Segmented Double-Stranded RNA Viruses

Structure and Molecular Biology

Edited by John T. Patton

Caister Academic Press

Caister Academic Press
Norfolk, UK

www.caister.com

British Library Cataloguing-in-Publication Data
A catalogue record for this book is available from the British Library

ISBN: 978-1-904455-21-9

Description or mention of instrumentation, software, or other products in this book does not imply endorsement by the author or publisher. The author and publisher do not assume responsibility for the validity of any products or procedures mentioned or described in this book or for the consequences of their use.

Printed and bound by Cromwell Press, Trowbridge, Wiltshire, Great Britain

Cover Stylized semi-transparent capsid shell protein structure of the cytoplasmic polyhedrosis virus (CPV, top) and shaded surface view of a CPV cryoEM reconstruction overlaid on a ribbon diagram rendering of the X-ray structure of rice dwarf virus P8 protein (bottom). Cover design by Xiaorui Zhang and Z. Hong Zhou.

Contents

Contributors

Stefan T. Arold
Centre de Biochimie Structurale
UMR CNRS 5048-INSERM 554-UM1
Montpellier
France

stefan.arold@cbs.cnr.fr

Matthew L. Baker
Department of Pathology Laboratory Medicine
University of Texas Medical School
Houston, TX
USA

Judith M. Ball
Department of Pathobiology
Texas A&M University
College Station, TX
USA

jball@cvm.tamu.edu

Dennis H. Bamford
Department of Biological and Environmental Sciences and Institute of Biotechnology
University of Helsinki
Finland

dennis.bamford@helsinki.fi

Michelle M. Becker
Virologie Moléculaire et Structurale
UMR CNRS 2472-INRA 1157
Gif-sur-Yvette
France

michelle.becker@vanderbilt.edu

Stephen K. Burley
Structural GenomiX Inc.
San Diego, CA
USA

<sburley@sgxpharma.com

Jacquelyn A. Campbell
Department of Microbiology and Immunology
Vanderbilt University School of Medicine
Nashville, TN
USA

jackie.campbell@vanderbilt.edu

José L. Carrascosa
Centro Nacional de Biotecnología
CSIC
Campus de Cantoblanco
Madrid
Spain

jlcarras@cnb.uam.es

José R. Castón
Centro Nacional de Biotecnología
CSIC
Campus de Cantoblanco
Madrid
Spain

Wah Chiu
National Center for Macromolecular Imaging
Baylor College of Medicine
Houston, TX
USA

wah@bcm.tmc.edu

Kevin M. Coombs
Department of Medical Microbiology and Infectious Diseases
University of Manitoba
Winnipeg
Manitoba
Canada

kcoombs@cc.umanitoba.ca

Sue E. Crawford
Department of Molecular Virology and Microbiology
Baylor College of Medicine
Houston, TX
USA

Rahul C. Deo
Department of Medicine
Brigham and Women's Hospital
Boston, MA
USA

rdeo@partners.org

Terence S. Dermody
Lamb Center for Pediatric Research
Vanderbilt University School of Medicine
Nashville, TN
USA

terry.dermody@vanderbilt.edu

Philip R. Dormitzer
Department of Pediatrics
Harvard Medical School
Boston, MA
USA

dormitze@crystal.harvard.edu

Kelly A. Dryden
Department of Cell Biology
The Scripps Research Institute
La Jolla, CA
USA

kdryden@scripps.edu

Mary K. Estes
Department of Molecular Virology and Microbiology
Baylor College of Medicine
Houston, TX
USA

Nora A. Gardner
The Laboratory of Biochemistry and Genetics
NIDDK
National Institutes of Health
Bethesda, MD
USA

noragardner@comcast.net

Caroline M. Groft
Weill Medical College of Cornell University
New York, NY
USA

cag2021@med.cornell.edu

Stephen C. Harrison
Department of Biological Chemistry and Molecular Pharmacology
Harvard Medical School
Boston, MA
USA

harrison@crystal.harvard.edu

Xiaofang Jiang
Verna and Marrs McLean Department of Biochemistry and Molecular Biology
Baylor College of Medicine
One Baylor Plaza
Houston
TX 77030
USA

John E. Johnson
Dept. of Molecular Biology
The Scripps Research Institute
La Jolla, CA
USA

jackj@scripps.edu

Wolfgang K. Joklik
Department of Molecular Genetics & Microbiology
Duke University Medical Center
Durham, NC
USA

Minni R. L. Koivunen
Institute of Biotechnology and Department of Biological and Environmental Sciences
University of Helsinki
Finland

minni.koivunen@gmail.com

Mukesh Kumar
Verna and Marrs McLean Department of Biochemistry and Molecular Biology
Baylor College of Medicine
Houston, TX
USA

Erika J. Mancini
Division of Structural Biology
The Henry Wellcome Building for Genomic Medicine
Oxford University
Oxford
UK

erika@strubi.ox.ac.uk

Leonard Mindich
Public Health Research Institute
Department of Microbiology
Newark, NJ
USA

mindich@phri.org

John T. Patton
Laboratory of Infectious Diseases
National Institute of Allergy and Infectious Diseases
National Institutes of Health
Bethesda, MD
USA

jpatton@niaid.nih.gov

Rebecca D. Parr
Department of Pathobiology
Texas A&M University
College Station, TX
USA

Didier Poncet
Virologie Moléculaire et Structurale
UMR CNRS 2472-INRA 1157
Gif-sur-Yvette
France

didier.poncet@vms.cnrs-gif.fr

Minna M. Poranen
Department of Biological and Environmental Sciences and Institute of Biotechnology
University of Helsinki
Finland

minna.poranen@helsinki.fi

B.V. Venkataram Prasad
Verna and Marrs McLean Department of Biochemistry and Molecular Biology
Baylor College of Medicine
Houston, TX
USA

vprasad@bcm.tmc.edu

Polly Roy
Department of Infectious & Tropical Diseases
London School of Hygiene & Tropical Medicine
London
UK

polly.roy@lshtm.ac.uk

José F. Rodríguez
Centro Nacional de Biotecnología
CSIC
Campus de Cantoblanco
Madrid
Spain

Michael R. Roner
Department of Biology
The University of Texas
Arlington, TX
USA

roner@uta.edu

L. Peter Sarin
Institute of Biotechnology and Department of Biological and Environmental Sciences
University of Helsinki
Finland

Pierre Schelling
Interfakultäres Institut für Biochemie
Eberhard-Karls Universität
Tübingen
Germany

Leslie A. Schiff
Department of Microbiology
University of Minnesota
Minneapolis, MN
USA

schif002@umn.edu

Clarence E. Schutt
Department of Chemistry
Henry H. Hoyt Laboratory
Princeton University
Princeton, NJ
USA

Thilo Stehle
Interfakultäres Institut für Biochemie
Eberhard-Karls Universität
Tübingen
Germany

thilo.stehle@uni-tuebingen.de

Jinghua Tang
Dept. of Molecular Biology
The Scripps Research Institute
La Jolla, CA
USA

Yizhi Jane Tao
Department of Biochemistry and Cell Biology
Rice University, TX
USA

ytao@rice.edu

Zenobia F. Taraporewala
Laboratory of Infectious Diseases
National Institute of Allergy and Infectious Diseases
National Institutes of Health
Bethesda, MD
USA

ztaraporewala@niaid.nih.gov

Roman Tuma
Department of Biological and Environmental Sciences and Institute of Biotechnology
University of Helsinki
Finland

roman.tuma@helsinki.fi

Damien Vitour
Virologie Moléculaire et Structurale
UMR CNRS 2472-INRA 1157
1 Avenue de la Terrasse
Bâtiment 14B
91198 Gif-sur-Yvette
France

vitour@pasteur.fr

Reed B. Wickner
The Laboratory of Biochemistry and Genetics
NIDDK
National Institutes of Health
Bethesda, MD
USA

wickner@helix.nih.edu

Mark Yeager
Division of Cardiovascular Diseases
Scripps Clinic
La Jolla, CA
USA

yeager@scripps.edu

Z. Hong Zhou
Department of Microbiology, Immunology and Molecular Genetics and The California NanoSystems Institute
University of California
Los Angeles, CA
USA

hongzhou@mednet.ucla.edu

Preface

The double-stranded (ds)RNA viruses are a diverse lot, varying widely in their host range (humans, animals, plants, fungi, bacteria), number of genome segments (one to twelve), and virion organization (T-number, capsid layers, turrets). Over the past several years, remarkable progress has been made in determining, at atomic and subnanometer levels, the capsid structures of several of these viruses, as well as the structures of a number of individual viral proteins. This information has provided invaluable insight into the events in the viral life cycle including attachment and entry, genome replication, gene expression, and capsid morphogenesis. Besides serving to connect the structure of the capsids and individual proteins with biological functions, these studies indicate significant parallels in the structure and replicative processes of many of the dsRNA viruses. This volume presents a collection of articles that highlight our current knowledge of the dsRNA viruses, chiefly from the perspective of the structure and related functions of their capsids and their individual viral protein products.

The world of dsRNA viruses can be divided into two groups, simply based on whether or not the virus capsid includes a pseudo T = 1 core. This structure, formed by 60 asymmetrical dimers of a single protein, serves as the platform to which the viral RNA polymerase and associated enzymes are anchored within the capsid. It is the platform-linked polymerases of these pseudo T = 1 cores that catalyze viral RNA synthesis. The vast majority of dsRNA viruses contain a pseudo T = 1 core, including those in the families *Reoviridae* (reovirus, rotavirus, bluetongue virus (BTV), cytoplasmic polyhedrosis virus (CPV), rice dwarf virus (RDV), wound tumor virus), *Totiviridae* (yeast L-A virus), and *Cystoviridae* (bacteriophage ϕ6). While nearly all these dsRNA viruses have segmented genomes, L-A virus is an exception, containing but a single molecule of dsRNA within its pseudo T = 1 core. *Penicillium chrysogenum* virus (PcV), a fungal virus of the *Chrysovirus* family, can also be placed into this group, since it too has a T = 1 core. However, the PcV core is different, representing a true T = 1 structure formed from 60 monomers of a single protein, unlike the 60 dimers that make up the pseudo T = 1 core. PcV is also unusual in that it has a multipartite dsRNA genome, with its four RNA segments separately encapsidated into virus particles.

Of the group of dsRNA viruses without a T = 1 core, the best studied belong to the *Birnaviridae*, a family that includes Infectious Bursal Disease Virus and Infectious Pancreatic Necrosis Virus. Instead of a T = 1 core, the bi-segmented dsRNA genome of the birnaviruses is encapsidated along with copies of the viral RNA polymerases into a T = 13 protein shell. The birnavirus genome contains 5′-linked VPg-like proteins and is translated into polyproteins, features more reminiscent of positive-strand RNA viruses, than to other segmented dsRNA viruses infecting higher animals.

By far, the most thoroughly studied dsRNA viruses are those of the *Reoviridae*. These viruses contain ten to twelve genome segments in a pseudo T = 1 core. However, two structurally distinct forms of the cores are noted, those with turrets protruding from their fivefold axes (reovirus, CPV) and those without (rotavirus, BTV). The turrets represent pentamers formed by the viral RNA-capping enzyme. Thus, the turreted core contains several-times as many copies of the capping enzyme as non-turreted cores. Regardless of the type of core, the number of polymerase molecules approximates one per vertex. The RNA polymerase of the *Reoviridae* seems to have an unusual structure, one that is perhaps dictated by the core-associated nature of its activity. In particular, the polymerase has two putative RNA exit channels, one a conduit for directing RNA products out of the core and the other a conduit for returning RNA templates back into the core.

In most cases, the *Reoviridae* virion is a triple-layered icosahedron, formed by the addition of two concentric layers of proteins over the T = 1 core. However, there are exceptions. Perhaps the most intriguing are CPV and the other cypoviruses, whose virions are single-layered capsids that are more-or-less the structural equivalents of turreted T = 1 cores. There are also examples of double-layered virions within the *Reoviridae* family, notably RDV, a non-turreted phytoreovirus. The second protein layer of *Reoviridae* virions is assembled from trimers arranged with T = 13 symmetry. When present, the projecting nature of turrets from the core causes the T = 13 layer to be incomplete at the fivefold axes. In contrast to the first and second protein layers, the third layer varies considerably in organization and protein content, leading to marked differences in the overall architecture of the *Reoviridae*. For instance, the outer protein layer of some non-turreted viruses (rotavirus, BTV) serves as the anchor for projecting attachment proteins that decorate the full surface of the particle. For some turreted viruses (reovirus), projecting attachment proteins are only anchored to the particle at the fivefold axes.

A common characteristic of cells infected by the *Reoviridae* is the appearance of cytoplasmic inclusion bodies, structures serving as viral factories in support of genome replication and core assembly. With all the viruses in this family similarly replicating their dsRNA genomes and forming pseudo T = 1 cores, it could be predicted that close structural and functional similarities would exist among their inclusion-forming proteins. However, only in the broadest sense have such relationships been noted, with no strong evidence as of yet of conserved structural elements or enzymatic motifs. Instead, the overall sense has to be that this family of viruses is remarkably diverse in the types of proteins that they encode, irrespective of the similarity of their core component. In no small part, this diversity likely reflects variations in the host ranges of these viruses as well as their mechanism of spread, both from cell-to-cell and organism-to-organism. The diversity of their protein products is also likely to correlate with differences in how these viruses enter and exit the cell, produce and assemble their outer capsid shells, and subvert host-specialized antiviral systems.

I am deeply indebted to the authors for their contributions to this book, including the many that updated their already-completed articles along the way. Hopefully the end result justifies their patience and efforts. My appreciation also goes to Max Nibert for his help in getting this endeavor launched, and to Sarah McDonald, Zenobia Taraporewala and Cindy Luongo for their assistance in editing the chapters of this volume. Finally, I would like to acknowledge the support of the Intramural Research Program of the National Institute of Allergy and Infectious Diseases, NIH.

John T. Patton

Part I

Virion Architecture and Assembly

The Structure of Orthoreoviruses

1

Kelly A. Dryden, Kevin M. Coombs, and Mark Yeager

Abstract

The orthoreoviruses (reoviruses) are the prototypic members of the virus *Reoviridae* family, and representative of the turreted members, which comprise about half the genera. Like other members of the family, the reoviruses are non-enveloped and characterized by concentric capsid shells that encapsidate a segmented dsRNA genome. In particular, reovirus has eight structural proteins and ten segments of dsRNA. A series of uncoating steps and conformational changes accompany cell entry and replication. High-resolution structures are known for almost all of the proteins of mammalian reovirus (MRV), which is the best-studied genotype. Electron cryomicroscopy (cryoEM) and X-ray crystallography have provided a wealth of structural information about two specific MRV strains, type 1 Lang (T1L) and type 3 Dearing (T3D). This review describes the icosahedral structures of intact virions, infectious subviral particles that bind to cell surface receptors, and core particles, which mediate RNA transcription. The size, shape, stoichiometry, interactions and atomic structures of the constituent proteins are described, as well as their functional properties. Such structural details about the individual proteins and the assembled particles are essential for a complete understanding of how the multiple proteins and genomic segments interact and participate in morphogenesis and infection.

Introduction

The *Reoviridae* family is characterized by virions that contain a genome of 10–12 segments of dsRNA encased by several proteins, organized as multiple concentric capsids. The family currently consists of 12 genera, which may be separated structurally as turreted (e.g. orthoreoviruses, aquareoviruses, cypoviruses) or non-turreted (e.g. rotaviruses, orbiviruses, phytoreoviruses) (Figure 1.1) (Mertens, 2004). The orthoreoviruses (reoviruses) are the prototypic members of the virus *Reoviridae* family and of the turreted members. Many of the *Reoviridae* genera can be further subdivided into genogroups. There are currently five such subdivisions among the orthoreovirus genus (reoviruses). These are the mammalian orthoreoviruses (MRV), avian reoviruses (ARV), Baboon reovirus (BRV), Nelson Bay virus (from flying fox) and a group that primarily affects reptiles (Duncan, 1999). The aquareoviruses (AqRV) are a distinct genus that appears to be the most closely related to the orthoreovirus genus (Attoui *et al.*, 2002).

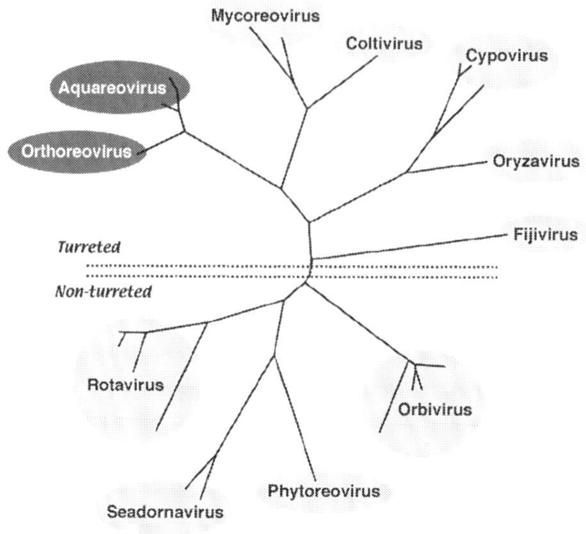

Figure 1.1 Dendrogram of the *Reoviridae*. Phylogenetic relationships of 11 family members, based on the sequences of the putative RNA-dependent RNA-polymerase. Adapted from Attoui *et al.*, 2006, with permission.

Although MRVs are not human pathogens, they have been studied extensively as a model system for virus entry, replication, and pathogenesis (Nibert and Schiff, 2001; Coombs, 2006). ARVs have also been studied increasingly due to their adverse impact on the poultry industry (Benavente and Martínez-Costas, 2007). Understanding the interactions that allow multiple proteins and genomic segments to participate in morphogenesis and infection is greatly aided by understanding the structures of individual proteins, as well as the complete particles. Much has been learned about the functions and roles of the viral proteins from studies of reassortants, temperature sensitive mutants, and recoating of subviral particles with expressed proteins. In addition, cryoEM and X-ray crystallography have provided a wealth of structural information about two specific MRV strains, type 1 Lang (T1L) and type 3 Dearing (T3D). Limited structural data are available for ARV, BRV, and AqRV. The detailed structures of other dsRNA viruses are reviewed in subsequent chapters.

Capsid morphology

Reoviruses are medium-sized (~850 Å diameter), non-enveloped viruses assembled from eight structural proteins that form two concentric capsids surrounding 10 dsRNA gene segments (Figure 1.2A). Virions serve as a gene delivery vehicle, and the outer capsid proteins in the virion recognize cell surface receptors and mediate entry of the viral core into the cell. In this process, virions are not static structures but rather undergo a series of uncoating steps and protein conformational changes. There are also two naturally occur-

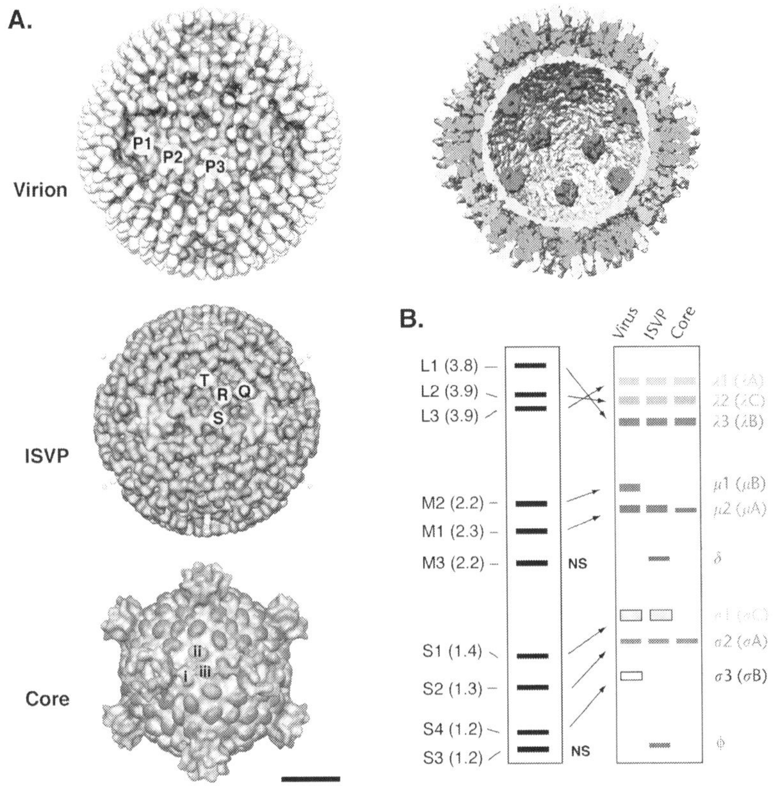

Figure 1.2 A) Surface-shaded, three-dimensional reconstructions of the mammalian reovirus (MRV) virion, intermediate subviral particle (ISVP), and core (left side from top to bottom, respectively), and a cross-section of the virion (top right). Features are colour-coded by protein as identified in B. Labels are shown for the P1, P2 and P3 channels on the virion, the Q, R, S and T positions of the μ1 trimers on the ISVP and the (i), (ii) and (iii) positions of the σ2 nodules on the core. Bar = 200 Å. B) Schematic representation of the electrophoretic patterns for the dsRNA segments with kilobase values (left), and the eight structural proteins (right). Homologous ARV proteins are indicated in parentheses (far right). (See also Plate 1.2.)

ring subviral particles, the infectious (or intermediate) subviral particle (ISVP) (~800 Å diameter), and the core (~540 Å diameter). The overall architecture of the virions consists of an outer capsid organized on a T = 13(*laevo*) icosahedral lattice that concentrically surrounds a T = 1 core capsid. The virion surface has 600 finger-like projections arranged as hexamers and incomplete hexamers around the fivefold vertices, such that pentameric turrets of the core participate in forming the outer surface of the virion. The ISVPs have lost the finger-like proteins, and can be generated extracellularly or *in vitro* by a variety of intestinal proteases (Sturzenbecker *et al.*, 1987; Bodkin *et al.*, 1989). Loss of the remaining outer capsid proteins generates core particles, which have a nodular surface distinguished by the turrets extending from the 12 fivefold vertices. Upon conversion to cores, the pentamers of protein λ2 at the fivefold vertex positions undergo a dramatic conformational change from a closed 'star-like' feature to the open 'crown-like' turret (compare the green pentamers in the ISVP and core particles in Figure 1.2A) (Dryden *et al.*, 1993). The intact core (five proteins plus all 10 dsRNA genes) is transcriptionally active and produces the viral mRNA during replication.

The outer capsid is formed by 600 copies of σ3, a 41 kDa protein that forms the finger-like projections (Metcalf *et al.*, 1991; Dryden *et al.*, 1993); 200 underlying trimers of the 76 kDa protein μ1; and 12 trimers of σ1, the 49 kDa cell attachment protein anchored at the distal centre of the turrets (Fraser *et al.*, 1990; Strong *et al.*, 1991; Chappell *et al.*, 2002) (Figures 1.2, 1.3, and 1.4). Reovirus has three distinct types of solvent channels that span the outer capsid layer (Metcalf *et al.*, 1991; Dryden *et al.*, 1993), which are identified as P1, P2, and P3 (Figure 1.2A). The 12 P1 channels are formed by the turrets at the fivefold vertices, which are blocked by σ1 in virions and ISVPs, but open when the particles are converted to cores. The 60 P2 and 60 P3 channels are surrounded by either four or six σ3 subunits, respectively. Although the σ3 molecules appear to be assembled as units around the channels, there is apparently little direct contact between the monomers (Dryden *et al.*, 1993; Liemann *et al.*, 2002). Rather, the μ1 trimers create the majority of capsid interactions, and σ3 merely decorates the trimers, forming heterohexameric structures. This protein arrangement generates the outer capsid with 600 copies of σ3 and μ1, rather than the 780 copies expected for a T = 13 lattice (as seen in non-turreted dsRNA viruses). This stoichiometry is a result of the twelve λ2 pentamers that disrupt the continuous lattice. The μ1/σ3 heterohexamers are located in four quasi-equivalent environments, which are identified as Q, R, S, and T, following the trimer convention for bluetongue virus (Grimes *et al.*, 1998) (Figure 1.2A). The T subunits are centred on the threefold axes, the Q subunits are closest to the fivefold axes, the S subunits are adjacent to the twofold axes, and the R subunits are the remainder. Transition to the ISVPs involves proteolysis and release of the σ3 molecules and a dramatic conformational change in the σ1 trimer. In some serotypes (e.g. T1L) σ1 is converted from a compact density centred on the fivefold axis (yellow fivefold density in Figure 1.2A, virion) to an extended, flexible fibre about 400 Å long (Furlong *et al.*, 1988; Fraser *et al.*, 1990; Dryden *et al.*, 1993) (Figure 1.3). Due to the flexibility of σ1, only the base of the fibre is visualized in icosahedral reconstructions (yellow beaded densities in Figure 1.2A, ISVP). In other serotypes (e.g. T3D) σ1 is digested and such ISVPs are non-infectious (Nibert *et al.*, 1995).

The core is constructed from a shell of 120 copies of the 142-kDa protein, λ1 (Harrison *et al.*, 1999), decorated with 150 monomers of σ2, a 47-kDa protein (Wiener *et al.*, 1989b), and 12 pentamers of λ2, the 144 kDa protein that forms the turrets (Reinisch *et al.*, 2000; Breun *et al.*, 2001) (Figure 1.2 and see Figures 1.5–1.8). The remaining core proteins, λ3, the RNA-dependent RNA polymerase (RdRp) (see Figure 1.9), and presumably μ2, are located in the interior beneath the λ2 turrets (Dryden *et al.*, 1998; Zhang *et al.*, 2003). λ1 assembles the core capsid as 60 asymmetric dimers with T = 1 lattice symmetry. The protein has two positions, designated A (or λ1.5f) for the subunits adjacent to the fivefold axes and B (or λ1.3f) for the subunits that interdigitate with A and interact at the threefold axes (Reinisch *et al.*, 2000; Zhang *et al.*, 2005a) (see Figure 1.7). The σ2 nodules are in three unique orientations; 60 each around the five- and threefold axes, and 30 on the twofold axes. The positions are identified as (i) (or σ2.5f) around the

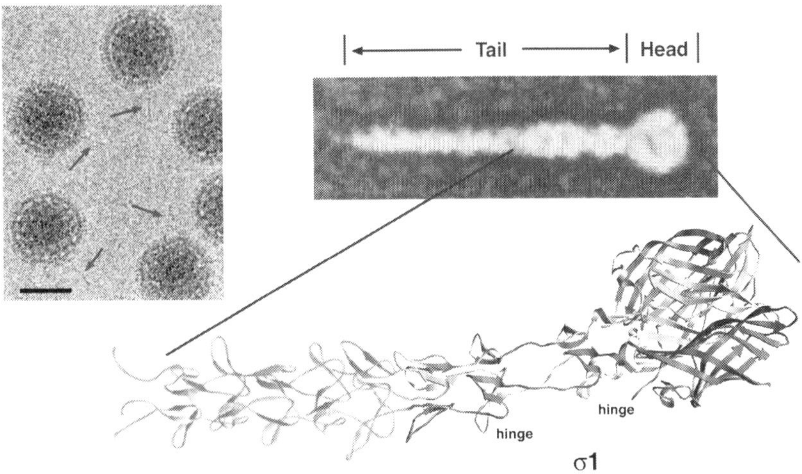

Figure 1.3 The cell attachment protein, σ1, mediates binding to receptors such as JAM and sialic acid. Cryo-electron micrograph of ISVPs (left) shows extended σ1 fibres (arrows). A two-dimensional average of negatively stained σ1 displays head and tail regions (upper right) (from Fraser *et al.*, 1990, with permission). The head domain and tail fragment that was determined by X-ray crystallography is coloured by monomer (pdb 1FN9), and the extra predicted β-spirals in grey (bottom). Hinges would confer conformational flexibility. The remainder of the protein is not modelled. Bar = 750 Å for the micrograph. (See also Plate 1.3.)

λ2 turret, (ii) (or σ2.3f) around the threefold axes, and (iii) (or σ2.2f) directly on the twofold axes (Reinisch *et al.*, 2000; Zhang *et al.*, 2005a) (see Figure 1.8). The pentamers of λ2 extend ~80 Å radially from the core capsid along each icosahedral fivefold axis (Luftig *et al.*, 1972; Metcalf *et al.*, 1991; Dryden *et al.*, 1993) (see Figures 1.5 and 1.6). One copy of the 142 kDa protein λ3 (RdRp) is positioned beneath the λ2 pentamers, but slightly off-centre of the fivefold axis (Zhang *et al.*, 2003) (see Figure 1.9).

Structure of the genome

The reovirus genome consists of 10 segments of dsRNA (Figure 1.2B); 3 large segments (L1, L2, L3), 3 medium segments (M1, M2, M3), and 4 small segments (S1, S2, S3, S4). Every particle contains a single copy of each segment, and each is proposed to associate with a separate λ3 polymerase (Nibert and Schiff, 2001). Particle images display an interior fingerprint pattern attributed to paracrystalline packing of the dsRNA genome. The ~26 Å spacing between the lines in the patterns suggests tight packing of the dsRNA that is comparable to the packing of DNA in bacteriophages (Harvey *et al.*, 1981).

All 10 genes of MRVs have been sequenced (Wiener *et al.*, 1989a; Breun *et al.*, 2001; Yin *et al.*, 2004). Each segment contains a single ORF encoding a single protein, except for the S1 gene that is bicistronic and encodes σ1 and an additional non-structural protein, σ1NS. Every MRV gene consists of a short 5′ non-translated region (NTR) of variable length, an ORF of variable length, and a 3′ NTR, also of variable length (Wiener *et al.*, 1989a; Breun *et al.*, 2001; Nibert and Schiff, 2001; Yin *et al.*, 2004). Each of the 10 genes has an m7GpppGm (type 1) cap structure and a consensus GCUA tetranucleotide at the

5′ end, a consensus UCAUC pentanucleotide at the 3′ end (of the plus-sense strand), and lack a 3′ poly(A) tail. Although still not known, the current idea is that signals directing genomic assembly reside in one or both of the NTRs, although some signals may also reside in the ORF.

Protein structures and functions

In reovirus particles, there are many complex interactions within protein multimers and between different subunits. Lower resolution three-dimensional (3D) reconstructions of the naturally occurring particles (Metcalf *et al.*, 1991; Dryden *et al.*, 1993) (Figure 1.2A) have been supplemented with recent higher resolution cryoEM reconstructions of virions (Zhang *et al.*, 2003; Zhang *et al.*, 2005a), high-resolution X-ray crystal structures of core particles (Reinisch *et al.*, 2000) and X-ray structures of all remaining MRV structural proteins except for μ2 (Chappell *et al.*, 2002; Liemann *et al.*, 2002; Tao *et al.*, 2002). Fitting of the known X-ray structures into a cryoEM map at 7.0 Å resolution has generated a pseudoatomic model for the entire virion. In addition, the 7.0 Å cryoEM map revealed densities ascribed to polypeptide regions that were not visualized in the X-ray structures (Zhang *et al.*, 2005a).

Most of the capsid proteins not only have a structural role in forming the shells of the virion but also have functional roles. Reassortants, temperature sensitive (ts) mutants, and redecorated core particles have all been used to elucidate the functions of individual proteins (Chandran *et al.*, 2001; Nibert and Schiff, 2001; Coombs, 2006).

σ1 is the cell attachment protein (reviewed by Schelling *et al.*, this volume)

σ1 is the outermost protein and the most variable reovirus protein of the T1L, T2J and T3D strains (Duncan *et al.*, 1990; Nibert *et al.*, 1990). σ1 is the cell attachment protein (Weiner *et al.*, 1980; Lee *et al.*, 1981) and is also the serotype determinant (Weiner *et al.*, 1978). In virions, σ1 forms a compact structure centred at the fivefold vertices. In T1L ISVPs, it has an extended conformation with a tail anchored by the lid domain of the λ2 pentamer, which closes the P1 channel, and a head at the distal end projecting away from the virion surface (Furlong *et al.*, 1988; Dryden *et al.*, 1993) (Figure 1.3). Several regions of flexibility have been proposed along the length of the tail (Fraser *et al.*, 1990). This flexibility accounts for the absence of σ1 density observed in cryoEM reconstructions, other than small beads of density projecting from the centre of the λ2 turrets in 3D reconstructions of ISVPs (Dryden *et al.*, 1993; Zhang *et al.*, 2005a) (Figure 1.2A, ISVP). Note that there is also a symmetry mismatch between the trimeric tail of σ1 and the λ2 pentamers.

The crystal structure of a receptor-binding fragment of σ1 was trimeric and contained the head and part of the tail of each subunit (Chappell *et al.*, 2002) (Figure 1.3). There are many sites of interaction along the monomers, which contribute to stabilizing the trimer. The head of σ1 is a compact eight-stranded β-barrel assembled from two Greek key motifs. The tail consists of repeating segments of two β-strands with connecting loops, forming a triple β-spiral in the trimer, which is similar to that seen in the adenovirus fibre (van Raaij *et al.*, 1999). A short α-helix links the β-spiral of the tail to the head domain. Although only three repeats are present in the crystal structure, five more repeats are predicted from the sequence. The remainder of the tail is predicted to be an α-helical coiled-coil (Bassel-Duby *et al.*, 1985). Significant bending of the trimer occurs in a hinge region within the tail, which results in a kink between the threefold axes of the head and the tail in the crystallized trimer (Chappell *et al.*, 2002). Conformational simulations of the trimer head by molecular dynamic studies suggest that the head is very flexible, with rotations up to 50° (Cavalli *et al.*, 2004). The extreme flexibility of the head and fibre, trimeric stoichiometry, and 1twofold valency of σ1 on the particles would maximize the opportunities for interaction with cell surface receptors.

σ1 can bind to several receptors, depending on the viral strain. All three strains bind to JAM, the junction adhesion molecule (Barton *et al.*, 2001). In T3D, σ1 also binds sialic acid and mediates haemagglutination (Weiner *et al.*, 1978; Gentsch and Pacitti, 1987). σ1 is also pivotal in determining viral infection patterns (Weiner *et al.*, 1980; Tyler *et al.*, 1986), most likely through the selective recognition of different cell-surface receptors. T3D induces apoptosis to a substantially greater extent than T1L, and this difference is determined primarily by the σ1-encoding S1 gene (Tyler *et al.*, 1995; Connolly *et al.*, 2000).

σ3 is the protector protein (reviewed by Schiff in Chapter 10)

The major capsid protein σ3 folds into a bi-lobed cylindrical structure with a length of ~76 Å and a diameter of ~47 Å, organized around a central α-helix, which is roughly perpendicular to the virion surface (Olland *et al.*, 2001) (Figure 1.4). The small lobe consists of the N-terminal 100 residues and a 40 residue α-helical hairpin from the large lobe. The protein contains a zinc finger motif (Giantini *et al.*, 1984) and binds a zinc ion (Schiff *et al.*, 1988; Olland *et al.*, 2001), which is required for correct folding (Danis *et al.*, 1992) and stability (Mabrouk and Lemay, 1994). Loss of the zinc-binding site decreases cytoplasmic stability of particles and eliminates the ability of σ3 to interact with the other major outer capsid protein, μ1, which is required for capsid assembly (Mabrouk and Lemay, 1994, Shepard *et al.*, 1996). A conserved, primarily α-helical region on the small lobe of σ3 mediates the interactions with μ1 (Olland *et al.*, 2001; Liemann *et al.*, 2002). The non-conserved side faces the P2 and P3 solvent channels and is composed largely of loops. The large lobe contains ~225 residues that fold into several α-helices, short β-sheets, and large loops. The loops are exposed on the virion surface and contain a proteolytically hypersensitive region (HSR) that is cleaved to initiate viral disassembly (Ebert *et al.*, 2002; Jané-Valbuena *et al.*, 2002). Apart from other reovirus outer capsid homologues, neither the complete protein nor either lobe is homologous to other proteins (Olland *et al.*, 2001).

σ3 plays an important role in particle stability (Drayna and Fields, 1982b; Drayna and Fields, 1982c). Temperature-sensitive (ts) mutants with lesions in the S4 gene, encoding σ3, are defective in outer capsid assembly at the restrictive temperatures (Fields *et al.*, 1971; Danis *et al.*, 1992). Indeed, at restrictive temperatures, mutant σ3 and μ1 fail to associate, and immunocoprecipitation experiments suggest that a μ1-σ3 interaction is required for their assembly onto cores (Shing and Coombs, 1996). The susceptibility of σ3 to proteolytic cleavage has consequences for virion stability and initial uncoating. Proteolytic digestion of σ3, but not necessarily complete removal, appears to be a prerequisite for cell entry (Chandran *et al.*, 2001; Nibert *et al.*, 2005). The digestion of σ3 has been studied extensively by SDS-PAGE and mass spectrometry (Mendez *et al.*, 2003; Hadzisejdic *et al.*, 2006). Cleavage is initiated in a central specific HSR (Ebert *et al.*, 2002; Jané-Valbuena *et al.*, 2002) located in different parts of the sequence for T1L and T3D strains (labelled

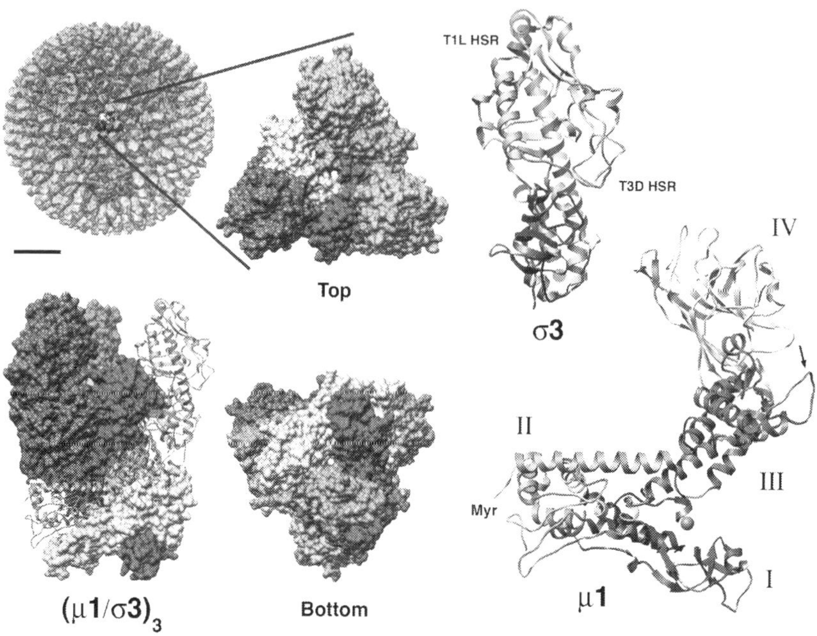

Figure 1.4 The major outer capsid proteins σ3 and μ1 confer stability to the virions. The left-hand panels are surface renderings of the μ1/σ3 heterohexamer, with an overview of the placement of one heterohexamer in the virion (upper left). The complex is viewed in different orientations with the μ1 subunits shown in red, orange and yellow and the σ3 subunits shown in dark blue, cyan, and green (pdb 1JMU). Note how each μ1 wraps completely around the complex. A closer view of the σ3 monomer (upper right) is coloured blue for the N-terminal small lobe, green for the C-terminal large lobe and loop to the small lobe, and yellow for the central helix. The proteolytically hypersensitive regions (HSR) are red. In μ1 (lower right), domains I to IV are coloured, blue, green, red and yellow, respectively. The μ1C/μ1N cleavage is identified with a magenta sphere at Asn42, and the arrow indicates the region for δ/ϕ cleavage. Bar = 200 Å for the virion; 25 Å for the crystal structures. (See also Plate 1.4.)

in Figure 1.4), and they differ in their cleavage patterns and susceptibility to proteases (Jané-Valbuena *et al.*, 1999). In most cases, digestion proceeds rapidly in both directions from the site of initiation until the entire protein has been digested (Mendez *et al.*, 2003). The cleavage products and rate of digestion also vary depending on the protease used, virus type, and the presence of detergents (Hadzisejdic *et al.*, 2006). A single lysine-to-histidine mutation near the carboxy-terminus of σ3 increases its susceptibility to proteolysis and causes a structural change that may mimic the protein in a low pH environment as would occur when exposed to gastric contents (Wilson *et al.*, 2002).

Transient or stable expression of σ3 alone does not influence cell growth or protein expression (Yue and Shatkin, 1996). However, σ3 does have non-structural functions in the viral life cycle by shutting down host protein synthesis (Sharpe and Fields, 1982) and counteracting host defences such as interferon activation (Imani and Jacobs, 1988; Giantini and Shatkin, 1989; Beattie *et al.*, 1995), possibly by binding dsRNA in the cytoplasm (Huismans and Joklik, 1976; Denzler and Jacobs, 1994). σ3 can bind dsRNA, and the site of binding has been localized to the C-terminal domain (Schiff *et al.*, 1988).

μ1 is the membrane penetration protein

In addition to σ3, the other major outer capsid protein, μ1, plays an important role in virion stability (Drayna and Fields, 1982c; Middleton *et al.*, 2002). μ1 is myristoylated at its N-terminus (Jayasuriya *et al.*, 1988; Nibert *et al.*, 1991) and undergoes several cleavage events during disassembly of virions (Jayasuriya *et al.*, 1988; Nibert and Fields, 1992). μ1 assembles as a trimer with dimensions of ~100 × ~80 × ~70 Å (Liemann *et al.*, 2002), and the subunits wrap around each other 360° as they twist counter clockwise from base to tip (Figure 1.4). There is a small, ~5 Å pore at the centre of the trimer, which would be permeable to water but not hydrated ions or signalling molecules such as ATP. Each μ1 subunit has a Z-shape with four domains. The three interior domains are predominantly α-helical (domains I, II, and III) and the fourth, most distal domain (domain IV), is a canonical jelly-roll, β-barrel as commonly seen in picornavirus, nodavirus, and tetravirus major capsid proteins (Liemann *et al.*, 2002). Domain I forms the base of the trimer that interacts with the surface of the underlying core. The crystal structure has disordered or poorly defined residues in this region, which are proposed to be flexible loops that would allow variable contacts with the core. The conserved autocatalytic cleavage site at residue 42 (Nibert *et al.*, 1991; Nibert *et al.*, 2005) resides in a pocket of domain I (Figure 1.4, magenta sphere). Domain II forms the central structure of μ1 with a long, horizontal α-helix that is roughly parallel to the particle surface and contains the presumptive myristoyl pocket (labelled Myr in Figure 1.4). Domain III is formed by five α-helices and a short β strand. Two helices and the bridging loop between them form a convex surface for attachment to σ3. The δ/ϕ cleavage sites reside in an exposed bridging loop of domain III (coloured orange in Figure 1.4). The jelly-roll, β-barrel motif of domain IV forms the head of the μ1 trimer. Domain IV has two antiparallel β sheets, and the extended surface loops participate in μ1-σ3 or μ1-μ1 interactions within the heterohexamer.

Each μ1 contacts two σ3 subunits and each σ3 contacts two μ1 subunits within the heterohexamer, which has dimensions of ~130 × ~100 × ~95 Å (Figure 1.4). Fitting of the crystallographic structure of the heterohexamer into a 7.0 Å resolution cryoEM map showed that the R, S, and T trimers are very similar. Not surprisingly, the Q position trimers vary from the others due to their proximity to λ2 (Zhang *et al.*, 2005a). Proposed contacts of μ1.Q with the λ2 pentamer may prevent conformational changes that block the binding pocket of the methyltransferase-1 domain of λ2 and restrict the elongation of mRNA. Similarly, proposed contacts of μ1.Q with λ2 may lead to altered protein–protein interactions in these local regions of this subset of μ1 molecules that might explain the lack of complete μ1 to δ/ϕ digestion (Coombs, 1998). The cryoEM analysis and molecular modelling also identified densities that may correspond to sequences not resolved in the crystal structure. For instance, within the P2 and P3 channels is a 'hub-and-spokes' structure, which is ascribed to 33 disordered amino acids of μ1 near the core and the myristoylated N-terminus in the pocket (Zhang *et al.*, 2005a).

When resolved by standard SDS-PAGE, about 95% of μ1 molecules in virions are cleaved to a large C-terminal fragment of 72 kDa, called μ1C. High-resolution SDS-PAGE identified an additional ~4 kDa N-terminal, myristylated peptide, called μ1N (Nibert *et al.*, 1991). This cleavage occurs between Asn42 and Pro43 (Figure 1.4, magenta sphere), and it was recently shown that a majority of the cleavage in virions takes place during sample preparation prior to SDS-PAGE (Nibert *et al.*, 2005). Nevertheless, the autocatalytic cleavage of μ1 is necessary for cell entry (Odegard *et al.*, 2004).

Following attachment and receptor-mediated endocytosis, virions are converted to ISVPs by proteolytic digestion of σ3 and cleavage of μ1 to the peptides, δ and ϕ (Figure 1.4, arrow). Thereafter, a transient particle, ISVP*, is a required intermediate on the pathway from ISVPs to cores that mediates membrane penetration (Chandran *et al.*, 2002). The ISVP* particles are distinguished by conformational rearrangements of μ1, including a separation of the head domains in the trimer (Zhang *et al.*, 2006), which makes the molecules more sensitive to proteolysis. In addition, the myristylated μ1N peptide is released from the particle, thereby conferring the ability to disrupt membranes (Odegard *et al.*, 2004). ISVP* also has a greater propensity to lose σ1, and unlike virions and ISVPs, the particles manifest transcriptase activity (Chandran *et al.*, 2002). Although the δ and ϕ peptides remain associated with ISVPs, they are subsequently lost and core-like particles enter the cytoplasm (Chandran *et al.*, 2003).

λ2 is the guanylyltransferase and capping protein

Although λ2 is a core protein, pentamers of λ2 span most of the outer capsid radii. The pentamers form turrets ~120 Å in diameter, ~80 Å in height, with a central cavity 70 Å in diameter (Figure 1.5). The five monomers of λ2 are closely entwined at an angle of 45° with respect to the fivefold axis (Reinisch *et al.*, 2000). The turrets do not apparently have tight contacts with the λ1 layer of the core, since removal of the turrets by heat, proteases

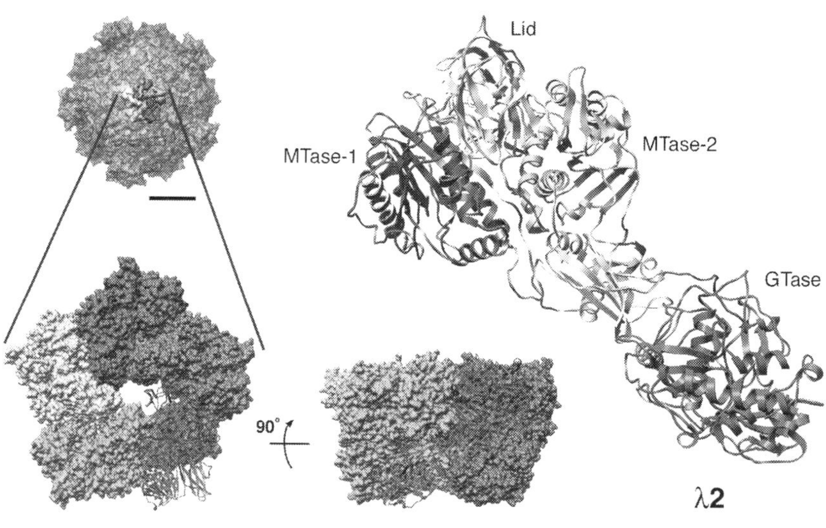

Figure 1.5 The λ2 turret protein is an RNA capping enzyme. The five subunits in the pentamer (bottom left) are coloured differently (pdb 1EJ6). A ribbon representation of the yellow monomer in the pentamer (right) shows the inner side of the protein. The seven domains are identified as red for the guanylyltransferase (GTase), blue and green for the methyltransferases (MTase), grey for the connecting domain, and yellow for the three Ig-like lid domains. Bar = 200 Å for the core particle; 40 Å for the crystal structures. (See also Plate 1.5.)

or ionic detergents does not significantly change the structure of the core (Luongo *et al.*, 2002), and core capsids that lack λ2 can be assembled by some ts mutants (Hazelton and Coombs, 1999) or *in vitro* (Kim *et al.*, 2002).

The domains of λ2 that form the star-like lid of the pentamer close the P1 channel in virions and ISVPs and provide a platform for attachment of σ1 (Dryden *et al.*, 1993). When the outer capsid proteins are lost during conversion to cores, the lid opens to various degrees depending on the reovirus strain (Figure 1.6A,B). In T1L cores, this C-terminal domain rotates up and out nearly 50° (Dryden *et al.*, 1993). T3D cores do not display such a large conformational change, rotating outward ~20° (Luongo *et al.*, 1997; Reinisch *et al.*, 2000). Attempts to crystallize T1L cores were unsuccessful, and reassortant experiments showed that the ability to crystallize T3D cores was mediated by T3D λ2 (Coombs *et al.*, 1990). From the differences in the T1L and T3D structures, the more protruding C-terminal domain in T1L may sterically hinder crystal packing. The transcriptionally active core particle is responsible for the production of the viral mRNA, which is extruded through channels in the λ2 turrets (Figure 1.6C,D) (Bartlett *et al.*, 1974; Yeager *et al.*, 1996). Full activation of the viral transciptase is dependent on removal of all the outer capsid proteins (σ3, μ1, and σ1) and conformational changes of the λ2 pentamers. The extent of turret opening may also influence the rate of exit of the capped mRNA from the core.

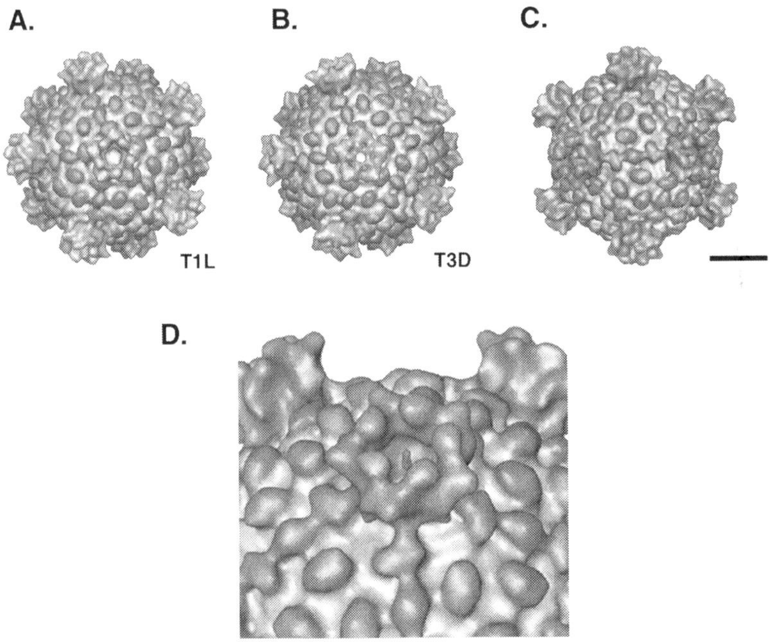

Figure 1.6 Surface-shaded 3D reconstructions of strain-specific conformational changes (A and B) and of transcriptionally active reovirus core particles (C and D). Two strains of MRV display a different amount of opening of the λ2 turrets. T1L (A) opens the P1 channel much wider than T3D (B) by swinging the λ2 lid domain almost perpendicular to the turret. A reconstruction of active core particles viewed down the twofold symmetry axis (C) shows that the particles remain intact during transcription. The oblique, close-up view of the fivefold vertex position (D) displays a thread of density (red) exiting the λ2 pentamers, ascribed to newly synthesized single-stranded RNA. Bar = 200 or 75 Å for A-C and D, respectively. (See also Plate 1.6.)

The high-resolution X-ray structure of the T3D reovirus core particles, coupled with mutagenesis studies, show that λ2 has seven compact domains that display mixed α/β folds (Reinisch *et al.*, 2000). Specific enzymatic activities of λ2 are important for mRNA processing during exit from the transcribing core (Luongo *et al.*, 2000; Qiu and Luongo, 2003). The pentamer binds GMP (Fausnaugh and Shatkin, 1990) and S-adenosyl-L-methionine (SAM) (Luongo *et al.*, 1998), possesses guanylyltransferase (GTase) (Cleveland *et al.*, 1986; Qiu and Luongo, 2003) and methyltransferase (MTase) activities, and caps the nascent mRNA (Shatkin, 1974; Luongo *et al.*, 2002). The active sites for the enzymatic domains face the interior of the turret (Figure 1.5). The spatial arrangement of the active sites may reflect the sequence of the capping reactions. The most N-terminal domain is at the base of the turret, and the most C-terminal domain is at the top. The first residues form a cup-like structure with a concave surface that faces the turret cavity. This domain (red in Figure 1.5) has been identified as a GTase so that guanylyl transfer occurs first, as the newly synthesized transcript enters the cavity of the turret. A small domain (grey in Figure 1.5) bridges the GTase and the MTases, and may provide structural support for the enzymatic domains. The first methyltransferase domain (MTase-1) is a variant of the SAM-binding domain seen in many other methyltransferases. The GTase site from one monomer is closer to the MTase-1 active site of one neighbour than to any other MTase-1 site, so the mRNA cap is more likely to pass from the GTase on one subunit to the MTase-1 site of another subunit. Active site sharing between two monomers also explains the absence of MTase activity in monomeric λ2 (Mao and Joklik, 1991). The second methyltransferase domain (MTase-2) has a fold that more closely resembles a consensus methyltransferase fold (seven-stranded β-sheet with flanking α-helices) plus an insertion forming a three-stranded parallel β-sheet and two α-helices. The lid at the top of the turret is formed from three Ig-like domains in the C-terminal portion of λ2 (coloured yellow in Figure 1.5) (Reinisch *et al.*, 2000).

λ1 forms the core capsid.

λ1 is the major capsid protein of the core, and 120 subunits form a thin, reasonably smooth shell (Figure 1.7), which then bind the σ2 nodules (Figure 1.8). The protein is a roughly triangular slab with dimensions ~135 Å in length, ~80 Å at the widest point (the triangular 'base'), and ~25 Å thick. The secondary structure is predominantly α-helical, with a β-sheet region distal from the fivefold axis (Reinisch *et al.*, 2000). λ1 assembles the core capsid as 60 asymmetric dimers with T = 1 lattice symmetry. Five λ1-A monomers interact in a 'pinwheel' fashion at the vertices, and three λ1-B monomers centred on the threefold axes interdigitate between the λ1-A subunits in the pinwheels. This packing pattern recapitulates the packing observed for VP3, the core shell protein of bluetongue virus (BTV) from the Orbivirus genus (Grimes *et al.*, 1998), although λ1 and VP3 have no sequence similarity. For BTV, the T = 13 shell of the VP7 trimers sits directly on the VP3 shell, without the intervening equivalent of the σ2 nodules (Figure 1.8). Although the overall secondary structure of the λ1-A and λ1-B subunits are similar, the bottom domain of λ1-B is pivoted ~5° counter-clockwise relative to the same region of the λ1-A subunit.

The first 240 residues of λ1 are largely disordered in the crystal structure, with three ordered segments observed in the λ1-B subunits, which appear to form a network of arms underneath the core shell around the threefold axes. Based on its position, one of these

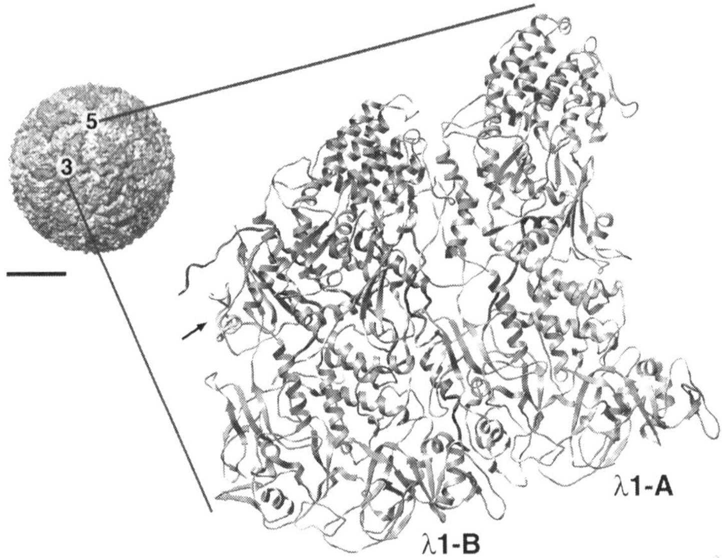

Figure 1.7 The λ1 core protein forms a fairly smooth shell for attachment of σ2. A surface-shaded view (left) showing the unique λ1-A (pink) and λ1-B (orange) subunits. Although the A and B subunits are very similar in structure (right), there is a subtle shift in the lower domains relative to each other (pdb 1EJ6). The additional amino acids observed in the crystal structure of λ1-B are in green, with the residues predicted to come from a neighbouring λ1-B subunit, 40–267, in dark green. Residues 130–159 (blue), which are essential for particle assembly, are located under the σ2 i and ii binding positions. The Zn-finger motif is at residues 181–208 (arrow). Bar = 200 Å. (See also Plate 1.7.)

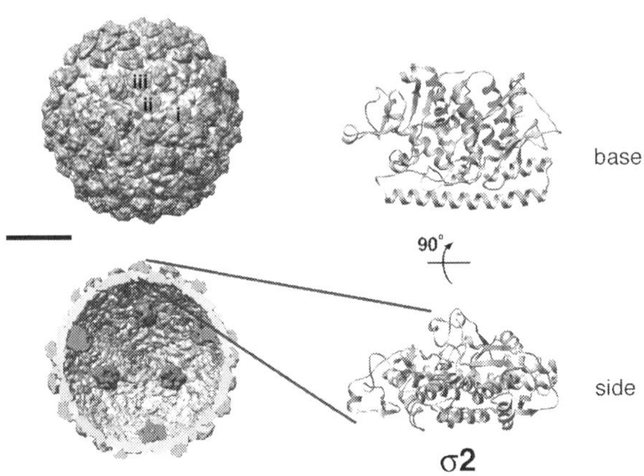

Figure 1.8 The σ2 nodules are required for core particle assembly. Surface-shaded view of the core capsid with the nodules viewed along the threefold axis (top left) and in cross-section (bottom left). Note that not all λ3 subunits display equally in the cross-section due to their off-axis positions. Close-up views of the protein from the base (top right) and side (bottom right) (pdb 1EJ6). Different residues on the base interact with the λ1 shell in three unique positions: (i) 60 around the fivefold axes, (ii) 60 around the threefold axes, and (iii) 30 on the twofold axis. Bar = 200 Å. (See also Plate 1.8.)

segments (residues 40–167) is most likely contributed from a neighbouring λ1-B subunit and may participate in stabilizing the λ1 shell of the core (Reinisch *et al.*, 2000). However, stable cores can be assembled *in vitro* when the hydrophilic N-terminal 230 residues are deleted, although residues 231-to-259 are required for core formation, possibly due to interactions with the σ2 nodules (Kim *et al.*, 2002).

λ1 contains three functional domains: (1) a zinc finger motif (Bartlett and Joklik, 1988), (2) a separate region that binds dsRNA (Lemay and Danis, 1994; Bisaillon and Lemay, 1997), and (3) a domain with NTPase activity, which is involved in transcriptional events (Bisaillon *et al.*, 1997; Noble and Nibert, 1997a). In the assembled core, the zinc finger in the N-terminal domain of λ1 is buried and probably not accessible for binding dsRNA (Reinisch *et al.*, 2000). However, the zinc finger is accessible in monomers and would be exposed during assembly, at which time it may interact with the RNA.

σ2 is the core clamp

The surface of the core is studded with σ2 nodules, which are compact, globular structures with a predominantly α-helical fold (Figure 1.8). The 150 nodules have three distinct asymmetric positions: (i) 60 around the fivefold axes, (ii) 60 around the threefold axes, and (iii) 30 on the twofold axis. The (i) and (ii) positions involve non-equivalent binding of λ1. The (i)-type subunits bind across the middle portions of a λ1-A and λ1-B protein, and (ii) bridges the middle of one λ1-B to the C-terminus of another λ1-B subunit. There is a single σ2 at site (iii), in one of two equally likely orientations. The twofold symmetric sides of site (iii) (on λ1-A) are the same as the C-terminal portion of site (ii) (on λ1-B). Both of the two unrelated (i) and (ii) type binding sites for σ2 have high affinity, as the σ2 subunit has only limited contacts within the core but no tendency to dissociate. Only modest conformational adjustments are needed to fit the non-equivalent positions. When λ1 is expressed in the absence of σ2, no icosahedral particles are assembled, indicating that σ2 is required for assembly of core capsids (Xu *et al.*, 1993; Coombs *et al.*, 1994; Kim *et al.*, 2002), as also inferred by a ts mutation in σ2 (Coombs *et al.*, 1994). σ2 also binds dsRNA (Schiff *et al.*, 1988; Dermody *et al.*, 1991) and plays a role in the induction of interferon-β in infected cultures of cardiac myocytes (Sherry *et al.*, 1998).

λ3 is the RNA-dependent RNA-polymerase (reviewed by Tao and Harrison, this volume)

λ3 is the RNA-dependant RNA polymerase (RdRp) (Drayna and Fields, 1982a; Koonin *et al.*, 1989; Starnes and Joklik, 1993). The λ3 subunits are situated interior to the core capsid near the fivefold axes, thereby positioning the newly synthesized mRNA for exit through the λ2 turrets (Figures 1.2 and 1.9). Since a λ3 monomer binds at each vertex, it is likely that a unique dsRNA segment is attached to each complex, probably via a cap-binding pocket identified on the surface (Tao *et al.*, 2002). The X-ray crystal structure showed that the protein is shaped roughly as a cube with a diameter of ~65 Å, containing several channels (Tao *et al.*, 2002) (Figure 1.9). The protein has the canonical three domains reported for many nucleic acid polymerases, but contains two additional regions, an N-terminal domain on one side and a C-terminal 'bracelet' domain that encircles one of the channels. Although a λ3 subunit binds below each λ2 turret (Coombs, 1998; Dryden

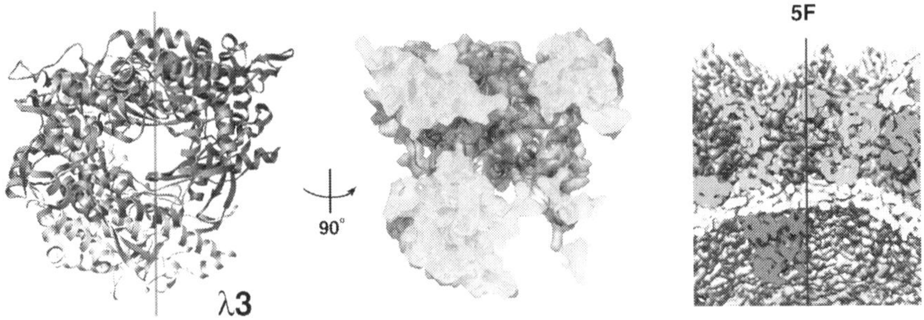

Figure 1.9 The λ3 RdRp mediates RNA synthesis. In the left panel, the three polymerase subdomains (red, blue and green) are bracketed by the N-terminal domain (yellow) and the C-terminal bracelet (pink) (pdb 1MUK). In the middle panel the monomer is rotated 90° and slabbed to display the channels in the same orientation as in the cross-section of the pseudoatomic model (right) (pdb 2CSE). The alignment of λ3 is based on fitting the crystal structure into a 7.6 Å resolution cryoEM density map (Zhang *et al.*, 2003) and positions λ3 slightly off centre from the fivefold axis. (See also Plate 1.9.)

et al., 1998), the best fit of the X-ray structure into the fivefold averaged density into the 7.6 Å cryoEM reconstruction suggested that λ3 is offset slightly from the fivefold axis (Figure 1.9, right). This position orients the protein such that transcripts exiting the polymerase domain would be directed through potential pores between adjacent λ1-A monomers (Zhang *et al.*, 2003).

μ2 may be a polymerase cofactor

μ2 is the only MRV structural protein for which there is currently no high-resolution structure, and its precise location remains unknown. It has been proposed that μ2 is present in 20 to 24 copies (Coombs, 1998) and may associate directly with the λ3 polymerase (Dryden *et al.*, 1998; Zhang *et al.*, 2003). Compared with the other reovirus structural proteins, the function of μ2 is somewhat uncertain. It may be an RdRp cofactor (Yin *et al.*, 1996; Noble and Nibert, 1997b) since it is involved in the *in vitro* transcription of ssRNA (Yin *et al.*, 1996), the conversion of mRNA to dsRNA (Coombs, 1996), and possesses nucleoside triphosphatase activity (Noble and Nibert, 1997b). μ2 also plays important roles in the pathogenesis of myocarditis (Sherry and Fields, 1989) and in the formation and morphology of viral inclusions (Parker *et al.*, 2002; Yin *et al.*, 2004; Kobayashi *et al.*, 2006), which may be mediated, in part, by its level of ubiquination (Miller *et al.*, 2004).

Structural comparison of MRV with its closest relatives, ARV and AqRV

Most of the orthoreovirus genogroups, except for MRV, are capable of inducing cell–cell fusion (syncytia formation), which is very unusual for non-enveloped viruses. This property has been observed for the ARVs (Shmulevitz and Duncan, 2000), BRVs (Duncan *et al.*, 1995) and reptilian reoviruses (Corcoran and Duncan, 2004). The closely related aquareoviruses, AqRV, are also fusogenic (Winton *et al.*, 1987), as are a number of other strains of *Reoviridae*. Unlike enveloped viruses, this activity is not triggered by a structural protein, but rather by a novel small non-structural peptide, which is encoded by an alternative reading frame in the RNA segment that encodes the σ1 homolog (Shmulevitz and Duncan, 2000).

Although MRVs have been most extensively studied, the structure of a related reovirus, ARV, has been examined by cryoEM and image analysis. The MRV and ARV M and S gene sequences exhibit 20–45% amino acid identity, so it is possible to infer homologous proteins (Table 1.1) (Martínez-Costas *et al.*, 1997; Duncan, 1999; Kim *et al.*, 2004; Noad *et al.*, 2006). ARV packages 10 dsRNA segments in a multilayered T = 13*l* capsid, very similar to that of MRV (Zhang *et al.*, 2005b) (Figure 1.10). ARV can also be converted into the MRV equivalent subviral particles, ISVPs and cores (Martínez-Costas *et al.*, 1997). As with MRV, the ISVPs are still infectious, and the cores are transcriptionally active (Benavente and Martínez-Costas, 2006).

There are 600 surface projections attributable to ARV σB, the protein that corresponds to σ3 in MRV. The σB subunits decorate a trimer base ascribed to μB, which, like its MRV counterpart, μ1, is myristoylated and must undergo cleavage during viral entry (Duncan, 1996). ARV has pentameric turrets formed by λC, which have a closed conformation in the virions and exhibit guanylyltransferase activity (Martínez-Costas *et al.*, 1995; Hsiao *et al.*, 2002). A small density at the centre of the turret along the fivefold axis is ascribed to σC, which is the ARV cell-attachment protein (Martínez-Costas *et al.*, 1997). Similar to the receptor-binding fragment of σ1, the X-ray crystal structure of the head domain and a portion of the tail for σC display equivalent tertiary structure to σ1 (Calvo *et al.*, 2005). The core capsid is a thin shell assigned to λA (Martínez-Costas *et al.*, 1997). As with MRV, there are 150 core nodules formed by monomers of protein σA, the homolog to σ2, which are also required for assembly (Xu *et al.*, 2004). There is density on the interior of the λA shell at the fivefold axes, ascribed to the transcriptase, likely formed by the λB RNA polymerase and probably μA as a cofactor, which has a predicted homologous NTPase domain (Noad *et al.*, 2006).

Table 1.1 Protein components of the *Reoviridae*

MRV	ARV	AqRV	Number of copies	Location	Function
λ1	λA	VP3	120	Core capsid	Binds RNA, NTPase?, helicase?
λ2	λC	VP1	60	Core turrets	Guanylyltransferase, methyltransferase
λ3	λB	VP2	~12	Core interior	RNA-dependent RNA polymerase
μ1	μB	VP4	600	Outer capsid	Membrane penetration
μ2	μA	VP5	~24	Core interior	Binds RNA, NTPase?
σ1	σC	–	~36	Outer capsid	Cell attachment protein
σ2	σA	VP6	150/120(AqRV)	Core nodules	Binds dsRNA
σ3	σB	VP7	600	Outer capsid	Protease sensitive, binds dsRNA

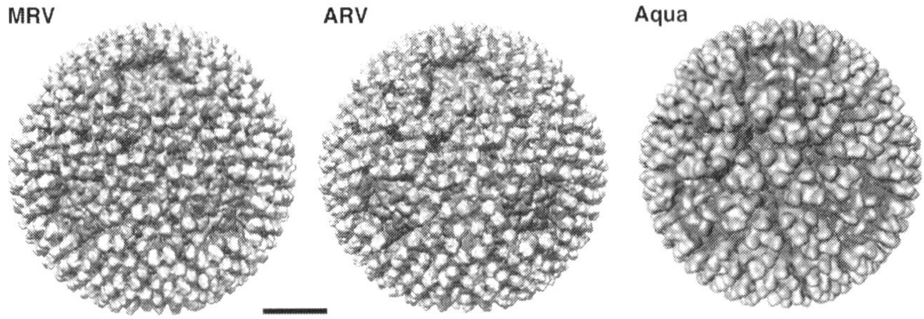

Figure 1.10 Comparison of reoviruses MRV with avian reovirus (ARV) and aquareovirus (AqRV) (Zhang *et al.*, 2005b). The structure of ARV (B) is very similar to MRV, except for the absence of density at low radii in the P1 and P2 channels (see Figure 1.2A). The closely related genus AqRV also has structural similarities (C; from Nason *et al.*, 2000, with permission), but the turrets appear narrower and have an open conformation. AqRV is coloured blue to radius 375 Å and pink for the outer radii. They are also missing the equivalent σ2 core nodules at the twofold symmetry axes (not shown). (See also Plate 1.10.)

At the 15 Å resolution of the cryoEM study, the greatest structural difference between ARV and MRV is the hub-and-spokes feature in MRV within the P2 and P3 channels (Zhang *et al.*, 2005b). The difference is not surprising because the C-terminal sequences within μ1 that presumably give rise to this density in MRV are absent from the homologous gene encoding μB (Noad *et al.*, 2006). Preliminary results for the structure of BRV, another fusogenic orthoreovirus, are very similar to ARV, including the absence of the hub-and-spokes density (T. Baker, R. Duncan, and M. Nibert, personal communication). BRV, however, also lacks an identifiable homolog to the cell attachment proteins of MRV and ARV (Dawe and Duncan, 2002).

AqRV, a distinct genus, are closely related to the orthoreoviruses, and have many structural similarities (Shaw *et al.*, 1996; Nason *et al.*, 2000). AqRV have homologous structural proteins to those of the reoviruses (Table 1.1) (Attoui *et al.*, 2002), but they package 11 instead of 10 segments of dsRNA. The cryoEM structure of the AqRV striped bass reovirus (SBR) has a multilayered capsid, with pentameric turrets extending from the core. There are surface structures similar to the μ1/σ3 heterohexamer, and the particles similarly undergo loss of the σ3 homolog, VP7, when treated with protease (Nason *et al.*, 2000). One structural difference, however, is that the turrets are partially open, even in the virion (Nason *et al.*, 2000). Another difference is that SBR has 120 core nodules, attributed to VP6, rather than the 150 observed in reoviruses. The nodules are at sites (i) and (ii), but absent across the twofold axes at site (iii) (see Figure 1.2A, core). Like BRV, AqRV does not have a cell-attachment homolog to σ1, and the SBR virion lacks density centred in the turret, which may also be a consequence of the more open turrets. Finally, the SBR virion lacks the hub-and-spokes density in its P2 or P3 channels, which makes it similar to ARV, with a truncation in the VP4 sequence corresponding to that of μ1.Q. Although phylogenetic relationships calculated for the M gene segment sequences place MRV, the fusogenic ARV, and AqRV similarly distant (Noad *et al.*, 2006), the structural similarities between studied members of the orthoreovirus genus are greater than structural similarities between orthoreoviruses and AqRV, other than the common absence of the hub-and-spokes feature in the fusogenic viruses.

During cell attachment and entry, reovirus incorporates a coordinated series of dramatic as well as subtle conformational changes and uncoating steps of the capsid proteins. Structural details about the individual proteins and the assembled particles are essential for a complete understanding of how the multiple proteins and genomic segments interact and participate in morphogenesis and infection.

Acknowledgments

We thank Timothy S. Baker for providing maps of MRV and ARV, and Barbie Ganser-Pornillos for helpful comments. Supported by NIH grant RO1 GM066087 (M.Y.) and CIHR grant 11630 (K.M.C.).

References

Attoui, H., Fang, Q., Jaafar, F.M., Cantaloube, J.-F., Biagini, P., de Micco, P., and de Lamballerie, X. (2002). Common evolutionary origin of aquareoviruses and orthoreoviruses revealed by genome characterization of Golden shiner reovirus, Grass carp reovirus, Striped bass reovirus and golden ide reovirus (genus *Aquareovirus*, family *Reoviridae*). J. Gen. Virol. *83*, 1941–1951.

Attoui, H., Jaafar, F.M., Belhouchet, M., de Micco, P., de Lamballerie, X., and Brussaard, C.P.D. (2006). *Micromonas pusilla* reovirus, a new member of the family Reoviridae assigned to a novel proposed genus (*Mimoreovirus*). J. Gen. Virol. *87*, 1375–1383.

Bartlett, J.A., and Joklik, W.K. (1988). The sequence of the reovirus serotype 3 L3 genome segment which encodes the major core protein λ1. Virology *167*, 31–37.

Bartlett, N.M., Gillies, S.C., Bullivant, S., and Bellamy, A.R. (1974). Electron microscopy study of reovirus reaction cores. J. Virol. *14*, 315–326.

Barton, E.S., Forrest, J.C., Connolly, J.L., Chappell, J.D., Liu, Y., Schnell, F.J., Nusrat, A., Parkos, C.A., and Dermody, T.S. (2001). Junction adhesion molecule is a receptor for reovirus. Cell *104*, 441–451.

Bassel-Duby, R., Jayasuriya, A., Chatterjee, D., Sonenberg, N., Maizel, J.V., Jr., and Fields, B.N. (1985). Sequence of reovirus haemagglutinin predicts a coiled-coil structure. Nature *315*, 421–423.

Beattie, E., Denzler, K.L., Tartaglia, J., Perkus, M.E., Paoletti, E., and Jacobs, B.L. (1995). Reversal of the interferon-sensitive phenotype of a vaccinia virus lacking E3L by expression of the reovirus S4 gene. J. Virol. *69*, 499–505.

Benavente, J., and Martínez-Costas, J. (2007). Avian reovirus: structure and biology. Virus Res. *123*, 105–119.

Bisaillon, M., Bergeron, J., and Lemay, G. (1997). Characterization of the nucleoside triphosphate phosphohydrolase and helicase activities of the reovirus λ1 protein. J. Biol. Chem. *272*, 18298–18303.

Bisaillon, M., and Lemay, G. (1997). Molecular dissection of the reovirus λ1 protein nucleic acids binding site. Virus Res. *51*, 231–237.

Bodkin, D.K., Nibert, M.L., and Fields, B.N. (1989). Proteolytic digestion of reovirus in the intestinal lumens of neonatal mice. J. Virol. *63*, 4676–4681.

Breun, L.A., Broering, T.J., McCutcheon, A.M., Harrison, S.J., Luongo, C.L., and Nibert, M.L. (2001). Mammalian reovirus L2 gene and λ2 core spike protein sequences and whole-genome comparisons of reoviruses type 1 Lang, type 2 Jones, and type 3 Dearing. Virology *287*, 333–348.

Calvo, P.G., Fox, G.C., Parrado, X.L.H., Llamas-Saiz, A.L., Costas, C., Martínez-Costas, J., Benavente, J., and van Raaij, M.J. (2005). Structure of the carboxy-terminal receptor-binding domain of avian reovirus fibre sigmaC. J. Mol. Biol. *354*, 137–149.

Cavalli, A., Prota, A.E., Stehle, T., Dermody, T.S., Recanatini, M., Folkers, G., and Scapozza, L. (2004). A Molecular Dynamics study of reovirus attachment protein σ1 reveals conformational changes in σ1 structure. Biophys. J. *86*, 3423–3431.

Chandran, K., Farsetta, D.L., and Nibert, M.L. 2002. Strategy for nonenveloped virus entry: a hydrophobic conformer of the reovirus membrane penetration protein μ1 mediates membrane disruption. J. Virol. *76*, 9920–9933.

Chandran, K., Parker, J.S.L., Ehrlich, M., Kirchhausen, T., and Nibert, M.L. (2003). The δ region of outer-capsid protein μ1 undergoes conformational change and release from reovirus particles during cell entry. J. Virol. 77, 13361–13375.

Chandran, K., Zhang, X., Olson, N.H., Walker, S.B., Chappell, J.D., Dermody, T.S., Baker, T.S., and Nibert, M.L. (2001). Complete *in vitro* assembly of the reovirus outer capsid produces highly infectious particles suitable for genetic studies of the receptor-binding protein. J. Virol. *75*, 5335–5342.

Chappell, J.D., Prota, A.E., Dermody, T.S., and Stehle, T. (2002). Crystal structure of reovirus attachment protein σ1 reveals evolutionary relationship to adenovirus fiber. EMBO J. *21*, 1–11.

Cleveland, D.R., Zarbl, H., and Millward, S. (1986). Reovirus guanylyltransferase is L2 gene product lambda 2. J. Virol. *60*, 307–311.

Connolly, J.L., Rodgers, S.E., Clarke, P., Ballard, D.W., Kerr, L.D., Tyler, K.L., and Dermody, T.S. (2000). Reovirus-induced apoptosis requires activation of transcription factor NF-κB. J. Virol. *74*, 2981–2989.

Coombs, K.M. (1996). Identification and characterization of a double-stranded RNA- reovirus temperature-sensitive mutant defective in minor core protein μ2. J. Virol. *70*, 4237–4245.

Coombs, K.M. (1998). Stoichiometry of reovirus structural proteins in virus, ISVP, and core particles. Virology *243*, 218–228.

Coombs, K.M. (2006). Reovirus structure and morphogenesis. Curr. Top. Microbiol. Immunol. *309*, 117–167.

Coombs, K.M., Fields, B.N., and Harrison, S.C. (1990). Crystallization of the reovirus type 3 Dearing core. Crystal packing is determined by the λ2 protein. J. Mol. Biol. *215*, 1–5.

Coombs, K.M., Mak, S.-C., and Petrycky-Cox, L.D. (1994). Studies of the major reovirus core protein σ2, Reversion of the assembly-defective mutant *ts*C447 is an intragenic process and involves back mutation of Asp-383 to Asn. J. Virol. *68*, 177–186.

Corcoran, J.A., and Duncan, R. (2004). Reptilian reovirus utilizes a small type III protein with an external myristylated amino terminus to mediate cell-cell fusion. J. Virol. *78*, 4342–4351.

Danis, C., Garzon, S., and Lemay, G. (1992). Further characterization of the ts453 mutant of mammalian orthoreovirus serotype 3 and nucleotide sequence of the mutated S4 gene. Virology *190*, 494–498.

Dawe, S., and Duncan, R. (2002). The S4 genome segment of baboon reovirus is bicistronic and encodes a novel fusion-associated small transmembrane protein. J. Virol. *76*, 2131–2140.

Denzler, K.L., and Jacobs, B.L. (1994). Site-directed mutagenic analysis of reovirus σ3 protein binding to dsRNA. Virology *204*, 190–199.

Dermody, T.S., Schiff, L.A., Nibert, M.L., Coombs, K.M., and Fields, B.N. (1991). The S2 gene nucleotide sequences of prototype strains of the three reovirus serotypes: characterization of reovirus core protein σ2. J. Virol. *65*, 5721–5731.

Drayna, D., and Fields, B.N. (1982a). Activation and characterization of the reovirus transcriptase: genetic analysis. J. Virol. *41*, 110–118.

Drayna, D., and Fields, B.N. (1982b). Biochemical studies on the mechanism of chemical and physical inactivation of reovirus. J. Gen. Virol. *63*, 161–170.

Drayna, D., and Fields, B.N. (1982c). Genetic studies on the mechanism of chemical and physical inactivation of reovirus. J. Gen. Virol. *63*, 149–159.

Dryden, K.A., Farsetta, D.L., Wang, G., Keegan, J.M., Fields, B.N., Baker, T.S., and Nibert, M.L. (1998). Internal structures containing transcriptase-related proteins in top component particles of mammalian orthoreovirus. Virology *245*, 33–46.

Dryden, K.A., Wang, G., Yeager, M., Nibert, M.L., Coombs, K.M., Furlong, D.B., Fields, B.N., and Baker, T.S. (1993). Early steps in reovirus infection are associated with dramatic changes in supramolecular structure and protein conformation: analysis of virions and subviral particles by cryoelectron microscopy and image reconstruction. J. Cell. Biol. *122*, 1023–1041.

Duncan, R. (1996). The low pH-dependent entry of avian reovirus is accompanied by two specific cleavages of the major outer capsid protein μ2C. Virology *219*, 179–189.

Duncan, R. (1999). Extensive sequence divergence and phylogenetic relationships between the fusogenic and nonfusogenic orthoreoviruses: a species proposal. Virology *260*, 316–328.

Duncan, R., Horne, D., Cashdollar, L.W., Joklik, W.K., and Lee, P.W.K. (1990). Identification of conserved domains in the cell attachment proteins of the three serotypes of reovirus. Virology *174*, 399–409.

Duncan, R., Murphy, F.A., and Mirkovic, R.R. (1995). Characterization of a novel syncytium-inducing baboon reovirus. Virology *212*, 752–756.

Ebert, D.H., Deussing, J., Peters, C., and Dermody, T.S. (2002). Cathepsin L and cathepsin B mediate reovirus disassembly in murine fibroblast cells. J. Biol. Chem. *277*, 24609–24617.

Fausnaugh, J., and Shatkin, A.J. (1990). Active site localization in a viral mRNA capping enzyme. J. Biol. Chem. *265*, 7669–7672.

Fields, B.N., Raine, C.S., and Baum, S.G. (1971). Temperature-sensitive mutants of reovirus type 3: defects in viral maturation as studied by immunofluorescence and electron microscopy. Virology *43*, 569–578.

Fraser, R.D.B., Furlong, D.B., Trus, B.L., Nibert, M.L., Fields, B.N., and Steven, A.C. (1990). Molecular structure of the cell-attachment protein of reovirus: correlation of computer-processed electron micrographs with sequence-based predictions. J. Virol. *64*, 2990–3000.

Furlong, D.B., Nibert, M.L., and Fields, B.N. (1988). Sigma 1 protein of mammalian reoviruses extends from the surfaces of viral particles. J. Virol. *62*, 246–256.

Gentsch, J.R., and Pacitti, A.F. (1987). Differential interaction of reovirus type 3 with sialylated receptor components on animal cells. Virology *161*, 245–248.

Giantini, M., Seliger, L.S., Furuichi, Y., and Shatkin, A.J. (1984). Reovirus type 3 genome segment S4: nucleotide sequence of the gene encoding a major virion surface protein. J. Virol. *52*, 984–987.

Giantini, M., and Shatkin, A.J. (1989). Stimulation of chloramphenicol acetyltransferase mRNA translation by reovirus capsid polypeptide σ3 in cotransfected COS cells. J. Virol. *63*, 2415–2421.

Grimes, J.M., Burroughs, J.N., Gouet, P., Diprose, J.M., Malby, R., Ziéntara, S., Mertens, P.P.C., and Stuart, D.I. (1998). The atomic structure of the bluetongue virus core. Nature *395*, 470–478.

Hadzisejdic, I., Cheng, K., Wilkins, J.A., Ens, W., and Coombs, K.M. (2006). High-resolution mass spectrometric mapping of reovirus digestion. Rapid Commun. Mass Spectrom. *20*, 438–446.

Harrison, S.J., Farsetta, D.L., Kim, J., Noble, S., Broering, T.J., and Nibert, M.L. (1999). Mammalian reovirus L3 gene sequences and evidence for a distinct amino-terminal region of the λ1 protein. Virology *258*, 54–64.

Harvey, J.D., Bellamy, A.R., Earnshaw, W.C., and Schutt, C. (1981). Biophysical studies of reovirus type 3. IV. Low-angle x-ray diffraction studies. Virology *112*, 240–249.

Hazelton, P.R., and Coombs, K.M. (1999). The reovirus mutant *tsA279* L2 gene is associated with generation of a spikeless core particle: implications for capsid assembly. J. Virol. *73*, 2298–2308.

Hsiao, J., Martínez-Costas, J., Benavente, J., and Vakharia, V.N. (2002). Cloning, expression, and characterization of avian reovirus guanylyltransferase. Virology *296*, 288–299.

Huismans, H., and Joklik, W.K. (1976). Reovirus-coded polypeptides in infected cells: isolation of two native monomeric polypeptides with affinity for single-stranded and double-stranded RNA, respectively. Virology *70*, 411–424.

Imani, F., and Jacobs, B.L. (1988). Inhibitory activity for the interferon-induced protein kinase is associated with the reovirus serotype 1 σ3 protein. Proc. Natl Acad. Sci. USA *85*, 7887–7891.

Jané-Valbuena, J., Breun, L.A., Schiff, L.A., and Nibert, M.L. (2002). Sites and determinants of early cleavages in the proteolytic processing pathway of reovirus surface protein σ3. J. Virol. *76*, 5184–5197.

Jané-Valbuena, J., Nibert, M.L., Spencer, S.M., Walker, S.B., Baker, T.S., Chen, Y., Centonze, V.E., and Schiff, L.A. (1999). Reovirus virion-like particles obtained by recoating infectious subvirion particles with baculovirus-expressed σ3 protein: an approach for analyzing σ3 functions during virus entry. J. Virol. *73*, 2963–2973.

Jayasuriya, A.K., Nibert, M.L., and Fields, B.N. (1988). Complete nucleotide sequence of the M2 gene segment of reovirus type 3 Dearing and analysis of its protein product μ1. Virology *163*, 591–602.

Kim, J., Tao, Y., Reinisch, K.M., Harrison, S.C., and Nibert, M.L. (2004). Orthoreovirus and aquareovirus core proteins: conserved enzymatic surfaces, but not protein-protein interfaces. Virus Res. *101*, 15–28.

Kim, J., Zhang, X., Centonze, V.E., Bowman, V.D., Noble, S., Baker, T.S., and Nibert, M.L. (2002). The hydrophilic amino-terminal arm of reovirus core shell protein λ1 is dispensable for particle assembly. J. Virol. *76*, 12211–12222.

Kobayashi, T., Chappell, J.D., Danthi, P., and Dermody, T.S. (2006). Gene-Specific inhibition of reovirus replication by RNA interference. J. Virol. *80*, 9053–9063.

Koonin, E.V., Gorbalenya, A.E., and Chumakov, K.M. (1989). Tentative identification of RNA-dependent RNA polymerases of dsRNA viruses and their relationship to positive strand RNA viral polymerases. FEBS Lett. *252*, 42–46.

Lee, P.W.K., Hayes, E.C., and Joklik, W.K. (1981). Protein σ1 is the reovirus cell attachment protein. Virology *108*, 156–163.

Lemay, G., and Danis, C. (1994). Reovirus λ1 protein: affinity for double-stranded nucleic acids by a small amino-terminal region of the protein independent from the zinc finger motif. J. Gen. Virol. *75*, 3261–3266.

Liemann, S., Chandran, K., Baker, T.S., Nibert, M.L., and Harrison, S.C. (2002). Structure of the reovirus membrane-penetration protein, μ1, in a complex with Its protector protein, σ3. Cell *108*, 283–295.
Luftig, R.B., Kilham, S.S., Hay, A.J., Zweerink, H.J., and Joklik, W.K. (1972). An ultrastructural study of vrions and cores of reovirus type 3. Virology *48*, 170–181.
Luongo, C.L., Contreras, C.M., Farsetta, D.L., and Nibert, M.L. (1998). Binding site for S-adenosyl-L-methionine in a central region of mammalian reovirus λ2 protein. Evidence for activities in mRNA cap methylation. J. Biol. Chem. *273*, 23773–23780.
Luongo, C.L., Dryden, K.A., Farsetta, D.L., Margraf, R.L., Severson, T.F., Olson, N.H., Fields, B.N., Baker, T.S., and Nibert, M.L. (1997). Localization of a C-terminal region of λ2 protein in reovirus cores. J. Virol. *71*, 8035–8040.
Luongo, C.L., Reinisch, K.M., Harrison, S.C., and Nibert, M.L. (2000). Identification of the guanylyltransferase region and active site in reovirus mRNA capping protein λ2. J. Biol. Chem. *275*, 2804–2810.
Luongo, C.L., Zhang, X., Walker, S.B., Chen, Y., Broering, T.J., Farsetta, D.L., Bowman, V.D., Baker, T.S., and Nibert, M.L. (2002). Loss of activities for mRNA synthesis accompanies loss of λ2 spikes from reovirus cores, An effect of λ2 on λ1 shell structure. Virology *296*, 24–38.
Mabrouk, T., and Lemay, G. (1994). Mutations in a CCHC zinc-binding motif of the reovirus σ3 protein decrease its intracellular stability. J. Virol. *68*, 5287–5290.
Mao, Z., and Joklik, W.K. (1991). Isolation and enzymatic characterization of protein λ2, the reovirus guanylyltransferase. Virology *185*, 377–386.
Martínez-Costas, J., Grande, A., Varela, R., García-Martínez, C., and Benavente, J. (1997). Protein architecture of avian reovirus S1133 and identification of the cell attachment protein. J. Virol. *71*, 59–64.
Martínez-Costas, J., Varela, R., and Benavente, J. (1995). Endogenous enzymatic activities of the avian reovirus S1133, Identification of the viral capping enzyme. Virology *206*, 1017–1026.
Mendez, I.I., She, Y.-M., Ens, W., and Coombs, K.M. (2003). Digestion pattern of reovirus outer capsid protein σ3 determined by mass spectrometry. Virology *311*, 289–304.
Mertens, P. (2004). The dsRNA viruses. Virus Res. *101*, 3–13.
Metcalf, P., Cyrklaff, M., and Adrian, M. (1991). The three-dimensional structure of reovirus obtained by cryo-electron microscopy. EMBO J. *10*, 3129–3136.
Middleton, J.K., Severson, T.F., Chandran, K., Gillian, A.L., Yin, J., and Nibert, M.L. (2002). Thermostability of reovirus disassembly intermediates (ISVPs) correlates with genetic, biochemical, and thermodynamic properties of major surface protein μ1. J. Virol. *76*, 1051–1061.
Miller, C.L., Parker, J.S.L., Dinoso, J.B., Piggott, C.D.S., Perron, M.J., and Nibert, M.L. (2004). Increased ubiquitination and other covariant phenotypes attributed to a strain- and temperature-dependent defect of reovirus core protein μ2. J. Virol. *78*, 10291–10302.
Nason, E.L., Samal, S.K., and Prasad, B.V.V. (2000). Trypsin-induced structural transformation in Aquareovirus. J. Virol. *74*, 6546–6555.
Nibert, M.L., Chappell, J.D., and Dermody, T.S. (1995). Infectious subvirion particles of reovirus type 3 Dearing exhibit a loss in infectivity and contain a cleaved σ1 protein. J. Virol. *69*, 5057–5067.
Nibert, M.L., Dermody, T.S., and Fields, B.N. (1990). Structure of the reovirus cell-attachment protein: a model for the domain organization of σ1. J. Virol. *64*, 2976–2989.
Nibert, M.L., and Fields, B.N. (1992). A carboxy-terminal fragment of protein μ1/μ1C is present in infectious subvirion particles of mammalian reoviruses and is proposed to have a role in penetration. J. Virol. *66*, 6408–6418.
Nibert, M.L., Odegard, A.L., Agosto, M.A., Chandran, K., and Schiff, L.A. (2005). Putative autocleavage of reovirus μ1 protein in concert with outer-capsid disassembly and activation for membrane permeabilization. J. Mol. Biol. *345*, 461–474.
Nibert, M.L., and Schiff, L.. (2001). Reoviruses and their replication. In: Fields Virology, D.M. Knipe, M. Howley, D.E. Griffin, R.A. Lamb, M.A. Martin, B. Roizman and S.E. Straus, eds. (Lippincott Williams & Wilkins, Philadelphia). pp. 1679–1728.
Nibert, M.L., Schiff, L.A., and Fields, B.N. (1991). Mammalian reoviruses contain a myristoylated structural protein. J. Virol. *65*, 1960–1967.
Noad, L., Shou, J., Coombs, K.M., and Duncan, R. (2006). Sequences of avian reovirus M1, M2 and M3 genes and predicted structure/function of the encoded μ proteins. Virus Res. *116*, 45–57.
Noble, S., and Nibert, M.L. (1997a). Characterization of an ATPase activity in reovirus cores and its genetic association with core-shell protein λ1. J. Virol. *71*, 2182–2191.

Noble, S., and Nibert, M.L. (1997b). Core protein μ2 is a second determinant of nucleoside triphosphatase activities by reovirus cores. J. Virol. *71*, 7728–7735.
Odegard, A.L., Chandran, K., Zhang, X., Parker, J.S.L., Baker, T.S., and Nibert, M.L. (2004). Putative autocleavage of outer capsid protein μ1, allowing release of myristoylated peptide μ1N during particle uncoating, is critical for cell entry by reovirus. J. Virol. *78*, 8732–8745.
Olland, A.M., Jané-Valbuena, J., Schiff, L.A., Nibert, M.L., and Harrison, S.C. (2001). Structure of the reovirus outer capsid and dsRNA-binding protein σ3 at 1.8 Å resolution. EMBO J. *20*, 979–989.
Parker, J.S.L., Broering, T.J., Kim, J., Higgins, D.E., and Nibert, M.L. (2002). Reovirus core protein μ2 determines the filamentous morphology of viral inclusion bodies by interacting with and stabilizing microtubules. J. Virol. *76*, 4483–4496.
Qiu, T., and Luongo, C.L. (2003). Identification of two histidines necessary for reovirus mRNA guanylyltransferase activity. Virology *316*, 313–324.
Reinisch, K.M., Nibert, M.L., and Harrison, S.C. (2000). Structure of the reovirus core at 3.6 Å resolution. Nature *404*, 960–967.
Schiff, L.A., Nibert, M.L., Co, M.S., Brown, E.G., and Fields, B.N. (1988). Distinct binding sites for zinc and double-stranded RNA in the reovirus outer capsid protein σ3. Mol. Cell Biol. *8*, 273–283.
Sharpe, A.H., and Fields, B.N. (1982). Reovirus inhibition of cellular RNA and protein synthesis: role of the S4 gene. Virology *122*, 381–391.
Shatkin, A.J. (1974). Methylated messenger RNA synthesis *in vitro* by purified reovirus. Proc. Natl Acad. Sci. USA *71*, 3204–3207.
Shaw, A.L., Samal, S.K., Subramanian, K., and Prasad, B.V.V. (1996). The structure of aquareovirus shows how the different geometries of the two layers of the capsid are reconciled to provide symmetrical interactions and stabilization. Structure *4*, 957–967.
Shepard, D.A., Ehnstrom, J.G., Skinner, P.J., and Schiff, L.A. (1996). Mutations in the zinc-binding motif of the reovirus capsid protein σ3 eliminate its ability to associate with capsid protein μ1. J. Virol. *70*, 2065–2068.
Sherry, B., and Fields, B.N. (1989). The reovirus M1 gene, encoding a viral core protein, is associated with the myocarditic phenotype of a reovirus variant. J. Virol. *63*, 4850–4856.
Sherry, B., Torres, J., and Blum, M.A. (1998). Reovirus induction of and sensitivity to beta interferon in cardiac myocyte cultures correlate with induction of myocarditis and are determined by viral core proteins. J. Virol. *72*, 1314–1323.
Shing, M., and Coombs, K.M. (1996). Assembly of the reovirus outer capsid requires μ1/σ3 interactions which are prevented by misfolded σ3 protein in temperature-sensitive mutant *tsG453*. Virus Res. *46*, 19–29.
Shmulevitz, M., and Duncan, R. (2000). A new class of fusion-associated small transmembrane (FAST) proteins encoded by the non-enveloped fusogenic reoviruses. EMBO J. *19*, 902–912.
Starnes, M.C., and Joklik, W.K. (1993). Reovirus protein λ3 is a poly(C)-dependent poly(G) polymerase. Virology *193*, 356–366.
Strong, J.E., Leone, G., Duncan, R., Sharma, R.K., and Lee, P.W.K. (1991). Biochemical and biophysical characterization of the reovirus cell attachment protein σ1: evidence that it is a homotrimer. Virology *184*, 23–32.
Sturzenbecker, L.J., Nibert, M., Furlong, D., and Fields, B.N. (1987). Intracellular digestion of reovirus particles requires a low pH and Ii an essential step in the viral infectious cycle. J. Virol. *61*, 2351–2361.
Tao, Y., Farsetta, D.L., Nibert, M.L., and Harrison, S.C. (2002). RNA synthesis in a cage: Structural studies of reovirus polymerase λ3. Cell. *111*, 733–745.
Tyler, K.L., McPhee, D.A., and Fields, B.N. (1986). Distinct pathways of viral spread in the host determined by Rrovirus S1 gene segment. Science *233*, 770–774.
Tyler, K.L., Squier, M.K.T., Rodgers, S.E., Schneider, B.E., Oberhaus, S.M., Grdina, T.A., Cohen, J.J., and Dermody, T.S. (1995). Differences in the capacity of reovirus strains to induce apoptosis are determined by the viral attachment protein σ1. J. Virol. *69*, 6972–6979.
van Raaij, M.J., Mitraki, A., Lavigne, G., and Cusack, S. (1999). A triple β-spiral in the adenovirus fibre shaft reveals a new structural motif for a fibrous protein. Nature *401*, 935–938.
Weiner, H.L., Powers, M.L., and Fields, B.N. (1980). Absolute linkage of virulence and central nervous system cell tropism of reoviruses to viral hemagglutinin. J. Infect Dis. *141*, 609–616.

Weiner, H.L., Ramig, R.F., Mustoe, T.A., and Fields, B.N. (1978). Identification of the gene coding for the hemagglutinin of reovirus. Virology *86*, 581–584.

Wiener, J.R., Bartlett, J.A., and Joklik, W.K. (1989a). The sequences of reovirus serotype 3 genome segments M1 and M3 encoding the minor protein μ2 and the major nonstructural protein μNS, respectively. Virology *169*, 293–304.

Wiener, J.R., McLaughlin, T., and Joklik, W.K. (1989b). The sequences of the S2 genome segments of reovirus serotype 3 and of the dsRNA-negative mutant ts447. Virology *170*, 340–341.

Wilson, G.J., Nason, E.L., Hardy, C.S., Ebert, D.H., Wetzel, J.D., Prasad, B.V.V., and Dermody, T.S. (2002). A single mutation in the carboxy terminus of reovirus outer-capsid protein σ3 confers enhanced kinetics of σ3 proteolysis, resistance to inhibitors of viral disassembly, and alterations in σ3 structure. J. Virol. *76*, 9832–9843.

Winton, J.R., Lannan, C.N., Fryer, J.L., Hedrick, R.P., Meyers, T.R., Plumb, J.A., and Yamamoto, T. (1987). Morphological and biochemical properties of four members of a novel group of reoviruses isolated from aquatic animals. J. Gen. Virol. *68*, 353–364.

Xu, P., Miller, S.E., and Joklik, W.K. (1993). Generation of reovirus core-like particles in cells infected with hybrid vaccinia viruses that express genome segments L1, L2, L3, and S2. Virology *197*, 726–731.

Xu, W., Patrick, M.K., Hazelton, P.R., and Coombs, K.M. (2004). Avian reovirus temperature-sensitive mutant *tsA12* has a lesion in major core PProtein σA and is defective in assembly. J. Virol. *78*, 11142–11151.

Yeager, M., Weiner, S.G., and Coombs, K.M. (1996). Transcriptionally active reovirus core particles visualized by electron cryo-microscopy and image reconstruction. Biophys. J. *70*, A116.

Yin, P., Cheang, M., and Coombs, K.M. (1996). The M1 gene is associated with differences in the temperature optimum of the transcriptase activity in reovirus core particles. J. Virol. *70*, 1223–1227.

Yin, P., Keirstead, N.D., Broering, T.J., Arnold, M.M., Parker, J.S.L., Nibert, M.L., and Coombs, K.M. (2004). Comparisons of the M1 genome segments and encoded μ2 proteins of different reovirus isolates. Virol. J. *1*, 6.

Yue, Z., and Shatkin, A.J. (1996). Regulated, stable expression and nuclear presence of reovirus double-stranded RNA-binding protein σ3 in HeLa cells. J. Virol. *70*, 3497–3501.

Zhang, L., Chandran, K., Nibert, M.L., and Harrison, S.C. (2006). Reovirus μ1 structural rearrangements that mediate membrane penetration. J. Virol. *80*, 12367–12376.

Zhang, X., Ji, Y., Zhang, L., Harrison, S.C., Marinescu, D.C., Nibert, M.L., and Baker, T.S. (2005a). Features of reovirus outer capsid protein μ1 revealed by electron cryomicroscopy and image reconstruction of the virion at 7.0 Å resolution. Structure *13*, 1545–1557.

Zhang, X., Tang, J., Walker, S.B., O'Hara, D., Nibert, M.L., Duncan, R., and Baker, T.S. (2005b). Structure of avian orthoreovirus virion by electron cryomicroscopy and image reconstruction. Virology *343*, 25–35.

Zhang, X., Walker, S.B., Chipman, P.R., Nibert, M.L., and Baker, T.S. (2003). Reovirus polymerase λ3 localized by cryo-electron microscopy of virions at a resolution of 7.6 Å. Nat. Struct. Biol. *10*, 1011–1018.

Cypovirus

2

Z. Hong Zhou

Abstract

Cytoplasmic polyhedrosis viruses (CPVs) form the genus *Cypovirus* of the family *Reoviridae*. CPVs are classified into 14 species based on the electrophoretic migration profiles of their genome segments. *Cypovirus* is unique among the *Reoviridae* in that it has only a single capsid shell, which is architecturally similar to the orthoreovirus inner core. Despite lacking protective outer shells, CPV exhibits striking capsid stability and is fully capable of endogenous RNA transcription and processing. The overall folds of CPV proteins are remarkably similar to those of their structural homologues in other reoviruses. However, CPV proteins have insertional domains and unique structures that contribute to their extensive intermolecular interactions. These differences, together with its unparalleled multiple modes of intermolecular complementarity, form the structural basis of the enhanced stability of CPV capsid. The CPV turret protein contains two methylase domains with a highly conserved helix-pair/β-sheet/helix-pair sandwich fold but lacks the β-barrel flap present in orthoreovirus λ2. The stacking of turret protein functional domains, the presence of constrictions and A spikes along the mRNA release pathway point to a mechanism that uses pores and channels to regulate the highly coordinated steps of RNA transcription, processing, and release.

Introduction

Cytoplasmic polyhedrosis viruses (CPVs) collectively form the genus *Cypovirus* of the family *Reoviridae* (Mertens, 1989). The CPVs are provisionally classified into 14 distinct species based on distinctive variations in the electrophoretic migration patterns of their genome segments (Payne and Rivers, 1976; Mertens *et al.*, 1989). The most extensively studied CPV is a member of the *Cypovirus*-1 subfamily that infects *Bombyx mori* silkworms (Mertens, 2000). Unlike most of the other members of the *Reoviridae*, which have double- or triple-layered capsid architecture, CPV particles have a single shell that is morphologically and biochemically comparable to the turreted cores of the other reoviruses (Nibert and Schiff, 2001).

CPV virion is composed of a single protein shell with 12 turret-like protrusions, enclosing 12 transcriptional enzyme complexes (TECs) and a dsRNA genome of 10 segments. Despite lacking the outer protein shells common to other members of the *Reoviridae*, CPV is unusually stable under various chemical conditions. It shares the same mechanism of replication and endogenous transcription with the other structurally more complex reoviruses. Recent findings from genome sequencing, molecular biology, and electron cryomicroscopy (cryoEM) have begun to shed light on some of the unusual properties of this simplest member of the *Reoviridae*.

Infection and replication

CPVs infect a variety of species of insects, most of which are pests (Smith, 1963). While its infection of silkworms can have a negative economic impact on silkworm farming in Asia, bioengineered CPV is also recognized as an emerging biocontrol agent, serving as an environmentally friendly pesticide for fruit and vegetable farming (Mertens *et al.*, 2000). In Asia, considerable efforts have been devoted to engineering CPV as a potential specific pesticide against various kinds of insects of agricultural significance.

CPV causes a chronic gut infection in insects (Smith, 1963). After replication and assembly, the CPV virions are usually embedded in variable numbers in cubic or polyhedral occlusion/inclusion bodies. These bodies, known as polyhedra, are formed by a crystalline arrangement of the viral protein polyhedrin (Smith, 1963; Mertens *et al.*, 2000; Coulibaly *et al.*, 2007). The shape of the polyhedra inside the host cell is characteristic of each type of CPV; they can be spherical, cuboid, hexahedral, rod shaped, or amorphous. The *Bombyx mori* CPV (*Bm*CPV) inclusion bodies are 1–3 μm in diameter. The virions are separated by approximately 500–1000 Å in the polyhedra (Xia, 1991; Mertens *et al.*, 2000; Coulibaly *et al.*, 2007), suggesting that about 10 000 virus particles can be occluded in a single polyhedral body. As illustrated in Figure 2.1, upon entering the alkaline midgut of an insect through the oral route, the matrix of polyhedra dissolves, resulting in the releasing of single-shelled CPV virions that, in turn, infect epithelial cells (Payne and Harrap, 1977). This type of CPV transmission is called vertical infection. Alternatively, non-occluded, single-shelled CPV capsids can be released to directly infect neighbouring cells in the gut of the same insect. This process is referred to as parallel infection.

Vertical infection is the means by which CPV is transmitted from one insect to another through oral ingestion of leaf tissues contaminated by polyhedra released from dead insects. The polyhedra containing the CPV capsids are dissolved in the alkaline environment of the insect gut, thereby releasing free viral particles. The exact mechanism by which the viruses enter the cells is not clear. Recent results suggest that intact virions enter columnar cells by means of direct penetration through the cell membrane (Tan, 2003). mRNA is synthesized by endogenous RNA transcription, capped, and released from the intact capsid. Upon protein synthesis, the viral capsid is assembled in virogenic stroma, a viral protein-rich region in the cytoplasm and a hallmark of active CPV infection. Virogenic stroma and progeny virions have been observed in columnar cells, but not in other midgut cells (Tan *et al.*, 2003).

Most studies of CPV infection have been restricted to the examination of tissues dissected from insect larvae infected with uncharacterized virus isolates. During CPV infection, the most pronounced effect on the host cell is the degradation of rough endoplasmic reticulum. Eventually, cell lysis occurs and CPV-containing polyhedra are released into the midgut lumen and excreted. Epithelial cells are regenerated within the tissue. This continual replacement of susceptible cells may account for the chronic, or slow-progressing nature of CPV disease in infected insects (Smith, 1963).

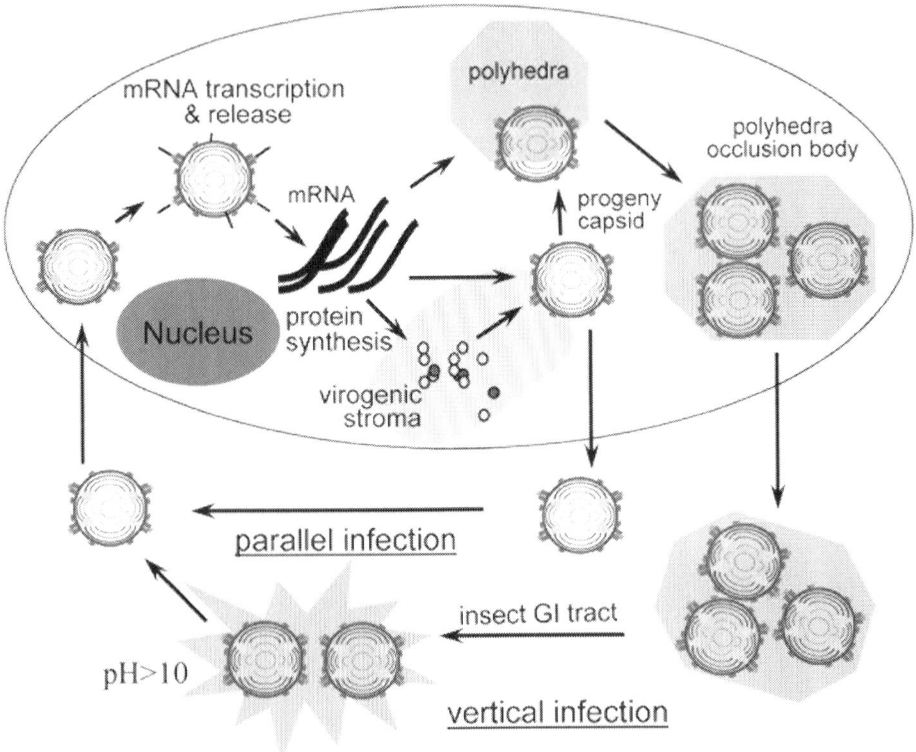

Figure 2.1 Schematic diagram of CPV replication cycle. Inside host cells, CPVs are either non-occluded or occluded in polyhedra, a crystalline inclusion body formed by viral protein polyhedrin. Non-occluded capsids can be released upon cell lysis to directly infect adjacent epithelial cells. This pathway is called parallel infection. Alternatively, occluded viruses are excreted and can be orally ingested by other insects to cause vertical infection. In this pathway, polyhedra dissolve in the alkaline environment of the midgut of the gastrointestinal (GI) tract, releasing 'naked' CPV to initiate infection by entering epithelial cells. mRNA synthesis takes place inside the cytoplasm. Protein synthesis and viral assembly occur inside virogenic stroma.

Genome structure and molecular composition

The 10 segments of the CPV dsRNA genome have a total estimated molecular mass of 20.4 MDa (Figure 2.2a) (Hayashi and Kawase, 1964). Similar to other reoviruses, the CPV genome is monocistronic; that is, each segment of the genome encodes a single, specific structural or non-structural protein (Table 2.1). The molecular masses of the genomic segments can be estimated by nucleic acid electrophoresis. The migration patterns of genome segments vary considerably among different cypoviruses (Payne and Rivers, 1976). These variations form the basis for classifying the cypoviruses into 14 distinct species (Payne and Rivers, 1976; Mertens *et al.*, 1989).

All segments of *Bm*CPV have been sequenced and their respective genes have been analyzed, as summarized in Table 2.1 (Hagiwara, 1998a; Hagiwara, 1998b; Hagiwara and Matsumoto, 2000; Hagiwara, 2001; Hagiwara, 2002). Of the gene products, five or six,

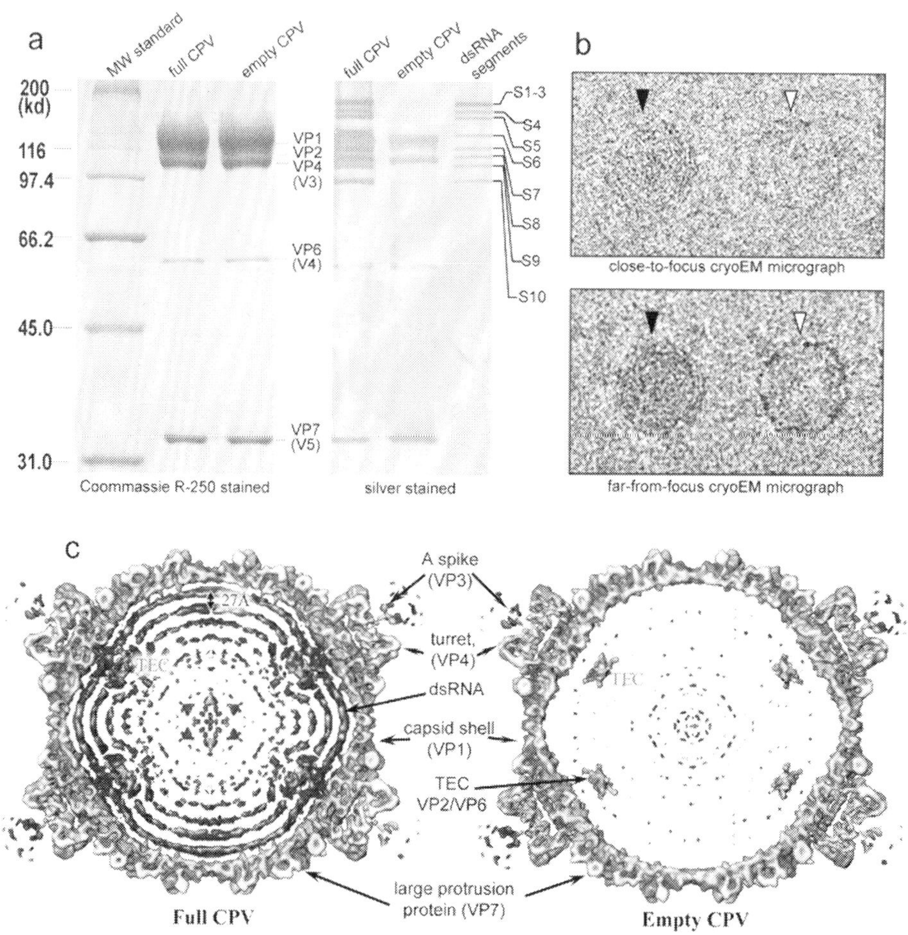

Figure 2.2 Genomic, protein and structural components of CPV. (a) 8.5% SDS PAGE of full and empty CPV particles and isolated dsRNA genome segments. Gels were stained by Coomassie blue R-250 to reveal only proteins, or by silver staining to reveal both proteins and genomic dsRNA segments. Protein names are based on modified nomenclature suggested recently (Hagiwara *et al.*, 2002; Mertens, 2004). (b) Focal-pair cryoEM micrographs of full (dark arrowhead) and empty (white arrowhead) CPV particles embedded in vitreous ice. Full CPV contains densely packed dsRNA, which gives rise to a characteristic fingerprint pattern inside the particle in its close-to-focus image. (c) Structural comparisons of full and empty CPV, with the structural components and locations of its constituent proteins indicated.

depending on the species or experimental conditions, are structural proteins and the others non-structural proteins (Table 2.1). For example, five structural proteins were shown to be associated with purified *Bm*CPV virions based on SDS gel electrophoresis. These structural proteins were originally named V1 through V5 based on the decreasing order of their molecular mass on the SDS gel (McCrae and Mertens, 1983; Payne and Mertens, 1983). These structural proteins (V1-V5) have since been renamed VP1, VP2, VP4, VP6, and VP7, respectively, with numbers designating their associated genome segments to avoid confusion arising from variations in protein mobility of different cypovirus species

Table 2.1 Genome segments and proteins of *Bm*CPV-1 (modified from Hagiwara *et al.*, 2002; Mertens *et al.*, 2004)

dsRNA segment number	Size (bp)	Protein[1](other names[2])	$M_r(\times 10^3)$	Copy number per virion	Function, putative location, and orthoreovirus homolog	Structural/functional homolog in reovirus
S1	4190	VP1 (V1)	148	120	Capsid shell protein (CSP)	λ1
S2	3854	VP2 (V2)	136	(12)	RNA-dependent RNA polymerase; forms TEC	λ3
S3	3846	VP3	140	(36)	Possibly forms A spikes	Possibly σ1
S4	3262	VP4 (V3, VP3)	115	60	Turret protein (TP), i.e. guanylyl transferase and methylases	λ2
S5	2852	NS5 NS5aNS5b	10180*23*	0	Non-structural protein, contains autocleavage sequences	Unknown
S6	1796	VP6 (V4)	64	(12)	ATP/GTP-binding protein, possibly in TEC	Possibly μ2
S7	1501	VP7 (V5)(p50)NSP7	50 (61*)58*31*	120	Non-structural proteins with structural cleavage products, *i.e.* large protrusion protein (LPP)	σ2 (note: 150 copies per reovirus)
S8	1328	NSP8 (P44)	44	unknown	Unknown	Unknown
S9	1186	NSP9 (NS5)	36	0	Non-structural,dsRNA binding	Unknown
S10	944	polyhedrin	28.5	0	forms crystalline polyhedra	None

Size of genome segments and encoded proteins was determined by sequence analysis of the genome segments.

*Sizes of some proteins estimated from electrophoretic migration.

[1]Protein names based on the new nomenclature (Hagiwara *et al.*, 2002).

[2]Protein names based on the old nomenclature (McCrae and Mertens, 1983).

(Hagiwara *et al.*, 2002). The biological functions of most of the non-structural proteins remain to be determined.

Like other members of the *Reoviridae*, the 5′ end of each of the CPV genomic segments is capped, but there is no poly-A tail at its 3′ end. The termini of the coding strands are identical across different genome segments within a species of CPV; the sequence in *Bm*CPV, for examples, is 5′-AGUAAA…GUUAGCC-3′. However, this terminal sequence seems to be virus- or species-specific and differs across cypovirus species.

When expressed in baculovirus, VP1 self-assembles and forms single-shelled, virus-like particles *in vitro* (Hagiwara and Naitow, 2003), indicating its structural role as the capsid shell protein (CSP) (Zhou, 2003). This is different from the orthoreovirus core, which requires co-expression of the CSP, λ1, and its nodule protein, σ2 (Xu, 1993). Interestingly, the recombinant VP1 virus-like particles are somewhat different from native virus particles in that the recombinant particles are more sensitive to acidic conditions (Hagiwara and Naitow, 2003).

Sequence analysis has suggested that segment 2 encoded VP2 is the RNA-dependent RNA polymerase of CPV (Rao, 2003). Sequencing and molecular biology studies have shown that segment 4 encodes VP4, which is incorporated into CPV polyhedra and released under alkaline conditions, suggesting a role for VP4 in CPV virion occlusion into the polyhedra through specific interactions with polyhedrin (Ikeda, 2001) (Note that the old 'VP3' designation was used by the authors of this study). Except for the CSP VP1, VP4 is the only major structural protein that is large and abundant enough to act as the turret protein (TP). Immunoblotting analysis indicates that VP6 is a minor component of the virion and is expressed at low levels *in vivo* (Hagiwara and Matsumoto, 2000). The low abundance and ATP/GTP binding activities suggest that VP6 might be a component of TEC. VP7 is a cluster of four variants produced from the original putative protein encoded by segment 7. These are thought to be incorporated into the process of virion maturation (Hagiwara and Matsumoto, 2000). The size and abundance of VP7 suggest that it is most likely to be the large protrusion protein (LPP) that connects adjacent VP1 molecules in the mature virion (Zhou *et al.*, 2003).

The matrix of the characteristic polyhedra of CPV is composed of a single protein, polyhedrin, which is synthesized in large amounts (up to one-third of the cell mass) during viral replication. The metabolic burden resulting from such large amounts of polyhedrin eventually has lethal effects on the infected insect. Polyhedrin is an alkaline-soluble protein that undergoes proteolysis at a pH higher than 10. It is encoded by the smallest segment of the viral genome, segment 10, and its expression appears to occur in the late infection stage (Mori, 1989). Its size varies from 25 to 31 kDa in different species of CPV. In *Bm*CPV, the polyhedrin gene has 942 base pairs and encodes a polypeptide of 248 amino acids (molecular mass of 28.5 kDa). Expression of the *Bm*CPV polyhedrin gene in a baculovirus expression vector revealed that it is capable of supramolecular assembly and that the assembly of *Bm*CPV polyhedrin occurs independently of other viral proteins (Mori, 1993).

Unparalleled capsid stability

Most of the reoviruses have two or three protein layers enclosing the dsRNA genome. The outer two layers of the capsid usually have two functional roles: to protect the viral RNA genome from the unfavourable environment outside the virus (Lawton, 2000) and to mediate specific viral attachment to host cells. Devoid of the outer protein layers, CPV particles are instead embedded in the polyhedra. Thus, the formation of occlusion (also referred to as 'inclusion') bodies is a specialized component of the CPV viral life cycle (Figure 2.1). The occlusion body is believed to have a role in virion protection during viral transfer between hosts or during vertical infection. Cypovirus is the only genus of the *Reoviridae* that has a single-shelled capsid, and is thus the structurally simplest member (Mertens *et al.*, 2000).

Despite lacking the outer protective layers found in other dsRNA viruses, CPV virions are resistant to chemical treatments, including highly concentrated cations, high pH, trypsin, chymotrypsin, ribonuclease A, deoxyribonuclease, phospholipase, and SDS, and they retain infectivity for weeks at –15 to 25°C (Mertens *et al.*, 2000; Zhang, 2002). The integrity of CPV capsids is not affected under high-pH conditions that disrupt the orthoreovirus inner core (Zhang *et al.*, 2002). This observation is consistent with the ability of CPV to maintain structural integrity in extremely alkaline environments during infection. Even after incubation at pH 12.4 for 30 minutes, it was rare (< 1%) to observe any spikeless capsids. Therefore, CPV capsids are more resistant to alkaline conditions than orthoreovirus cores. However, CPV is sensitive to low pH, detergents, and urea, and it disintegrates after 5 minutes of incubation in as little as 0.01% NP-40 and Triton X-100 or 4.5 M urea (Zhang *et al.*, 2002).

The enhanced stability of CPV capsids may reflect the ability of the virus to establish productive infection in its natural hosts. A progeny CPV capsid, whether directly released as a free virion or embedded in viral polyhedra, can establish highly productive infection in an alkaline environment (pH ~10.5), whether in the gut of the same insect infected by its parental virus or in that of a different, previously uninfected insect (Figure 2.1). Although it is more resistant to alkaline conditions than the orthoreovirus core, the CPV capsid is sensitive to acidic environments, urea, and detergent treatment (Zhang *et al.*, 2002). These observations suggest that CPV has evolved specifically to adapt to the high-pH environment encountered during its normal infection life cycle. It is interesting that most of the broken particles observed in the chemical treatments retain prominent spikes or turrets, indicating that the molecular interactions between the turret and the shell proteins are at least as strong as those among the shell proteins. The enhanced connectivity between the turret and the shell in CPV is probably required to compensate for its lack of protective outer shells.

Empty and full CPV particles and genome packing

Purification of polyhedral materials by differential centrifugation or density gradients yields two types of CPV particles: full, which contain dsRNA, and empty, which do not (Zhang, 1999) (Figure 2.2b). The origin and biological significance of the empty particles are not known. Repeated freezing and thawing or chemical treatment of purified full particles can result in accumulation of empty particles in solution. This suggests that empty particles, rather than being the precursor of full particles, probably originate from full particles that have released their dsRNA genome during the viral isolation and purification process.

The three-dimensional (3D) cryoEM structures of both empty and full CPV show identical T = 1 capsid shells consisting of 60 dimers of the CSP. The capsids are decorated by prominent surface nodules or protrusions, as well as 12 turrets (B spikes) at the icosahedral fivefold vertices. The capsid has a diameter of 570 or 715 Å when the turrets are included (but excluding the poorly defined A spikes) (Figure 2.2c). This overall architecture is similar to that of the inner core of orthoreovirus (Hill, 1999; Zhang *et al.*, 1999), confirming early suggestions that CPV is structurally and biochemically comparable to the virus cores of other turreted reoviruses (Lewandowski and Traynor, 1972). However, CPV capsid differs from orthoreovirus inner core in two aspects. First, CPV has 12 loosely attached A spikes that extend radially from the inner rims of the turrets, or B spikes (Yazaki and Miura, 1980; Hatta and Francki, 1982) (Figure 2.2c). Second, CPV has only 120 copies of LPP, in contrast to the 150 copies of its structural homolog σ2 in orthoreovirus (Dryden, 1993; Hill *et al.*, 1999; Zhang *et al.*, 1999).

3D structural comparisons of the full and empty CPV particles by cryoEM revealed the structural organization of the TEC and the packing of the dsRNA genome inside the capsid (Zhang *et al.*, 1999; Xia, 2003) (Figure 2.2c,d). The empty CPV capsid has twelve TEC densities attached to the inside of the capsid shell underneath the turret, which serves as the mRNA-capping enzyme complex. The full capsid shell is identical but contains additional internal densities attributed to genomic dsRNA. The RNA densities proximal to the capsid shell are organized into layers with a dodecahedral appearance, suggesting that the 10 genome segments are each organized as a cone-shaped spool surrounding a TEC (Figure 2.2c) (Xia *et al.*, 2003). The 27-Å interduplex spacing between dsRNA segments in full CPV particles indicates that the CPV genome is more densely packed than that of blue-tongue virus (BTV) (Gouet, 1999) and L-A virus (Caston, 1997) but less densely packed than those of bacteriophages, which have interduplex spacing of 22–24 Å (Harrison, 1996). The RNA densities conform to the angular shape of the surrounding capsid shell, suggesting that the capsid shell serves as a scaffold for appropriate positioning of the RNA genome and TEC. Furthermore, it has been shown that nascent mRNA is released through the constricted central channel of the turret (Yazaki and Miura, 1980). Based on these observations, a detailed moving-template transcription mechanism has been proposed to explain the highly efficient endogenous RNA transcription of dsRNA viruses (Xia *et al.*, 2003).

Molecular organization of the CPV capsid shell

The determination of the 3D structure of full CPV to subnanometer resolution has permitted the analyses of molecular interactions among its constituent structural proteins and the identification of secondary structural elements within each protein (Zhou *et al.*, 2003) (Figure 2.3a). The capsid shell of CPV is composed of three major structural proteins: VP1, the CSP and structural homolog of BTV VP3 and orthoreovirus λ1; VP4, the TP homolog of orthoreovirus λ2; and VP7, the LPP homolog of orthoreovirus σ2 (Zhou *et al.*, 2003) (Table 2.1). Each CPV capsid shell consists of 60 copies of TP and 120 copies each of CSP and LPP (Figure 2.3).

Each asymmetric unit of the CPV capsid shell contains one TP monomer and two pairs of quasi-equivalent but non-icosahedrally related molecules, CSP-A/CSP-B and LPP-5/LPP-3 (near fivefold and threefold axes, respectively) (Figure 2.3b,c). The two non-icosahedrally related molecules in each of these pairs are similar in both their overall shapes and secondary structure elements. The framework of the CPV capsid is formed by 120 plate-like CSP molecules. A pair of CSP-A molecules spans the edge across each

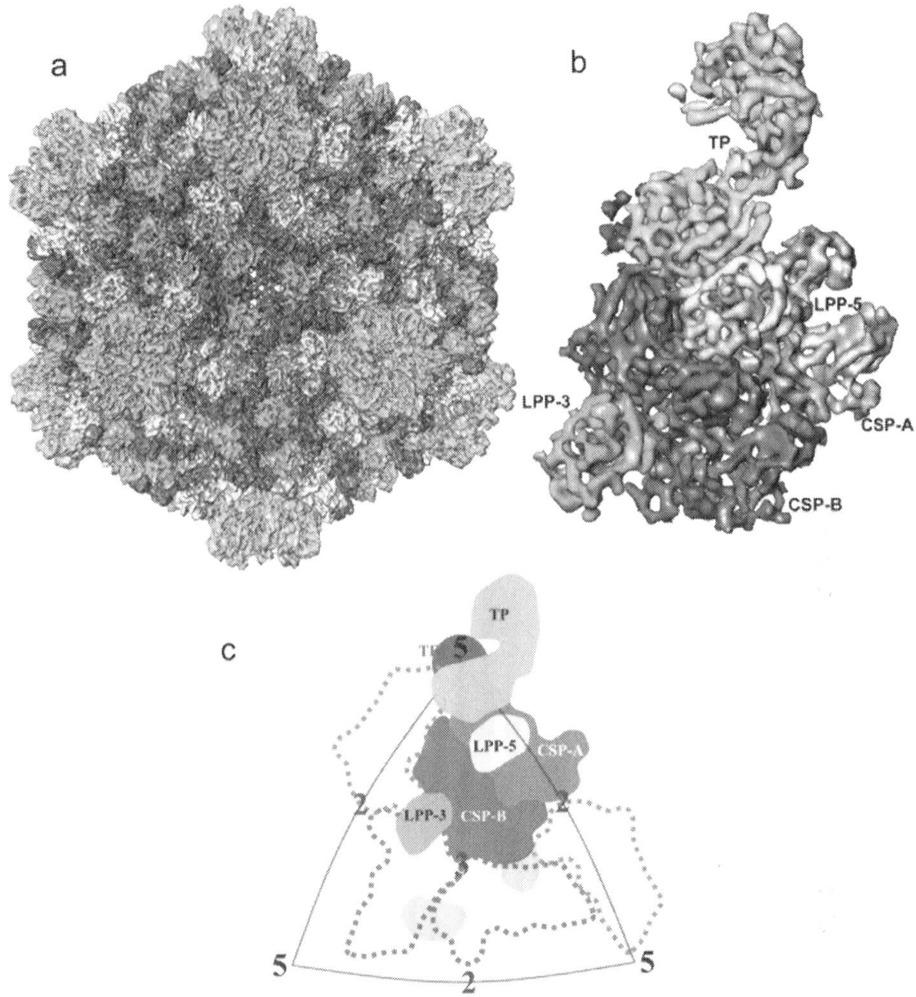

Figure 2.3 Structure organization and surface complementarity of CPV. (a) Shaded surface view of a CPV reconstruction at about 7-Å resolution as viewed along a threefold axis. Different molecules are coloured differently. (b) Close-up view of an asymmetric unit, which is made up of one TP (VP4), two copies of the CSP (VP1), and two copies of the LPP (VP7) (fivefold-proximal LPP-5 is in yellow; threefold-proximal LPP-3 is in green). (c) Schematic diagram illustrating the positions and interactions among the structural components with respect to the icosahedral five-, three-, and twofold axes. The blue triangle represents one triangular face of the icosahedron. (b) and (c) are adapted from (Zhou *et al.*, 2003) with permission from the publisher. (See also Plate 2.3.)

icosahedral twofold axis, and three CSP-B molecules seal the central space of the icosahedral triangular face near each icosahedral threefold axis (Figure 2.3c). LPP-5 and LPP-3 further strengthen this basic framework by serving as molecular clamps joining two (one CSP-A and one CSP-B) or three (one CSP-A and two CSP-B) adjacent CSP molecules, respectively. Each turret, a pentamer of TP located on each icosahedral vertex of the capsid, provides additional strength for this framework by 'sewing' together five CSP-A molecules and by making two contacts with the tip of each nearby CSP-B (Figure 2.3).

Structural basis of the enhanced stability of CPV

CSP is shaped like a thin, planar wedge with the dimensions 150 × 60 × 25–40 Å (Figure 2.4). When viewed from the outside of the capsid, it resembles an inverted 'F' and can be divided into four domains, three of which are similar to the domains identified in the inner capsid proteins in other related viruses such as BTV VP3 (Grimes, 1998), orthoreovirus λ1 (Reinisch, 2000), and rice dwarf virus P3 (Zhou, 2001; Nakagawa, 2003). These three domains are designated as apical, carapace, and dimerization domains following the nomenclature of BTV VP3 (Grimes *et al.*, 1998). CPV CSP contains a unique fourth domain, the small protrusion domain, which is involved in interactions between CSP molecules (c.f. Figure 2.4a and b). Unlike the structural homologues in BTV and animal reovirus, where large domain twists and radial curvature are observed between CSP-A and CSP-B, the quasi-equivalent CSP molecules in CPV are almost superimposable across all domains except for a 20° twist between the two small protrusion domains (Zhou *et al.*, 2003). Despite a general lack of sequence homology (identity ~15%) between CPV CSP and its structural homologues in other reoviruses, most of the α-helices and β-sheets identified in the 8-Å map of CPV CSP are topologically conserved with those revealed in the X-ray crystallographic models of reovirus λ1 (Reinisch *et al.*, 2000), BTV VP3 (Grimes *et al.*, 1998), and rice dwarf virus P3 (Nakagawa *et al.*, 2003). That is, the same types of secondary structure elements appear at roughly similar locations with only slightly different orientations, though their exact locations and orientations differ (Zhou *et al.*,

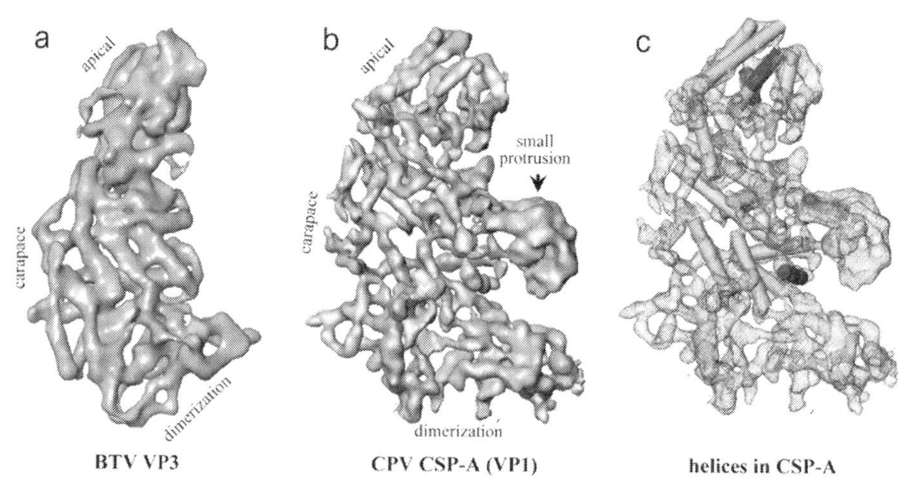

Figure 2.4 Structure of CSP. (a) The atomic structure of BTV VP3 was generated by Gaussian-filtering the atomic model (Grimes *et al.*, 1998) to 6 Å. (b) CPV CSP-A as viewed from outside of the capsid. Except for the unique small protrusion domain, the domains are labelled by analogy to those used in BTV VP3 (Grimes *et al.*, 1998). (c) CSP-A shown using wire-frame representation, superimposed with the identified helices, which were modelled as cylinders 5 Å in diameter. CPV CSP contains four unique helices (dark grey).

2003) (Figure 2.4). In addition, CPV CSP has a β-sheet-rich protrusion domain (Figure 2.4b) and four unique α-helices, two of which are located in the small protrusion domain and the other two in the apical domain (Zhou *et al.*, 2003) (Figure 2.4c).

The mechanism by which the 120 CSP (VP1) molecules associate with each other to form the thin and contiguous capsid shell represents a remarkable example of multiple modes of interaction and geometric complementarity (Zhou *et al.*, 2003). First, there are no apparent pores or gaps larger than 10 Å across the entire capsid shell, indicating perfect surface complementarity among all inter-CSP contacts. Remarkably, CPV CSP has achieved this perfect surface complementarity without introducing major domain twists and curvature adjustments. Second, there is no similarity in the molecular contacts made by the corresponding regions of CSP-A and CSP-B with neighbouring molecules, except for a small region (< 20 Å) in the dimerization domain across the dyad of a CSP-A and a CSP-B (near LPP-3 in Figure 2.3c). For example, the carapace domain of CSP-A interacts with the opposite side of a single CSP-B molecule, while the carapace domain of CSP-B interacts with a different domain of a CSP-A as well as the dimerization domain of another CSP-B (Figure 2.3c). These multiple modes of molecular interaction apply not only to inter-CSP interactions, but also to the interactions between CSP and LPP (Zhou *et al.*, 2003).

The 120 LPP (VP7) molecules in each CPV capsid act as molecular clamps, tying together neighbouring CSP molecules on the capsid shell (Figure 2.3b,c). The quasi-equivalent LPP-5 and LPP-3 have essentially identical rectangular shapes and dimensions, and similar secondary structural elements when viewed from above. However, LPP-3 and LPP-5 adopt completely different structures in the regions that interact with underlying CSP molecules (Zhou *et al.*, 2003).

Most of the reoviruses contain one or two outer protein capsid shells that stabilize the thin inner shell. In addition to its conserved roles in enclosing the dsRNA genome and transcriptional enzymes, the CPV capsid has evolved to assume the structural roles that are served by the outer shells of other dsRNA viruses. Although CPV particles are protected during transmission between insects by occlusion inside viral polyhedra, this matrix is dissolved upon entering the alkaline insect midgut, releasing 'naked' CPV capsids to infect epithelial cells. Non-occluded CPV can also be released to directly infect neighbouring cells in the parallel infection pathway (Figure 2.1) (Payne and Harrap, 1977). To adequately protect its genome, it is conceivable that a less porous and more structurally stable capsid shell is required for CPV during its attachment and entry into host cells, as it lacks the protective outer shells of other dsRNA viruses. As illustrated in Figure 2.3c, the 120 LPPs on each CPV capsid are perfectly positioned to join adjacent CSPs together either at the intra- or inter-dimer interfaces. The substantial interactions between LPPs and the underlying CSPs, the high surface complementarity of CSP, the extra interactions provided by the small protrusion domain, and the extensive TP-CSP interactions might collectively contribute to the enhanced capsid stability of CPV. On the other hand, the fact that the capsid shell lacks pores to accommodate the release of mRNA suggests that conformational changes of CSP must occur during active mRNA synthesis and release. The unusual molecular interactions used by CPV, characterized by a multivalent, variable, or promiscuous nature, may facilitate such conformational transitions required during transcription activation.

The mRNA capping complex

Each CPV turret (B spike) is a multifunctional complex involved in the highly coordinated steps of transcriptional processing and release of nascent mRNA, similar to its structural analogue in the orthoreovirus core (Bartlett, 1974; Furuichi, 1974; Bellamy and Harvey, 1976; White and Zweerink, 1976; Yazaki and Miura, 1980; Zhang *et al.*, 1999; Reinisch *et al.*, 2000). 3D structural comparisons of intact CPV capsids with spikeless (lacking turrets) CPV capsids generated by chemical treatment have allowed the determination of the molecular boundaries between the turret and the CPV capsid (Figure 2.5a) (Zhang *et al.*, 2002). Each turret is a hollow cylinder, 150 Å in diameter and 80 Å in height, with an axial channel leading to the inside of the capsid (Figure 2.5b,c). Under each turret is a pore that is plugged by a small spherical density of about 10 Å. This plug is formed by a cluster of five pairs of α-helices located in the apical domains of the CSP-A pentamer. This plugged pore can be triggered to open by chemical treatments, suggesting a possible route of nascent mRNA release (Zhang *et al.*, 2002). The plug and the two constrictions inside the turret channel result in the formation of two cylindrical cavities (~35 Å tall, 60 Å in

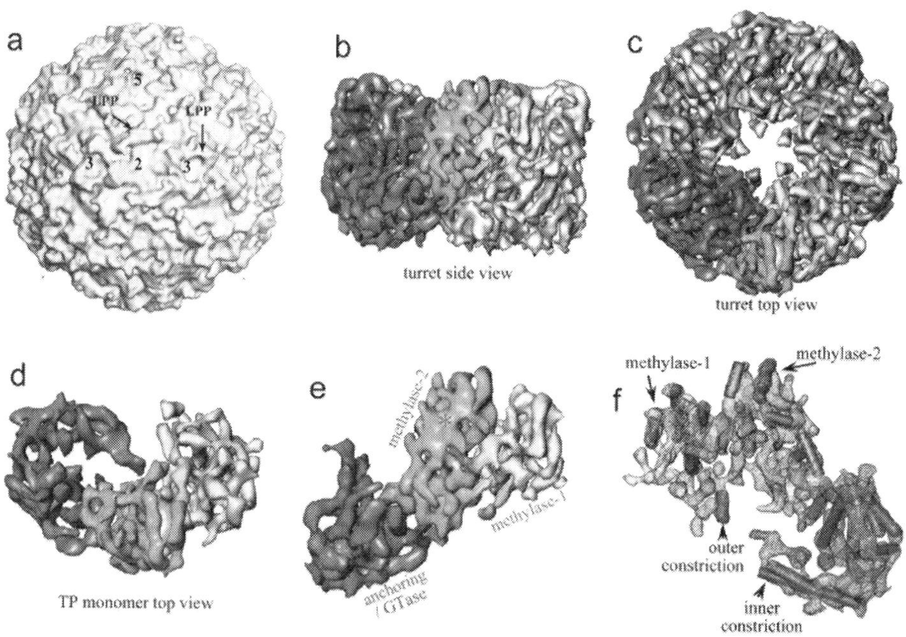

Figure 2.5 Structure of the mRNA capping complex, or turret. (a) Shaded surface representation of CPV with its turrets removed by chemical treatment (Zhang *et al.*, 2002). (b,c) Side (perpendicular to fivefold axis) (b) and top (parallel to fivefold axis) (c) views of a computationally isolated turret. Two adjacent subunits are highlighted in colour. (d,e) Top (d) and side (e) views of an extracted TP monomer. The asterisk indicates the large, twisted β-sheet in the characteristic methylase-like, helix-pair/β-sheet/helix-pair fold of the methylase-2 domain. (f) Wire-frame representation of TP as viewed from the inside the turret channel looking outward. The superimposed 5-Å blue cylinders represent the α-helices identified in each TP. The helices contributing to the formation of the inner and outer channel constrictions are indicated. Adapted from (Zhang *et al.*, 2002) (a) and (Zhou *et al.*, 2003) (b-f) with permission from the publishers. (See also Plate 2.5.)

diameter) along each turret channel and a V-shaped opening in the outermost region of the turret (Zhou *et al.*, 2003).

CPV turrets have methyl-transferring activities and contain *S*-adenosyl-L-methionine (SAM)-binding sites (Furuichi, 1974) similar to those of orthoreovirus λ2 (Nibert and Schiff, 2001). When viewed from the side (Figure 2.5e), each TP monomer (VP4) is y-shaped and slanted such that its longer side lies approximately 45° relative to the fivefold axis of the capsid. The y-shaped TP subunit can be divided into three domains based on the geometrical disposition of its densities and on considerations of their putative functional roles (Figure 2.5d-f). First, in the outermost region of TP, a large and slightly twisted β-sheet density is flanked by a pair of helices (13 and 20 Å long, respectively) on either side, forming a characteristic helix-pair/β-sheet/helix-pair sandwich fold. This is a highly conserved fold that forms the SAM-binding domain of numerous methyltransferases (Schluckebier, 1995; Hodel, 1996; Reinisch *et al.*, 2000; Zhou *et al.*, 2003). An almost identical sandwich fold has also been identified in another domain of TP (methylase-1 domain in Figure 2.5). The identification of this highly conserved fold in these two domains of the 8-Å CPV map has their assignments as the corresponding methylase-1 and methylase-2 domains in CPV TP. Finally, a putative guanylyltransferase (GTase) domain was also proposed based on the similarities of secondary structure arrangements between the CSP-proximal (shell-anchoring) domain of the CPV TP and the reovirus λ2 GTase domain (Zhou *et al.*, 2003) (Figure 2.5e,f). However, the GTase domain of CPV TP contains a unique set of three roughly parallel helices on its external surface (Zhou *et al.*, 2003).

The subnanometer-resolution CPV structure has provided some insight into the structural basis and possible mechanism by which mRNA processing and release are regulated. Nascent mRNA is released from the TEC through the turret channel along a path that contains three elements of constriction. The first is the plug that blocks a pore on the capsid shell, which is formed by the clustering of five pairs of long helices in the CSP-A pentamer, in a topology similar to that of a closed iris diaphragm in a camera (Zhou *et al.*, 2003). The other two constrictions are in the middle of the turret channel, with opening diameters of only 25 and 10 Å, respectively. These two constrictions are stacked such that the constriction helices in the putative GTase domain of one TP subunit are directly under the constriction helix in the methylase-1 domain of the neighbouring TP subunit (Figure 2.5b,c). The iris diaphragm-like configurations of the constrictions and the alternating stacking of functional domains from neighbouring subunits provide a mechanism to facilitate conformational transitions and regulate RNA processing and release during endogenous RNA transcription. The initiation of mRNA transcription by the TEC at the apical domain of CSP-A may trigger a conformational change in this region (Zhang *et al.*, 2002; Xia *et al.*, 2003), resulting in the opening of the shell pore constriction in a manner similar to the movement of the plates in an iris diaphragm. The 5′ end of the nascent mRNA is then modified by guanylyl addition at the putative GTase domain, followed by two consecutive steps of methylation at the two methylase domains of an adjacent TP positioned directly above the guanylyltransferase domain. An atomic model will be necessary to accurately assign the m^7G (cap 0) and the 2′-*O*-ribose (cap 1) methyltransferase activities to the methylase 1 and 2 domains, as has been accomplished for those of the orthoreovirus λ2 (Reinisch *et al.*, 2000; Bujnicki and Rychlewski, 2001).

Nevertheless, the inner and outer turret constrictions provide the physical barriers to separate the enzymes involved in this successive RNA processing, thus ensuring the proper sequence of actions. Such a mechanism may also augment a ratchet-like, unidirectional movement of the processed mRNA through the turret channel to achieve the optimal efficiency observed in *in vitro* transcription experiments (Furuichi, 1974).

Comparison with other dsRNA viruses

All reoviruses and other dsRNA viruses whose 3D structures have been examined have a thin shell formed by 60 dimers of the CSP on a T = 1 lattice (Lawton, 1997; Grimes *et al.*, 1998; Reinisch *et al.*, 2000; Zhou *et al.*, 2001). The most striking examples of these are the much smaller (430 Å) L-A virus, which infects yeast (Caston *et al.*, 1997), and bacteriophage ϕ6, which has only three dsRNA segments (Butcher, 1997). Moreover, the general α+β folds of the CSPs, with their helix-rich apical and carapace domains and β-sheet-rich dimerization domains, are also conserved among all dsRNA viruses whose secondary structure is known. Beyond this innermost capsid shell, however, structures diverge substantially among dsRNA viruses, with variations in the presence or absence and the number of copies of LPP, as well as the presence or absence of additional protein layers. The conservation of the innermost capsid shell and diversification of external protein layers among the dsRNA viruses argue strongly for a common remote ancestor of these viruses that have evolved to infect an extremely broad spectrum of hosts, including bacteria, plants, yeast, and mammals. The conservation of CSP structure reflects the conserved role for forming the capsid shell, which is to provide an appropriate enclosure for the genome and replication enzymes that facilitate endogenous RNA transcription.

Despite the general similarity among the overall folds of the CSPs of the dsRNA viruses, it is interesting to note that no significant sequence homology has been detected among these CSPs using currently available multiple or pairwise sequence alignment tools, indicating that CSP molecules have also undergone substantial modifications during the viral evolution process. The most notable modifications are large domain deletions/insertions in the CSP molecules. In L-A virus, substantial sequence deletion exists, reflecting the fact that its genome is much smaller and a smaller capsid shell is needed (Caston *et al.*, 1997). In contrast, in CPV, domain insertion has been used to achieve additional inter-CSP interactions, probably related to the requirement for achieving additional capsid stability (see above). Thus, while the structural divergences of the outer shell proteins are related to the evolutionary pressure of host adaptability, modifications to the inner capsid shell protein CSP are mostly likely a consequence of structural adaptability driven by the autonomous needs of different viruses.

While the overall folds of the two methylase domains and the shell GTase domain in CPV TP are similar to those observed in orthoreovirus λ2, there are several marked differences between CPV TP and λ2. First, CPV TP completely lacks the C-terminal three β-sheet-rich immunoglobulin domains making up the flap of the orthoreovirus λ2. The λ2 flap of orthoreovirus is involved in the association with the cell-attachment protein σ1 and can undergo drastic conformational changes at different states of viral infection and maturation (Dryden *et al.*, 1993; Chandran, 1999). A similar flap is seen in the aquareovirus turret, which exhibits similar conformational changes upon trypsin treatment (Nason, 2000). The lack of this flap domain in CPV suggests a different mechanism for host specificity in CPV, possibly through the A spike, which interacts with the inner rims of the turret (Hill *et al.*, 1999; Zhang *et al.*, 1999). Second, the three characteristic parallel helices found on the outer surface of the GTase domain of CPV TP are not present in λ2. Finally, in contrast to the good match between the locations of secondary structure elements in the shell-anchoring and methylase-2 domains of CPV TP and λ2, the CPV methylase-1 domain is rotated ~40° inward toward the channel, making the CPV turret a more compact structure (Zhou *et al.*, 2003).

Acknowledgments

I thank Dr Xuekui Yu for providing Figure 2.2a and for critical reading of the manuscript; Dr Yuyao (Mario) Liang for assistance with preparation of Figures 2.3 and 2.4; Professor Jiang-Qiang Zhang for providing CPV-containing polyhedra samples; and Ms Pierrette Lo for editorial assistance. Research in my group has been supported by grants from the National Institutes of Health (GM071940) and the Welch Foundation.

References

Bartlett, N., Gillies, S., Bullivant, S., and Bellamy, A. (1974). Electron microscopy study of reovirus reaction cores. J. Virol. *14*, 315–326.

Bellamy, A., and Harvey, J. (1976). Biophysical studies of reovirus type 3. III. A laser light-scattering study of the RNA transcriptase reaction. Virology. *70*, 28–36.

Bujnicki, J.M., and Rychlewski, L. (2001). Reassignment of specificities of two cap methyltransferase domains in the reovirus lambda 2 protein. Genome Biol. *2*, RESEARCH0038.0031.

Butcher, S.J., Dokland, T., Ojala, P.M., Bamford, D.H., and Fuller, S.D. (1997). Intermediates in the assembly pathway of the double-stranded RNA virus phi6. EMBO J. *16*, 4477–4487.

Caston, J.R., Trus, B.L., Booy, F.P., Wickner, R.B., Wall, J.S., and Steven, A.C. (1997). Structure of L-A virus, a specialized compartment for the transcription and replication of double-stranded RNA. J. Cell Biol. *138*, 975–985.

Chandran, K., Walker, S.B., Chen, Y., Contreras, C.M., Schiff, L.A., Baker, T.S., and Nibert, M.L. (1999). *In vitro* recoating of reovirus cores with baculovirus-expressed outer-capsid proteins mu1 and sigma3. J. Virol. *73*, 3941–3950.

Coulibaly, F., Chiu, E., Ikeda, K., Gutmann, S., Haebel, P.W., Schulze-Briese, C., Mori, H., and Metcalf, P. (2007). The molecular organization of cypovirus polyhedra. Nature *446*, 97–101.

Dryden, K.A., Wang, G.J., Yeager, M., Nibert, M.L., Coombs, K.M., Furlong, D.B., Fields, B.N., and Baker, T.S. (1993). Early steps in reovirus infection are associated with dramatic changes in supramolecular structure and protein conformations. J. Cell Biol. *122*, 1023–1041.

Furuichi, Y. (1974). "Methylation-coupled" transcription by virus-associated transcriptase of cytoplasmic polyhedrosis virus containing double-stranded RNA. Nucleic Acids Res. *1*, 809–822.

Gouet, P., Diprose, J.M., Grimes, J.M., Malby, R., Burroughs, J.N., Zientara, S., Stuart, D.I., and Mertens, P.P. (1999). The highly ordered double-stranded RNA genome of bluetongue virus revealed by crystallography. Cell *97*, 481–490.

Grimes, J.M., Burroughs, J.N., Gouet, P., Diprose, J.M., Malby, R., Zientara, S., Mertens, P.P.C., and Stuart, D.I. (1998). The atomic structure of the bluetongue virus core. Nature. *395*, 470–478.

Hagiwara, K., Tomita, M., Kobayashi, J., Miyajima, S., and Yoshimura, T. (1998a). Nucleotide Sequence of Bombyx mori Cytoplasmic Polyhedrosis Virus Segment 8. Biochemical and Biophysical Research Communications 247.

Hagiwara, K., Tomita, M., Nakal, K., Kobayashi, J., Mayajima, S., and Yoshimura, T. (1998b). Determination of the nucleotide sequence of Bombyx mori Cytoplasmic Polyhedrosis Virus segment 9 and its expression in BmN4 cells. J. Virol. July, 5762–5768.

Hagiwara, K., and Matsumoto, T. (2000). Nucleotide sequences of genome segments 6 and 7 of Bombyx mori cypovirus 1, encoding the viral structural proteins V4 and V5, respectively. J. Gen. Virol. *81*, 1143–1147.

Hagiwara, K., Kobayashi, J., Tomita, M., and Yoshimura, T. (2001). Nucleotide sequence of genome segment 5 from Bombyx mori cypovirus 1. Arch Virol. *146*, 181–187.

Hagiwara, K., Rao, S., Scott, S.W., and Carner, G.R. (2002). Nucleotide sequences of segments 1, 3 and 4 of the genome of Bombyx mori cypovirus 1 encoding putative capsid proteins VP1, VP3 and VP4, respectively. J. Gen. Virol. *83*, 1477–1482.

Hagiwara, K., and Naitow, H. (2003). Assembly into single-shelled virus-like particles by major capsid protein VP1 encoded by genome segment S1 of Bombyx mori cypovirus 1. J. Gen. Virol. *84*, 2439–2441.

Harrison, S., Wiley, D.C., and Skehel, J.J. (1996). Virus structure. In, Virology B. Field, D.M. Knipe, and P.M. Howley, ed. Lippincott-Raven, Philadelphia. p. 59–100.

Hatta, T., and Francki, R.I.B. (1982). Similarity in the structure of cytoplasmic polyhefrosis virus, leafhopper A virus and Fiji disease virus particles. Intervirol. *18*, 203–208.

Hayashi, Y., and Kawase, S. (1964). Base pairing in ribonucleic acid extracted from the cytoplasmic polyhedra of the silkworm. Virology *23*, 611–614.

Hill, C.L., Booth, T.F., Prasad, B.V., Grimes, J.M., Mertens, P.P., Sutton, G.C., and Stuart, D.I. (1999). The structure of a cypovirus and the functional organization of dsRNA viruses. Nat. Struct. Biol. *6*, 565–568.

Hodel, A.E., Gershon, P.D., Shi, X., and Quiocho, F.A. (1996). The 1.85 A structure of vaccinia protein VP39, a bifunctional enzyme that participates in the modification of both mRNA ends. Cell. *85*, 247–256.

Ikeda, K., Nagaoka, S., Winkler, S., Kotani, K., Yagi, H., Nakanishi, K., Miyajima, S., Kobayashi, J., and Mori, H. (2001). Molecular characterization of Bombyx mori cytoplasmic polyhedrosis virus genome segment 4. J. Virol. *75*, 988–995.

Lawton, J.A., Zeng, C.Q., Mukherjee, S.K., Cohen, J., Estes, M.K., and Prasad, B.V.V. (1997). Three-dimensional structural analysis of recombinant rotavirus-like particles with intact and amino-terminal-deleted VP2, implications for the architecture of the VP2 capsid layer. J. Virol. *71*, 7353–7360.

Lawton, J.A., Estes, M.K., and Prasad, B.V. (2000). Mechanism of genome transcription in segmented dsRNA viruses. Adv. Virus Res. *55*, 185–229.

Lewandowski, L.J., and Traynor, B.L. (1972). Comparison of the structure and polypeptide composition of three double-stranded ribonucleic acid-containing viruses (diplornaviruses), cytoplasmic polyhedrosis virus, wound tumor virus, and reovirus. J. Virol. *10*, 1053–1070.

McCrae, M.A., and Mertens, P.P.C. (1983). *In vitro* translation studies on and RNA coding assignments for cytoplasmic polyhedrosis viruses. In, Double-stranded RNA Viruses, R.W. Compans, and D.H.L. Bishop, ed. Elsevier Science, New York. p. 35–41.

Mertens, P.P.C., Crook, N.E., Rubinstein, R., Pedley, S., and Payne, C.C. (1989). Cytoplasmic polyhedrosis virus classification by electropherotype; validation by serological analyses and agarose gel electrophoresis. J. Gen. Virol. *70*, 173–185.

Mertens, P.P.C., Arella, M., Attoui, H., Belloncik, S., Bergoin, M., Boccardo, G., Booth, T.F., Chiu, W., Diprose, J.M., Duncan, R., Estes, M.K., Gorziglia, M., Gouet, P., Gould, A.R., Grimes, J.M., Hewat, E., Homes, I.A., Hosshino, Y., Hill, C., Joklik, W.K., Knowles, N., Ferber, M.L., Malby, R., Marcachi, C., McCrae, M.A., Milne, R.G., Nibert, M., Nunn, M., Omura, T., Prasad, B.V.V., Pritchard, I., Roy, P., Samal, S.K., Schoehn, G., Shikata, E., Stoltz, D.B., Stuart, D.I., Suzuki, N., Upadhyaya, N., Uyeda, I., Waterhouse, P., Williams, C.F., Winton, J.R., and Zhou, Z.H. (2000). Family *Reoviridae*. In, Virus Taxonomy, M.H.V. van Regenmortel, C.M. Fauguet, D.H.L. Bishop, E.B. Carstens, M.K. Estes, S.M. Lemon, J. Maniloff, M.A. Mayo, D.J. McGeoch, C.R. Pringle, and R.B. Wickner, ed. Academic Press, San Diego, CA, USA p. 395–480.

Mertens, P.P.C., Rao, S., and Zhou, Z.H. (2004). Cypovirus, *Reoviridae*. In, Virus Taxonomy, VIIIth Report of the ICTV, C.M. Fauquet, M.A. Mayo, J. Maniloff, U. Desselberger, and L.A. Ball, ed. Elsevier/Academic Press, London. p. 522–533.

Mori, H., Minobe, Y., Sasaki, T., and Kawase, S. (1989). Nucleotide sequence of the polyhedrin gene of Bombyx mori cytoplasmic polyhedrosis virus A strain with nuclear localization of polyhedra. J. Gen. Virol. *70*, 1885–1888.

Mori, H., Ito, R., Nakazawa, H., Sumida, M., Matsubara, F., and Minobe, Y. (1993). Expression of Bombyx mori cytoplasmic polyhedrosis virus polyhedrin in insect cells by using a baculovirus expression vector, and its assembly into polyhedra. J. Gen. Virol. *74*, 99–102.

Nakagawa, A., Miyazaki, N., Taka, J., Naitow, H., Ogawa, A., Fujimoto, Z., Mizuno, H., Higashi, T., Watanabe, Y., Omura, T., Cheng, R.H., and Tsukihara, T. (2003). The atomic structure of rice dwarf virus reveals the self-assembly mechanism of component proteins. Structure *11*, 1227–1238.

Nason, E.L., Samal, S.K., and Venkataram Prasad, B.V. (2000). Trypsin-induced structural transformation in aquareovirus. J. Virol. *74*, 6546–6555.

Nibert, M.L., and Schiff, L.A. (2001). Reoviruses and their replication. In, Fields Virology 4 edn, D.M. Knipe, P.M. Howley, D.E. Griffin, R.A. Lamb, M.A. Martin, B. Roizman, and S.E. Straus, ed. Lippincott Williams & Wilkins, Philadelphia. p. 1679–1728.

Payne, C.C., and Rivers, C.F. (1976). A provisional classification of cytoplasmic polyhedrosis viruses based on the sizes of the RNA genomes segments. J. Gen. Virol. *33*, 71–85.

Payne, C.C., and Harrap, K.A. (1977). Cytoplasmic polyhedrosis viruses. In, The Atlas of Insect and Plant Viruses, K. Maramorosch, ed. Academic Press, New York. p. 105–129.

Payne, C.C., and Mertens, P.P.C. (1983). Cytoplasmic polyhedrosis viruses. In, The *Reoviridae*, W.K. Joklik, ed. Academic Press, New York. p. 425–504.

Rao, S., Carner, G.R., Scott, S.W., Omura, T., and Hagiwara, K. (2003). Comparison of the amino acid sequences of RNA-dependent RNA polymerases of cypoviruses in the family *Reoviridae*. Arch Virol. *148*, 209–219.

Reinisch, K.M., Nibert, M.L., and Harrison, S.C. (2000). Structure of the reovirus core at 3.6 Å resolution. Nature *404*, 960–967.

Schluckebier, G., O'Gara, M., Saenger, W., and Cheng, X. (1995). Universal catalytic domain structure of AdoMet-dependent methyltransferases. J. Mol. Biol. *247*, 16–20.

Smith, K.M. (1963). The cytoplasmic virus diseases. In, Insect Pathology, E.A. Steinhaus, ed. Academic Press, New York. p. 457–497.

Tan, Y.R., Sun, J.C., Lu, X.Y., Su, D.M., and Zhang, J.Q. (2003). Entry of Bombyx mori cypovirus 1 into midgut cells *in vivo*. J Electron Microsc (Tokyo). *52*, 485–489.

White, C., and Zweerink, H. (1976). Studies on the structure of reovirus cores, selective removal of polypeptide lambda 2. Virology *70*, 171–180.

Xia, D., Sun, Y.-k., McCrae, M.A., and Rossmann, M.G. (1991). X-ray powder pattern analysis of cytoplasmic polyhedrosis virus inclusion bodies. Virology. *180*, 153–158.

Xia, Q., Jakana, J., Zhang, J.Q., and Zhou, Z.H. (2003). Structural comparisons of empty and full cytoplasmic polyhedrosis virus, protein–RNA interactions and implications for endogenous RNA transcription mechanism. J. Biol. Chem. *278*, 1094–1100.

Xu, P., Miller, S.E., and Joklik, W.K. (1993). Generation of reovirus core-like particles in cells infected with hybrid vaccinia viruses that express genome segments L1, L2, L3, and S2. Virology *197*, 726–731.

Yazaki, K., and Miura, K.-I. (1980). Relation of the structure of cytoplasmic polyhedrosis viruses and the synthesis of its messenger RNA. Virology *105*, 467–479.

Zhang, H., Zhang, J., Yu, X., Lu, X., Zhang, Q., Jakana, J., Chen, D.H., Zhang, X., and Zhou, Z.H. (1999). Visualization of protein–RNA interactions in cytoplasmic polyhedrosis virus. J. Virol. *73*, 1624–1629.

Zhang, H., Yu, X.K., Lu, X.Y., Zhang, J.Q., and Zhou, Z.H. (2002). Molecular interactions and viral stability revealed by structural analyses of chemically treated cypovirus capsids. Virology *298*, 45–52.

Zhou, Z.H., Baker, M.L., Jiang, W., Dougherty, M., Jakana, J., Dong, G., Lu, G., and Chiu, W. (2001). Electron cryomicroscopy and bioinformatics suggest protein fold models for rice dwarf virus. Nat. Struct. Biol. *8*, 868–873.

Zhou, Z.H., Zhang, H., Jakana, J., Lu, X.-Y., and Zhang, J.-Q. (2003). Cytoplasmic polyhedrosis virus structure at 8 Å by electron cryomicroscopy, structural basis of capsid stability and mRNA processing regulation. Structure *11*, 651–663.

Rotavirus Structure

3

Xiaofang Jiang, Sue E. Crawford, Mary K. Estes, and B. V. Venkataram Prasad

Abstract

This chapter describes our current understanding of the rotavirus structure and how structural studies have provided some insights into capsid-associated functions including cell entry, antibody neutralization, trypsin-enhanced infectivity and endogenous transcription. In the last two decades, several dsRNA viruses have been structurally characterized using cryo-EM and X-ray crystallographic techniques. Despite some distinctive structural features reflecting their host specifies, these studies have revealed several common mechanistic themes particularly in regard to endogenous transcription of the genome. The possible evolutionary implications of the common features among these viruses are also discussed.

Introduction

Rotavirus is the most common cause of acute gastroenteritis in infants and young children worldwide. This virus contains a double-stranded RNA (dsRNA) genome and is a member of the *Reoviridae* family. Viruses with dsRNA genomes are currently grouped into six families: *Reoviridae, Birnaviridae, Totiviridae, Partitiviridae, Hypoviridae*, and *Cystoviridae* (reviewed in Fields, 2001; Kapikian, Hoshino *et al.*, 2001). Of these six families, the *Reoviridae* is the largest and most diverse in terms of host range. *Reoviridae* are currently classified into nine genera (Table 3.1, adapted from Lawton, Estes *et al.*, 2000). The genomes of these viruses consist of 10 to 12 segments of dsRNA, each generally encoding one protein. The mature virions are non-enveloped. Their capsids, formed by multiple proteins, have icosahedral symmetry and are arranged generally in concentric layers. A distinguishing feature of the dsRNA viruses, irrespective of their family association, is their ability to carry out transcription of the dsRNA segments, under appropriate conditions, within the capsid. In all these viruses, the enzymes required for endogenous transcription are thus part of the virion structure. This chapter focuses on our current understanding of the rotavirus structure and mechanistic details of some capsid-related functions with a brief discussion on the comparison of rotavirus structural features with other dsRNA viruses and evolutionary implications.

Table 3.1 Genera within the family of *Reoviridae*

Genus	Primary host	Prototype virus	Genome segments
Orthoreovirus	Mammals	Reovirus 3	10
Orbivirus	Mammals	Bluetongue virus 1	10
Rotavirus	Mammals	Simian rotavirus SA11	11
Coltivirus	Mammals	Colorado tick fever virus	12
Aquareovirus	Fish	Golden shiner virus	11
Cypovirus	Insects	*Bombyx mori* cypovirus 1	10
Fijivirus	Plants	Fiji disease virus	10
Phytoreovirus	Plants	Wound tumour virus	12
Oryzavirus	Plants	Rice ragged stunt virus	10

Rotavirus architecture

The genome of rotavirus consists of eleven segments of dsRNA (Figure 3.1). Each genome segment codes for one protein with the exception of segment 11, which codes for two proteins (Estes 2001). Among the twelve proteins, six are structural (VPs) (Table 3.2, adapted from Estes, Fields Virology 2001) and six are non-structural proteins (NSPs) (Table 3.3). Rotavirus is a triple layered, icosahedral particle approximately 1000 Å in diameter (Prasad, Wang *et al.*, 1988; Prasad, Burns *et al.*, 1990; Yeager, Dryden *et al.*, 1990; Prasad, Rothnagel *et al.*, 1996) (Figure 3.2A).

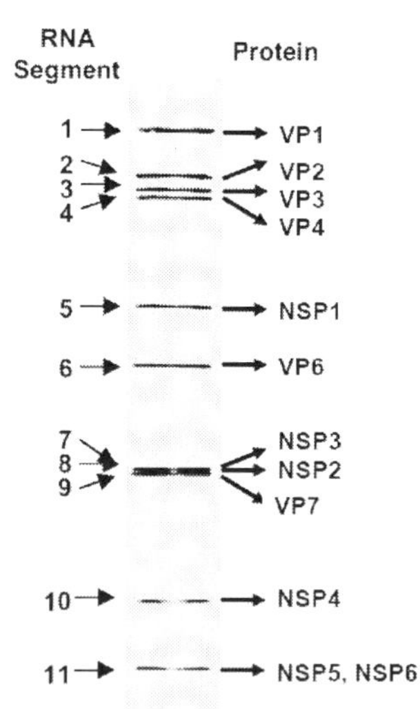

Figure 3.1 Rotavirus genome. RNA gel showing rotavirus gene products. The RNA segments are numbered in order of a polyacrylamide gel migration along with their corresponding protein products. Gene 11 is alternatively translated to produce NSP5 and NSP6 (Torres-Vega, Gonzalez *et al.*, 2000; Welch, Crawford *et al.*, 1989). For colour figure see http://ncmi.bcm.tmc.edu/~chin/dsRNA_chapter3_figs.pdf

Table 3.2 Rotavirus structural proteins

Protein	Gene segment number	Mw (kD)[a]	Location in virus particles	Functions and properties
VP1	1	125	Core	RNA-dependent RNA polymerase, ssRNA binding, complex with VP3, interacts with NSP5
VP2	2	103	Core	RNA binding, required for replicase activity of VP1, dimer, interacts with NSP5
VP3	3	98	Core	Guanylyltransferase, methyltransferase, ssRNA binding, complex with VP1
VP4 (VP5*/8*)	4	88 (60/28)	Outermost layer	Haemagglutinin, neutralization antigen, protease-enhanced infectivity, virulence, putative fusion region, cell attachment, protection
VP6	6	45	Middle layer	Hydrophobic, trimer, subgroup antigen
VP7	9	37	Outermost layer	RER integral membrane glycoprotein, trimer, neutralization antigen, two hydrophobic N-terminal regions, binds Ca^{2+}

a Predicted Mw of simian rotavirus SAII proteins.

Table 3.3 Non-structural proteins of rotavirus

Protein	Gene segment number	Mw (kD)[a]	Localization	Functions and properties
NSP1	5	58	Cytoplasm	Slightly basic, two conserved cysteine-rich zinc-finger motifs, RNA binding, interacts with IRF3 and induces IRF3 degradation
NSP2	8	37	Viroplasm	Basic, octamer, non-specific ssRNA binding, NTPase activity, binds NSP5 and VP1, involved in viroplasm formation, upregulates NSP5 phosphorylation
NSP3	7	35	Cytoplasm	Slightly acidic, homodimer, specifically binds 3′ end of rotavirus mRNA, interacts with eIF4GI, enhances translation of viral mRNA
NSP4	10	20	Endoplasmic reticulum	RER transmembrane glycoprotein, enterotoxin, receptor for budding of double-layer particle through ER membrane, regulates RNA synthesis, multifunctional
NSP5	11	22	Viroplasm	Slightly basic, glycosylated, rich in serine and threonine, phosphoprotein, interacts with NSP2 to form viroplasm, interacts with VP1, VP2, and NSP6
NSP6	11	11	Viroplasm	Product of second, out-of-frame ORF, interacts with NSP5

a Predicted Mw of simian rotavirus SAII proteins.

The outermost VP7 layer and VP4 spikes

The outermost layer consists of VP4 (~88 kD) and VP7 (~37 kD). VP7, a glycoprotein, is the major constituent of the outermost layer (Figure 3.2A). There are 260 VP7 trimers located at all the local and strict threefold axes of the T = 13 icosahedral lattice. A distinctive feature of the rotavirus structure is the presence of 132 aqueous channels at all the fivefold and quasi sixfold positions of the T = 13 lattice. Based on their locations in the T = 13 lattice, these aqueous channels are grouped into three types. Type I channels are at the fivefold axes of the icosahedral capsid, type II channels are at the six-coordinated

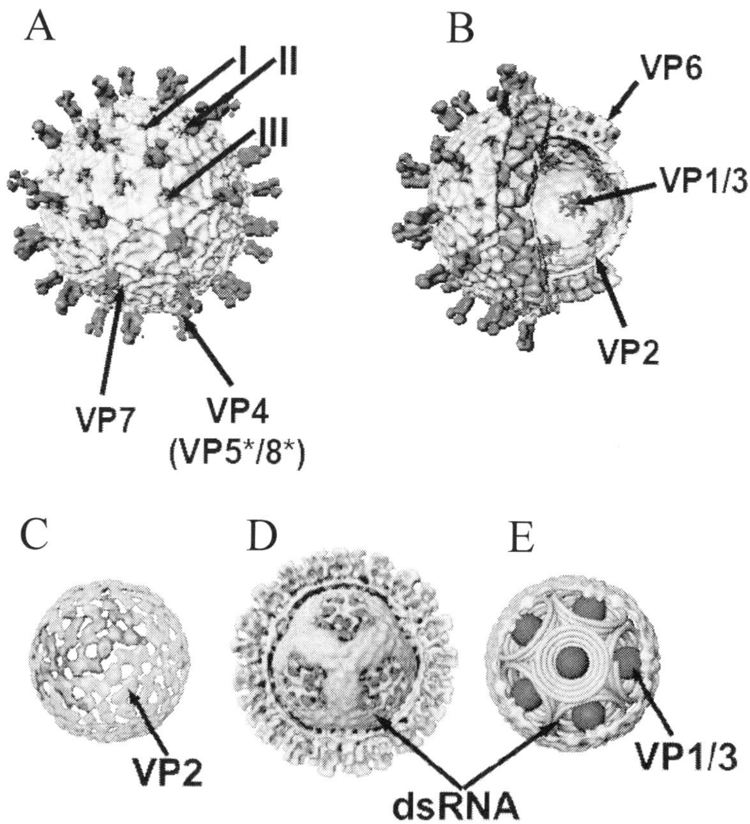

Figure 3.2 Rotavirus architecture. A. The three types of aqueous channels are labelled I, II, and III. The VP4 spikes are coloured red and the outer capsid layer protein, VP7, is shown in yellow. B. The internal VP6 protein layer is in blue and the core VP2 layer in green. The flower-shaped VP1/VP3 transcription complex is attached to the inside of the VP2 layer at the fivefold icosahedral axes directly below the type I channels and is coloured red (adapted from Pesavento *et al.*, 2006). C. Organization of the VP2 layer with 2 subunits in the icosahedral asymmetric unit shown in red and purple around one of the fivefold axes. D. Manifestation of the ordered layer of the genomic RNA in the icosahedrally averaged reconstruction of the DLP. E. Model for the structural organization of the dsRNA segments in rotavirus. The dsRNA segments are represented as inverted cones surrounding the transcriptions enzyme complex (VP1 and VP3) shown as red balls attached to the VP2 layer shown in green. For colour figure see http://ncmi.bcm.tmc.edu/~chin/dsRNA_chapter3_figs.pdf

positions surrounding type I channels. The type III channels are those on the other six-coordinated positions that surround the icosahedral threefold axes.

There are 60 spikes that emanate from the VP7 layer. (Figure 3.2A) In mature rotavirus, grown in the presence of trypsin, these spikes are distinctly bilobed and are situated at the edge of the type II channels. The spikes extend from the virion surface by approximately 90 Å. Cryo-EM studies of the mature triple-layered particles (TLPs) using VP4 specific antibody have shown that these spikes are formed by VP4 (Prasad, Burns *et al.*, 1990; Tihova, Dryden *et al.*, 2001). In these studies, two Fab molecules are seen bound to each spike indicating that each spike is a dimer of VP4. In addition to the exposed portion, the spike extends inward with a large globular portion buried inside the type II channels.

The rotavirus spike protein, VP4, is the major determinant of host cell entry and neutralization. Proteolytic cleavage of VP4, resulting in VP5* and VP8* (an asterisk denotes proteolytic fragments), enhances viral infectivity several fold and facilitates virus entry into cells (Estes, Graham *et al.*, 1981; Lopez, Arias *et al.*, 1985). The consensus opinion that has emerged from several recent studies is that rotavirus cell entry is a coordinated multistep process involving sequential interactions with sialic acid (SA)-containing receptors in the initial cell attachment step followed by post-attachment interactions with hsc70 (Zarate, Cuadras *et al.*, 2003), and integrins such as $\alpha v\beta 3$ (Guerrero, Mendez *et al.*, 2000), $\alpha 4\beta 1$ (Coulson, Londrigan *et al.*, 1997), and $\alpha 2\beta 1$ (Hewish, Takada *et al.*, 2000; Zarate, Espinosa *et al.*, 2000). In the entry process, the VP8* domain is involved in interactions with SA (Isa, Lopez *et al.*, 1997), whereas VP5* is implicated in interactions with integrins (Coulson, Londrigan *et al.*, 1997; Guerrero, Mendez *et al.*, 2000; Zarate, Espinosa *et al.*, 2000). How does VP4 facilitate such multistep entry processes in rotavirus? One possibility is that VP4 undergoes distinct conformational changes at various stages during cell entry to mask certain interaction domains and reveal others in order to allow optimal interactions with different receptors and the cellular membrane (Pesavento, Crawford *et al.*, 2006). Such distinct conformational states during cell entry processes are also known for viruses such as influenza viruses, alphavirus, and picornaviruses.

Unique entry-associated conformational changes

X-ray structures of VP8* and a portion of VP5* show that the distal globular domain of the VP4 spike is composed of VP8* and the remaining body of the spike consists of VP5* (Dormitzer, Sun *et al.*, 2002; Dormitzer, Nason *et al.*, 2004) (Figure 3.3A). Cryo-EM and biochemical studies of TLPs grown in the absence of trypsin show that VP4 is structurally disordered (Figure 3.3B) until it undergoes trypsin cleavage (Crawford, Mukherjee *et al.*, 2001). Trypsinization renders the spike conformationally stable and structurally ordered, perhaps enabling it to interact more efficiently with cellular receptors.

In addition to this unique disorder-to-order transition, recent studies have indicated further conformational flexibility inherent in the spikes. Cryo-EM studies of trypsinized TLPs at elevated pH have shown that the bilobed 90 Å long VP4 spike undergoes a dramatic irreversible conformational change and becomes stunted, with a pronounced trilobed appearance (Pesavento, Crawford *et al.*, 2005) (Figure 3.3C). These pH-treated particles are non-infectious and have lost the ability to haemagglutinate red blood cells. In addition, they exhibit SA-independent cell binding in contrast to the SA-dependent cell binding of the untreated particles. This surprising observation suggests that a pH-induced

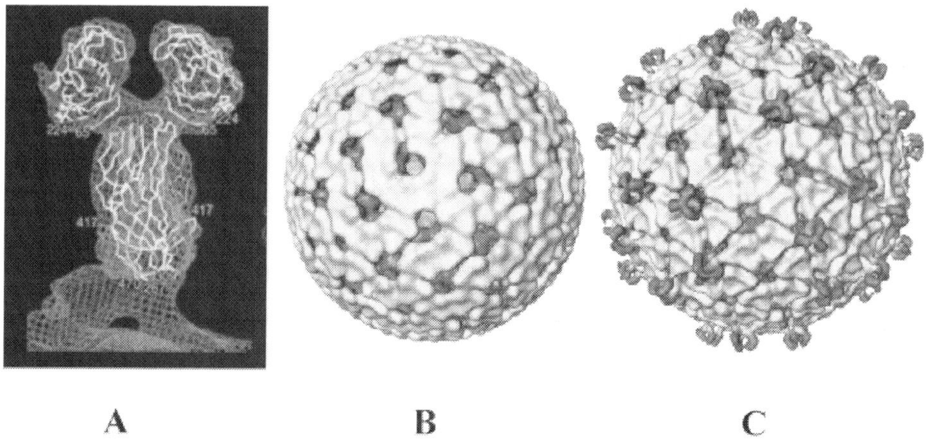

Figure 3.3 VP4 spikes of rotavirus. (A) X-ray structures of VP5* and VP8* (shown in the backbone representation) fitted into the cryo-EM envelope of the VP4 spike derived from a 12 Å-resolution map. (B) Structure of the rotavirus particles grown in the absence of trypsin showing no spikes (Crawford, Mukherjee *et al.*, 2001). Biochemical analysis indicates that VP4 in these particles is present at the same stoichiometric levels as in the virus grown in the presence of trypsin (Figure 3.2A). (C) Treatment of rotavirus with ~pH 11 induces a conformational change in the spike resulting in a tri-lobed stunted spike. (Pesavento, Crawford *et al.*, 2006). For colour figure see http://ncmi.bcm.tmc.edu/~chin/dsRNA_chapter3_figs.pdf

conformational change exposes an epitope that switches the SA-dependent virus to bind to cells without the requirement of SA. Despite large-scale conformational changes, neither the stoichiometry nor the chemical composition of the spike is altered in the pH-treated particles. Cryo-EM analysis with a VP5-specific monoclonal antibody shows that three Fabs bind to each of the altered spikes, thus suggesting the unique and remarkable possibility that the VP4 spike undergoes a dimer-trimer transition. In the crystal structure, VP5* forms a trimer with substantial intersubunit interactions, indicating VP5* has a propensity to form strong trimers by itself (Dormitzer, Nason *et al.*, 2004).

An interesting result from the experiments at elevated pH is that the high pH-induced conformational change is completely protected by a VP5*-specific neutralizing antibody. The structural integrity of the VP4 spike is preserved as are its functional properties such as SA-dependent cell binding and haemagglutination when particles are bound to this antibody. The neutralizing antibody also remains bound to the particles even after drastic changes in pH (from pH 7 to 11 and back to 7) (Pesavento, Crawford *et al.*, 2005). Based on these studies, it is proposed that high pH treatment triggers a conformational change that mimics a post SA-attachment step to expose an epitope recognized by the downstream receptor(s) in the rotavirus cell entry process, which as described above involves sequential interactions with multiple receptors. The mechanism by which this antibody neutralizes infectivity is proposed to be by preventing this conformational change.

Taken together, the available structural studies on rotavirus spikes suggest a model for entry associated conformational changes in the VP4 spike. According to this model, each spike is possibly a trimer of VP4, and upon trypsinization, two molecules form the visible spike as seen in the cryo-EM reconstructions of the trypsinized rotavirus particles, with the other monomer being floppy and not visible in reconstructions (Dormitzer, Nason *et al.*, 2004). During cell entry, by a yet unknown entry associated event, the floppy VP4 monomer trimerizes together with the other two molecules of VP4, as indicated by the high pH-treated particles and X-ray crystallographic analysis. While this hypothesis looks attractive, further studies are necessary to unequivocally prove that VP4 spikes are indeed trimers.

The intermediate VP6 layer

Directly underneath the capsid layer formed by the VP7 trimers and VP4 spikes is a layer formed by VP6 (Yeager, Berriman *et al.*, 1994; Shaw, Rothnagel *et al.*, 1996) (Figure 3.2B). This protein is the major protein of the rotavirus by mass. Similar to the VP7 layer, 780 copies of VP6 are arranged as 260 trimers on a T = 13 icosahedral lattice. The VP6 trimers are located directly below the VP7 trimers such that the aqueous channels in the outer VP7 and in the intermediate VP6 layers are in register (Pesavento, Crawford *et al.*, 2006). Because the VP7 trimers are wider, the channel opening is narrower in the mature virion than in the particles without the VP7 layer, which are referred to as double layered particles (DLPs). Due to its location, VP6 interacts with the outer layer proteins VP7, and VP4 (Shaw, Rothnagel *et al.*, 1993), and VP2, which forms the innermost layer of the virion.

X-ray crystallographic analysis of VP6 shows that VP6 crystallizes as trimers with strong intersubunit interactions consistent with the observations from the cryo-EM structure of the virion (Mathieu, Petitpas *et al.*, 2001) (Figure 3.4). Each subunit structure has two domains linked by a loop. One of the domains exhibits a classical 8-stranded

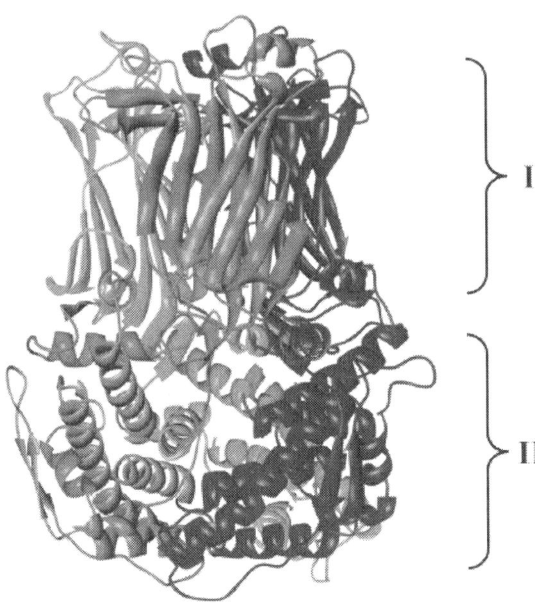

Figure 3.4 X-ray structure of VP6 (Mathieu, Petitpas *et al.*, 2001; PDB ID: 1QHD). The three subunits of the trimer are coloured green, red, and blue. Distal 8-stranded β-barrel domain (I) that interacts with VP7 and VP4 and the lower α–helical domain (II) that interfaces with VP2 layer are indicated. For colour figure see http://ncmi.bcm.tmc.edu/~chin/dsRNA_chapter3_figs.pdf

antiparallel β-strand structure whereas the other domain contains several α-helices. Fitting the X-ray structure of VP6 into cryo-EM structures of DLPs and TLPs has provided some insight into how VP6 interacts with other structural proteins in the virion (Mathieu, Petitpas *et al.*, 2001). Interaction with VP7 predominantly involves the top portion of the β-barrel domain of VP6, whereas interactions with VP2 involve the α-helical domain. The VP6 trimers that surround the type II channels interact with the buried globular domain of the VP4 spike through their sides. All the interacting surfaces (VP6-VP4, VP6-VP7, and VP6-VP2) contain the most conserved residues of VP6. VP6 interactions with VP2 are predominantly hydrophobic, whereas interactions with VP7 and VP4 involve several charged residues. The fitting also shows that the VP6 trimers interact laterally to form the T = 13 layer. When the VP6 protein is expressed alone, it readily forms two dimensional crystalline arrays and helical tubes with pseudo-hexagonal packing, but rarely T = 13 icosahedral shells as in the virus. Thus, the VP6 alone does not contain the information to make a closed icosahedral shell as seen in the virion structure (Lepault, Petitpas *et al.*, 2001).

The innermost VP2 layer and the transcription enzymes

The innermost capsid layer of rotavirus is formed by VP2 (Prasad, Rothnagel *et al.*, 1996) (Figure 3.2B and C). Cryo-EM studies have suggested that 120 copies, as 60 dimers, interact closely, similar to that seen in the structure of the bluetongue virus capsid, to form a T = 1 icosahedral shell of ~30 Å thickness (Lawton, Zeng *et al.*, 1997). There are no prominent holes in the VP2 shell structure unlike in the outer two layers of the virus. The aqueous channels of VP7 and VP6 layers terminate at the surface of the VP2 layer. Among all the structural proteins of rotavirus, only VP2 has the ability to form a native-like icosahedral shell upon expression (Labbe, Charpilienne *et al.*, 1991; Prasad, Rothnagel *et al.*, 1996). Co-expression of VP2 with other structural proteins in insect cells results in the formation of virus-like particles with identical capsid organization as in native virions (Prasad, Rothnagel *et al.*, 1996). Thus, it is likely that VP2 is what directs the correct assembly of the virion, and VP6 by interacting with both the inner VP2 and the outer VP7 layers provides an appropriate organization to integrate the two principal processes of the virus, namely cell entry and endogenous transcription (Jayaram, Estes *et al.*, 2004).

The remaining two structural proteins, VP1, the viral polymerase (Valenzuela, Pizarro *et al.*, 1991), and VP3, the capping enzyme (Liu, Mattion *et al.*, 1992; Chen, Luongo *et al.*, 1999) essential for endogenous transcription and subsequent capping of the transcripts, are present in very small amounts. Comparative cryo-EM studies of the VLPs formed by the co-expression of VP2 and VP6 with and without these enzymes have shown that VP1 and VP3 are attached to the inner surface of the VP2 layer at the fivefold axes possibly as heterodimers (Prasad, Rothnagel *et al.*, 1996) (Figure 3.2B and 3.2C). Based on cryo-EM reconstructions and also biochemical data, it is likely that 12 molecules of VP1 and VP3 are present in each virion. Although these studies indicate the locations of these proteins, structural features of these proteins and the details of how they interact between themselves and with VP2 are lost in reconstructions because of the imposition of icosahedral averaging. However, the flower-shaped complex that is attributed to the VP1 and VP3 complex in the cryo-EM structure of the VP1/2/3/6 particles are not seen when the N-terminal

92 residues of VP2 are deleted, thereby suggesting that this region of VP2 is essential for interactions with VP1/3 (Lawton, Zeng *et al.*, 1997; Zeng, Estes *et al.*, 1998).

Structural basis of endogenous transcription

One of the most fascinating aspects of dsRNA viruses in general and rotavirus in particular is the endogenous transcription. In rotaviruses, DLPs are the transcriptionally competent part of the virion structure. During the cell entry process, the outer layer is removed, and the resulting DLPs in the cytoplasm become transcriptionally active (Lawton, Estes *et al.*, 1997). None of the protein components of the DLPs, namely VP6, VP2, VP1, VP3, is by itself capable of transcribing the dsRNA segments. The single-layered particles (SLPs), obtained by the removal of VP6, are also transcriptionally inactive. Thus, the structural integrity of the DLP is essential for endogenous transcription. The DLP is a molecular machine capable of continuous transcription of the dsRNA segments limited only by the pool of available nucleotides. Cryo-EM studies of transcriptionally active DLPs have shown that during the process of transcription, DLPs maintain their structural integrity and the type I channels in the VP6 layer are used as conduits for exit of the nascent transcripts (Lawton, Estes *et al.*, 1997) (Figure 3.5). Fitting the X-ray structure of VP6 into the cryo-EM structure of the DLP suggests that the net concentration of negative charges on the walls lining the type I channel may facilitate the extrusion of the mRNA transcript by increasing its fluidity due to electrostatic repulsion between the channel wall and the exiting mRNA transcript (Mathieu, Petitpas *et al.*, 2001; Greig, Berriman *et al.*, 2006).

One interesting question is why TLPs are transcriptionally inactive and how the removal of the outer layer activates endogenous transcription. Interestingly, certain VP6-specific antibodies, which attach to the distal end of the VP6 trimers in an analogous

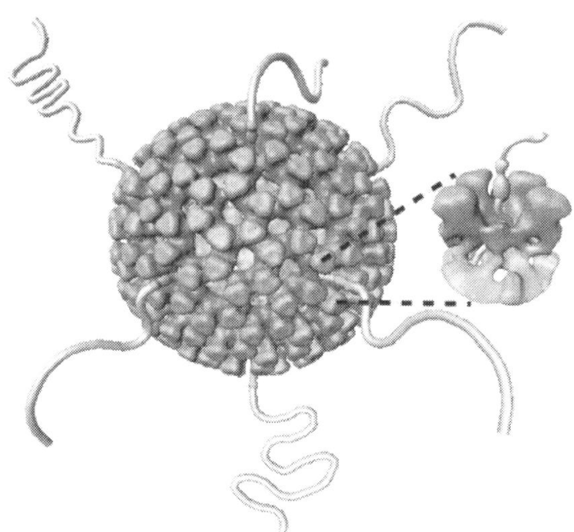

Figure 3.5 Proposed model from cryo-EM reconstruction of actively transcribing DLPs. The endogenous transcription results in the simultaneous release of the transcribed mRNAs from channels located at the fivefold vertex of the icosahedral DLP. (From Lawton, J. *et al.*, 1997). For colour figure see http://ncmi.bcm.tmc.edu/~chin/dsRNA_chapter3_figs.pdf

fashion to VP7, also inhibit transcription (Lawton, Estes *et al.*, 1999). How the binding of ligands such as VP7 or the monoclonal antibodies to the distal end of VP6 almost 180 Å from the site of transcription affects transcription remains unclear. The entire process of endogenous transcription can be thought of as consisting of three steps: initiation of mRNA synthesis, transcript elongation, and translocation. Biochemical studies have shown that in the TLPs, as well as in DLPs decorated with antibodies that inhibit transcription, initiation as well as limited elongation including mRNA capping occurs, indicating that the enzymatic activities of VP1 and VP3 are not affected by the binding of VP7 or the antibodies (Lawton, Estes *et al.*, 1999; Lawton, Estes *et al.*, 2001). When the outer VP7 layer is removed by adding EDTA, pulse-chase experiments show that transcript elongation resumes leading to full-length transcripts (Lawton, Estes *et al.*, 2001). One possible model is that removal of the outer layer induces conformational changes in the VP2 layer allowing emerging transcripts to exit through the type I channel without interruption during transcription. Although there are some hints of these conformational changes from the comparative cryo-EM analysis of TLPs, DLPs, and transcriptionally active DLPs at 20 Å resolution (Lawton, Estes *et al.*, 1999; Feng, Lawton *et al.*, 2002), further studies at subnanometer resolution are necessary to obtain a more precise molecular understanding of these changes. Formation of 7–8 nucleotide long capped transcripts observed in the TLPs (and in antibody decorated DLPs) upon transcriptional activation indicate that the capping enzyme VP3 is in close proximity to VP1, further supporting the suggestion from the cryo-EM studies that VP1 and VP3 are attached to the VP2 layer as heterodimers (Lawton, Estes *et al.*, 2001).

Structural organization of the genome

Given that rotaviruses transcribe the eleven dsRNA segments simultaneously and repeatedly within the confines of the DLP structure, interesting questions are how the genomic segments are organized within the VP2 layer and how these segments interact with VP1 and VP3. Unless the genome adheres to strict icosahedral symmetry, the precise details of how the genome is organized inside the VP2 layer and how it interacts with VP1 and VP3 would be lost in reconstructions of rotavirus particles, in which icosahedral symmetry is implicitly assumed. Neither the genome nor VP1 and VP3, is expected to obey strict icosahedral symmetry. Comparative structural analyses of virions and VLPs, which lack the genome, clearly indicate that the rotavirus genome is enclosed within the VP2 layer. By comparing reconstructions of native DLPs and recombinant baculovirus-expressed empty virus-like particles (VLPs), Prasad *et al.* (Prasad, Rothnagel *et al.*, 1996) showed that the viral genome in the icosahedrally-averaged reconstructions manifests as concentric shells in which the dsRNA helices, interacting closely with the VP2 layer, are packed around the transcription enzyme complex located at the icosahedral fivefold axes (Pesavento, Crawford *et al.*, 2006) (Figure 3.2D). A similar structural organization of the genome is seen in the X-ray structures of bluetongue virus (BTV) (Grimes, Burroughs *et al.*, 1998; Gouet, Diprose *et al.*, 1999), and the orthoreovirus core (Reinisch, Nibert *et al.*, 2000). Earlier low angle X-ray scattering studies of orthoreovirus particles indicate that the dsRNA genome is tightly packed as parallel helices in a semicrystalline array (Harvey, Bellamy *et al.*, 1981). Recently, through biochemical and cryo-EM analysis of rotavirus subjected to high pH conditions, the rotavirus genome has been observed to exhibit a remarkable ability to undergo reversible condensation and expansion without affecting the integrity of the surrounding capsid layers (Pesavento, Lawton *et al.*, 2001). From the previous biochemical and structural data, a plausible model for the structural organization of the rotavirus genome has been proposed (Gouet, Diprose *et al.*, 1999; Pesavento, Estes *et al.*, 2003). In this model, each dsRNA genome segment is spooled around a transcription complex that is anchored to the inner surface of the VP2 layer at the fivefold axis (Figure 3.2E). This model allows for up to 12 independent transcription complexes, each attached to an individual dsRNA segment for simultaneous, independent transcription (Pesavento, Crawford *et al.*, 2006). Consistent with this model, to date no dsRNA virus with more than 12 segments has been observed. This model is also consistent with the isometric and concentric condensation of the genome seen in rotavirus, which is achieved simply by reducing the inter-strand separation in each of the cones (Pesavento, Lawton *et al.*, 2001).

Comparison of rotavirus structural organization with other dsRNA viruses

In the last two decades, several dsRNA viruses within the *Reoviridae* such as orbivirus (Grimes, Jakana *et al.*, 1997; Grimes, Burroughs *et al.*, 1998; Gouet, Diprose *et al.*, 1999), orthoreovirus (Reinisch, Nibert *et al.*, 2000; Zhang, Ji *et al.*, 2005), aquareovirus (Shaw, Samal *et al.*, 1996; Nason, Samal *et al.*, 2000), phytoreovirus (Zhou, Baker *et al.*, 2001; Nakagawa, Miyazaki *et al.*, 2003), and cypovirus (Hill, Booth *et al.*, 1999), and outside this family such as totivirus (Naitow, Tang *et al.*, 2002), birnavirus (Coulibaly, Chevalier *et al.*, 2005), and cystovirus (Huiskonen, de Haas *et al.*, 2006), have been structurally characterized using either or both cryo-EM and X-ray crystallographic techniques.

One necessary and common function inherent in all these viruses is the ability to carry out endogenous transcription of the genome. Although these viruses vary substantially in the organization of their outer capsid layer reflecting specific cell recognition requirements, they exhibit significant similarities in the structural elements that are required for endogenous transcription. In all these viruses, it appears that the viral protein responsible for transcription is attached to their inner-most layer that surrounds the genome at the fivefold positions similar to that seen in the case of rotavirus. The innermost layer with 120 subunits also is strikingly duplicated in many dsRNA viruses, from totiviruses of yeast origin which consists of two dsRNA segments to orthoreoviruses with twelve segments. This structural similarity is seen in the overall organization and in subunit structure, despite lacking any noticeable protein sequence similarity, as indicated by X-ray crystallographic structures of some of the dsRNA viruses. The only exception is birnavirus, in which the entire capsid exhibits T = 13 organization (Bottcher, Kiselev *et al.*, 1997; Coulibaly, Chevalier *et al.*, 2005), perhaps because of a slightly different genome organization and replication strategy from that usually seen in other dsRNA viruses (Birghan, Mundt *et al.*, 2000). Although transcription of the genome is capsid-dependent as in other dsRNA viruses, birnavirus exhibits picornavirus-like characteristics including protein-primed replication (Magyar, Chung *et al.*, 1998), and proteolytic processing of the precursor proteins (Birghan, Mundt *et al.*, 2000; Da Costa, Chevalier *et al.*, 2002). The capsid protein VP2 of birnavirus shares significant similarities not only with capsid proteins of dsRNA viruses which form a T = 13 shell, such as VP6 of rotavirus (Mathieu, Petitpas *et al.*, 2001), VP7 of BTV (Grimes, Jakana *et al.*, 1997) and μ1 of reovirus (Liemann, Chandran *et al.*, 2002;

Zhang, Ji *et al.*, 2005) but also with the capsid proteins of T = 3 and T = 4 positive-sense ssRNA viruses (Ahlquist 2005; Coulibaly, Chevalier *et al.*, 2005).

The icosahedral organization with 120 subunits, also referred to as T = 1 icosahedral structure (Grimes, Burroughs *et al.*, 1998), is unique and has only been observed in dsRNA viruses. This unique assembly of the innermost capsid layer in these viruses is thought to have evolved to serve a dual purpose: to properly position the transcription enzyme complex and organize the genome to facilitate endogenous transcription (Prasad and Prevelige, 2003). In this icosahedral assembly, the asymmetric unit of the icosahedral lattice consists of two subunits. One of these subunits is closer to the icosahedral fivefold axis, and the other subunit is slightly offset from the fivefold axis. Based on the location of the polymerase at the fivefold axis, the subunit closer to the fivefold axis perhaps provides a proper anchoring point for the polymerase, while the other subunit may provide substantial interactions with the underlying layer of dsRNA. In rotavirus, it has been shown that the N-terminal 92 residues of VP2 are required for interaction with the polymerase and also the RNA (Labbe, Baudoux *et al.*, 1994; Zeng, Estes *et al.*, 1998). Cryo-EM studies of recombinant particles with deletion in this region suggests that the N-termini of VP2 subunits protrudes inward (Lawton, Zeng *et al.*, 1997) consistent with the X-ray structures of other members of *Reoviridae* such as BTV (Gouet, Diprose *et al.*, 1999) (Grimes, Jakana *et al.*, 1997), orthoreovirus (Reinisch, Nibert *et al.*, 2000), and phytoreovirus (Nakagawa, Miyazaki *et al.*, 2003).

While the structural organization of the innermost layer and the location of the polymerase are conserved in all these viruses, there are differences in the location of the capping enzyme. In rotavirus and BTV, it appears that the capping enzyme is located inside the innermost layer in close proximity to the polymerase (Prasad, Rothnagel *et al.*, 1996; Nason, Rothagel *et al.*, 2004). However, in orthoreovirus, aquareovirus, and cypovirus, the capping enzyme is located on the outer surface of the innermost layer at the fivefold axes forming a distinct turret like structure (Hill, Booth *et al.*, 1999; Nason, Samal *et al.*, 2000; Reinisch, Nibert *et al.*, 2000). Based on this difference, dsRNA viruses are grouped into two classes, turreted and non-turreted (Hill, Booth *et al.*, 1999). Rotavirus, BTV, and phytoreoviruses belong to the non-turreted class, whereas orthoreovirus, aqauereovirus, and cypoviruses belong to the turreted class. In the turreted viruses, because of the prominent turrets at the fivefold axes, the outer capsid layer(s) that surround the common capsid layer with 120 subunits exhibit incomplete T = 13 lattice with 600 subunits (200 trimers) instead of the expected 780 subunits (260 trimers) as found in the non-turreted viruses.

Among the simplest of the dsRNA viruses are the totiviruses, which are intracellular yeast viruses with a single capsid layer consisting of the 120 subunits, followed by the cypoviruses, which incorporate turrets at their fivefold axes, perhaps indicating that the more complex non-turreted and turreted dsRNA viruses diverged from these ancestral capsid architectures, by incorporating additional capsid layers. Birnaviruses perhaps represent an evolutionary link between the positive-sense RNA viruses and dsRNA viruses (Ahlquist, 2005; Coulibaly, Chevalier *et al.*, 2005), and the emergence of more complex dsRNA viruses with T = 13 capsid layers with diverse host range.

Conclusion

Structural studies using cryo-EM and X-ray crystallography techniques, together with a variety of biochemical studies, have provided a detailed picture of rotavirus architecture along with mechanistic insights into various functions such as cell entry, trypsin-enhanced infectivity, antibody interaction, and endogenous transcription (Pesavento, Crawford *et al.*, 2006). While uncovering several common themes with other dsRNA viruses, these studies have also underscored unique features of rotaviruses. One of the most interesting aspects of the rotavirus structure is the possibility of unique entry-related structural transformations in the VP4 spike protein (Dormitzer, Nason *et al.*, 2004; Pesavento, Crawford *et al.*, 2005), which undoubtedly will be the subject of further structural and biochemical studies. Emerging techniques such as electron tomography are predicted to become useful in establishing the oligomeric status of the VP4 spike in trypsinized particles. Further structural and biochemical studies are expected to include: (1) the high resolution structures of VP7, and the internal globular domain of VP4, and how these proteins interact with one another and with VP6 in coordinating the conformational changes of VP4 during cell entry; (2) receptor interactions with VP4; (3) structural alterations that follow the removal of the outer capsid layer during cell entry, which render the resulting DLPs transcriptionally competent; (4) mechanistic details of how the capsid facilitates the dynamic process of simultaneous and continuous genome transcription; (5) high resolution structural analysis of VP1, the viral polymerase, and VP3, the capping enzyme, and how they interact with each other and with VP2 to coordinate endogenous transcription, and how this ternary complex interacts with non-structural proteins such as NSP2 (Jayaram, Taraporewala *et al.*, 2002) and NSP5 (Taraporewala and Patton 2004) (Jiang, Jayaram *et al.*, 2006) to coordinate genome packaging and capsid assembly. With further advances in the reverse genetics for rotavirus, following preliminary successful attempts (Komoto, Sasaki *et al.*, 2006), more in-depth understanding of mechanistic details of rotavirus replication is anticipated.

Acknowledgements

This work was supported by NIH grants AI-36040 (B.V.V.P.) and DK-30144 (M.K.E.) and a grant from R. Welch foundation (B.V.V.P).

References

Ahlquist, P. (2005). Virus evolution, fitting lifestyles to a T. Curr. Biol. *15*(12), R465–7.

Birghan, C., E. Mundt, *et al.* (2000). A non-canonical lon proteinase lacking the ATPase domain employs the ser-Lys catalytic dyad to exercise broad control over the life cycle of a double-stranded RNA virus. EMBO J. *19*(1), 114–23.

Bottcher, B., N.A. Kiselev, *et al.* (1997). Three-dimensional structure of infectious bursal disease virus determined by electron cryomicroscopy. J. Virol. *71*(1), 325–30.

Chen, D., C.L. Luongo, *et al.* (1999). Rotavirus open cores catalyze 5′-capping and methylation of exogenous RNA, evidence that VP3 is a methyltransferase. Virology *265*(1), 120–30.

Coulibaly, F., C. Chevalier, *et al.* (2005). The birnavirus crystal structure reveals structural relationships among icosahedral viruses. Cell *120*(6), 761–72.

Coulson, B.S., S.L. Londrigan, *et al.* (1997). Rotavirus contains integrin ligand sequences and a disintegrin-like domain that are implicated in virus entry into cells. Proc. Natl Acad. Sci. USA *94*(10), 5389–94.

Crawford, S.E., S.K. Mukherjee, *et al.* (2001). Trypsin cleavage stabilizes the rotavirus VP4 spike. J. Virol. *75*(13), 6052–61.

Da Costa, B., C. Chevalier, *et al.* (2002). The capsid of infectious bursal disease virus contains several small peptides arising from the maturation process of pVP2. J. Virol. *76*(5), 2393–402.
Dormitzer, P.R., E.B. Nason, *et al.* (2004). Structural rearrangements in the membrane penetration protein of a non-enveloped virus. Nature *430*(7003), 1053–8.
Dormitzer, P.R., Z.Y. Sun, *et al.* (2002). The rhesus rotavirus VP4 sialic acid binding domain has a galectin fold with a novel carbohydrate binding site. EMBO J. *21*(5), 885–97.
Estes, M.K. (2001). Rotaviruses and their replication. Philadelphia, Lippincott Willimas & Wilkins.
Estes, M.K., D.Y. Graham, *et al.* (1981). Proteolytic enhancement of rotavirus infectivity, molecular mechanisms. J. Virol. *39*(3), 879–88.
Feng, N., J.A. Lawton, *et al.* (2002). Inhibition of rotavirus replication by a non-neutralizing, rotavirus VP6-specific IgA mAb. J. Clin. Invest *109*(9), 1203–13.
Gouet, P., J.M. Diprose, *et al.* (1999). The highly ordered double-stranded RNA genome of bluetongue virus revealed by crystallography. Cell *97*(4), 481–90.
Greig, S.L., J.A. Berriman, *et al.* (2006). Structural determinants of rotavirus subgroup specificity mapped by cryo-electron microscopy. J. Mol. Biol. *356*, 209–221.
Grimes, J.M., J.N. Burroughs, *et al.* (1998). The atomic structure of the bluetongue virus core. Nature *395*(6701), 470–8.
Grimes, J.M., J. Jakana, *et al.* (1997). An atomic model of the outer layer of the bluetongue virus core derived from X-ray crystallography and electron cryomicroscopy. Structure *5*(7), 885–93.
Guerrero, C.A., E. Mendez, *et al.* (2000). Integrin alpha(v)beta(3) mediates rotavirus cell entry. Proc. Natl Acad. Sci. USA *97*(26), 14644–9.
Harvey, J.D., A.R. Bellamy, *et al.* (1981). Biophysical studies of reovirus type 3. IV. Low-angle x-ray diffraction studies. Virology *112*(1), 240–9.
Hewish, M.J., Y. Takada, *et al.* (2000). Integrins alpha2beta1 and alpha4beta1 can mediate SA11 rotavirus attachment and entry into cells. J. Virol. *74*(1), 228–36.
Hill, C.L., T.F. Booth, *et al.* (1999). The structure of a cypovirus and the functional organization of dsRNA viruses. Nat. Struct Biol. *6*(6), 565–8.
Huiskonen, J.T., F. de Haas, *et al.* (2006). Structure of the bacteriophage phi6 nucleocapsid suggests a mechanism for sequential RNA packaging. Structure *14*(6), 1039–48.
Isa, P., S. Lopez, *et al.* (1997). Functional and structural analysis of the sialic acid-binding domain of rotaviruses. J. Virol. *71*(9), 6749–56.
Jayaram, H., M.K. Estes, *et al.* (2004). Emerging themes in rotavirus cell entry, genome organization, transcription and replication. Virus Res. *101*(1), 67–81.
Jayaram, H., Z. Taraporewala, *et al.* (2002). Rotavirus protein involved in genome replication and packaging exhibits a HIT-like fold. Nature *417*(6886), 311–5.
Jiang, X., H. Jayaram, *et al.* (2006). Cryoelectron microscopy structures of rotavirus NSP2-NSP5 and NSP2-RNA complexes, implications for genome replication. J. Virol. *80*(21), 10829–35.
Kapikian, A.Z., Y. Hoshino, *et al.* (2001). Fields Virology. Philadelphia, Lippincoot Williams & Wilkins.
Komoto, S., J. Sasaki, *et al.* (2006). Reverse genetics system for introduction of site-specific mutations into the double-stranded RNA genome of infectious rotavirus. Proc. Natl Acad. Sci. USA *103*(12), 4646–51.
Labbe, M., P. Baudoux, *et al.* (1994). Identification of the nucleic acid binding domain of the rotavirus VP2 protein. J. Gen. Virol. 75 (Pt 12), 3423–30.
Labbe, M., A. Charpilienne, *et al.* (1991). Expression of rotavirus VP2 produces empty corelike particles. J. Virol. *65*(6), 2946–52.
Lawton, J.A., M.K. Estes, *et al.* (1997). Three-dimensional visualization of mRNA release from actively transcribing rotavirus particles. Nat. Struct Biol. *4*(2), 118–21.
Lawton, J.A., M.K. Estes, *et al.* (1999). Comparative structural analysis of transcriptionally competent and incompetent rotavirus-antibody complexes. Proc. Natl Acad. Sci. USA *96*(10), 5428–33.
Lawton, J.A., M.K. Estes, *et al.* (2000). Mechanism of genome transcription in segmented dsRNA viruses. Adv. Virus Res. *55*, 185–229.
Lawton, J.A., M.K. Estes, *et al.* (2001). Identification and characterization of a transcription pause site in rotavirus. J. Virol. *75*(4), 1632–42.
Lawton, J.A., C.Q. Zeng, *et al.* (1997). Three-dimensional structural analysis of recombinant rotavirus-like particles with intact and amino-terminal-deleted VP2, implications for the architecture of the VP2 capsid layer. J. Virol. *71*(10), 7353–60.

Lepault, J., I. Petitpas, *et al.* (2001). Structural polymorphism of the major capsid protein of rotavirus. EMBO J. *20*(7), 1498–507.

Liemann, S., K. Chandran, *et al.* (2002). Structure of the reovirus membrane-penetration protein, Mu*1*, in a complex with is protector protein, Sigma3. Cell *108*(2), 283–95.

Liu, M., N.M. Mattion, *et al.* (1992). Rotavirus VP3 expressed in insect cells possesses guanylyltransferase activity. Virology *188*(1), 77–84.

Lopez, S., C.F. Arias, *et al.* (1985). Primary structure of the cleavage site associated with trypsin enhancement of rotavirus SA11 infectivity. Virology *144*(1), 11–9.

Magyar, G., H.K. Chung, *et al.* (1998). Conversion of VP1 to VPg in cells infected by infectious pancreatic necrosis virus. Virology *245*(1), 142–50.

Mathieu, M., I. Petitpas, *et al.* (2001). Atomic structure of the major capsid protein of rotavirus, implications for the architecture of the virion. EMBO J. *20*(7), 1485–97.

Naitow, H., J. Tang, *et al.* (2002). L-A virus at 3.4 A resolution reveals particle architecture and mRNA decapping mechanism. Nat. Struct Biol. *9*(10), 725–8.

Nakagawa, A., N. Miyazaki, *et al.* (2003). The atomic structure of rice dwarf virus reveals the self-assembly mechanism of component proteins. Structure *11*(10), 1227–38.

Nason, E.L., R. Rothagel, *et al.* (2004). Interactions between the inner and outer capsids of bluetongue virus. J. Virol. *78*(15), 8059–67.

Nason, E.L., S.K. Samal, *et al.* (2000). Trypsin-induced structural transformation in aquareovirus. J. Virol. *74*(14), 6546–55.

Pesavento, J., S. Crawford, *et al.* (2006). Rotavirus Proteins–Structure and Assembly. New York, Springer.

Pesavento, J.B., S.E. Crawford, *et al.* (2005). pH-induced conformational change of the rotavirus VP4 spike, implications for cell entry and antibody neutralization. J. Virol. *79*(13), 8572–80.

Pesavento, J.B., M.E. Estes, *et al.* (2003). Structural organization of the genome in rotavirus. London, Elsevier Science B.V.

Pesavento, J.B., J.A. Lawton, *et al.* (2001). The reversible condensation and expansion of the rotavirus genome. Proc. Natl Acad. Sci. USA *98*(4), 1381–6.

Prasad, B.V., J.W. Burns, *et al.* (1990). Localization of VP4 neutralization sites in rotavirus by three-dimensional cryo-electron microscopy. Nature *343*(6257), 476–9.

Prasad, B.V., and P. E. Prevelige, Jr. (2003). Viral genome organization. Adv. Protein Chem. *64*, 219–58.

Prasad, B.V., R. Rothnagel, *et al.* (1996). Visualization of ordered genomic RNA and localization of transcriptional complexes in rotavirus. Nature *382*(6590), 471–3.

Prasad, B.V., G.J. Wang, *et al.* (1988). Three-dimensional structure of rotavirus. J. Mol. Biol. *199*(2), 269–75.

Reinisch, K.M., M.L. Nibert, *et al.* (2000). Structure of the reovirus core at 3.6 A resolution. Nature *404*(6781), 960–7.

Shaw, A.L., R. Rothnagel, *et al.* (1993). Three-dimensional visualization of the rotavirus hemagglutinin structure. Cell 74(4), 693–701.

Shaw, A.L., R. Rothnagel, *et al.* (1996). Rotavirus structure, interactions between the structural proteins. Arch. Virol. Suppl *12*, 21–7.

Shaw, A.L., S.K. Samal, *et al.* (1996). The structure of aquareovirus shows how the different geometries of the two layers of the capsid are reconciled to provide symmetrical interactions and stabilization. Structure 4(8), 957–67.

Taraporewala, Z.F., and J. T. Patton (2004). Nonstructural proteins involved in genome packaging and replication of rotaviruses and other members of the *Reoviridae*. Virus Res. *101*(1), 57–66.

Tihova, M., K.A. Dryden, *et al.* (2001). Localization of membrane permeabilization and receptor binding sites on the VP4 hemagglutinin of rotavirus, implications for cell entry. J. Mol. Biol. *314*(5), 985–92.

Torres-Vega, M.A., R.A. Gonzalez, *et al.* (2000). The C-terminal domain of rotavirus NSP5 is essential for its multimerization, hyperphosphorylation and interaction with NSP6. J. Gen. Virol. *81*(Pt 3), 821–30.

Valenzuela, S., J. Pizarro, *et al.* (1991). Photoaffinity labeling of rotavirus VP1 with 8-azido-ATP, identification of the viral RNA polymerase. J. Virol. *65*(7), 3964–7.

Welch, S.K., S.E. Crawford, *et al.* (1989). Rotavirus SA11 genome segment 11 protein is a nonstructural phosphoprotein. J. Virol. *63*(9), 3974–82.

Yeager, M., J.A. Berriman, *et al.* (1994). Three-dimensional structure of the rotavirus haemagglutinin VP4 by cryo-electron microscopy and difference map analysis. EMBO J. *13*(5), 1011–8.

Yeager, M., K.A. Dryden, *et al.* (1990). Three-dimensional structure of rhesus rotavirus by cryoelectron microscopy and image reconstruction. J. Cell Biol. *110*(6), 2133–44.
Zarate, S., M.A. Cuadras, *et al.* (2003). Interaction of rotaviruses with Hsc70 during cell entry is mediated by VP5. J. Virol. *77*(13), 7254–60.
Zarate, S., R. Espinosa, *et al.* (2000). Integrin alpha2beta1 mediates the cell attachment of the rotavirus neuraminidase-resistant variant nar3. Virology *278*(1), 50–4.
Zeng, C.Q., M.K. Estes, *et al.* (1998). The N. terminus of rotavirus VP2 is necessary for encapsidation of VP1 and VP3. J. Virol. *72*(1), 201–8.
Zhang, X., Y. Ji, *et al.* (2005). Features of reovirus outer capsid protein mu1 revealed by electron cryomicroscopy and image reconstruction of the virion at 7.0 Angstrom resolution. Structure *13*(10), 1545–57.
Zhou, Z.H., M.L. Baker, *et al.* (2001). Electron cryomicroscopy and bioinformatics suggest protein fold models for rice dwarf virus. Nat. Struct Biol. *8*(10), 868–73.

Structure and Function of Bluetongue Virus and its Proteins

4

Polly Roy

Abstract

The members of Orbivirus genus within the *Reoviridae* family are arthropod-borne viruses and are responsible for high morbidity and mortality in ruminants. Bluetongue virus (BTV) which causes disease in livestock (sheep, goat, cattle) has been in the forefront of molecular studies for the last three decades and now represents the best understood orbivirus at the molecular and structural levels. BTV, like other members of the family, is a complex non-enveloped virus with seven structural proteins and an RNA genome consisting of 10 variously sized double-stranded (ds) RNA segments. This article will be centred on the molecular dissection of BTV with a view to understanding the role of each protein in the virus replication cycle. Data obtained from studies over a number of years has defined the key players in BTV entry, replication, assembly and exit and has increasingly found roles for host proteins at each stage. Specifically, it has been possible to determine the complex nature of the virion through three-dimensional cryo-electron microscopy (EM) reconstructions (diameter ~800 Å); the atomic structure of proteins and the internal capsid (~700 Å, the first large highly complex structure ever solved); the definition of the virus encoded enzymes required for RNA replication; the ordered assembly of the capsid shell and the protein sequestration required for it; and the role of host proteins in virus entry and virus release. These areas are important in themselves for BTV replication but they also indicate the pathways that related viruses, which includes viruses that are pathogenic to man and animals, might also use providing an informed starting point for intervention or prevention.

Introduction

Bluetongue virus (BTV) is the prototype virus of the *Orbivirus* genus within the family *Reoviridae*. BTV and other orbiviruses (14 serogroups within the genus) are distinct from reovirus and rotaviruses in a number of ways. The most distinctive feature of these viruses are that they infect and multiply in arthropod vectors (e.g. gnats, mosquitoes and ticks) and are transmitted by these blood-feeding vectors to a variety of vertebrates, including man (sometimes causing severe infection). BTV consists of 24 different serotypes (BTV-I, –2, etc.) and is transmitted by *Culicoides* spp. (gnats) infecting field and domestic ruminants (e.g. sheep and cattle), often with high morbidity and mortality causing considerable economical consequences in many parts of the world. Other orbiviruses that may cause disease in animals with serious economic consequences in some regions of the world, include African horse sickness virus (AHSV; nine serotypes, AHSV-1, -2, etc.) and epizootic haemorrhagic disease virus (EHDV; seven serotypes, EHDV-1, -2, etc.) of deer, also transmitted by *Culicoides*. Although BTV and AHSV were isolated in 1900, and initial morphological and biochemical characterization were reported as early as 1969 (see review article, Gorman, 1990), much of the current knowledge on the molecular biology, genome structure and encoded products has been accumulated only in the last two decades. This has been driven by the fact that the spread of *Culicoides* species from endemic to non-BTV (and also related AHSV and EHDV of deer) regions of the world has heightened concern that these viruses are a threat to regions of the world that are presently free from viral infection. Indeed recently BTV has emerged in many European countries for the first time. Due to the economic significance of BTV in many parts of the worlds, much of the research interests have been focused on BTV and thus to date it not only represents the best characterized orbivirus but is now also one of the most well characterized animal viruses.

BTV and other orbiviruses are structurally different from reoviruses and rotaviruses and have proteins in their organization that bear no similarity in primary sequence to those of reovirus or rotaviruses (Roy *et al.*, 1990). The virions are more fragile than reovirus, BTV infectivity being lost in mildly acidic conditions and on exposure to lipid solvents and detergents, although there is no evidence for the presence of a lipid component in purified virions (Gorman *et al.*, 1983). BTV virions (550S) are architecturally complex structures composed of 7 discrete proteins in a specific but non-equimolar ratio that are organized into two shells, the inner and outer capsids (Hewat *et al.*, 1992b; Prasad *et al.*, 1992; Roy, 1996) (Figure 4.1). The virion proteins encapsidate a genome of 10 dsRNA segments. The outer capsid which is composed of two major structural proteins (VP2 and VP5) is

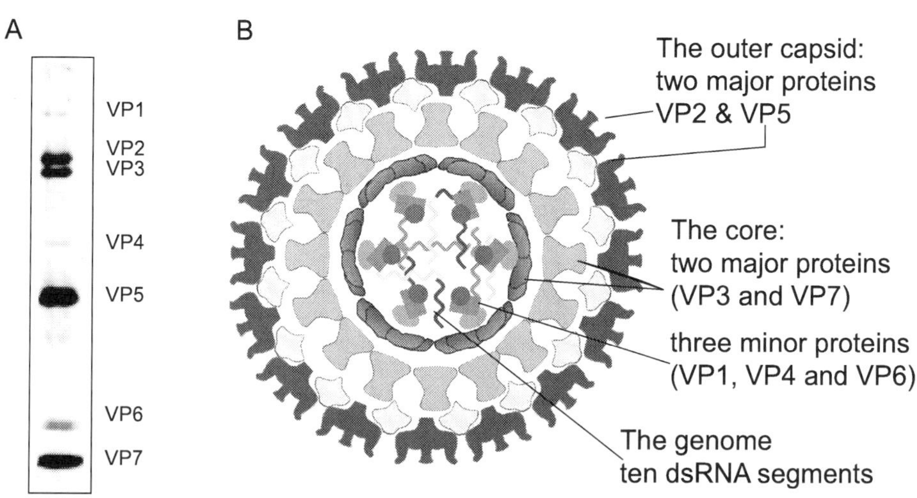

Figure 4.1 Bluetongue virus particles and proteins. (A) Migration patterns of seven structural proteins on a 10% PAGE. (B) Schematic diagram of bluetongue virus (BTV) showing the positions and structural organizations of BTV components.

removed (virion uncoating) shortly after infection, to yield a transcriptionally active 470S core particle. The core is composed of two major proteins (VP7 and VP3) and three minor proteins (VP1, VP4 and VP6) in addition to the dsRNA genome (Roy, 1996). There is no evidence that any trace of the outer capsid remains associated with these cores, as has been described for reovirus. The cores may be further uncoated to form 390S subcore particles that lack VP7, also in contrast to reovirus (Nibert *et al.*, 1996). The subviral particles are probably akin to cores derived *in vitro* from virions by physical or proteolytic treatment that removes the outer capsid and causes activation of genome transcription (Huismans *et al.*, 1987). In addition to the seven structural proteins, three non-structural (NS) proteins, NS1, NS2, NS3 (and a related NS3A) are made in BTV-infected cells, each of which appears to be involved either in virus replication, assembly or egress.

Over the past years, significant advances have been made in the understanding of the structure–function relationships of the BTV proteins, protein and capsid structure, the assembly process for the formation of virus particles and virus morphogenesis. This article will mainly focus on BTV proteins, their structural topography in the virus, with particular references to virus penetration, the virion assembly process and what is known about BTV egress from infected cells.

BTV outer capsid proteins, structural organization and their involvement in cell entry

The non-glycosylated outer capsid proteins of BTV are involved in virus entry to the host cells. A combination of three-dimensional (3D) structural determination and molecular/biochemical analysis of the BTV outer capsid has given some insight into how the two outer capsid proteins could function in cell entry/membrane penetration processes in an analogous manner to enveloped virus proteins. Cryo-electron microscopy (Cryo-EM) and image reconstruction of BTV particles have been used to deduce the general relationships of the two outer capsid proteins within the virus particle. An initial low-resolution (45 Å) 3D reconstruction followed by a higher resolution (22 Å) reconstruction of BTV particles revealed a well-ordered morphology of the virion particle with icosahedral symmetry and provided details of the essential organization of the two outer capsid proteins (Hewat *et al.*, 1992b; Nason *et al.*, 2004). An accurate position of the two proteins, VP2 and VP5, in relation to each other and to the underlying protein layer of the core was clearly defined (Figure 4.2). In addition, their juxtaposition and structural features gave an insight into their role in virus attachment to the host cell receptor and membrane penetration that are supported further by biochemical data. The outer capsid of BTV has an icosahedral configuration, and exhibits a unique organization made up of two distinct shapes, triskelion and globular. The outer layer does not conform to a T = 13 icosahedral symmetry like rotaviruses and reoviruses and consists of a total of sixty triskelion densities and 120 globular densities (Figure 4.2).

VP2, its structure and function

The most external part of the outer capsid is the propeller-shaped triskelion motif. At the tip of each blade of the propeller, the molecule broadens out and bends upwards perpendicular to the plane of the virus. These bent tips give the entire virion a diameter of ~880 Å and extend from the main body of the particle by 4 nm (Hewat *et al.*, 1992b,1994;

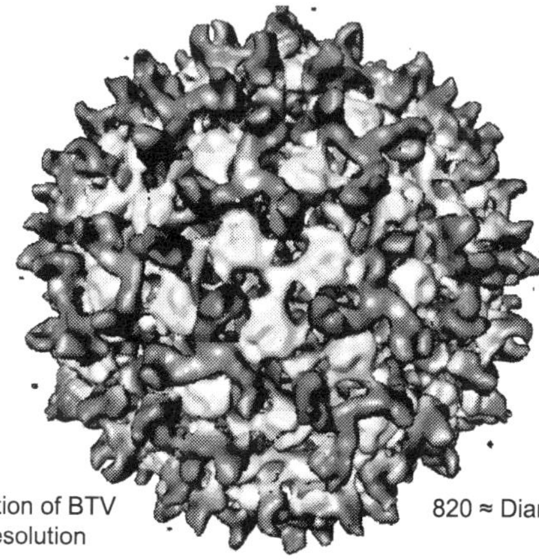

Figure 4.2 Surface representations of the 3D cryo-EM structures (22 Å resolution) of BTV. (A) Whole particle showing sail-shaped triskelion propellers (VP2 trimers) in red and globular domains (VP5 trimers) in yellow (Nason *et al*., 2004).(See also Plate 4.2.)

Nason *et al.*, 2004). The estimated volume of each triskelion structure closely matches that of three VP2 molecules. In solution recombinant VP2 oligomerises into a trimer readily, supporting the structural data (Hassan and Roy, 1999). The structural features of VP2 are of a sailed-shaped, extended propeller blade on the virion surface and correlate well with the functional features of VP2. It is the most variable of the BTV proteins, eliciting virus neutralizing antibody, and is responsible both for serotype-specificity and virus haemagglutination activity indicating that VP2 is the cellular receptor binding protein (Huismans and Erasmus, 1981; Kahlon *et al.*, 1983; Fukusho *et al.*, 1987; Ghiasi *et al.*, 1987; Inumaru and Roy, 1987; Gould, 1988; French and Roy, 1990; Roy *et al.*, 1990; Roy, 1992;). Direct evidence that VP2 indeed mediates attachment of the virion to the cell surface and its subsequent internalization has been obtained subsequently by biochemical studies using virion particles, recombinant particles and soluble VP2 (Eaton and Crameri, 1989; Hassan and Roy, 1999).

VP5, its structure and function

Interspersed between the VP2 triskelions and lying more internally are the VP5 globular densities that are also entirely exposed in the virion. Three molecules of VP5 (~60 kDa) conform to the globular density observed and as recombinant VP5 in solution also forms trimers, the 120 globular densities per virus particle comprise a total of 360 VP5 molecules (Hassan *et al.*, 2001). At the sequence level VP5 is divided into an amino terminal coiled-coil domain and a carboxyl terminal globular domain with a flexible hinge region in between. The N-terminus of VP5 also includes a membrane-inserting hydrophobic region followed by a long heptad repeat region analogous to the other fusion proteins (Purdy *et al.*, 1986; Hassan *et al.*, 2001; Forzan *et al.*, 2004). These features, as well as its location, support its involvement in membrane permeabilizing activity. In the absence of other BTV proteins, VP5 is highly cytotoxic and purified VP5 protein was shown to have intrinsic membrane 'pore-formation' ability (Hassan *et al.*, 2001), an activity that maps to the amphipathic helices.

That VP5 is indeed the BTV membrane penetration protein has been formally demonstrated by an experimental system where VP5 was expressed on the cell-surface (Forzan *et al.*, 2004). By this system it was possible to show that VP5 has the ability to interact with host cell membranes and to induce cell–cell fusion in a manner similar to that seen for the fusion proteins of enveloped viruses. VP5 exhibits its membrane-interacting properties only after it has undergone a low-pH triggered activation step, which presumably alters the conformation to fusion-competence. Deletion of the first 40 residues (the two predicted amphipathic helices) abolished fusion activity and, notwithstanding the precise mechanism, it is clear that VP5 can functionally substitute for a typical viral fusion protein. Although it is not yet clear if fusion induced by VP5 follows the same mechanisms proposed for enveloped viruses (Peisajovich and Shai, 2002), it seems probable that the amphipathic helices insert into the target membrane. This early interaction and the formation of the initial pore might be mechanistically different for VP5 when compared with other fusion proteins as the amphipathic helices of VP5 more closely resemble pore-forming antimicrobial peptides than typical fusion peptides (Merrifield *et al.*, 1982; Putsep *et al.*, 1999).

Thin section EM studies exhibited the presence of BTV particles in coated pits, suggesting clathrin-mediated endocytosis may be used for BTV entry (Eaton *et al.*, 1990). A recent comprehensive investigation using biochemical and confocal microscopy studies in combinations of various specific inhibitors and RNA interference conclusively demonstrated that indeed BTV enters cells by clathrin-mediated endocytosis (Figure 4.3) and requires acidic pH-dependent penetration (Forzan *et al.*, 2007).

Notably, in the BTV particle the VP5 protein makes strong contact with the underlying VP7 layer and also with the spike protein VP2 which protrudes above VP5

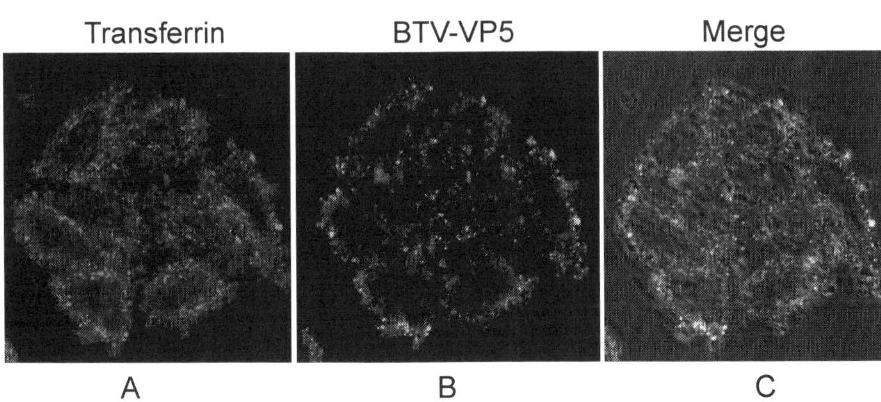

Figure 4.3 BTV entry into the cells via endosome. HeLa cells were infected with BTV and cellular uptake of BTV was monitored by confocal microscopy. BTV particles were stained with a polyclonal anti-VP5 primary antibody and localized by a FITC-conjugated secondary antibody (panel B). Transferrin-TRITC conjugate (Tf-TRITC) was monitored as endosomal marker (panel A). Note that VP5 co-localises with transferrin in the endosomes of infected HeLa cells (Panel C) (Forzan *et al.*, 2007).

and has direct contacts with underlying VP7 as well (see Figure 4.2). Thus, a plausible hypothesis would be that the attachment of VP2 with receptor and subsequent low pH inside endosomes induce a rearrangement and conformational changes in VP2 and VP5 which result in loosening the interactions of VP5 and VP7 and simultaneously allowing VP5 to form a protein layer with intrinsic outside-in curvature. Proteolytic removal or pH induced conformational change of VP2 then allows the VP5 amphipathic helices to freely interact with membranes and initiate the permeabilization process. This allows core particles that now lack both VP2 and VP5 proteins, to be released into the cytoplasm from the endosomal vesicles.

BTV core proteins, viral genome and their structural organization

The released uncoated, core particle in the host cytoplasm is a highly stable structure providing a robust compartment for the viral genome that remains within the core throughout the infection cycle. Thus, the core is the endpoint of virus disassembly within which the ten dsRNA genome segments can repeatedly and simultaneously be transcribed by the core-associated enzymes, resulting in the extrusion of the newly synthesized, capped and methylated mRNA. The extruded transcripts in turn program the synthesis of viral structural and NS proteins and also act as templates for synthesis of genomic RNAs. *In vitro,* BTV mRNA synthesis can be activated by proteolytic digestion of the outer capsid of the virus (Van Dijk and Huismans, 1980; Van Dijk and Huismans, 1988). The structural integrity of the core particle appears to be essential for maintaining an efficient transcriptional activity which like other members of the family, occurs by a fully conservative process (Bannerjee and Shatkin, 1970) which requires the efficient co-ordination of a series of enzyme activities. Using individual recombinant proteins and *in vitro* assay systems, it has been possible to delineate the specific catalytic activities provided solely by the three core-associated minor proteins, VP1, VP4 and VP6 (discussed later) that are encapsidated by the two major core proteins, VP3 and VP7.

BTV core structure and its two major proteins

Unlike the whole virion particle, the BTV core appears as an icosahedral capsomeric structure by negative staining. Considerable information is available on core structures and individual proteins due to both Cryo-EM analysis and X-ray crystallography analysis (Hewat *et al.*, 1992a, 1992b, 1994; Prasad *et al.*, 1992; Grimes *et al.*, 1997, 1998). The 470S core with a diameter of 70 nm consists of two concentric protein layers surrounding a central density. The surface layer, which is made up of clusters of 260 VP7 (38 kDa) trimers, conforms to an icosahedral symmetry with a triangulation number of 13 (T = 13) in a left-handed configuration and occurs between 260 Å and 345 Å from the centre of the core. The VP7 trimers sit on a thin layer formed by 120 copies of VP3 (100 kDa) and are easily visible as protrusions by negative staining of the particles under EM, giving a bristly appearance to cores described by Els and Verwoerd as capsomers arranged in ring-like structures (Els and Verwoerd, 1969). These ring structures were alluded to in the genus name 'Orbis' (ring or circle). These trimers are arranged around 132 distinctive channels as six-member rings, with five-member rings at the vertices of the icosahedrons (Figure 4.4). The organization of the five quasi-equivalent trimers (T13) that form a protomeric unit

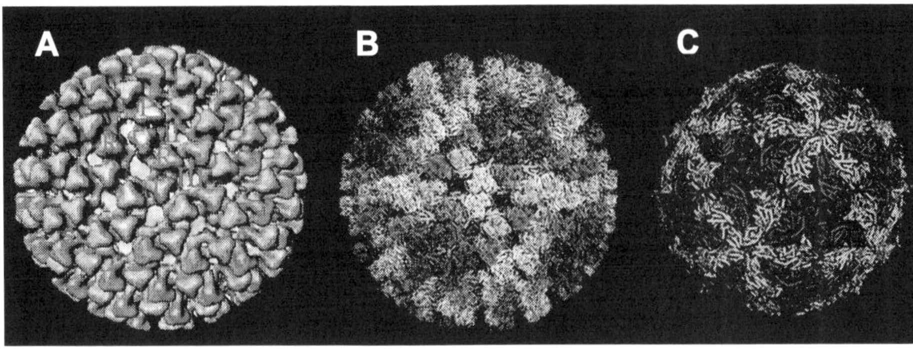

Figure 4.4 3D structures of BTV core. (A). Surface representations of the 3D Cryo-EM structures of BTV-10 core (700 Å in diameter) viewed along the icosahedral threefold axis showing the trimers of VP7 (in blue). The five quasi-equivalent trimers (P, Q, R, S, and T) and the locations of channels II and III are marked. (Grimes *et al.*, 1997). (B). The X-ray structure of the BTV-1 core surface showing the detail arrangement of 260 trimers. The structurally equivalent trimers have been given the same colour. (C). The structure of the VP3 layer showing the arrangement of 120 copies of two conformationally distinct types of VP3 molecules, 'A' shown in green and 'B' shown red. 60 copies of each of the two molecules together form 12 decamers that are linked together forming a thin protein shell. The icosahedral fivefold, threefold and symmetry axes are indicated (Grimes *et al.*, 1998).(See also Plate 4.4.)

(P, Q, R, S, T) could be seen clearly. The aqueous channels that are surrounded by these trimers can be grouped into three types; type I channels run along the icosahedral fivefold axes, type II channels surround the fivefold axes, and the type III channels are located around the icosahedral threefold axes. Some of these channels extend inward, penetrating through to the inner layer. The threefold axis of the trimer (85 Å long) is perpendicular to the core surface, and the broader base of the trimer (65 Å) contacts the internal network of the VP3 shell (subcore) which has a disc-shaped appearance that serves as the scaffold for the deposition of VP7 layer. The 120 molecules of the VP3 shell form this scaffold organized with icosahedral symmetry, on a pseudo T = 2 lattice. The channels or 'pores' are small (7 Å diameter) at the icosahedral threefold axes (type III) and slightly larger (9 Å diameter) at the fivefold axes (type I), and the latter act as 'portals' to release the newly synthesized transcripts. While the 'T' trimers of VP7 sit tightly on the 'pores' at the threefold axes, the 'P' trimers are located loosely above the pores at fivefold axes.

The monomer of VP7 has a 2 domain structure composed of an outer (distal from core centre) and an inner domain (Basak *et al.*, 1992, 1996, 1997; Grimes *et al.*, 1995, 1998). The outer domain consists of about 130 residues representing the central portion (aa 121–249) of the VP7 polypeptide chain and has a 'jelly-roll' (folded into an anti-parallel β-sandwich) with topology common to many viral structural proteins (Stuart, 1993). The inner domain is entirely α-helical consisting of a total of nine α-helices and long extended loops and is formed by both the N (aa 1–120) and the C (aa 250–349) termini of the molecule. The relative disposition of these domains is such that they wrap around the molecular threefold axis in a right handed sense (Grimes *et al.*, 1995). The packing of different trimers onto the VP3 layer requires minimal distortion and the two layers interact through flat, largely hydrophobic surfaces.

Each VP3 molecule (110 kDa) contains many α helices as well as β strands and resembles a triangular wedge made up of three distinct domains: an apical domain (aa 298–587) that sits nearest to the icosahedral fivefold axis; a carapace domain (aa 7–297, aa 588–698 and aa 855–901), which forms a rigid plate; and a dimerization domain (aa 699–854) that is involved in forming the quasi-twofold interaction between the different molecules of VP3 (Grimes *et al.*, 1998). The 120 VP3 molecules are arranged as two structurally distinct sets (A and B) of 60 subunits, five of which ('A' group) are arranged as a five-pointed star around each fivefold axis and the other five ('B' group) molecules are positioned, one between each of the points of the star, but at a further distance from the fivefold axis. Thus, VP3 molecules form a convex disc shape in which twelve decamers, positioned one at each of the fivefold axes, are arranged end to end, so as to form the complete VP3 shell. The molecular contacts between the A and B subunits within the VP3 decamers surrounding the fivefold axes of the VP3 shell are very extensive and it seems plausible that this is the primary building block for virus assembly. This notion has been further supported by mutational analysis (Kar *et al.*, 2004). The VP3 shell has a maximum diameter of 59 nm, forming a central space with an internal diameter of 38 nm that contains both the genome as well as the three minor core proteins.

The genome organization of the core particle

The core structure also exhibited electron density distributed throughout the internal space of the VP3 shell (Prasad *et al.*, 1992; Grimes *et al.*, 1998; Gouet *et al.*, 1999). The density map showed features consistent with the layers of RNA organized in a highly ordered fashion (Gouet *et al.*, 1999). Approximately 80% of the entire genome of 19219 base pairs (Fukusho *et al.*, 1989; Roy, 1989) can be modelled as four distinct concentric layers that have centre to centre spacing between RNA strands of 26–30 Å. As a consequence of the spacing of the 'A' and 'B' forms of VP3, the grooves in the inner surface of the subcore shell form a spiral around the fivefold axis. The outer dsRNA layer appears to interact with these grooves. However, it is not clear how the RNA molecules pass inwards to the next and subsequent concentric shells. The density detected in the inner layers of RNA gets progressively weaker, suggesting that its organization and icosahedral symmetry is progressively less rigid, although each layer maintains an overall spiral organization (Gouet *et al.*, 1999).

Protein–protein interactions drive the assembly of core-like and virus-like particles in a heterologous expression system

While expression of individual proteins by use of the baculovirus expression system have made possible an understanding of the structure–function relationship of each of the BTV proteins, the discovery that the high level expression of VP3 and VP7 or VP2, VP3, VP5 and VP7 by recombinant baculoviruses could spontaneously assemble into core-like and virus-like particles (CLPs and VLPs) in insect derived *Spodoptera frugiperda* cells (*Sf* cells) made significant advances in orbivirus research (French and Roy, 1990). These have included the capsid assembly process and pathways, the structural organization of certain capsid proteins and the development of a novel vaccine technology. Initially a dual baculovirus expression vector was utilized to express and assemble the VP3 and VP7 of core particle and it was demonstrated that these two proteins have inherent interacting properties for each other (French and Roy, 1990). CLPs produced in *Sf* cells were empty, lacking the genetic materials but had similar in size and appearance to cores prepared from BTV (Figure 4.5).

The assembly of CLPs had given the notion that it might be possible to assemble the outer capsid proteins onto these CLPs. Therefore additional expression cassettes were generated; one that expressed VP2 and VP5 simultaneously and the other that expressed VP2, VP3, VP4 and VP7 proteins by a single recombinant baculovirus cell (Belyaev and Roy, 1993; French *et al.*, 1990). Co-infections of insect cells with two dual expression vectors (namely VP3 and VP7 and VP2 and VP5) yielded VLPs that contained different amounts of the outer capsid proteins, depending on the experiment while the expressed proteins from the quadruple vector assembled into virtually homogenous double-capsid particles (Figure 4.5). The formation of CLPs or complete VLPs in the absence of non-structural proteins or the internal minor proteins implies that these proteins are not necessary for the assembly of BTV capsids.

Further, the 3D reconstruction of VLPs determined from cryo-electron micrographs, revealed an icosahedral structure of 86 nm in diameter with essentially the same features as for the native particle. The VLP was thus shown to contain the four constituent proteins as the native virus particle (Hewat *et al.*, 1994), with each of the protein positions highly occupied. In contrast, CLPs made by VP3 and VP7, although having an identical organization of VP7 and VP3 as obtained from core structure, lacked five VP7 trimers around each of the fivefold axes (Hewat *et al.*, 1992a, 1992b; Nason *et al.*, 2004). Thus CLPs had only 200 trimers instead of 260 trimers, arranged on a T = 13 lattice. Only the VP7 trimers which have the strongest affinity for the VP3 inner core and are involved in maintaining the structural integrity of the CLPs were incorporated. It is likely that the VP7 trimers that were missing are much more distant from the VP3 layer than the other trimers and that the outer capsid proteins VP2 and VP5 are necessary for the adhesion of these VP7 trimers around the fivefold axes. Analysis of the adhesion surfaces of VP7 suggests an order for the trimer assembly process whereby trimers attach first to the icosahedral threefold axes, with the peripentonal trimers being the last to attach and the first to be released (Grimes *et al.*, 1998; Gouet *et al.*, 1999). This model is further supported by biochemical and mutagenesis studies (Limn *et al.*, 2000; Limn and Roy, 2003) and also appears to hold true for other dsRNA viruses (Lu *et al.*, 1998; Nakagawa *et al.*, 2003).

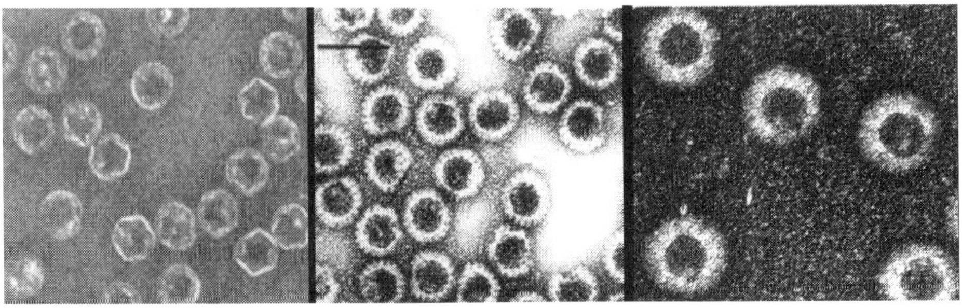

Figure 4.5 Recombinant Baculovirus expressed BTV-like particles. (Left panel): Expression of VP3 assembled into a single layer subcore-like particles. (Middle panel): Expression of VP3 and VP7 assembled as doubled-shelled core-like particles. (Right panel): Expression of VP2, VP3, VP5 and VP7 assembled of triple-layered virus-like particles.

Structural organization of VP1 and VP4, the two minor proteins within CLPs

The assembly of recombinant proteins in capsid formation was further exploited to investigate the protein–protein interactions in internal protein encapsidation mechanism as well as the structural organization of these proteins within the CLPs (Le Blois *et al.*, 1992; Loudon and Roy, 1991) . It was clear that while the two minor proteins, VP1 and VP4 were readily encapsidated by VP3-VP7 CLPs, VP6 was not easily incorporated into CLPs by the same assay system indicating that unlike the VP1 and VP4, VP6 might not have direct contact with VP3 under surface. Cryo-EM reconstruction of CLPs with and without the internal VP1 and VP4 proteins have revealed a flower-shaped density formed by VP1 and VP4 within the VP3 shell of particles generated with VP1, VP3, VP4 and VP7 (Nason *et al.*, 2004). The VP1 and VP4 flower-shaped density, similar to that of rotavirus particle, was clearly observed directly beneath the icosahedral fivefold axes and attached to the underside of the VP3 layer (Prasad *et al.*, 1996). Although the individual contribution of VP1 and VP4 to the flower-shaped density was difficult to delineate from the reconstruction of CLPs, based on recent results from reovirus (Zhang *et al.*, 2003), it is likely that the central mass is predominantly due to the VP1 protein and that some of the sideward projecting density is due to VP4. By co-immunoprecipitation studies it was further confirmed that VP1 and VP4 not only interact with one another strongly but also interact with VP3.

The location of the VP1 and VP4 complex is consistent with the model proposed for genome organization in BTV suggested from X-ray crystallographic studies (Gouet *et al.*, 1999). It is noteworthy that in the presence of magnesium ions and NTP substrates the BTV core-associated enzymes (i.e. the minor proteins, see below) are activated and conformational changes can be seen around the fivefold axis of the BTV core. This results in outward movement of VP3 and VP7 allowing opening of a pore in the VP3 layer at the fivefold axis. The mRNAs synthesized by the core are extruded through this pore and subsequently pass out through the central space in the pentameric rings of VP7 trimers ('P' trimers). There are also pores in the VP3 layer at the threefold axes but in the intact core particle these are completely blocked by the 'T' trimer of VP7 as discussed above.

VLPs as novel vaccine

VLPs generated in insect cells were not only structural mimic of the virion particles, but were also demonstrated as functional equivalents to the native virions at the biological and immunological levels. VLPs exhibited high levels of haemagglutination activity similar to those of authentic BTV (French *et al.*, 1990). As expected, antibodies raised in guinea pigs, to the expressed particles contained high titres of neutralizing activity against the homologous BTV serotype emphasizing their authenticity at functional level (French *et al.*, 1990). Subsequently, a series of vaccine trials were performed to examine the protective efficacy of VLPs (10–200 μg per sheep) in 1-year old BTV free sheep divided in groups (4–8 sheep/group) (Roy *et al.*, 1992, 1994). All sheep were challenged by subcutaneous inoculation of 1 ml of infective sheep blood containing virulent virus at day 117 or at 14 months. The challenged sheep developed neither clinical signs nor viraemia, indicating the suppressed replication of BTV. The post-challenge blood samples of the sheep that only received saline were viraemic and these sheep developed high neutralizing antibody titres indicative of a primary infection. Protective immunity to BTV disease was obtained by vaccinating sheep with doses of 10 μg or more of BTV VLPs and showed long-lasting protection against homologous BTV challenge. Some evidence was also obtained for cross-protection, depending on the challenge virus and the amounts of antigen used for vaccination (Roy *et al.*, 1994; Roy, 1995).

Use of recombinant proteins and particles to delineate the function of the three enzymatic core proteins and their structural organization

Although the location and the predicted amino acid sequences of the three internal proteins suggested their possible roles in genome replication (Roy *et al.*, 1988), direct evidence for their catalytic activities was only possible to obtain due to the availability of purified soluble proteins and the suitable *in vitro* assay systems.

RNA dependent RNA polymerase activity

The first direct evidence of polymerase activity of the largest viral protein, VP1 (Mr of 149.5 kDa) was obtained from an *in vitro* poly(A) polymerase activity assay from cell extracts infected with a recombinant baculovirus expressing VP1 protein (Urakawa *et al.*, 1989). Subsequently, replicase activity of the protein was demonstrated by an *in vitro* replicase assay using purified soluble protein form (Boyce *et al.*, 2004). The data showed that soluble BTV VP1 possesses processive replicase activity in the absence of any other viral proteins (Boyce *et al.*, 2004), unlike the rotavirus VP1 polymerase where the polymerase protein is only active in the context of a subviral particle (Chen *et al.*, 1994, 2001; Patton *et al.*, 1997; Chen and Patton, 1998; Tortorici *et al.*, 2003). In addition to elongating the nascent minus strand, the synthesis of a unit length labelled RNA shows that purified VP1 initiates BTV minus strand synthesis *de novo*, rather than extending a fortuitous secondary structure present at the 3′ end of the 'plus' strand template. The polymerase in the absence of other core components exhibited little sequence specificity for BTV 'plus' strand template, suggesting that specificity for viral template probably comes from location within the viral core. The lack of specificity for the template 3′ end of VP1 is consistent with a lack of specificity reported using crystals of reovirus λ3 complexed with a short oligonucleotide template (Tao *et al.*, 2002). Purified recombinant reovirus λ3 has been shown to be a poly (C)-dependent poly (G) polymerase, but does not possess replicase activity (Starnes and Joklik, 1993). Also, although purified rotavirus VP1 has been shown to bind to the 3′ end of *in vitro* synthesized rotavirus 'plus' strand RNA, replicase activity was only detected when the remaining core proteins were present or in the context of subviral particles (Patton, 1996; Patton *et al.*, 1997). It can therefore be hypothesized that template specificity may be added to the replicase activity of BTV VP1 by viral proteins present in the assembling virus core. It is plausible that after dsRNA synthesis is complete viral core proteins may trigger the switch from replicase activity to further transcriptase activity.

The X-ray structure of BTV VP1 is not available. However, based on reovirus λ3 polymerase together with two other viral RNA dependent RNA polymerases (RdRps) it has been possible to generate a model for the structure of BTV VP1 (Wehrfritz *et al.*, 2007). Similar to the other polymerases, three distinct domains were identified from the model, an N-terminal domain (NTD), a C-terminal (CTD) domain and a central polymerase domain (PD). The experimental data obtained from a mutant VP1 protein, (in which the aspartate residues of the catalytic site, a characteristic GDD motif, aa 763–765 were mutated to alanine residues) confirmed that the PD is indeed the active domain of the protein. It is therefore highly likely that this fragment will be folded into a typical polymerase structure as the modelling data predicts. When each of the three domains were synthesized individually, only the purified PD fragment but not the NTD or CTD fragment was able to bind NTP similar to that of the full-length protein. None of the domains had catalytic activity when tested individually or in tandem but when all three were mixed together *in vitro*, the RdRp activity was reconstituted. It is possible that although PD possesses the catalytic activity, the other two domains are needed to stabilise the protein. It is noteworthy that in reovirus λ3, the large N and C-terminal domains form a cage around the centrally placed polymerase domain holding it rigid and preventing movement during the catalytic cycle (Tao *et al*., 2002). It is likely therefore that the VP1 and λ3 share similar structural organization as seen in the VP1 model.

BTV mRNA capping activity

The transcripts produced by BTV cores have 'cap' structures at their 5′ termini. By using recombinant purified VP4 protein (76.4 kDa), it had been possible to demonstrate that VP4, in the absence of any other viral proteins, was sufficient for the generation of a cap structure (type 1) at the 5′ end of BTV transcripts. It is remarkable that although the protein comprises only 644 amino acid residues it can efficiently execute a series of catalytic reactions *in vitro* resulting in the 'cap' structure (Ramadevi *et al*., 1998a). Further it was shown that the soluble VP4 possesses RNA 5′-triphosphatase activity which catalyses the first step in the RNA-capping sequence and also covalently binds the G of GTP via a phosphoamide linkage and subsequently catalyses a GTP-PPi exchange reaction. These are essential and characteristic features of a guanylytransferase enzyme. Moreover encapsidated VP4 in recombinant CLPs reacted with GTP and covalently bind G similar to that of soluble protein indicating that the soluble VP4 is functional equivalent of the VP4 within the viral core. Both the soluble form and the encapsidated form were capable of catalysing the conversion of unmethylated GpppG or *in vitro*-produced uncapped BTV RNA transcripts to m7GpppGm in the presence of *S*-adenosyl-L-methionine indicating that VP4 is also a methyltransferase enzyme. Analysis of the methylated products identified both methyltransferase type 1 and type 2 activities associated with VP4, demonstrating that the complete BTV cap reaction is associated with this single core protein (Martínez-Costas *et al*., 1998; Ramadevi *et al*., 1998a; Ramadevi and Roy, 1998). These data are particularly significant as the capping activities associated with vaccinia virus had shown that an enzyme complex of 3 proteins was required to achieve the 'cap' formation (Venkatesan *et al*., 1980). The close association of VP4 with the polymerase protein within the core structure is however, necessary since VP4 must cap the nascent mRNAs as they emerge from the polymerase.

How a single viral protein may orchestrate these series of reactions in a consecutive manner became clear from a recent 2.5 Å resolution crystal structure of the protein (Sutton *et al*., 2007). The atomic structure revealed an elongated, slightly eccentric egg-timer molecule with various active sites organized in multidomain architecture. The overall layouts of active sites on the elongated VP4 molecule are arranged in a largely linear fashion along the molecule (Figure 4.6). The polypeptide chain folds into four discrete domains, the N-terminal 108 residues form a kinase-like (KL) domain which is followed by part of the second domain, the N-terminal part of the N7MTase domain. The polypeptide chain then forms the entire 2′OMTase domain before completing the N7MT domain and finally forming the guanylyltransferase domain (GT). The KL, N7MT and 2′OMT domains are mixed α/β structures with a central β sheet and associated helices. The C-terminal GT domain is almost entirely α-helical and includes a number of relatively poorly ordered loops. There are extensive interaction surfaces between some of the domains, as shown in Figure 4.6. The GT and N7MT domains are closely associated, whilst the KL domain is reasonably strongly associated with both of these domains. In contrast, the 2′OMT domain is rather flexibly attached to the remainder of the molecule. These interactions are reflected in the flexibility of the molecule as judged from the flexing of domains observed in different crystals of VP4.

Within the crystal a potential dimer is formed by a crystallographic twofold axis via 56 amino acids (located at both N and C terminus) from each monomer. It is noteworthy that through gel filtration and sucrose gradient analyses it was shown that the purified VP4 is also dimeric in solution (Ramadevi *et al.*, 1998b).

VP6 as a helicase protein

It is logical to presume that a helicase protein is necessary either to unwind the dsRNA ahead of the transcriptase protein or to separate the parental and newly synthesized RNAs

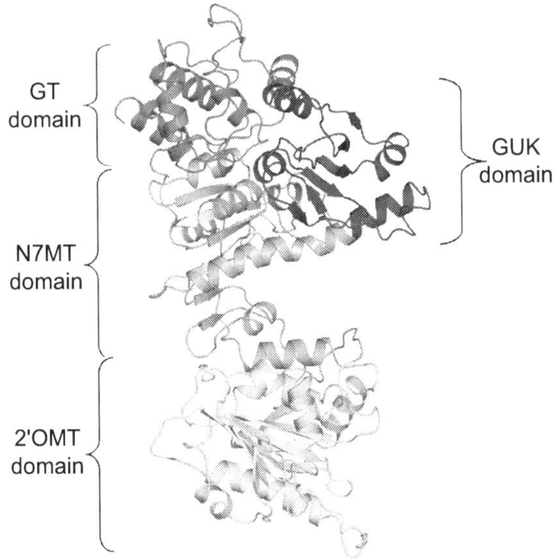

Figure 4.6 Cartoon representation of the structure of the VP4 molecule. Ribbon diagram of the VP4 structure with enzymatic domains coloured: guanylyl transferase (GTase, red) domain, kinase-like (kl, blue) domain, guanine-N7-methyltransferase (N7MTase, orange) domain and nucleoside-2′-O-methyltransferase (2′OMT, green) domain (Sutton *et al.*, 2007).(See also Plate 4.6.)

following transcription. BTV is the only member of the *Reoviridae* where helicase activity has been located to a minor structural component of the core, the smallest 35.7 kDa VP6 protein. It was found that incubation of purified soluble VP6 with dsRNA in the presence of ATP and magnesium ions, leads to the unwinding of the dsRNA template (Stauber *et al.*, 1997). Unusually, this activity was detected not only for templates which were complete duplexes but also for those which had short 5′ or 3′ overhangs (Stauber *et al.*, 1997). Most models explaining helicase activity require two features: ATP driven conformational rearrangements of the protein that, in turn, generate kinetic energy, and possession of at least two nucleic acid binding sites in order to maintain contact with nucleic acid (Lohman *et al.*, 1998). This requirement seems indispensable for the unwinding reaction, since it allows the protein to bind both single strand and duplex regions of a polynucleotide simultaneously or two complementary single strands in an unwinding fork. VP6 possesses these features and were shown to be necessary for its activity (Kar *et al.*, 2004).

VP6 also forms readily a hexamer in solution, a common feature shared by many helicase proteins (Kar *et al.*, 2004). Furthermore, incubation of purified VP6 with RNA resulted in the formation of protein–RNA complexes that are visible as a ring-like structure by electron microscopy (Kar *et al.*, 2004).

Protein synthesis and involvement of BTV non-structural protein NS2 in virus replication/core assembly

In infected cells BTV specific proteins are detectable as early as at 2 hours post infection, and the rate of protein synthesis increases rapidly until 11 to 13 hours post infection. Most viral proteins are synthesized throughout the infection cycle and accumulate during infection until cell lysis (Huismans, 1979). Infection of mammalian cells with BTV leads to a rapid inhibition of cellular macromolecular synthesis and the induction of an apoptotic response. The latter response appears to be triggered by the combination of two signals involving attachment of the virus to the cell and the activity of the membrane permeabilizing protein, VP5 (Mortola *et al.*, 2004). Shortly after release from the endosome core particles become very rapidly associated with a matrix that gradually surrounds the particles to form virus inclusion bodies (VIBs). As the infection progresses, these VIBs increase both in size and in number. The newly synthesized viral transcripts, all the four subcore proteins (VP1, VP3, VP4 and VP6) as well as both the cores and subcores can be identified within these VIBs; thus, they appear to be the site of BTV replication and of early viral assembly (Eaton *et al.*, 1988; Hyatt and Eaton, 1988; Brookes *et al.*, 1993).

The formation of inclusion bodies (IBs) similar to the VIBs could also be obtained by synthesizing the NS2 phosphoprotein in absence of other viral proteins (Thomas *et al.*, 1990; Modrof *et al.*, 2005). Thus, it seems that NS2 is the driving force for IB formation in the infected cells. However, it seems that their make-up is dependent only on a full-length NS2 with the C terminus of the protein being the major functional domain involved (Kar *et al.*, 2007). Co-expression of core structural proteins has shown that while VP1, VP3, VP4, and VP6 all co-localise with NS2 inclusions without other viral proteins, VP7 requires the co-expression of VP3 to be recruited in to these structures. This may be an early indication that partially assembled core structures in which VP3 and VP7 are already interacting are the building blocks of core formation.

NS2 is a highly charged protein and binds to any ssRNA molecules readily (Fukusho *et al.*, 1989; Roy *et al.*, 1990). However, by careful designing of *in vitro* competition assays together with electrophoretic mobility shift assays it was demonstrated that BTV RNAs were preferentially bound over non-specific RNAs via RNA structures that are unique in each RNA segment (Lymperopoulos *et al.*, 2003, 2006). These observations support the suggestion that NS2 may have a role in the recruitment of mRNA for replication. They are significant since the mechanism by which the 10 dsRNA segments that make up the viral genome are selectively recruited and packaged into newly assembling virus particles is unresolved and represents one of the most enduring questions in the field. The sequences that represent the putative binding partners for NS2 are neither the conserved 5′ terminal octanucleotides nor 3′ terminal hexanucleotides of all BTV RNA segments or the potential panhandle structures (formed by the partial complimentary sequences of the 5′ and 3′ termini). Instead they are distributed throughout the coding and non-coding regions of the different genome segments (Lymperopoulos *et al.*, 2003). The confirmation that NS2 indeed binds at unique hairpin-loop secondary structures of BTV RNA transcripts were first provided by chemical and enzymatic structure probing of regions bound preferentially by NS2 and then subsequently by using hairpin mutants which alter the sequence or the structure of the RNA independently (Figure 4.7). No other RNA binding protein of any other member of *Reoviridae* so far has been shown to have similar RNA structural specificity. However, rotavirus NSP3 appears to bind a linear sequence found at the 3′ end of all rotavirus RNA segments (Poncet *et al.*, 1993) and rotavirus VP1 also binds to the 3′end of rotavirus RNA (Patton, 1996). The significance of these interactions for genome packaging is, at this stage, unclear and it is worth pointing out that, other than BTV, there are several systems where viral proteins recognise packaging signals that are contained in hairpin structures with no apparent sequence homology (Bae *et al.*, 2001; Beasley and Hu, 2002). Although the RNA binding activity of NS2 can explain how BTV mRNAs are selected from the pool of cellular messages for incorporation into assembling virus particles, it remains unclear how only a single copy of each genome segment is included in newly formed core particles. NS2 could bring the viral RNAs into close proximity, and inter-segment RNA interactions allow the formation of an RNA complex that is the basis of core assembly or that different RNA subsets that have already interacted may be bound by NS2. The latter hypothesis is consistent with the observations that NS2 may form decameric complexes (Butan *et al.*, 2004) and that each NS2 protein subunit may have several RNA binding domains (Fillmore *et al.*, 2002).

Although inclusion bodies have obvious advantages in providing a site for core assembly and genome packaging as well as preventing premature shut-off of core transcription by preventing the addition of viral outer capsid proteins they may also represent a 'trap' for newly assembled core particles. In order for newly assembled core to be released from inclusions it is therefore necessary that these structures must be dynamic and easily disassembled as well as assembled. NS2 is the only BTV-encoded phosphoprotein. Recent data suggest that protein kinase casein kinase II (CKII), interacts with NS2 and is able to phosphorylate it *in vitro* (Modrof *et al.*, 2005). Mutagenesis of two normally phosphorylated serine residues (aa 249 and 259) to alanine in NS2 abrogates the formation of distinct VIBs whilst engineered mutation of the same residues to aspartic acid, mimicking

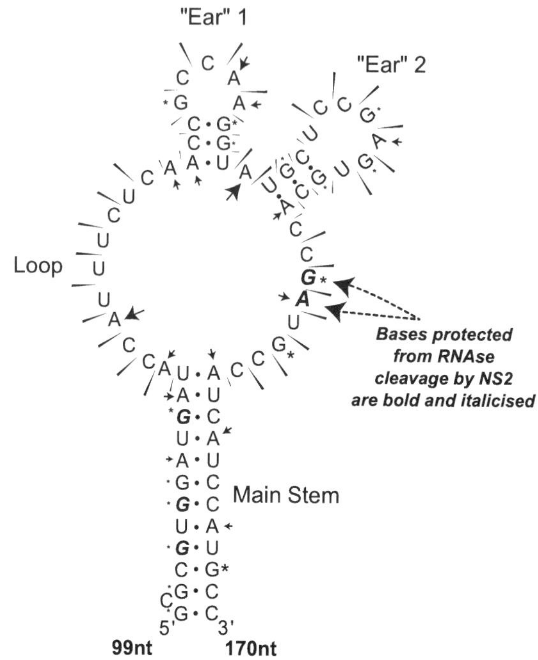

Figure 4.7 Secondary structure of BTV ssRNA that binds NS2. Chemical and enzymatic structure probing of regions bound preferentially by NS2 has shown that the NS2 binds BTV RNA transcripts at unique hairpin-loop secondary structures. Note that the NS2-hairpin interaction has been confirmed by using hairpin mutants which alter the sequence or the structure of the RNA independently (Lymperopoulos *et al.*, 2003, 2006).

a constitutively phosphorylated state, results in the formation of normal inclusion body structures (Modrof *et al.*, 2005). Furthermore the expression of the non-phosphorylated mutant of NS2 was sufficient to disrupt the formation of VIBs in virus-infected cells, demonstrating the potential of the mutant to act in a dominant negative fashion (Modrof *et al.*, 2005). The sequence context of the phosphorylated serine 249 ($_{248}$LSDDDDQ$_{254}$) in NS2 conforms to a conserved sequence which generally stabilises helix unfolding (Zetina, 2001) suggesting that phosphorylation of NS2 in this position could be involved in stabilizing protein folding within its C-terminal domain. Thus, phosphorylation may act to control the propensity of NS2 to form large viral inclusion bodies and thus provide a mechanism by which newly assembled core particles can be released from VIBs. The newly assembled cores that are released from VIBs would then be free to acquire the outer layers (VP5 and VP2) outside of the VIB structure. It is noteworthy that none of the two outer capsid proteins are associated with NS2 or NS2 IBs. A cartoon summary of the role of NS2 in the various stages of core assembly is shown in Figure 4.8.

Assembly mechanism of BTV capsids

The assembly of BTV capsids is especially intriguing as it requires a complex highly ordered series of protein–protein interactions. Recent studies have focussed on the use of

CLPs and VLPs together with structure-based mutagenesis to reveal the key principles that drive the dynamic assembly of this structure.

Assembly of the core particle

VP3 shell appears to play the major role in core assembly as it interacts with all the core components and serves as a scaffold for surface protein VP7 trimers. An important question in VP3 assembly is whether the decamer present in the assembled particle is an identifiable intermediate of the assembly process or only arises upon assembly. The requirement of decamer formation in assembly of the VP3 layer of the core was demonstrated by deleting the dimerization domain of VP3 which abolished the assembly of the subcore (Kar *et al.*, 2004). The deletion of this domain, however, did not perturb the stable decamer formation which was still able to interact with VP1 and VP4 as did the complete VP3 shell (Kar *et al.*, 2004). By contrast, BTV RNAs while interacting with the intact VP3 *in vitro* very efficiently, failed to associate with VP3 decamers under same conditions. These data suggest that assembly of the BTV core may start with the complex formed by polymerase and capping enzyme with the VP3 decamers and these assembly intermediates subsequently recruit the viral RNA and possibly VP6 complexes prior to completion of the assembly of the VP3 subcore.

The assembly of VP3 subcore is stabilized by subsequent attachment of the VP7 trimer. The mismatch between the number of subunits in the VP3 and VP7 layers of the core poses an interesting problem as to how these layers interact to form an intact icosahedral

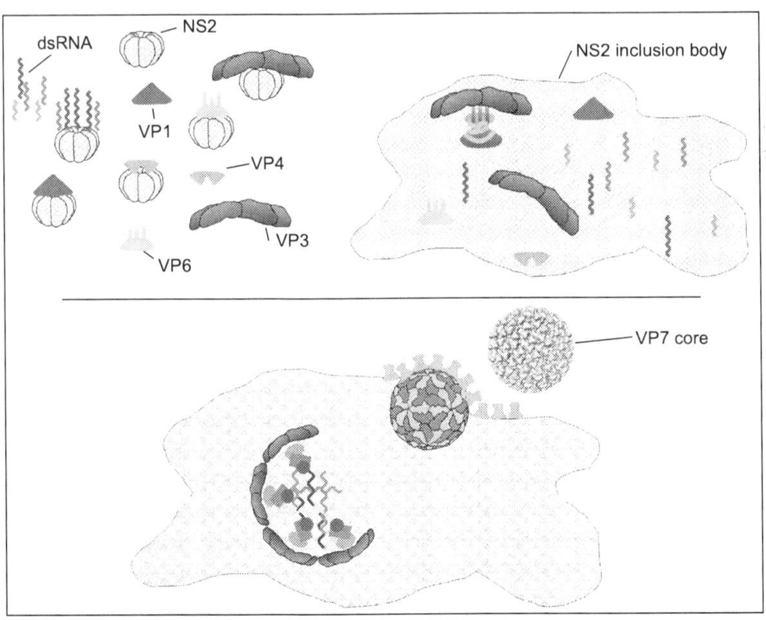

Figure 4.8 Interaction of NS2 and viral components and core assembly. Schematic showing the interaction of NS2 oligomers/VIBs with BTV proteins and the BTV ssRNA and subsequent assembly of VP3 layer into subcore that eventually recruits VP7 trimers prior to completion of the core formation.

structure. Extensive site-directed mutagenesis experiments in combination with the use of various assembly assay systems have suggested the order of the assembly pathway of VP7 onto the subcore (VP3) layer during stable core formation. Mutations in VP7 directed at perturbing monomer-monomer, trimer-trimer, and VP7-VP3 interactions have been generated. The effects of these mutations on VP7 solubility, ability to trimerize and form CLPs in the presence of VP3 have revealed critical information on protein–protein interaction in core formation and also on the order of core assembly (Limn *et al.*, 2000; Limn and Roy, 2003). Stable core assembly depends not only on trimer formation, but more importantly the precise 'shape' of the trimers is necessary to drive the formation of the tight lattice of 260 VP7 trimers on the core surface. The data also suggest that, initially, multiple sheets of VP7 form around different nucleation sites instead of a cascade of trimer associations originating from a single nucleation site and thus it is likely that a number of strong VP7 trimer-VP3 contacts act as multiple equivalent initiation sites and that a second set of weaker interactions then 'fill the gaps' to complete the outer layer of the core.

Assembly of the outer capsid

Although our understanding of the physical contacts between the outer layer of the core, VP7 and the viral outer capsid proteins, VP2 and VP5, has increased recently, (Nason *et al.*, 2004), our understanding of how this process is controlled is less clear. Since the addition of VP2 and VP5 abolishes the transcription activity of the core, addition of the outer capsid is likely to be a highly regulated process to prevent premature shut-off of transcription. Unlike rotavirus, there is no evidence that assembly of the outer capsid is associated with the ER. Instead, within BTV infected cells virus particles are found associated with the vimentin intermediate filaments (Eaton *et al.*, 1990). Furthermore the presence of both VP2 and VP5 was found to be necessary to direct VLPs to the cytoskeleton of insect cells (Hyatt *et al.*, 1993). By contrast, VP2 alone associates with vimentin intermediate filaments, and disruption of these filaments drastically reduces the normal release of virus particles from infected cells (Bhattacharya *et al.*, 2007). Given the importance of these structures for the assembly and egress of many other viruses (Arcangeletti *et al.*, 1997; Nedellec *et al.*, 1998) it may be that intermediate filaments are used to control the addition of the VP2 and possibly VP5 to the assembled core.

NS1 protein and its assembly as tubules in infected cells

The 64 kDa protein NS1 of BTV is synthesized in abundance, up to 25% of the virus-specified proteins, in virus infected cells and readily assembles as tubular structures within the cytoplasm that are biochemically and morphologically distinct from the microtubules and neurofilaments of normal cells (Huismans, 1979). Indeed tubules are characteristic features in all orbivirus infection. These high-molecular-weight tubular structures are formed by helically coiled ribbons of NS1 dimers with a diameter of 52 nm and lengths of up to 1000 nm. The early and abundant synthesis of NS1 and tubules suggests their involvement in virus replication and/or virus translocation.

In a recent study intracellular expression of a single chain antibody to the viral NS1 protein (scFv-αNS1) demonstrated that high levels of NS1 in BTV infected cells are critical to the morphology and trafficking of virus particles (Owens *et al.*, 2004). Four major changes were apparent: first, there was a tremendous reduction in virus-induced cytopathic effect (CPE); second, there was a more than 10-fold increase in the amount of virus released into the culture medium; third, there was a shift from lytic release of virus to budding from the plasma membrane; and fourth, NS1 tubule formation was completely inhibited by scFv-αNS1 expression. Each of these changes, except for the lack of tubule formation, is reminiscent of what occurs during BTV infection of insect cells in culture. Based on these findings it can be hypothesized that the NS1 protein is a major determinant of pathogenesis in the vertebrate host, and that its mechanism of action is the augmentation of virus-cell association (but not transport of virus to the cell surface), which ultimately leads to lysis of the infected cell. It is also possible that NS1 is somehow involved in virus trafficking. Differential virus release in different cells suggests the involvement of host proteins. Indeed a yeast two-hybrid analysis has identified NS1 interaction with the cellular protein SUMO-1 (Noad and Roy, unpublished data). This protein is conjugated to cellular targets as a post-translational modification involved in the intracellular trafficking. In addition it appears that sumoylation of cellular proteins is highly enhanced in BTV infected cells. Upregulation of SUMO-1 has recently been shown to regulate dynamin-dependent protein trafficking within the cell. Thus, it could be speculated that the role of NS1 is to control the trafficking of immature or mature virus particles in infected cells through the interaction with cellular proteins such as SUMO-1.

Non-structural protein NS3 and BTV egress from host cells

Recent evidence has implicated the viral non-structural protein NS3 in virus release from the host cells (Beaton *et al.*, 2002; Wirblich *et al.*, 2006). Although the majority of mature virus particles remain cell associated in mammalian cells like the other members of the family and thus cause substantial cytopathic effect, the virus titre in the infected cell supernatant increases significantly before the onset of dramatic CPE. Thus it is clear that there are defined mechanisms that traffic newly synthesized progeny virus particles out of infected cells. Indeed particles can be seen either to bud through the cell membrane or groups of virus particles move through a local disruption of plasma membrane and both types of particles have been observed to be released from some cell surfaces (Hyatt *et al.*, 1989).

NS3 (229 aa) and its shorter form NS3A (216 aa) are the only membrane proteins encoded by BTV (Lee and Roy, 1986; French *et al.*, 1989;). Both proteins can be found associated with smooth-surfaced, intracellular vesicles (Hyatt *et al.*, 1991) but do not form part of the stable structure of the mature virus. NS3/NS3A proteins have long N-terminal and a shorter C-terminal cytoplasmic domains connected by two transmembrane domains and a short extracellular domain (Figure 4.9A) (Bansal *et al.*, 1998; Beaton *et al.*, 2002; Wu *et al.*, 1999). A single glycosylation site is present in the extracellular domain of BTV NS3 but is missing in NS3 of AHSV, a closely related orbivirus, and therefore does not seem to be essential for the function of the protein (Bansal *et al.*, 1998; van Staden and Huismans, 1991). The NS3/NS3A proteins accumulate to only very low levels in BTV-infected mammalian cells but in invertebrate vector cells the expression levels of these proteins are high (French *et al.*, 1989; Guirakhoo *et al.*, 1995). The correlation between high NS3/NS3A expression and non-lytic virus release has suggested a significant functional role for NS3 in virus egress from invertebrate cells. In addition NS3 and NS3A facilitate the release of baculovirus-expressed VLPs in heterologous insect cells. The NS3 protein is localized at the site of the membrane where VLPs are released (Hyatt *et al.*, 1993), confirming that the integration of NS3/NS3A into the plasma membrane may trigger the release of mature virions. The NS3 protein of BTV exhibit cytoxicity when expressed individually in mammalian or insect cells (French *et al.*, 1989; van Staden *et al.*, 1995). The cytoxicity of NS3 requires membrane association of the protein and depends on the presence of the N terminal transmembrane domain, suggesting that NS3 might function as a viroporin, and facilitate virus release by inducing membrane permeabilization (Han and Harty, 2004). It is possible that this permeabilization activity causes local disruption

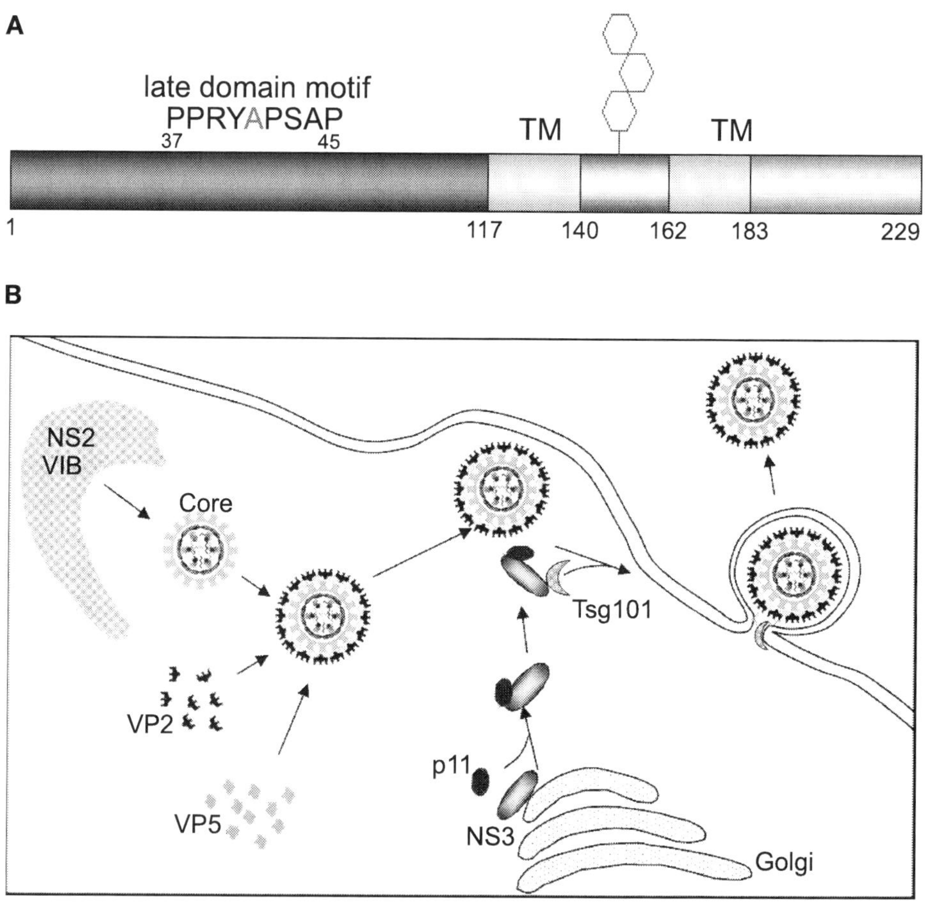

Figure 4.9 Schematics showing the organization of NS3 protein domains and how NS3 facilitate BTV release. A: BTV NS3 is an integral membrane protein with two transmembrane domain (TM) and contains the two most commonly observed late domain motifs PPXY and PS/TAP that recruit the components of cellular ESCRT (endosomal complex required for transport) complexes. B: To facilitate BTV budding the N-terminus (cytoplasmic) interacts with cellular release factors calpactin p11 and Tsg101 (via late domain motifs) while its C-terminus interacts with the BTV outer capsid protein VP2. The interactions with calpactin and p11 are important for proper release of newly assembled virus particles from infected cells. (Beaton *et al.*, 2002; Wirblich *et al.*, 2006).

of the plasma membrane allowing virus particles to be extruded through a membrane pore without acquiring a lipid envelope.

The first 13 amino acids of NS3 have potential to form an amphipathic helix. These amino terminal residues (not present on NS3A) of NS3 interact with the calpactin light chain (p11) of the cellular annexin II complex (Beaton *et al.*, 2002), a complex that has been implicated in membrane related events along the endocytic and regulated secretory pathways including the trafficking of vesicles (for reviews see (Creutz, 1992; Raynal and Pollard, 1994). The interaction between NS3 with p11 is highly specific and an NS3 peptide (a mimic of the sequences of the p11 binding domain inhibits the progeny virions from BTV-infected insect cells. Although the exact physiological role of this interaction is still unknown it is likely that interaction of p11 with NS3 may direct NS3 to sites of active cellular exocytosis, or that NS3 could become part of an active extrusion process. There is some indications that cytoskeletal material is released at sites of BTV egress (Hyatt *et al.*, 1991) which may be Annexin II being drawn through the membrane during the extrusion process, while it is still associated with NS3. The significance of this interaction to BTV egress becomes more apparent in the light of the observation that the other cytoplasmic domain of the protein, situated at the C-terminal end interacts specifically with the BTV outer capsid protein VP2.

In addition to its interaction with the p11 component of the annexin II complex, NS3 is also capable of interaction with Tsg101, a cellular protein implicated in the intracellular trafficking and release of a number of enveloped viruses (Freed, 2004). It is intriguing that NS3 and NS3A of both BTV and AHSV both bind *in vitro* with similar affinity to human Tsg101 and also to its orthologue from *Drosophila melanogaster* (Wirblich *et al.*, 2006). This interaction is mediated by the conserved PSAP motif in NS3 of BTV (Figure 4.9A) and seems to play a role in virus release as knockdown of Tsg101 with siRNA inhibits BTV release from mammalian cells (Wirblich *et al.*, 2006). Like most other viral proteins which recruit Tsg101, NS3 also binds host NEDD4-like ubiquitin ligases *in vitro* via a resident PPXY late domain motif. However, the late domain motifs in NS3 do not function as effectively as the late domains of other enveloped viruses (Wirblich *et al.*, 2006). This appears to be mainly due to the presence of a unique arginine at position 3 of the PPXY motif in NS3. The low activity of the NS3 late domain motifs can be reversed by converting the PPRY motif of NS3 into a more universal PPPY, rendering NS3 as effective as Gag in facilitating retroviral VLP release as the late domains of enveloped viruses. The ability to usurp the vacuolar protein sorting pathway is likely to be more important in insect hosts as orbiviruses establish persistent infections in insect cells without causing significant cytopathic effect. Identification of insect proteins that interact with NS3 should shed more light on the question of whether NS3 is better adapted to engage insect proteins so facilitating improved virus release. Current data suggest that NS3 makes use of host proteins and acts as intermediate to facilitate the release of newly synthesis progeny virus across the cell membrane (see the schematic, Figure 4.9B).

Concluding remarks

Bluetongue disease became an emerging disease in Europe recently and is becoming one of the major threats in livestock industry world-wide. Although our knowledge of host-virus interactions and viral pathogenesis in animals is limited, BTV is well characterized both genetically (the sequence was completed in 1989) and structurally. Understanding of the molecular biology of the virus and mapping the role of each protein in virus life cycle has benefited significantly through the availability of recombinant BTV proteins and sub-viral particles. In addition the structure of BTV proteins, core and virion particles have contributed greatly to understanding the mechanism of protein–protein interaction in the virus assembly pathway of BTV and other orbiviruses. Most importantly, information gained from these studies has laid sound foundation for the generation of safe BTV vaccines with the possibility of use in animals in the near future. Latterly, studies have concentrated on the fundamental mechanisms that are used by the virus to invade, replicate in and escape from susceptible host cells. Progress has been made in understanding the structure and entry of intact virus particles, the role of each enzymatic protein in the transcription complex, the critical interactions that occur between the viral non-structural proteins and viral RNA and the role of cellular proteins in non-enveloped virus egress.

Despite these advances, some critical questions remain unanswered for the BTV life cycle and a more complete understanding of the interactions between the virus and the host cell is required for these to be addressed. For example, although progress has been made in the identification of signals for the recruitment of BTV RNA segments into the virion assembly site in the host cell cytoplasm, it has not been possible yet to determine how exactly each genome segment is packaged into the progeny virus. It is also not apparent when and how these genome segments wrap around the polymerase complex once the RNA has been recruited. One of the major drawbacks of research with BTV and other members of *Reoviridae* has been the lack of availability of a suitable system for genetic manipulation of the virus. This has been a major obstacle in understanding the replication processes of these viruses. However, one of the recent developments in the field of BTV research has been to rescue live virus from transfection of BTV transcripts (Boyce and Roy, 2007). There is no doubt that this will be soon extended to establish *in vitro* manipulative genetic system and will be utilized to address some of these remaining questions.

Very little is known of the intracellular trafficking of newly generated virions although there are some indications of involvement of the cytoskeleton, intermediate filaments and vimentin during BTV morphogenesis. Host–virus interactions during virus trafficking will be one of the future areas needing intense attention. Recent work has revealed unexpected and striking parallels between the entry and release pathways of BTV and pathways involved in entry and release of enveloped viruses. These parallels may be the result of an enveloped ancestor virus or because there are a limited number of cellular pathways that can be useful for the egress of large protein complexes from cells. It is notable that the NS3 glycoprotein of BTV is an integral membrane protein that is functionally involved in virus egress by bridging between the outer capsid protein VP2 and the cellular export machinery. Although no cell-free enveloped form of BTV has been isolated, budding of BTV particles from infected cells at the plasma membrane are quite apparent. The exact role of NS3 in this process and the role host proteins (Annexin II and p11, Tsg101 and MVB) and their contribution in the release of non-enveloped viruses, such as BTV, remains to be clarified.

References

Arcangeletti, M.C., Pinardi, F., Missorini, S., De Conto, F., Conti, G., Portincasa, P., Scherrer, K., and Chezzi, C. (1997). Modification of cytoskeleton and prosome networks in relation to protein synthesis in influenza A virus-infected LLC-MK2 cells. Virus Res. *51*, 19–34.

Bae, S.H., Cheong, H.K., Lee, J.H., Cheong, C., Kainosho, M., and Choi, B.S. (2001). Structural features of an influenza virus promoter and their implications for viral RNA synthesis. Proc. Natl Acad. Sci. USA *98*, 10602–10607.

Bannerjee, A.K., and Shatkin, A.J. (1970). Transcription *in vitro* by reovirus associated ribonucleic acid-dependent polymerase. J. Virol. *6*, 1–11.

Bansal, O.B., Stokes, A., Bansal, A., Bishop, D.H.L., and Roy, P. (1998). Membrane organization of bluetongue virus non-structural glycoprotein NS3. J. Virol. *72*, 3362–3369.

Basak, A., Grimes, J., Burroughs, N., Mertens, P.P.C., Roy, P., and Stuart, D. (1992). Crystallographic studies on the structure of the Bluetongue virus inner capsid, In Bluetongue, African Horse Sickness and Related Orbiviruses., T.E. Walton, and B.I. Osburn, eds. (Boca Raton, FL: CRC Press), pp. 483–490.

Basak, A.K., Gouet, P., Grimes, J., Roy, P., and Stuart, D. (1996). Crystal structure of the top domain of African horse sickness virus VP7: comparisons with bluetongue virus VP7. J. Virol. *70*, 3797–3806.

Basak, A.K., Grimes, J.M., Gouet, P., Roy, P., and Stuart, D.I. (1997). Structures of orbivirus VP7: implications for the role of this protein in the viral life cycle. Structure *5*, 871–883.

Beasley, B.E., and Hu, W.S. (2002). cis-Acting Elements Important for Retroviral RNA Packaging Specificity. J. Virol. *76*, 4950–4960.

Beaton, A.R., Rodriguez, J., Reddy, Y.K., and Roy, P. (2002). The membrane trafficking protein calpactin forms a complex with bluetongue virus protein NS3 and mediates virus release. Proc. Natl Acad. Sci. USA *99*, 13154–13159.

Belyaev, A.S., and Roy, P. (1993). Development of baculovirus triple and quadruple expression vectors: co-expression of three or four bluetongue virus proteins and the synthesis of bluetongue virus-like particles in insect cells. Nucleic Acids Research *21*, 1219–1223.

Bhattacharya, B., Noad, R.J., and Roy, P. (2007). Interaction between Bluetongue virus outer capsid protein VP2 and vimentin is necessary for virus egress. Virol. J. *Jan 15; 4:7.*

Boyce, M., Wehrfritz, J., Noad, R., and Roy, P. (2004). Purified recombinant bluetongue virus VP1 exhibits RNA replicase activity. J. Virol. *78*, 3994–4002.

Boyce, M., and Roy, P. (2007). Reovery of infectious Bluetongue virus from RNA. J. Virol. *81*, 2179–2186.

Brookes, S.M., Hyatt, A.D., and Eaton, B.T. (1993). Characterization of virus inclusion bodies in bluetongue virus infected cells. J. Gen. Virol. *74*, 525–530.

Butan, C., Van Der Zandt, H., and Tucker, P.A. (2004). Structure and assembly of the RNA binding domain of bluetongue virus non-structural protein 2. J. Biol. Chem. *279*, 37613–37621.

Chen, D., Barros, M., Spencer, E., and Patton, J.T. (2001). Features of the 3′ Consensus Sequence of Rotavirus mRNAs Critical to Minus Strand Synthesis. Virology *282*, 221–229.

Chen, D., and Patton, J.T. (1998). Rotavirus RNA replication requires a single-stranded 3′ end for efficient minus-strand synthesis. J. Virol. *72*, 7387–7396.

Chen, D., Zeng, C.Q., Wentz, M.J., Gorziglia, M., Estes, M.K., and Ramig, R.F. (1994). Template-dependent, *in vitro* replication of rotavirus RNA. J. Virol. *68*, 7030–7039.

Creutz, C.E. (1992). The annexins and exocytosis. Science *258*, 924–931.

Eaton, B.T., and Crameri, G.S. (1989). The site of bluetongue virus attachment to glycophorins from a number of animal erythrocytes. J. Gen. Virol. *70*, 3347–3353.

Eaton, B.T., Hyatt, A.D., and Brookes, S.M. (1990). The replication of bluetongue virus. Curr. Top Microbiol. Immunol. *162*, 89–118.

Eaton, B.T., Hyatt, A.D., and White, J.R. (1988). Localization of the nonstructural protein NS1 in bluetongue virus-infected cells and its presence in virus particles. Virology *163*, 527–537.

Els, H.J., and Verwoerd, D.W. (1969). Morphology of bluetongue virus. Virology *38*, 213–219.

Fillmore, G.C., Lin, H., and Li, J.K.-K. (2002). Localization of the single-stranded RNA-binding domains of bluetongue virus nonstructural protein NS2. J. Virol. *76*, 499–506.

Forzan, M., Wirblich, C., and Roy, P. (2004). A capsid protein of nonenveloped Bluetongue virus exhibits membrane fusion activity. Proc. Natl Acad. Sci. USA *101*, 2100–2105.

Forzan, M., and Roy, P. (2007). Bluetongue virus entry into the cells. Journal of Virology *2007 Jan 31; [Epub ahead of print]*
Freed, E.O. (2004). Mechanisms of enveloped virus release. Virus Res. *106*, 85–86.
French, T.J., Inumaru, S., and Roy, P. (1989). Expression of two related nonstructural proteins of bluetongue virus (BTV) type 10 in insect cells by a recombinant baculovirus: production of polyclonal ascitic fluid and characterization of the gene product in BTV-infected BHK cells. J. Virol. *63*, 3270–3278.
French, T.J., Marshall, J.J., and Roy, P. (1990). Assembly of double-shelled, virus-like particles of bluetongue virus by the simultaneous expression of four structural proteins. J. Virol. *64*, 5695–5700.
French, T.J., and Roy, P. (1990). Synthesis of bluetongue virus (BTV) corelike particles by a recombinant baculovirus expressing the two major structural core proteins of BTV. J. Virol. *64*, 1530–1536.
Fukusho, A., Ritter, G.D., and Roy, P. (1987). Variation in the bluetongue virus neutralization protein VP2. J. Gen. Virol. *68*, 2967–2973.
Fukusho, A., Yu, Y., Yamaguchi, S., and Roy, P. (1989). Completion of the sequence of bluetongue virus serotype 10 by the characterization of a structural protein, VP6, and a non-structural protein, NS2. J. Gen. Virol. *70*, 1677–1689.
Ghiasi, H., Fukusho, A., Eshita, Y., and Roy, P. (1987). Identification and characterization of conserved and variable regions in the neutralization VP2 gene of bluetongue virus. Virology *160*, 100–109.
Gorman, B.M. (1990). The bluetongue viruses. Curr. Top Microbiol. Immunol. *162*, 1–19.
Gorman, B.M., Taylor, J., and Walker, P.J. (1983). Orbiviruses, In The *Reoviridae*, W.K. Joklik, ed. (New York: Plenum), pp. 287–357.
Gouet, P., Diprose, J.M., Grimes, J.M., Malby, R., Burroughs, J.N., Zientara, S., Stuart, D.I., and Mertens, P.P. (1999). The highly ordered double-stranded RNA genome of bluetongue virus revealed by crystallography. Cell *97*, 481–490.
Gould, A.R. (1988). Nucleotide sequence of the Australian bluetongue virus serotype 1 RNA segment 10. J. Gen. Virol. *69*, 945–949.
Grimes, J., Basak, A.K., Roy, P., and Stuart, D. (1995). The crystal structure of bluetongue virus VP7. Nature *373*, 167–170.
Grimes, J.M., Burroughs, J.N., Gouet, P., Diprose, J.M., Malby, R., Zientara, S., Mertens, P.P.C., and Stuart, D.I. (1998). The atomic structure of the bluetongue virus core. Nature *395*, 470–478.
Grimes, J.M., Jakana, J., Ghosh, M., Basak, A.K., Roy, P., Chiu, W., Stuart, D.I., and Prasad, B.V. (1997). An atomic model of the outer layer of the bluetongue virus core derived from X-ray crystallography and electron cryomicroscopy. Structure *5*, 885–893.
Guirakhoo, F., Catalan, J.A., and Monath, T.P. (1995). Adaptation of bluetongue virus in mosquito cells results in overexpression of NS3 proteins and release of virus particles. Arch. Virol. *140*, 967–974.
Han, Z., and Harty, R.N. (2004). The NS3 protein of bluetongue virus exhibits viroporin-like properties. J. Biol. Chem. *279*, 43092–43097.
Hassan, S.H., and Roy, P. (1999). Expression and functional characterization of bluetongue virus VP2 protein: Role in cell entry. J. Virol. *73*, 9832–9842.
Hassan, S.H., Wirblich, C., Forzan, M., and Roy, P. (2001). Expression and functional characterization of bluetongue virus VP5 protein: role in cellular permeabilization. J. Virol. *75*, 8356–8367.
Hewat, E.A., Booth, T.F., Loudon, P.T., and Roy, P. (1992a). Three-dimensional reconstruction of baculovirus expressed bluetongue virus core-like particles by cryo-electron microscopy. Virology *189*, 10–20.
Hewat, E.A., Booth, T.F., and Roy, P. (1992b). Structure of bluetongue virus particles by cryoelectron microscopy. J. Struct Biol. *109*, 61–69.
Hewat, E.A., Booth, T.F., and Roy, P. (1994). Structure of correctly self-assembled bluetongue virus-like particles. J. Struct Biol. *112*, 183–191.
Huismans, H. (1979). Protein synthesis in bluetongue virus-infected cells. Virology *92*, 385–396.
Huismans, H., and Erasmus, B.J. (1981). Identification of the serotype-specific and group-specific antigens of bluetongue virus. Onderstepoort J. Vet. Res. *48*, 51–58.
Huismans, H., van Dijk, A.A., and Bauskin, A.R. (1987). *In vitro* phosphorylation and purification of a nonstructural protein of bluetongue virus with affinity for single-stranded RNA. J. Virol. *61*, 3589–3595.
Hyatt, A.D., and Eaton, B.T. (1988). Ultrastructural distribution of the major capsid proteins within bluetongue virus and infected cells. J. Gen. Virol. *69*, 805–815.

Hyatt, A.D., Eaton, B.T., and Brookes, S.M. (1989). The release of bluetongue virus from infected cells and their superinfection by progeny virus. Virology *173*, 21–34.
Hyatt, A.D., Gould, A.R., Coupar, B., and Eaton, B.T. (1991). Localization of the non-structural protein NS3 in bluetongue virus-infected cells. Journal of General Virology *72*, 2263–2267.
Hyatt, A.D., Zhao, Y., and Roy, P. (1993). Release of bluetongue virus-like particles from insect cells is mediated by BTV nonstructural protein NS3/NS3A. Virology *193*, 592–603.
Inumaru, S., and Roy, P. (1987). Production and characterization of the neutralization antigen VP2 of bluetongue virus serotype 10 using a baculovirus expression vector. Virology *157*, 472–479.
Kahlon, J., Sugiyama, K., and Roy, P. (1983). Molecular basis of bluetongue virus neutralization. J. Virol. *48*, 627–632.
Kar, A.K., Bhattacharya, B., and Roy, P. (2007). Bluetongue virus RNA binding protein NS2 is a modulator of viral replication and assembly. BMC Mol. Biol. *8*, 4.
Kar, A.K., Ghosh, M., and Roy, P. (2004). Mapping the assembly of Bluetongue virus scaffolding protein VP3. Virology *324*, 387–399.
Le Blois, H., French, T., Mertens, P.P., Burroughs, J.N., and Roy, P. (1992). The expressed VP4 protein of bluetongue virus binds GTP and is the candidate guanylyl transferase of the virus. Virology *189*, 757–761.
Lee, J.W., and Roy, P. (1986). Nucleotide sequence of a cDNA clone of RNA segment 10 of bluetongue virus (serotype 10). J. Gen. Virol. *67*, 2833–2837.
Limn, C.-H., Staeuber, N., Monastyrskaya, K., Gouet, P., and Roy, P. (2000). Functional dissection of the major structural protein of bluetongue virus: identification of key residues within VP7 essential for capsid assembly. J. Virol. *74*, 8658–8669.
Limn, C.K., and Roy, P. (2003). Intermolecular interactions in a two-layered viral capsid that requires a complex symmetry mismatch. J. Virol. *77*, 11114–11124.
Lohman, T.M., Thom, K., and Vale, R.D. (1998). Staying on track: common features of DNA helicases and microtubule motors. Cell *93*, 9–12.
Loudon, P.T., and Roy, P. (1991). Assembly of five bluetongue virus proteins expressed by recombinant baculoviruses: inclusion of the largest protein VP1 in the core and virus-like proteins. Virology *180*, 798–802.
Lu, G., Zhou, Z.H., Baker, M.L., Jakana, J., Cai, D., Wei, X., Chen, S., Gu, X., and Chiu, W. (1998). Structure of double-shelled rice dwarf virus. J. Virol. *72*, 8541–8549.
Lymperopoulos, K., Noad, R., Tosi, S., Nethisinghe, S., Brierley, I., and Roy, P. (2006). Specific binding of Bluetongue virus NS2 to different viral plus-strand RNAs. Virology *353*, 17–26.
Lymperopoulos, K., Wirblich, C., Brierley, I., and Roy, P. (2003). Sequence specificity in the interaction of Bluetongue virus non-structural protein 2 (NS2) with viral RNA. J. Biol. Chem. *278*, 31722–31730.
Martínez-Costas, J., Sutton, G., and Roy, P. (1998). Guanylyltranferase and RNA5-triphosphatase activities of the purified expressed VP4 protein of bluetongue virus. J. Mol. Biol. *280*, 859–866.
Merrifield, R.B., Vizioli, L.D., and Boman, H.G. (1982). Synthesis of the antibacterial peptide cecropin A (1–33). Biochem. *21*, 5020–5031.
Modrof, J., Lymperopoulos, K., and Roy, P. (2005). Phosphorylation of Bluetongue Virus Nonstructural Protein 2 Is Essential for Formation of Viral Inclusion Bodies. J. Virol. *79*, 10023–10031.
Mortola, E., Noad, R., and Roy, P. (2004). Bluetongue virus outer capsid proteins are sufficient to trigger apoptosis in mammalian cells. J. Virol. *78*, 2875–2883.
Nakagawa, A., Miyazaki, N., Taka, J., Naitow, H., Ogawa, A., Fujimoto, Z., Mizuno, H., Higashi, T., Watanabe, Y., Omura, T., *et al.* (2003). The atomic structure of *Rice dwarf* virus reveals the self-assemby mechanism of component proteins. Structure *11*, 1227–1238.
Nason, E., Rothnagel, R., Muknerge, S.K., Kar, A.K., Forzan, M., Prasad, B.V.V., and Roy, P. (2004). Interactions between the Inner and Outer Capsids of Bluetongue Virus. J. Virol. *78*, 8059–8067.
Nedellec, P., Vicart, P., Laurent-Winter, C., Martinat, C., Prevost, M.C., and Brahic, M. (1998). Interaction of Theiler's virus with intermediate filaments of infected cells. J. Virol. *72*, 9553–9560.
Nibert, M., Schiff, L., and Fields, B. (1996). Reoviruses and their replication, In Fields Virology F.B.N, D. Knipe, and P. M. Howley, eds. (New York: Lippincott-Raven).
Owens, R.J., Limn, C., and Roy, P. (2004). Role of an arbovirus nonstructural protein in cellular pathogenesis and virus release. Journal of Virology *78*, 6649–6656.
Patton, J.T. (1996). Rotavirus VP1 alone specifically binds to the 3′ end of viral mRNA, but the interaction is not sufficient to initiate minus-strand synthesis. J. Virol. *70*, 7940–7947.

Patton, J.T., Jones, M.T., Kalbach, A.N., He, Y.W., and Xiaobo, J. (1997). Rotavirus RNA polymerase requires the core shell protein to synthesize the double-stranded RNA genome. J. Virol. *71*, 9618–9626.

Peisajovich, S.G., and Shai, Y. (2002). New insights into the mechanism of virus-induced membrane fusion. Trends Biochem. Sci. *27*, 183–190.

Poncet, D., Aponte, C., and Cohen, J. (1993). Rotavirus protein NSP3 (NS34) is bound to the 3′ end consensus sequence of viral mRNAs in infected cells. J. Virol. *67*, 3159–3165.

Prasad, B.V.V., Rothnagel, R., Zeng, C.Q., Jakana, J., Lawton, J.A., Chiu, W., and Estes, M.K. (1996). Visualization of ordered genomic RNA and localization of transcriptional complexes in rotavirus. Nature *382*, 471–473.

Prasad, B.V.V., Yamaguchi, S., and Roy, P. (1992). Three-dimensional structure of single-shelled bluetongue virus. J. Virol. *66*, 2135–2142.

Purdy, M.A., Ritter, G.D., and Roy, P. (1986). Nucleotide sequence of cDNA clones encoding the outer capsid protein, VP5, of bluetongue virus serotype 10. J. Gen. Virol. *67*, 957–962.

Putsep, R.A., Branden, C.I., Boman, H.G., and Normark, S. (1999). Antibacterial peptide from H. pylori. Nature *398*, 671–672.

Ramadevi, N., Burroughs, J.N., Mertens, P.P.C., Jones, I.M., and Roy, P. 1998a. Capping and methylation of mRNA by purified recombinant VP4 protein of Bluetongue virus. Proc. Natl Acad. Sci. USA 95: 13537–13542.

Ramadevi, N., Rodriguez, J., and Roy, P. 1998b. A leucine zipper-like domain is essential for dimerization and encapsidation of bluetongue virus nucleocapsid protein VP4. J. Virol. 72: 2983–2990.

Ramadevi, N., and Roy, P. (1998). Bluetongue virus core protein VP4 has nucleoside triphosphate phosphohydrolase activity. J. Gen. Virol. *79*, 2475–2480.

Raynal, P., and Pollard, H.B. (1994). Annexins: the problem of assessing the biological role for a gene family of multifunctional calcium- and phospholipid-binding proteins. Biochem. Biophys Res. Acta *1197*, 63–93.

Roy, P. (1989). Bluetongue virus genetics and genome structure. Virus Res. *13*, 179–206.

Roy, P. (1992). Bluetongue virus proteins. J. Gen. Virol. *73*, 3051–3064.

Roy, P., French, T., and Erasmus, B.J. (1992). Protective efficacy of virus-like particles for bluetongue disease. Vaccine *10*, 28–32.

Roy, P., Bishop, D.H., LeBlois, H., and Erasmus, B.J. (1994). Long-lasting protection of sheep against bluetongue challenge after vaccination with virus-like particles: evidence for homologous and partial heterologous protection. Vaccine *12*, 805–811.

Roy, P. (1995). Synthesis of particulate structures as Bluetongue virus vaccines and their use as multiple immunogen delivery systems. Bulletin de I'Institut Pasteur *93*, 3–20.

Roy, P. (1996). Orbivirus structure and assembly. Virology *216*, 1–11.

Roy, P., Fukusho, A., Ritter, G.D., and Lyon, D. (1988). Evidence for genetic relationship between RNA and DNA viruses from the sequence homology of a putative polymerase gene of bluetongue virus with that of vaccinia virus: conservation of RNA polymerase genes from diverse species. Nucleic Acids Res. *16*, 11759–11767.

Roy, P., Marshall, J.J., and French, T.J. (1990). Structure of the bluetongue virus genome and its encoded proteins. Current Topics in Microbiol. Immunol. *162*, 43–87.

Roy, P., Adachi, A., Urakawa, T., Booth, T.F., and Thomas, C.P. (1990). Identification of bluetongue virus VP6 protein as a nucleic acid-binding protein and the localization of VP6 in virus-infected vertebrate cells. Journal of Virology *64*, 1–8.

Starnes, M.C., and Joklik, W.K. (1993). Reovirus protein lambda 3 is a poly(C)-dependent poly(G)polymerase. Virology *193*, 356–366.

Stauber, N., Martínez-Costas, J., Sutton, G., Monastyrskaya, K., and Roy, P. (1997). Bluetongue virus VP6 protein binds ATP and exhibits an RNA-dependent ATPase function and a helicase activity that catalyze the unwinding of double-stranded RNA substrates. J. Virol. *71*, 7220–7226.

Stuart, D.I. (1993). Viruses. Curr. Opin. Struct Biol. 3, 167–174.

Sutton, G., Grimes, J.M., Stuart, D., and Roy, P. (2007). The structure of bluetongue virus VP4; a multifunctional RNA-capping production-line. Nat. Struct. Mol. Biol. *14*, 449–451.

Tao, Y., Farsetta, D.L., Nibert, M.L., and Harrison, S.C. (2002). RNA Synthesis in a Cage-Structural Studies of Reovirus Polymerase lambda3. Cell *111*, 733–745.

Thomas, C.P., Booth, T.F., and Roy, P. (1990). Synthesis of bluetongue virus-encoded phosphoprotein and formation of inclusion bodies by recombinant baculovirus in insect cells: it binds the single-stranded RNA species. J. Gen. Virol. *71*, 2073–2083.

Tortorici, M.A., Broering, T.J., Nibert, M.L., and Patton, J.T. (2003). Template recognition and formation of initiation complexes by the replicase of a segmented double-stranded RNA virus. J. Biol. Chem. 3, 3.

Urakawa, T., Ritter, D.G., and Roy, P. (1989). Expression of largest RNA segment and synthesis of VP1 protein of bluetongue virus in insect cells by recombinant baculovirus: association of VP1 protein with RNA polymerase activity. Nucleic Acids Res. *17*, 7395–7401.

Van Dijk, A.A., and Huismans, H. (1980). The *in vitro* activation and further characterization of the bluetongue virus-associated transcriptase. Virology *104*, 347–356.

Van Dijk, A.A., and Huismans, H. (1988). *In vitro* transcription and translation of bluetongue virus mRNA. J. Gen. Virol. *69*, 573–581.

van Staden, V., and Huismans, H. (1991). A comparison of the genes which encode non-structural protein NS3 of different orbiviruses. J. Gen. Virol. *72*, 1073–1079.

van Staden, V., Stoltz, M.A., and Huismans, H. (1995). Expression of nonstructural protein NS3 of African horsesickness virus (AHSV): evidence for a cytotoxic effect of NS3 in insect cells, and characterization of the gene products in AHSV infected Vero cells. Arch. Virol. *140*, 289–306.

Venkatesan, S., Gershowitz, A., and Moss, B. (1980). Modification of the 5′ end of mRNA. Association of RNA triphosphatase with the RNA guanylyltransferase-RNA (guanine-7-)methyltransferase complex from vaccinia virus. J. Biol. Chem. *255*, 903–908.

Wehrfritz, J., Boyce, M., Mirza, S., and Roy, P. (2007). Reconstitution of bluetongue virus polymerase activity from isolated domains based on a three dimensional structural model. Biopolymers *In Press*.

Wirblich, C., Bhattacharya, B., and Roy, P. (2006). Nonstructural protein 3 of bluetongue virus assists virus release by recruiting ESCRT-I protein Tsg101. J. Virol. *80*, 460–473.

Wu, C.L., Leu, T.S., Chang, T.T., and Shiau, A.L. (1999). Hepatitis C virus core protein fused to hepatitis B virus core antigen for serological diagnosis of both hepatitis C and hepatitis B infections by ELISA. J. Med. Virol. *57*, 104–110.

Zetina, C.R. (2001). A conserved helix-unfolding motif in the naturally unfolded proteins. Proteins *44*, 479–483.

Zhang, X., Walker, S.B., Chipman, P.R., Nibert, M.L., and Baker, T.S. (2003). Reovirus polymerase lambda3 localized by cryo-electron microscopy of virions at a resolution of 7.6 A. Nat. Struct Biol. *10*, 1011–1018.

Structures of Phytoreoviruses 5

Matthew L. Baker, Z. Hong Zhou, and Wah Chiu

Abstract

Phytoreoviruses are non-turreted reoviruses that are major agricultural pathogens, particularly in Asia. One member of this family, rice dwarf virus (RDV), has been extensively studied by electron cryomicroscopy and X-ray crystallography. From these analyses, atomic models of the capsid proteins and a plausible model for capsid assembly have been derived. While the structural proteins of RDV share no sequence similarity to other proteins, their folds and the overall capsid structure are similar to those of other *Reoviridae*.

Introduction

The *Reoviridae* infect a wide variety of vertebrate, invertebrate and plant hosts (Sabin, 1959). This family was first designated as a subgroup of respiratory and enteric viruses linked not by a common human disease, but rather by distinct characteristics. *Reoviridae* can be divided into two distinct subgroups – turreted and non-turreted – based on viral architecture. These may be further divided into nine unique genera. Each of these genera has distinct biological, genetic and structural differences, which in turn are reflected in their host ranges and pathogenesis. Most of these genera have been well studied, with the best characterized being orthoreovirus, which contains reovirus, and orbivirus, with the notable member, bluetongue virus (BTV). Members in *Reoviridae* all have a segmented dsRNA genome, with 10–12 unique segments, encapsulated by concentric icosahedral protein layers.

Phytoreoviruses, which belong to the non-turreted type, are unique among *Reoviridae* in that there are 12 dsRNA segments in each virus. Like all other *Reoviridae*, they contain their own RNA-dependant RNA polymerase, allowing for endogenous synthesis of viral mRNA within the virus (Nibert, 1996). Although less well studied than orthoreovirus and orbivirus, considerable structural and biochemical studies have been undertaken to characterize phytoreoviruses, which in general infect plants.

There are three recognized phytoreoviruses: wound tumour virus (WTV), rice dwarf virus (RDV) and rice gall dwarf virus (RGDV). Possible additional family members include tobacco leaf enation virus (TLEF), rice bunchy stunt virus and sweet potato virus (ICTV, VIDE) (Mertens *et al.*, 2000). There are also several RDV isolates, all of which share over 90% sequence identity (Mertens *et al.*, 2000). Of these phytoreoviruses, RDV has been extensively studied and is the best characterized member. Most of our knowledge about phytoreovirus assembly and structure is based on structural data accumulated in RDV studies, which will be the focus of the discussion of this chapter.

Infection and pathogenesis

RDV and other phytoreoviruses represent a major agricultural pathogen in Asia, particularly southern China and Japan (RDV, RGDV), as well as Africa (TLEV) and North America (WTV). They infect dicotyledons (WTV) or gramnacious plants (RDV and RGDV), such as rice and wheat. Transmission of phytoreoviruses does not occur through mechanical transmission, but rather through propagative transmission as the virus replicates in an insect vector (Mertens *et al.*, 2000). More specifically, RDV can only be vectored by replication in cicadellid leafhoppers (*Neopotettix cinticeps, N. nigropictus, Recilia dorsalis*) (Nault, 1989; Suzuki *et al.*, 1994). Viral latency in these insects is 10 to 20 days, from which point viral transmittance is maintained for the insect's life. Transmission occurs when the leafhopper, carrying a propagating virus, chews on an uninfected plant.

In most phytoreoviruses, infection of susceptible plants results in tumours. However with RDV, infection fails to produce neoplasia, instead causing chlorotic flecks on the leaves. Infection also results in the stunting of the plant bodies and failure to bear seeds. Interestingly, phytoreovirus infection produces no known disease in the host insects, although RDV infection can result in abnormalities in mycetocytes and fat body cells. Additionally, only viruses that have passed through these insect vectors become infectious to the plant host cells. As with other *Reoviridae*, replication is believed to occur in the cytoplasm with the associated viroplasm in infected cells. Unlike RDV, which can be found in many areas of the infected plant, WTV and RGDV are restricted to the phloem.

Genome and proteins

The phytoreovirus genome consists of 12 unique dsRNA segments. RDV contains 25.7 million base pairs (bp) of dsRNA, accounting for ~22% of the total mass of the viral particle. Each of the 12 unique dsRNA strands contains a conserved terminal 5′-oligonucleotide sequence and has one large open reading frame (ORF), except for segments 11 and 12 of RDV and segment 9 of WTV and RGDV. RDV segment 11 contains two ORFs from two in-frame initiation codons. The remaining multiple ORF segments contain a short ORF downstream of the large ORF that encodes for the non-structural protein Pns12 in RDV and Pns9 in WTV and RGDV. Unique to RDV are several short ORFs in segment S12. Table 5.1 summarizes the proteins encoded by the RDV genome.

Although awaiting experimental confirmation, it is generally believed, based on other reoviruses, that one copy of each of the genome segments is packaged into a virion, one underneath each of the 12 icosahedral vertices of the capsid (Lawton *et al.*, 1997; Pesavento *et al.*, 2001; Prasad *et al.*, 1996). Each of these segments is transcribed, producing a single-stranded, plus-sense transcript by the viral-associated RNA-dependent RNA polymerase. Transcription occurs at the fivefold vertex, interior to the inner capsid, where the RNA-dependent RNA polymerase and associated RNA-binding protein and guanylyltransferase are believed to be located. Based on structural and biochemical studies of other reoviruses, it is likely that the individual RNA segments are essentially wrapped around this fivefold complex in a 'coned-coil' fashion, where each unique genome segment occupies a separate fivefold vertex (Pesavento *et al.*, 2001; Xia *et al.*, 2003).

Table 5.1 RDV genome

Genome segment	Protein name	dsRNA (bp)/ length (aa)	Function
1	P1 (Pol)	4423/1444	RNA Polymerase
2	P2	3512/	Minor outer capsid
3	P3	3195/1019	Major inner core protein (vector transmission)
4	Pns4	2468/727	Non-structural
5	P5	2570/801	Guanylyltransferase
6	Pns6	1699/509	Non-structural
7	P7	1696/506	Nucleic acid binding protein
8	P8	1427/421	Major outer capsid protein
9	Pns9	1305/351	Non-structural
10	Pns10	1321/353	Non-structural
11	Pns11a Pns11b	1067/181	Non-structural
12	Pns12 Pns12OPa Pns12OPb	1066/312	Non-structural Phosphoprotein

Morphology

All non-turreted reoviruses have multiple, concentric, icosahedral protein layers encapsulating the segmented linear dsRNA genome (Fields, 1996). Most phytoreoviruses have a distinct angular appearance (Figure 5.1), with two such layers and no capsid spikes or turrets (WTV is reportedly the exception, with a third, outer amorphous layer) (Boccardo *et al.*, 1985; Hagiwara *et al.*, 2003; Inoue, 1985; Lu *et al.*, 1998; Nakagawa *et al.*, 2003; Uyeda, 1982; Wu *et al.*, 2000; Zhou *et al.*, 2001). All phytoreoviruses contain one copy of each of 12 supercoiled dsRNA segments, resulting in a total molecular weight of ~75 MDa (Anzola *et al.*, 1987). Six to seven of these segments encode structural proteins, two of which form the majority of the inner and outer protein capsid layers (Murao *et al.*, 1994; Suzuki *et al.*, 1994).

In RDV, three copies of P8 (780 copies, 46kDa) form trimers arranged on a T = 13l icosahedral lattice with an outer diameter of ~700 Å, while the inner protein capsid layer, with a diameter of ~570 Å, is composed of P3 (120 copies, 114 kDa) dimers arranged on a T = 1 icosahedral lattice (Figure 5.2A). A minor capsid protein, P2, is also involved in virus infection of vector cells and is required for oral entry of the virus into the insect vector (Tomaru *et al.*, 1997; Yan *et al.*, 1996). In addition to these capsid proteins, RDV and other phytoreoviruses contain an RNA-dependent RNA polymerase (P1, 164 kDa), mRNA-guanylyltransferase (P5, 91 kDa) and a non-specific nucleic acid binding protein (P7, 55 kDa) within the core of the virus capsid. Organization of the capsid proteins is

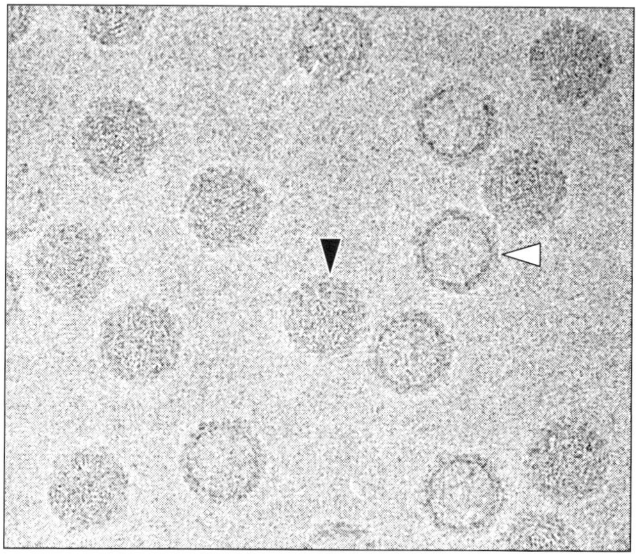

Figure 5.1 CryoEM images of RDV particles embedded in ice. One full and one empty viral particle are indicated by solid and open arrowhead, respectively. The image was recorded in a 400 kV JEM4000 electron cryomicroscope (courtesy of Joanita Jakana, Baylor College of Medicine).

slightly different in WTV, with the outermost amorphous layer and inner core each containing two structural proteins and three sub-core proteins.

Structure of a phytoreovirus

Many of the members of *Reoviridae* have been the subjects of extensive structural analyses. The structure of human orthoreovirus (Chappell *et al.*, 2002; Liemann *et al.*, 2002; Olland *et al.*, 2001; Reinisch *et al.*, 2000; Tao *et al.*, 2002) and BTV (Basak *et al.*, 1996; Grimes *et al.*, 1998; Grimes *et al.*, 1997) have all been solved by X-ray crystallography. Like RDV, the structures of rotavirus (e.g. Prasad *et al.*, 1990; Prasad *et al.*, 1988), cytoplasmic polyhedrosis virus (CPV) (e.g. Zhang *et al.*, 1999; Zhou *et al.*, 2003) and aquareovirus (e.g. Shaw *et al.*, 1996) have been extensively studied by electron cryomicroscopy (cryoEM). These two techniques have been both very useful for studying the basic architecture of these viruses (Grimes *et al.*, 1998; Grimes *et al.*, 1997; Lu *et al.*, 1998; Mathieu *et al.*, 2001; Reinisch *et al.*, 2000; Zhang *et al.*, 2003). In particular, our understanding of RDV has benefited from the combination of a medium-resolution cryoEM structure (Figure 5.2A) (Zhou *et al.*, 2001) and a complete X-ray crystal structure (Nakagawa *et al.*, 2003).

Capsid shells

The outer shell of RDV is 69 Å thick and is quite angular, with triangular capsomeres formed by trimeric complexes of P8 protein, similar in appearance to those of BTV (Figure 5.2A,B) (Lu *et al.*, 1998; Zhou *et al.*, 2001). As with other reoviruses, the arrangement of this outer capsid was determined to be T = 13l, meaning that there are 4 and 1/3 unique capsomeres per asymmetric unit. Slight protrusions (~15 Å) of the trimeric capsomeres at

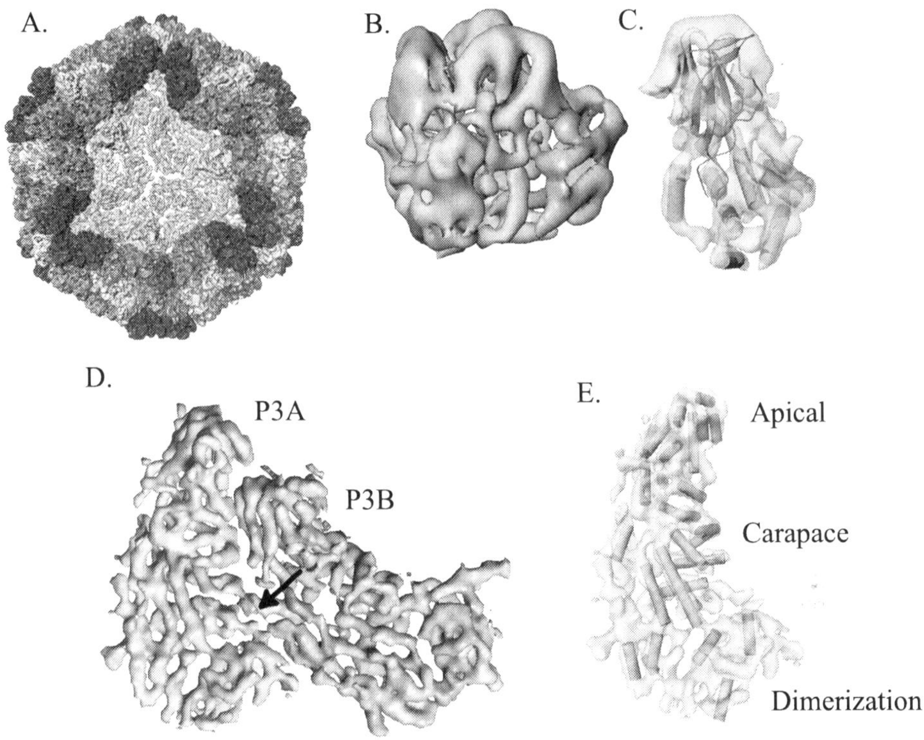

Figure 5.2 CryoEM structure of RDV. (A) Structure of the double-layered RDV at 6.8 Å resolution as determined by cryoEM. An asymmetric unit from the outer capsid shell has been computationally extracted. Individual subunits within the capsid layers are coloured (P3A light green, P3B light red, P trimer red, Q trimer orange, S trimer yellow, R timer green, T trimer blue) (B) Computationally averaged P8 trimer of RDV. A single P8 monomer is highlighted. (C) Pseudo-atomic model of P8 derived from structural analysis. (D) P3 dimer (P3A and P3B). Helix 1 of P3B (light red) is marked with an arrow. (E) Locations of alpha helices of P3A derived from structural analysis. (See also Plate 5.2.)

the fivefold vertices add to the overall angular appearance of the capsid. Dotting the surface of the structure are 132 holes, 12 at the fivefold vertices and 120 at the local sixfold axes.

The inner capsid layer is relatively thin (25 Å), smooth and round (Figure 5.2A,D) (Lu *et al.*, 1998; Zhou *et al.*, 2001). In contrast to the relatively compact P8 outer capsid layer, the inner layer protein, P3, is relatively expansive and porous, with a large hole (~20 Å) located directly along the fivefold vertex. The inner layer is composed of 60 P3 dimers arranged on a T = 1 lattice, creating a symmetry mismatch between the two layers.

Atomic models of the capsid proteins

The first pseudo-atomic models of the two major capsid proteins, P3 and P8, were generated from the 6.8-Å cryoEM structure of RDV using a combination of structural analysis and bioinformatics techniques (Chiu *et al.*, 2002; Zhou *et al.*, 2001). These models were subsequently confirmed by the crystal structure of the whole virion (Nakagawa *et al.*, 2003), which validated the structure analysis and models in the 6.8-Å cryoEM map.

However, of the other RDV structural proteins, only a small portion of the P7 protein could be visualized in the X-ray structure.

The cryoEM structure revealed that RDV P8 contains a lower α-helical domain and an upper β-sheet domain (Figure 5.2C), similar to that of BTV VP7, although differing in relative orientations in their spatial arrangements. The N-terminal 150 amino acids of P8 forms at least five major α-helices, which account for slightly more than half the mass of the lower domain mass. The next ~150 amino acids form three distinct β-sheets that combine to form the upper domain β-sandwich fold, reminiscent of BTV VP7. The final 120 amino acids form the final four major α-helices of the lower domain, with the C-terminus also located in the lower domain. In forming the trimer, one copy of P8 is twisted such that the lower domain of one P8 molecule resides underneath the upper domain of the neighbouring P8 molecule. The resulting trimer appears to have two triangular domains when viewed down the threefold axis, where each side is ~90 Å long with an overall height of nearly 70 Å. As with the upper domain, the overall architecture of the lower domain of P8 appears structurally homologous to the corresponding BTV inner capsid protein, VP3.

The cryoEM structure showed that the overall fold of RDV P3 is highly helical (Figure 5.2E), with three distinct domains like its counterparts in BTV and other reovirus inner cores. The apical domain is almost entirely helical and is orientated towards the fivefold vertices. It forms the thickest region of the inner capsid, which results in the slight protrusion of the fivefold P8 trimers and, thus, the overall angular appearance of the capsid. The second P3 domain, the carapace domain, is also mostly helical and is characterized by five long helices. Three are nearly parallel to each other, and lie above and perpendicular to two other parallel helices. Sandwiching this domain are the aforementioned apical domain and the third domain, the dimerization domain. Unlike the previous two domains, this domain contains several large β-sheets and relatively few helices. As indicated by its name, the dimerization domain is the site of interactions between adjacent P3 molecules and a point of twofold symmetry.

The two P3 molecules making up a dimer are designated as P3A and P3B. While P3A and P3B share the same polypeptide sequence and much of the overall structure, many differences in arrangement of secondary structure elements were observed in the 6.8 Å cryoEM map and the crystal structure. The outermost region of the apical domain of P3A forms holes at the fivefold vertices. In contrast, the apical domain of P3B, which sits adjacent to P3A, is not exposed at all and is covered on two sides by neighbouring P3A apical domains. Additionally, the most N-terminal helix, designated as helix 1, is visible in P3B but not in P3A (arrow Figure 5.2D). Helix 1 in P3B appears to extend back to its corresponding helix in P3A to form a connection to the underside of P3A.

Symmetry mismatch and protein–protein interactions

While symmetry mismatch is often found in the rotor/stator mechanism of molecular motors, reoviruses use it to provide the necessary mechanism for viral stability and replication. The cryoEM structure of RDV showed that the symmetry mismatch, coupled with relatively promiscuous interactions between the two layers, allows the two layers to effectively shield the genomic RNA from the surrounding environment. At the fivefold axes, holes in both capsid layers align and have been shown in rotavirus to be the points for RNA exit during viral transcription (Lawton *et al.*, 1999).

While nearly every P8 molecule is identical, changes in small regions of P8 are apparent at various positions due to interactions with P3. Interactions between these regions appear to be primarily through hydrogen bonds and electrostatic interactions, with secondary interactions coming from surface complementarities. These interactions are strongest around the threefold axes (T trimers, blue Figure 5.2A) and weakest at the fivefold axes (P trimers, red, Figure 5.2A). To accommodate the symmetry mismatch, the P8 trimers cover the 'joints' between P3 subunits. The resulting double-layered capsid allows the virus to shield its genome, creating an elaborate defence mechanism.

The X-ray structure of RDV further elucidated the intricate interactions of the capsid proteins at an atomic detail (Figure 5.3). Residues 1–49 are disordered in P3A due to blockade by the 10 N-terminal residues of P3B, which are not impeded from forming helix 1 in P3B. The N-terminal arms of P3A instead point inward, and thus interact with the genomic RNA or the RNA-dependent RNA polymerase or its associated proteins. Additional stabilization of the P3 dimers comes from extensive hydrogen bonding and salt bridges between the pair of subunits.

As with P3, a variety of interactions stabilize the P8 outer capsid (Figure 5.3). Neighbouring P8 trimers are stabilized by electrostatic complementarity due to the patches of positive and negative charges found on the surface of the trimer. Interestingly, residues 274–276 (GGG) interact very closely and are very near to the corresponding residues related by the icosahedral twofold axis. This triplet is semi-conserved in other phytoreoviruses, and might thus be crucial in forming the bilateral interactions in the P8 outer capsid layer.

Minor capsid proteins

To date, no structure has been reported for any of the minor structural proteins of phytoreoviruses. The X-ray crystallographic structure of RDV does show a portion of the P7 protein, but no densities could be attributed to P1 and P5 in the inner shell. Studies of other reoviruses by cryoEM have shown that a density resides internal to the P3 capsid

A. B.

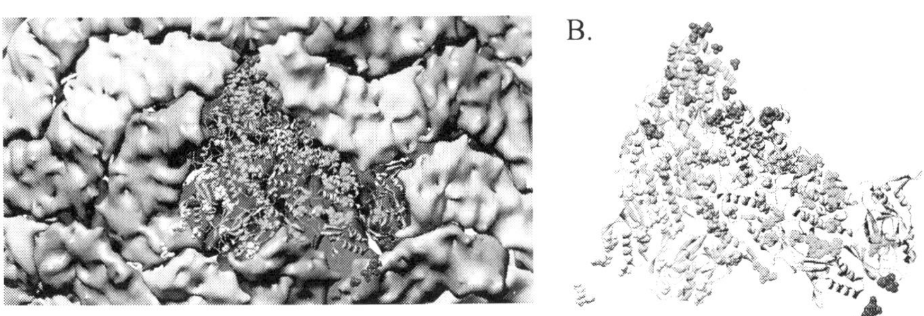

Figure 5.3 RDV protein shell Interactions. Interactions, determined from the X-ray structure, of the outer capsid layer protein, P8, with the inner shell protein, P3, are shown in top views. The atoms shown represent the interacting residues, coloured keyed (same colour scheme as in Figure 5.2) to the corresponding P8 protein trimers. (See also Plate 5.3.)

layer at the icosahedral fivefold vertices. This density has been proposed to be the RNA polymerase and its associated proteins. The observation of mRNA transcripts exiting from the fivefold vertices in rotavirus is consistent with this proposed location of the RNA polymerase (Lawton *et al.*, 1999). X-ray crystallographic studies have also recently yielded the structure of the RNA-dependent RNA polymerase from human orthoreovirus (Tao *et al.*, 2002). In combination with the 7.6 Å resolution structure of reovirus, a model for RNA transcription and exit from the virion was proposed and serves as the general model for reovirus transcription (Zhang *et al.*, 2003).

Interpretation of the putative RNA polymerase density in cryoEM reconstructions of RDV have been hampered because only one copy each of P1, P5 and P7 resides at the fivefold vertices. Icosahedral symmetry is imposed during structure determination of icosahedral viruses, thus averaging out the combined densities associated with these proteins. Thus, only the most basic analysis of this density has been possible. However, improvements in image processing have provided a method for further analyzing 'noisy' data. By applying a bilateral filter (Jiang *et al.*, 2003), which removes noise without disrupting the signal, to the subnanometer-resolution structures of the full (6.8 Å resolution) and empty (9.0 Å resolution) RDV virions, it was possible to enhance the visualization of the genome and other minor capsid proteins (Figure 5.4). As a result, a mushroom-shaped density was visible in both structures of RDV. There was also a noticeable change in the relative vertical position of the polymerase complex between the empty and full capsids. Although no high-resolution structure exists for any of the components of the phytoreovirus polymerase complex, parallels to the structure, function and arrangement of the orthoreovirus polymerase are likely to hold true for most members of the *Reoviridae*.

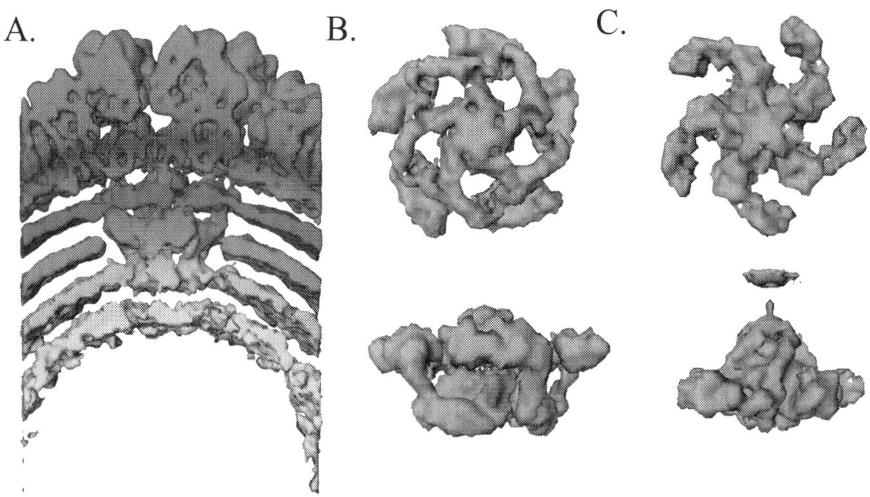

Figure 5.4 Minor proteins of RDV based on cryoEM structures. (A) Radially coloured density at a fivefold vertex. The outer capsid layer is coloured magenta, while the inner layer is coloured blue. The concentric rings of RNA forming the coiled coil structure and the polymerase complex are shown in green. RNA-dependant RNA polymerase density at a fivefold vertex after bilateral filtering in the full (B) and empty (C) RDV cryoEM reconstructions. (See also Plate 5.4.)

Assembly model

After establishing the structure of the major capsid proteins of RDV, it became possible to propose a model for assembly of the capsid (Figure 5.5). It has been suggested (Nakagawa *et al.*, 2003) that the strong interactions between the P3A and P3B molecules found in one dimer initiate the self-assembly of the RDV core particle. From the P3 dimer, either three dimers assemble about the icosahedral threefold axis, or five dimers assemble about the icosahedral fivefold axis. These subunits then begin to assemble the inner capsid, at which time P7 may be involved in transporting the nucleic acid into the viral particles. After formation of the inner capsid layer, the strong interactions between a P8 trimer and the icosahedral threefold axis perpetuate the binding of the P8 T-trimer. The R trimers then attach and form the complementary electrostatic interactions between the P8 trimers. This is then followed by the Q and S trimers, and finished by the weak attachment of the P trimers.

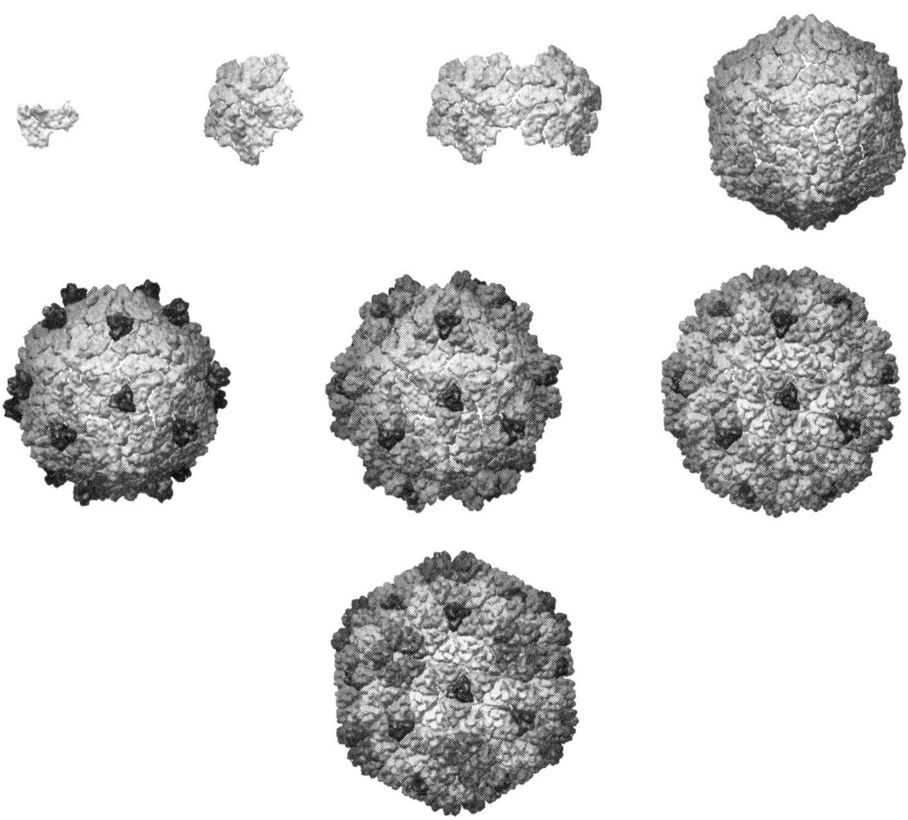

Figure 5.5 Assembly pathway. Assembly of RDV based on a model proposed by Nakagawa and co-workers (Nakagawa *et al.*, 2003). Proteins are coloured as stated in Figure 5.2. (See also Plate 5.5).

Conservation and diversity among *Reoviridae*: sequence vs. structure

Among the *Reoviridae*, some basic characteristics are consistent throughout all the genera. Every member appears to have a T = 1 icosahedral core that encloses a segmented dsRNA genome and the necessary proteins for transcription of the viral genome. Of course, the different genera have specific characteristics, as do members within the genera. However, differences in primary, secondary and tertiary structure still give rise to very similar virus architectures.

As mentioned previously, there are three major phytoreovirus: WTV, RDV and RGDV. A search for protein sequences similar to either of the two major capsid proteins (P8 and P3) yields only the analogous proteins from other phytoreoviruses (BLAST, E-value < 0.01, M.L. Baker, Unpublished). It should also be noted that total sequence length of the P3 and P8 protein analogues may vary up to 25% between different genera of *Reoviridae*. However, a sequence search using the RDV RNA-dependent RNA polymerase revealed similarity to RNA polymerases of two orbiviruses (African horse sickness virus and Chuzan virus). This level of sequence similarity is likely to indicate an evolutionary path that emphasizes the ability of reoviruses to diversify their host range and pathogenesis while maintaining a common method for viral propagation. The absence of sequence similarity might suggest little structural conservation within *Reoviridae*. However, there is a distinct structural similarity between the overall architecture of the major capsid proteins in RDV and those of other *Reoviridae* members. A search for homologous structures revealed that RDV P8 shares structural similarity to BTV VP7 (Z-score 4.8, RMSD 5.1) and African horse sickness virus (Z-score 4.6, RMSD 3.4), both members of the orbivirus genus of *Reoviridae*.

RDV P8, rotavirus VP6 and BTV VP7 all share a common domain arrangement despite the absence of detectable sequence similarity; each contains a lower domain with N- and C- terminal helix-rich segments, and an upper domain with a β-sheet-rich jelly roll. The two domains are connected by a relatively small, flexible region. Additionally, RDV P8 shares partial structural similarity to several other proteins containing the Nucleoplasmin-like/VP fold, common to other viral structural proteins (Table 5.2). This fold is similar to the viral protein domain fold of BTV VP7, AHSV VP7, rotavirus VP6 and RDV P8 upper domain, differing in the total number strands in the two sheets. While the lower helical domain cannot be identified using structural searches, such as DALI (Holm and Sander, 1995), it is evident that RDV P8 has a similar virus capsid protein alpha-helical

Table 5.2 DALI results for RDV P8

PDB ID	Z-score	RMSD (Å)	Protein
1bvp-1	4.8	5.1	Bluetongue virus 10 (usa) vp7
1phm	4.7	3.3	Peptidylglycine α-hydroxylating monooxygenase
1ahs-A	4.6	3.4	African horse sickness virus (serotype 4) vp7
1aym-1	3.6	5.9	Human rhinovirus 16 coat protein (hrv16)
1aoh-A	3.4	3.2	Cellulosome-integrating protein cipa fragment

domain fold, similar to that of BTV VP7 and rotavirus VP6. In all three viruses, the exact position and orientation of the nine major helices varies, but overall similarity in architecture and topology is maintained. In all cases, their morphological unit is always a trimer of the major capsid protein, which forms the T = 13l icosahedral layer of the virus (Figure 5.6 A–C, Table 5.3).

As described previously, RDV P3 contains three domains: a mostly compact helical apical domain, a mostly helical carapace domain, and a mainly β-sheet dimerization domain (Figure 5.2E). Although a DALI search for structural homologues revealed similarity only to BTV VP3 (Z-score: 7.2, RMSD: 4.7), a nearly identical structural arrangement

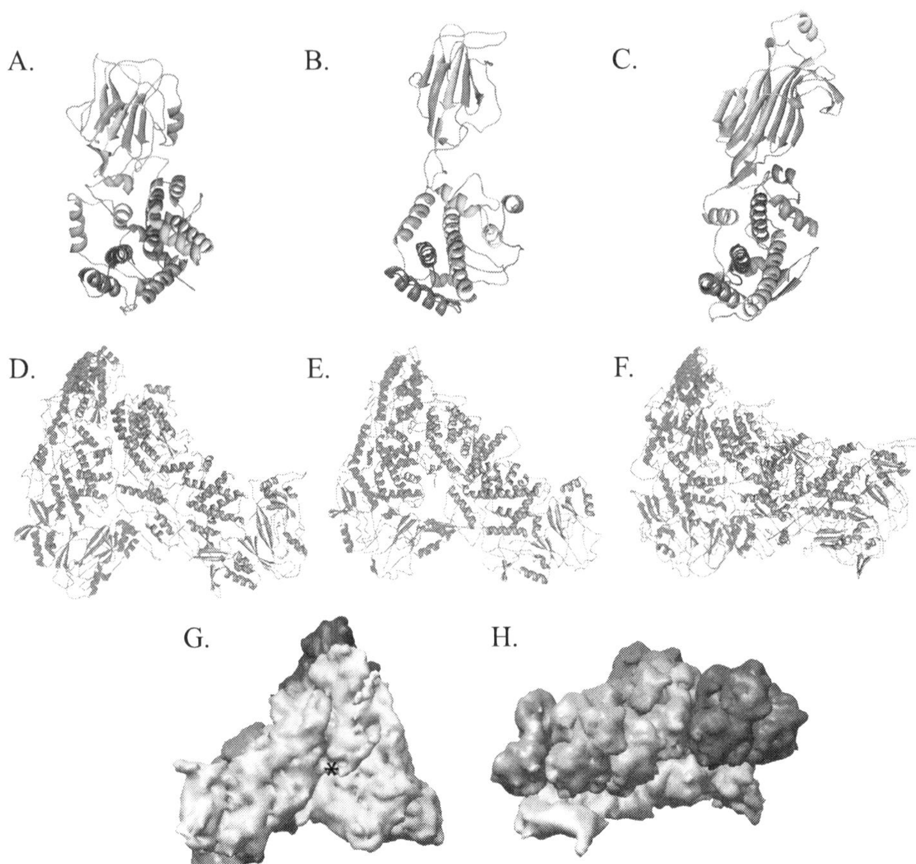

Figure 5.6 Structural comparisons of *Reoviridae* major capsid proteins. (A-C) Comparison of the outer capsid proteins: RDV P8 (A), BTV VP7 (B) and rotavirus VP6 (C). The ribbon representations of the structures are coloured from the N-terminus (blue) to the C-terminus (red). (D-F) Structural comparison of the dimer of the inner capsid proteins: RDV P3 (D), BTV VP3 (E) and reovirus (F). The subunit directly adjacent to the fivefold axis is coloured green, while the other subunit is coloured red. Asymmetric unit of RDV in a bottom view (G) and side view (H). Helix 1 of P3B is marked with an asterisk in (G). (See also Plate 5.6.)

Table 5.3 Structures of *Reoviridae* structural proteins

Organism	Genus	PDB ID
Major outer capsid protein		
Rice dwarf virus	Phytoreovirus	1UF2 (EBI: 1060)
Bluetongue virus	Orbivirus	2BTV
African horse sickness virus	Orbivirus	1AHS
Rotavirus	Rotavirus	1QHD
Major inner capsid protein		
Rice dwarf virus	Phytoreovirus	1UF2 (EBI: 1060)
Bluetongue virus	Orbivirus	2BTV
Orthoreovirus	Orthoreovirus	1EJ6
RNA dependent RNA polymerase		
Orthoreovirus	Orthoreovirus	1MUK

can be found in the orthoreovirus λ1 structure (Figure 5.6D–F, Table 5.3). In all three structures, the apical domain is populated with nearly parallel helices whose helical axes point towards the fivefold vertices. A single helix in RDV P3 (residues 425–437) extends outward, nearly perpendicular to the other helices in the apical domain, and is not visible in orthoreovirus and BTV. Additionally, orthoreovirus contains four sheets, two of which are not seen in either RDV or BTV. Perhaps the most conserved domain of P3 is the carapace domain, which is predominated by three long, parallel helices arranged orthogonal to two large helices just below. The dimerization of the three inner capsid proteins is relatively diverse. RDV and BTV have similar architectures of the dimers, although there appears to be relative rotation of the domain. In contrast, RDV and orthoreovirus share a similar domain orientation but less similar architecture. RDV contains an additional unique feature: the N-terminal 30 amino acids of P3B extend beneath the P3A subunit, terminating near the start of the P3A structure (residue 50). Neither orthoreovirus nor BTV contain such a feature, although the N-termini of the RDV P3A, BTV VP3A and orthroreovirus λ1b structures are on the inner surface of the inner capsid and are oriented toward the fivefold vertices. Like the major outer capsid protein, the major inner capsid proteins of the *Reoviridae* accomplish their structural, and presumably functional, roles in nearly the same manner in the absence of sequence similarity.

The *Reoviridae* are a diverse group of viruses with a wide range of host specificities and pathogenesis, as well as a unique structural organization. Phytoreoviruses share many of the same folds and architectures seen in other *Reoviridae,* despite having almost no sequence similarity. Such conservation underpins the common structural framework used by the *Reoviridae,* while retaining the flexibility of novel features that allow for functional diversity.

Acknowledgements

This research was supported by grants from the National Institutes of Health (P41RR02250 and R01AI38469) and the National Science Foundation (EIA-0325004) and Welch Foundation. M.L.B. was supported by a training fellowship from the Keck Center for Computational and Structural Biology of the Gulf Coast Consortia (NLM Grant No. 5T15LM07093). We would also like to thank Guangying Lu and Wen Jiang for their work on the cryoEM structure of RDV. Figures 5.3–5.6 were generated using UCSF Chimera (Pettersen *et al.*, 2004).

References

Anzola, J.V., Xu, Z.K., Asamizu, T., and Nuss, D.L. (1987). Segment-specific inverted repeats found adjacent to conserved terminal sequences in wound tumor virus genome and defective interfering RNAs. Proc. Natl Acad. Sci. USA *84*, 8301–8305.

Basak, A.K., Gouet, P., Grimes, J., Roy, P., and Stuart, D. (1996). Crystal structure of the top domain of African horse sickness virus VP7, comparisons with bluetongue virus VP7. J. Virol. *70*, 3797–3806.

Boccardo, G., Milne, R.G., Disthaporn, S., Chettanachit, D., and Putta, M. (1985). Morphology and nucleic acid of rice gall dwarf virus. InterVirology *23*, 167–171.

Chappell, J.D., Prota, A.E., Dermody, T.S., and Stehle, T. (2002). Crystal structure of reovirus attachment protein sigma1 reveals evolutionary relationship to adenovirus fiber. EMBO J. *21*, 1–11.

Chiu, W., Baker, M.L., Jiang, W., and Zhou, Z.H. (2002). Deriving folds of macromolecular complexes through electron cryomicroscopy and bioinformatics approaches. Curr. Opin. Struct Biol. *12*, 263–269.

Fields, B.N. (1996). *Reoviridae*. In Fields Virology B.N. Fields, Knipe, D.M., and Howley, P.M., ed. (Philadelphia, Lippincott-Raven), pp. 1553–1555.

Grimes, J.M., Burroughs, J.N., Gouet, P., Diprose, J.M., Malby, R., Zientara, S., Mertens, P.P., and Stuart, D.I. (1998). The atomic structure of the bluetongue virus core. Nature. *395*, 470–478.

Grimes, J.M., Jakana, J., Ghosh, M., Basak, A.K., Roy, P., Chiu, W., Stuart, D.I., and Prasad, B.V. (1997). An atomic model of the outer layer of the bluetongue virus core derived from X-ray crystallography and electron cryomicroscopy. Structure. *5*, 885–893.

Hagiwara, K., Higashi, T., Namba, K., Uehara-Ichiki, T., and Omura, T. (2003). Assembly of single-shelled cores and double-shelled virus-like particles after baculovirus expression of major structural proteins P3, P7 and P8 of Rice dwarf virus. J. Gen. Virol. *84*, 981–984.

Holm, L., and Sander, C. (1995). Dali, a network tool for protein structure comparison. Trends Biochem. Sci. *20*, 478–480.

Inoue, H., and Timmins, P.A. (1985). The structure of rice dwarf virus determined by small angle scattering measurements. Virology 214–216.

Jiang, W., Baker, M.L., Wu, Q., Bajaj, C., and Chiu, W. (2003). Applications of a bilateral denoising filter in biological electron microscopy. J. Struct Biol. *144*, 114–122.

Lawton, J.A., Estes, M.K., and Prasad, B.V. (1997). Three-dimensional visualization of mRNA release from actively transcribing rotavirus particles. Nat. Struct Biol. *4*, 118–121.

Lawton, J.A., Estes, M.K., and Prasad, B.V. (1999). Comparative structural analysis of transcriptionally competent and incompetent rotavirus-antibody complexes. Proc. Natl Acad. Sci. USA *96*, 5428–5433.

Liemann, S., Chandran, K., Baker, T.S., Nibert, M.L., and Harrison, S.C. (2002). Structure of the reovirus membrane-penetration protein, Mu1, in a complex with is protector protein, Sigma3. Cell. *108*, 283–295.

Lu, G., Zhou, Z.H., Baker, M.L., Jakana, J., Cai, D., Wei, X., Chen, S., Gu, X., and Chiu, W. (1998). Structure of double-shelled rice dwarf virus. J. Virol. *72*, 8541–8549.

Mathieu, M., Petitpas, I., Navaza, J., Lepault, J., Kohli, E., Pothier, P., Prasad, B.V., Cohen, J., and Rey, F.A. (2001). Atomic structure of the major capsid protein of rotavirus, implications for the architecture of the virion. EMBO J. *20*, 1485–1497.

Mertens, P.P.C., Arella, M., Attoui, H., Belloncik, S., Bergoin, M., Boccardo, G., Booth, T.F., Chiu, W., Diprose, J.M., Duncan, R., *et al.* (2000). Family *Reoviridae*. In Virus Taxonomy, M.H.V. van Regenmortel, C.M. Fauguet, D.H.L. Bishop, E.B. Carstens, M.K. Estes, S.M. Lemon, J. Maniloff, M.

Mayo, D.J. McGeoch, C.R. Pringle, and R.B. Wickner, eds. (San Diego, CA, USA, Academic Press), pp. 395–480.
Murao, K., Suda, N., Uyeda, I., Isogai, M., Suga, H., Yamada, N., Kimura, I., and Shikata, E. (1994). Genomic heterogeneity of rice dwarf phytoreovirus field isolates and nucleotide sequences of variants of genome segment 12. J. Gen. Virol. 75 (Pt 8), 1843–1848.
Nakagawa, A., Miyazaki, N., Taka, J., Naitow, H., Ogawa, A., Fujimoto, Z., Mizuno, H., Higashi, T., Watanabe, Y., Omura, T., *et al.* (2003). The atomic structure of rice dwarf virus reveals the self-assembly mechanism of component proteins. Structure (Camb). *11*, 1227–1238.
Nault, L.R., and Ammar, D.E. (1989). Leafhopper and planthopper transmission of plant viruses. Annu. Rev. Entomol, 503–539.
Nibert, M.L., Schiff, L.A., and Fields, B.N. (1996). Reoviruses and their replication. In Fields Virology B.N. Fields, Knipe, D.M., and Howley, P.M., ed. (Philadelphia, Lippincott-Raven), pp. 1557–1586.
Olland, A.M., Jané-Valbuena, J., Schiff, L.A., Nibert, M.L., and Harrison, S.C. (2001). Structure of the reovirus outer capsid and dsRNA-binding protein sigma3 at 1.8 Å resolution. EMBO J. *20*, 979–989.
Pesavento, J.B., Lawton, J.A., Estes, M.E., and Venkataram Prasad, B.V. (2001). The reversible condensation and expansion of the rotavirus genome. Proc. Natl Acad. Sci. USA *98*, 1381–1386.
Pettersen, E.F., Goddard, T.D., Huang, C.C., Couch, G.S., Greenblatt, D.M., Meng, E.C., and Ferrin, T.E. (2004). UCSF Chimera--a visualization system for exploratory research and analysis. J. Comput Chem. *25*, 1605–1612.
Prasad, B.V., Rothnagel, R., Zeng, C.Q., Jakana, J., Lawton, J.A., Chiu, W., and Estes, M.K. (1996). Visualization of ordered genomic RNA and localization of transcriptional complexes in rotavirus. Nature. *382*, 471–473.
Prasad, B.V.V., Burns, J.W., Marietta, E., Estes, M.K., and Chiu, W. (1990). Localization of VP4 neutralization sites in rotavirus by three-dimensional cryo-electron microscopy. Nature (London). *343*, 476–479.
Prasad, B.V.V., Wang, G.J., Clerx, J.P.M., and Chiu, W. (1988). Three-dimensional structure of rotavirus. Journal of Molecular Biology. *199*, 269–275.
Reinisch, K.M., Nibert, M.L., and Harrison, S.C. (2000). Structure of the reovirus core at 3.6 Å resolution. Nature. *404*, 960–967.
Sabin, A.B. (1959). Reoviruses. A new group of respiratory and enteric viruses formerly classified as ECHO type 10 is described. Science. *130*, 1387–1389.
Shaw, A.L., Samal, S.K., Subramanian, K., and Prasad, B.V. (1996). The structure of aquareovirus shows how the different geometries of the two layers of the capsid are reconciled to provide symmetrical interactions and stabilization. Structure. *4*, 957–967.
Suzuki, N., Sugawara, M., Kusano, T., Mori, H., and Matsuura, Y. (1994). Immunodetection of rice dwarf phytoreoviral proteins in both insect and plant hosts. Virology *202*, 41–48.
Tao, Y., Farsetta, D.L., Nibert, M.L., and Harrison, S.C. (2002). RNA synthesis in a cage--structural studies of reovirus polymerase lambda3. Cell. *111*, 733–745.
Tomaru, M., Maruyama, W., Kikuchi, A., Yan, J., Zhu, Y., Suzuki, N., Isogai, M., Oguma, Y., Kimura, I., and Omura, T. (1997). The loss of outer capsid protein P2 results in nontransmissibility by the insect vector of rice dwarf phytoreovirus. J. Virol. *71*, 8019–8023.
Uyeda, I.a.S., E. (1982). Ultrastructure of rice dwarf virus. Annals of the Phyopathological Society of Japan, 295–300.
Wu, B., Hammar, L., Xing, L., Markarian, S., Yan, J., Iwasaki, K., Fujiyoshi, Y., Omura, T., and Cheng, R.H. (2000). Phytoreovirus T = 1 core plays critical roles in organizing the outer capsid of T = 13 quasi-equivalence. Virology *271*, 18–25.
Xia, Q., Jakana, J., Zhang, J.Q., and Zhou, Z.H. (2003). Structural comparisons of empty and full cytoplasmic polyhedrosis virus. Protein-RNA interactions and implications for endogenous RNA transcription mechanism. J. Biol. Chem. *278*, 1094–1100.
Yan, J., Tomaru, M., Takahashi, A., Kimura, I., Hibino, H., and Omura, T. (1996). P2 protein encoded by genome segment S2 of rice dwarf phytoreovirus is essential for virus infection. Virology *224*, 539–541.
Zhang, H., Zhang, J., Yu, X., Lu, X., Zhang, Q., Jakana, J., Chen, D.H., Zhang, X., and Zhou, Z.H. (1999). Visualization of protein–RNA interactions in cytoplasmic polyhedrosis virus. J. Virol. *73*, 1624–1629.

Zhang, X., Walker, S.B., Chipman, P.R., Nibert, M.L., and Baker, T.S. (2003). Reovirus polymerase lambda 3 localized by cryo-electron microscopy of virions at a resolution of 7.6 A. Nat. Struct Biol. *10*, 1011–1018.

Zhou, Z.H., Baker, M.L., Jiang, W., Dougherty, M., Jakana, J., Dong, G., Lu, G., and Chiu, W. (2001). Electron cryomicroscopy and bioinformatics suggest protein fold models for rice dwarf virus. Nat. Struct Biol. *8*, 868–873.

Zhou, Z.H., Zhang, H., Jakana, J., Lu, X.-Y., and Zhang, J.-Q. (2003). Cytoplasmic polyhedrosis virus structure at 8 Å by electron cryomicroscopy, structural basis of capsid stability and mRNA processing regulation. Structure. *11*, 651–663.

The Yeast dsRNA Virus L-A Resembles Mammalian dsRNA Virus Cores

6

Reed B. Wickner, Jinghua Tang, Nora A. Gardner and John E. Johnson

Abstract

The L-A dsRNA virus of the yeast *Saccharomyces cerevisiae* has a single 4.6 kb genomic segment that encodes its major coat protein, Gag (76 kDa) and a Gag–Pol fusion protein (180 kDa) formed by a −1 ribosomal frameshift. L-A can support the replication and encapsidation in separate viral particles of any of several satellite dsRNAs, called M dsRNAs, each of which encodes a secreted protein toxin (the killer toxin) and immunity to that toxin. L-A and M are transmitted from cell to cell by the cytoplasmic mixing that occurs in the process of mating. Neither is naturally released from the cell nor enters cells by other mechanisms, but the high frequency of yeast mating in nature results in the wide distribution of these viruses in natural isolates. Moreover, the structural and functional similarities with dsRNA viruses of mammals has made it useful to consider these entities as viruses. The common structure of L-A dsRNA virions and the cores of dsRNA viruses of higher eukaryotes may reflect similarities in their replication cycles. Specifically, both types of viruses transcribe and replicate their genomes within nucleocapsids, processes requiring movement of the RNA template. The L-A major coat protein has a unique mRNA decapping activity essential for expression of the capless viral messages.

Viral replication cycle: *in vivo* and *in vitro*

Both the (+) and the (−) strands of L-A are synthesized within viral particles (Figure 6.1). Mature particles containing L-A dsRNA synthesize (+) strands (transcription) by a conservative mechanism, and extrude the new strand from the particles. These (+) strands serve as mRNA, to make coat proteins, and as the species encapsidated by those proteins to make new nucleoprotein particles. Encapsidated (+) strands are converted to dsRNA form (replication). The absence of both 5′ cap and 3′ poly(A) on the viral mRNA is the basis of complex interactions of virus and host in translation, as is discussed below (reviewed in Wickner, 2001; Wickner *et al.*, 2003).

Exposure of isolated viral particles to very low salt conditions results in rupture of particles with loss of the dsRNA. These opened empty particles specifically bind viral (+) strands in a reaction that reflects the *in vivo* RNA packaging reaction (Esteban *et al.*, 1988). Indeed, the RNA packaging site (Figure 6.2), determined by *in vitro* RNA binding to the opened particles, directs *in vivo* packaging of an arbitrary mRNA into L-A particles. This RNA packaging site of L-A is a stem–loop with a critical A residue protruding from

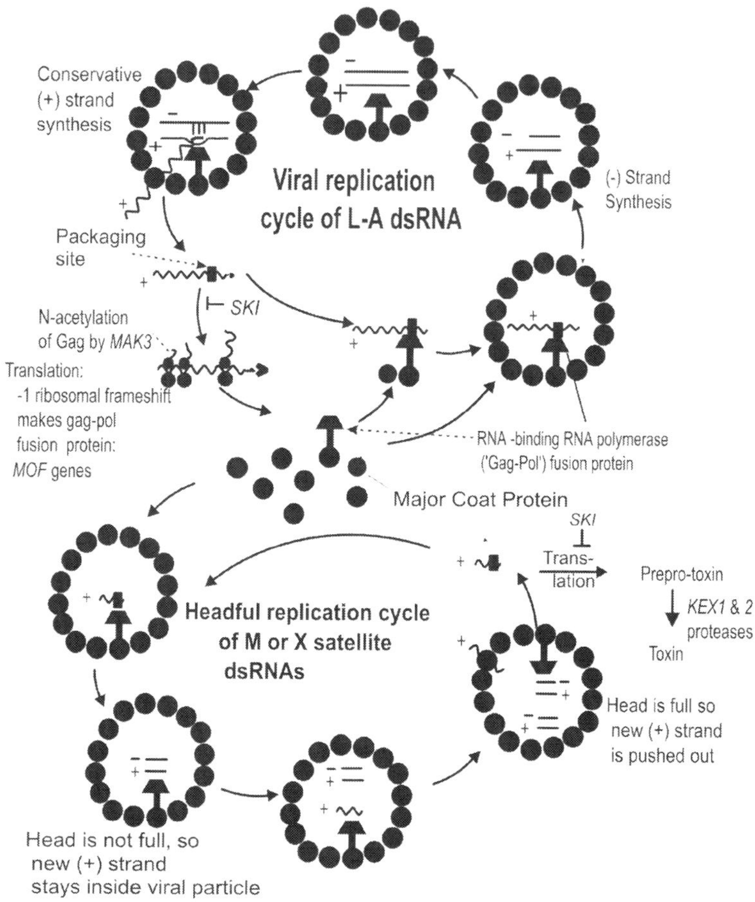

Figure 6.1 Replication cycles of the yeast L-A dsRNA virus and its killer toxin-encoding satellites, M.

the 5′ side of the stem (Esteban *et al.*, 1989). The sequence of the loop is also important. Two similar sites were identified in the sequence of M_1 (+) ssRNA, and each is likewise able to direct packaging *in vivo* (Esteban *et al.*, 1989; Shen and Bruenn, 1993).

Conversion of (+) ssRNA to dsRNA form requires sequences at both the 3′ end of the (+) strand and an internal site that largely overlaps with the packaging site (Figure 6.2). The polymerase binding to the internal (packaging) site apparently increases the frequency with which it encounters and recognizes the 3′ end, but this does not involve either a tracking or an allosteric effect (Fujimura and Wickner, 1992).

The *in vitro* transcription reaction is likewise specific for L-A viral or M satellite dsRNA (Fujimura and Wickner, 1989). The affinity of the polymerase for template dsRNA is much lower than for (+) ssRNA, perhaps reflecting the fact that while (+) strands must be fished out of a sea of cellular mRNAs, the dsRNA is confined to the viral particles with the polymerase.

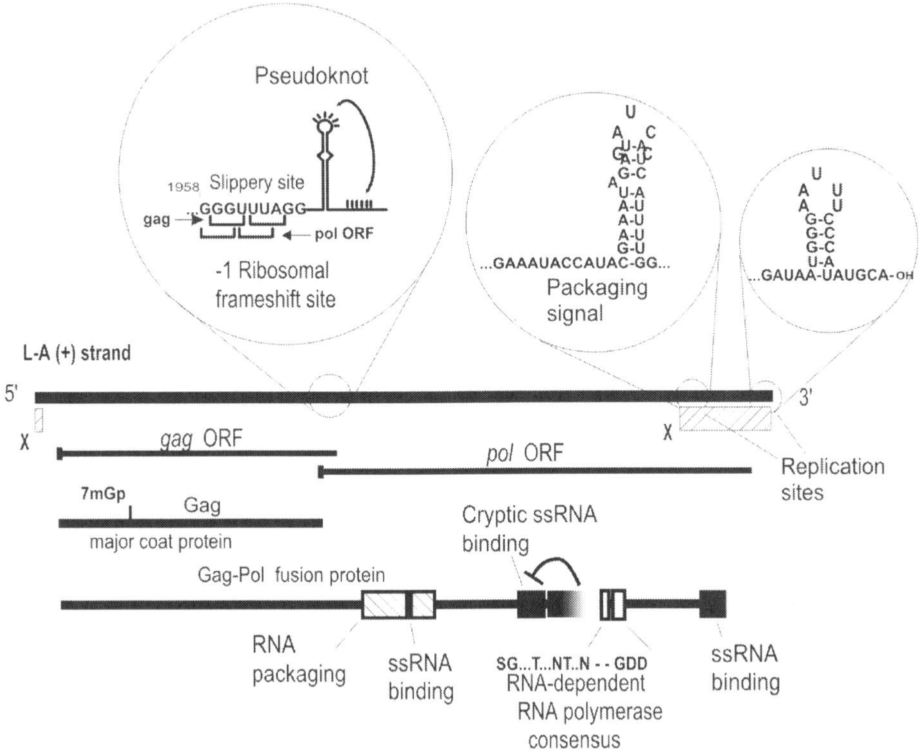

Figure 6.2 L-A coding information. The absence of 5′ cap and 3′ poly(A) structures on L-A (+) strands leads to the decapping activity of the Gag protein and the sensitivity of L-A mRNA to translation limitation by the Ski proteins.

Asymmetric dimer of Gag forms viral particles

The structure of the L-A virus particles was determined by X-ray crystallography at 3.4 angstroms resolution (Naitow *et al.*, 2002). Consistent with earlier cryoelectron microscopic results (Cheng *et al.*, 1994; Caston *et al.*, 1997), the L-A virus particles have 120 Gag molecules per particle, with 60 molecules in each of two different environments (Figure 6.3). The 'A' Gag molecules contact the fivefold and twofold axes, while the B molecules contact only the threefold axes. This icosahedral T = 1 architecture with a dimer of Gag as the asymmetric unit, also referred to as 'T = 2' because there are 120 molecules of Gag/particle, is unique to the cores of dsRNA viruses. In contrast to the quasi-equivalence observed in most icosahedral virus capsid structures, the two identical protein molecules in the asymmetric unit of L-A are located in distinctly non-equivalent structural environments. This requires different conformations in several regions to accommodate non-equivalent contacts with adjacent subunits. Conformational flexibility of this type is seen in a number of regions of the subunit. A region not involved with subunit contacts, but a chemical function, is on the outermost surface of the Gag molecule (Figure 6.4) and the particle

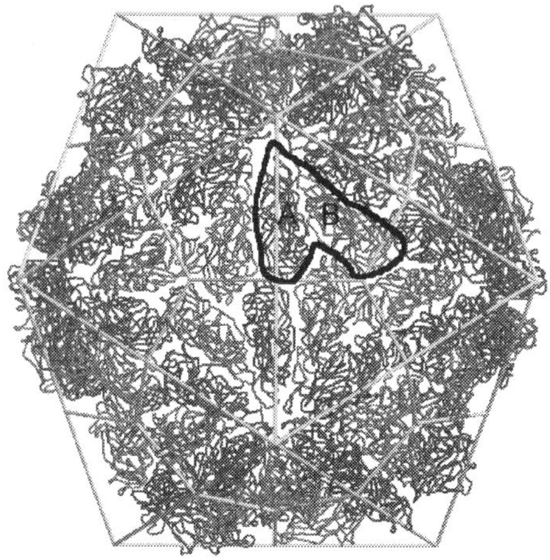

Figure 6.3 Gag monomers are found in two distinct environments: 'A' monomers contact the fivefold and twofold axes while 'B' monomers contact the threefold axes.

also shows differences between A and B subunits. This trench structure, around His 154, has been implicated in the decapping activity discussed below. The upper rim of the trench contains several β-strands connected by flexible loops (Figure 6.4). The flexibility of this region is shown by high temperature factors in the refined structure and large differences between the two monomers in the asymmetric unit. It is possible that crystal lattice effects cause these differences since interactions between these regions and neighbouring particles in the crystal will be different.

Comparing L-A cores (Naitow *et al.*, 2002) with other dsRNA viruses is an interesting exercise since the overall organization is immediately recognizable as similar to the others, but it is impossible to make a detailed comparison. All the inner core subunits are predominantly helical with a hinge region that allows the identical subunit to accom-

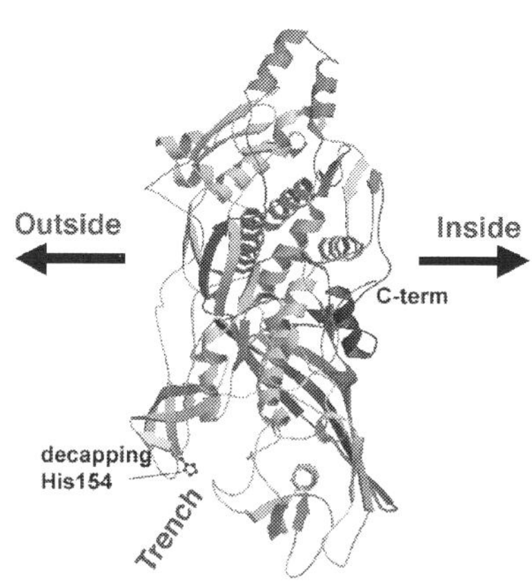

Figure 6.4 Gag monomer showing the decapping site (His154) in the 'trench', the C-terminus (where Pol attaches), and the inside-outside orientation.

modate two environments. They all occupy the same relative volume elements of the particle and each subunit interacts with neighbouring subunits in similar ways. However, as shown in Table 6.1 the L-A gag is roughly two-thirds the size of the other inner core subunits. There is not detectable sequence similarity among any of the core subunits and it is impossible to make a meaningful structural superposition. These subunits have clearly evolved to maintain maximum flexibility for interactions while conserving their relative positions in the capsid. Since most dsRNA viruses have comparable cores, the cores must be optimized for allowing entry of the metabolites for RNA replication and the exit of newly synthesized RNA.

The polymerase molecules, present in only two copies per particle, are not visible in the electron density because icosahedral symmetry was used for phase refinement. Since the outside surface of the particle determines the crystal packing, the polymerase molecules are likely to be distributed in 60 different locations in the crystal lattice, but their occupancy will be 1/30 that of the coat protein and therefore invisible. As expected, the visible C-termini of the subunits are internal and positioned to connect with the polymerase molecules (Figure 6.4).

The dsRNA is organized in striking concentric shells that are clearly visible in the low resolution electron density map. Although the RNA does not have icosahedral symmetry, it is roughly spherical at low resolution and therefore averages to non-zero values when icosahedral symmetry is applied to the density. The centre to centre distance between the shells is approximately 30 Å, the distance expected between two dsRNA segments. Similar RNA organization was seen in other dsRNA viruses, however, the packing of RNA within the L-A protein shell is less dense than in the others.

Compared to dsDNA, ssRNA viruses, or other dsRNA viruses, the density of dsRNA packing in the L-A virus is substantially lower (Caston *et al.*, 1997). This may be because the activity of a single RNA polymerase molecule covalently fixed to the wall requires a great deal of movement of the template RNA during transcription.

Translation of viral mRNA

The absence of both 5′ cap and 3′ poly(A) structures on viral mRNA is expected to limit the efficiency of translation. Indeed, selection of mutants expressing satellite (M) mRNA at higher levels (superkillers, *SKI2, 3, 6, 7, 8* genes) produced mutations that are specifically derepressed for expression of non-poly(A) mRNA (Masison *et al.*, 1995) (Benard

Table 6.1 Comparison of L-A with other dsRNA viruses

	L-A virus (whole)	BTV core (VP3 shell)	Reovirus core (λ1 shell)
Radius (angstroms)	200	270	300
Surface	Rough	Smooth	Smooth
No. of subunits	120	120	120
No. of residues	681	901	1275
Genome segments	1	10	10
Total RNA (kb)	4.6	19	23.5

et al., 1998; Benard et al., 1999). A homologue of Ski2p, called Slh1p, has a similar action (Searfoss and Wickner, 2000). Kinetic data and specific tests of mRNA degradation indicate that the primary effect of these mutations on the killer system is on translation (Masison et al., 1995; Searfoss et al., 2001; Searfoss and Wickner, 2000). In cells blocked for the major 5′→3′ mRNA turnover system, an effect of the Ski2, 3, 7, 8 proteins on a secondary mRNA turnover pathway can also be detected (Jacobs-Anderson and Parker, 1998). Ski2p, Ski3p and Ski8p form a complex located in the cytoplasm (Brown et al., 2000).

Using electroporated mRNA, the role of poly(A) in translation appears to be in promoting a Fun12p-dependent step (Searfoss et al., 2001). Fun12p is involved in 60S subunit joining (Pestova et al., 2000), and poly(A) has previously been implicated in this process (Munroe and Jacobson, 1990). Since there is little or no poly(A) requirement without Ski2p and Slh1p, it is proposed that these proteins act in 60S ribosome subunit joining (Searfoss et al., 2001). Consistent with this picture is the finding that a deficiency of 60S subunits results in lowered copy number of L-A and loss of M dsRNA (Ohtake and Wickner, 1995). Like its effect on the non-poly(A) viral mRNAs, lowering free 60S subunits adversely affects expression of electroporated non-poly(A) reporter mRNAs more than it does poly(A)+ mRNA (Edskes et al., 1998).

Ribosomal frameshifting makes a Gag–Pol fusion protein

L-A encodes its major coat protein (Gag) and the RNA-dependent RNA polymerase (Pol) in overlapping reading frames, a scheme reminiscent of retroviruses (Icho and Wickner, 1989), Figure 6.2). The 5′ reading frame encoding Gag is traversed by most ribosomes to the terminator. However, about 2% of ribosomes translating Gag shift into the −1 reading frame about 30 codons from the C-terminus, and continue translation in this (Pol) frame to make a Gag–Pol fusion protein (Dinman et al., 1991). The mechanism of this −1 frameshift is essentially identical to the simultaneous slippage mechanism described in retroviruses (Jacks et al., 1988).

A stable mRNA structure, a pseudoknot, makes the ribosomes pause just over a 'slippery' sequence, G GGT TTA (Gag reading frame shown; Figure 6.2). The Gly and Leu tRNAs paired to these two codons (in the P and A sites, respectively) can simultaneously slip back one base on the mRNA, and still have their non-wobble bases (the most important ones) correctly paired.

A host of translation factors and ribosomal components are responsible for maintaining the proper reading frame (Plant et al., 2003), and this is of course critical for proper translation. Mutants selectively altering the frameshifting efficiency at the viral slippery site have been described (Dinman and Wickner, 1994), and it has been suggested that drugs with the same effects may be a means to approach antiretroviral therapy.

mRNA decapping reaction

In a search for proteins able to bind the 5′ 7methyl-G cap structure of eukaryotic mRNAs, the covalent attachment of this structure to the L-A Gag protein was found (Blanc et al., 1992). The virus coat protein attaches 7mGMP from the cap structure to His154 of Gag in a reaction shown to occur both *in vitro* and *in vivo* (Blanc et al., 1992, 1994). Mutation of His154 to Arg, Asn or Ser abolishes the reaction *in vitro* and prevents expression of viral mRNA *in vivo* (Blanc *et al.*, 1994). This mutation does not affect the efficiency of dsRNA replication as long as viral proteins are supplied from an L-A cDNA clone, showing that this reaction has no role in the (+) or (−) strand synthesis reactions (Blanc *et al.*, 1994). The 5′ cap structure is important both for mRNA translation and for its stability. Efforts to detect the transfer of the cap to viral mRNA were unsuccessful (Masison *et al.*, 1995), and L-A (+) strands are reported to lack a cap (Bruenn and Keitz, 1976).

The instability of uncapped mRNA is due to its degradation by a 5′→3′ exoribonuclease specific for uncapped mRNAs, called Xrn1p/Ski1p (Hsu and Stevens, 1993; Stevens, 1980; Stevens, 1993; Stevens and Maupin, 1987). Indeed, *xrn1Δ* cells no longer need Gag His154 to express viral mRNAs, and the level of viral mRNA, which is lowered by His154Asn, is restored to normal levels in an *xrn1Δ* strain (Masison *et al.*, 1995) (Figure 6.5).

The X-ray structure of L-A viral particles shows that His154 is located on the outer surface of the particle, in a cleft or trench (Figure 6.4). The cap analogue 7mGDP binds to viral particles (Blanc *et al.*, 1992; Blanc *et al.*, 1994) and occupies a site in the trench near His154 (Tang *et al.*, 2005). Analysis of mutations of residues near His 154 and the bound cap analogue showed that the hydroxyls of Tyr452 and Tyr150, and the ring of Tyr538, as well as several nearby basic residues, are important in catalysis (Tang *et al.*, 2005).

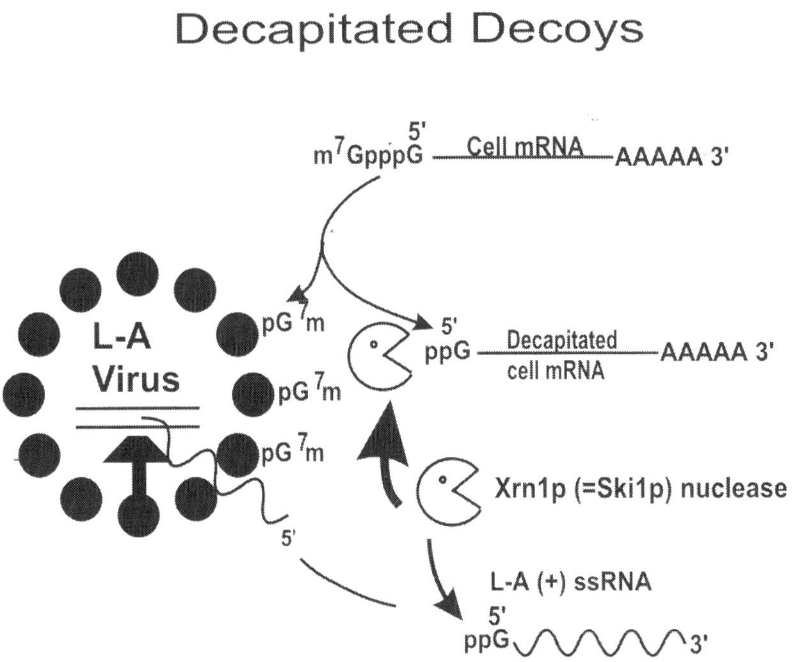

Figure 6.5 The decapping activity of Gag and its role in L-A virus expression. By removing caps from some cellular mRNAs, the Xrn1/Ski1 exoribonuclease is partially distracted from concentrating on viral mRNA degradation.

Satellite-encoded killer toxins

The various M dsRNAs each encode a secreted protein toxin, called the killer toxins, and a factor giving cells immunity to the toxin produced (reviewed in Wickner *et al.*, 2003). The toxins are each heterodimers, processed from a pre-protoxin by a series of proteolytic cleavages. The protoxin acts as the immunity factor. Studies of the mechanisms of protoxin cleavages led to the discovery of mammalian prohormone processing enzymes, the Kex2-like enzymes that cleave after dibasic residues in proinsulin, pro-opiomelanocortin and other prohormones. The killer toxins bind first to $\beta 1 \rightarrow 6$ glucan, a cell wall component, then to the membrane protein Kre1p, and finally kill by making ion pores in the cell membrane of sensitive cells.

Similarity of L-A to animal dsRNA virus cores

Animal and plant dsRNA viruses have up to three capsid layers, with structures adapted to protect the virus in its extracellular travels, to facilitate uptake by specific target cells and to engineer the uncoating of the outer layers after uptake. The yeast L-A virus can dispense with much of this apparatus because of its strictly intracellular life cycle. However, both L-A and plant and animal dsRNA viruses synthesize both (+) and (−) viral RNA strands within particles. The core particles in the case of the animal viruses and the structures of these inner nucleoprotein particles are similar as described above.

In spite of the similarities, there are differences between L-A and the animal viruses even in their intracellular lives. While animal viruses synthesize a 5′ cap for their mRNAs, L-A decaps cellular mRNAs to distract the mRNA degradation system. L-A is also less intent on maximizing translation of its mRNA because it must live with its host cell.

Problems for future work

The decapping reaction explains why viral mRNA is at least partially stable, but it remains a mystery how this uncapped, non-poly(A) mRNA is translated. Efforts to detect an internal ribosome entry site have been unsuccessful (Russell *et al.*, 1991), although an IRES has been found in a closely related dsRNA virus of *Giardia lamblia* (De-Chao and Wang, 1996).

Although we know the structure of the capsid, the RNA dependent RNA polymerase is an integral part of the viral structure, and its disposition within the particle and detailed structure are still unknown. Efforts to make soluble active L-A polymerase have not succeeded, perhaps because of an inherent requirement for Gag for activity, as has been found in rotaviruses (Tortorici *et al.*, 2003).

A model proposed for the assembly of L-A suggests initiation by the Gag–Pol fusion protein (Fujimura and Wickner, 1988), but determining the detailed mechanism of this process will require establishing an *in vitro* assembly system for this virus.

References

Benard, L., Carroll, K., Valle, R.C.P., and Wickner, R.B. (1998). Ski6p is a homolog of RNA-processing enzymes that affects translation of non-poly(A) mRNAs and 60S ribosomal subunit biogenesis. Mol. Cell. Biol. *18*, 2688–2696.

Benard, L., Carroll, K., Valle, R.C.P., and Wickner, R.B. (1999). The Ski7 antiviral protein is an EF1-α homolog that blocks expression of non-poly(A) mRNA in *Saccharomyces cerevisiae*. J. Virol. 73, 2893–2900.

Blanc, A., Goyer, C., and Sonenberg, N. (1992). The coat protein of the yeast double-stranded RNA virus L-A attaches covalently to the cap structure of eukaryotic mRNA. Mol. Cell. Biol. *12*, 3390–3398.

Blanc, A., Ribas, J.C., Wickner, R.B., and Sonenberg, N. (1994). His154 is involved in the linkage of the *Saccharomyces cerevisiae* L-A double-stranded RNA virus *gag* protein to the cap structure of mRNAs and is essential for M_1 satellite virus expression. Mol. Cell. Biol. *14*, 2664–2674.

Brown, J.T., Bai, X., and Johnson, A.W. (2000). The yeast antiviral proteins Ski2p, Ski3p and Ski8p exist as a complex *in vivo*. RNA *6*, 449–457.

Bruenn, J., and Keitz, B. (1976). The 5′ ends of yeast killer factor RNAs are pppGp. Nucleic Acids Res *3*, 2427–2436.

Caston, J.R., Trus, B.L., Booy, F.P., Wickner, R.B., Wall, J.S., and Steven, A.C. (1997). Structure of L-A virus, a specialized compartment for the transcription and replication of double-stranded RNA. J. Cell Biol. *138*, 975–985.

Cheng, R.H., Caston, J.R., Wang, G.-j., Gu, F., Smith, T.J., Baker, T.S., Bozarth, R.F., Trus, B.L., Cheng, N., Wickner, R.B., and Steven, A.C. (1994). Fungal virus capsids are cytoplasmic compartments for the replication of double-stranded RNA formed as icosahedral shells of assymetric Gag dimers. J. Mol. Biol. *244*, 255–258.

De-Chao, Y., and Wang, C.C. (1996). Identification of *cis*-acting signals in the giardiavirus (GLV) genome required for expression of firefly luciferase in *Giardia lamblia*. RNA *2*, 824–834.

Dinman, J.D., Icho, T., and Wickner, R.B. (1991). A −1 ribosomal frameshift in a double-stranded RNA virus of yeast forms a gag-pol fusion protein. Proc. Natl Acad. Sci. USA *88*, 174–178.

Dinman, J.D., and Wickner, R.B. (1994). Translational maintenance of frame, mutants of *Saccharomyces cerevisiae* with altered −1 ribosomal frameshfiting efficiencies. Genetics *136*, 75–86.

Edskes, H.K., Ohtake, Y., and Wickner, R.B. (1998). Mak21p of *Saccharomyces cerevisiae*, a homolog of human CAATT-binding protein, is essential for 60S ribosomal subunit biogenesis. J. Biol. Chem. *273*, 28912–28920.

Esteban, R., Fujimura, T., and Wickner, R.B. (1988). Site-specific binding of viral plus single-stranded RNA to replicase-containing open virus-like particles of yeast. Proc. Natl Acad. Sci. USA *85*, 4411–4415.

Esteban, R., Fujimura, T., and Wickner, R.B. (1989). Internal and terminal cis-acting sites are necessary for *in vitro* replication of the L-A double-stranded RNA virus of yeast. EMBO J. *8*, 947–954.

Fujimura, T., and Wickner, R.B. (1988). Gene overlap results in a viral protein having an RNA binding domain and a major coat protein domain. Cell 55, 663–671.

Fujimura, T., and Wickner, R.B. (1989). Reconstitution of template-dependent *in vitro* transcriptase activity of a yeast double-stranded RNA virus. J. Biol. Chem. *264*, 10872–10877.

Fujimura, T., and Wickner, R.B. (1992). Interaction of two cis sites with the RNA replicase of the yeast L-A virus. J. Biol. Chem. *267*, 2708–2713.

Hsu, C.L., and Stevens, A. (1993). Yeast cells lacking 5′→3′ exoribonuclease 1 contain mRNA species that are poly(A) deficient and partially lack the 5′ cap structure. Mol. Cell. Biol. *13*, 4826–4835.

Icho, T., and Wickner, R.B. (1989). The double-stranded RNA genome of yeast virus L-A encodes its own putative RNA polymerase by fusing two open reading frames. J. Biol. Chem. *264*, 6716–6723.

Jacks, T., Madhani, H.D., Masiarz, F.R., and Varmus, H.E. (1988). Signals for ribosomal frameshifting in the Rous sarcoma virus gag-pol region. Cell 55, 447–458.

Jacobs-Anderson, J.S., and Parker, R. (1998). The 3′ to 5′ degradation of yeast mRNAs is a general mechanism for mRNA turnover that requires the SKI2 DEVH box protein and 3′ to 5′ exonucleases of the exosome complex. EMBO J. *17*, 1497–1506.

Masison, D.C., Blanc, A., Ribas, J.C., Carroll, K., Sonenberg, N., and Wickner, R.B. (1995). Decoying the cap- mRNA degradation system by a dsRNA virus and poly(A)- mRNA surveillance by a yeast antiviral system. Mol. Cell. Biol. *15*, 2763–2771.

Munroe, D., and Jacobson, A. (1990). mRNA poly(A) tail, a 3′ enhancer of translation initiation. Mol. Cell. Biol. *10*, 3441–3455.

Naitow, H., Canady, M.A., Wickner, R.B., and Johnson, J.E. (2002). L-A dsRNA virus at 3.4 Angstroms resolution reveals particle architecture and mRNA decapping mechanism. Nature, Struct. Biol. 9, 725-728.

Ohtake, Y., and Wickner, R.B. (1995). Yeast virus propagation depends critically on free 60S ribosomal subunit concentration. Mol. Cell. Biol. *15*, 2772–2781.

Pestova, T.V., Lomakin, I.B., Lee, J.H., Choi, S.K., Dever, T.E., and Hellen, C.U.T. (2000). The ribosomal subunit joining reaction in eukaryotes requires eIF5B. Nature *403*, 332–335.

Plant, E.P., Jacobs, K.L., Harger, J.W., Meskauskas, A., Jacobs, J.L., Baxter, J.L., Petrov, A.N., and Dinman, J.D. (2003). The 9-A solution, how mRNA pseudoknots promote efficient programmed –1 ribosomal frameshifting. RNA *9*, 168–174.

Russell, P.J., Hambidge, S.J., and Kirkegaard, K. (1991). Direct introduction and transient expression of capped and non-capped RNA in *Saccharomyces cerevisiae*. Nucleic Acids Res. *19*, 4949–4953.

Searfoss, A., Dever, T.E., and Wickner, R.B. (2001). Linking the 3′ poly(A) Tail to the Subunit Joining Step of Translation Initiation, Relations of Pab1p, eIF5B (Fun12p) and Ski2p-Slh1p. Mol. Biol. Cell. *21*, 4900–4908.

Searfoss, A.M., and Wickner, R.B. (2000). 3′ poly(A) is dispensable for translation. Proc. Natl Acad. Sci. USA *97*, 9133–9137.

Shen, Y., and Bruenn, J.A. (1993). RNA structural requirements for RNA binding, replication, and packaging in the yeast double-stranded RNA virus. Virology *195*, 481–491.

Stevens, A. (1980). Purification and characterization of a *Saccharomyces cerevisiae* exoribonuclease which yeilds 5′ mononucleotides by a 5′→3′ mode of hydrolysis. J. Biol. Chem. *255*, 3080–3085.

Stevens, A. (1993). Eukaryotic nucleases and mRNA turnover. In Control of messenger RNA stability, J.Belasco, and G. Brawerman, eds. (San Diego, Academic Press), pp. 449–471.

Stevens, A., and Maupin, M.K. (1987). A 5′→3′ exoribonuclease of *Saccharomyces cerevisiae*, size and novel substrate specificity. Arch. Biochem. Biophys. *252*, 339–347.

Tang, J., Naitow, H., Gardner, N.A., Kolesar, A., Tang, L., Wickner, R.B., and Johnson, J.E. (2005). The structural basis of recognition and removal of cellular mRNA 7-methyl G 'caps', by a viral capsid protein: a unique viral response to host defense. J. Mol. Recognit. *18*, 158–168.

Tortorici, M.A., Broering, T.J., Nibert, M.L., and Patton, J.T. (2003). Template recognition and formation of initiation complexes by the replicase of a segmented double-stranded RNA virus. J. Biol. Chem. *278*, 32673–32682.

Wickner, R.B. (2001). Viruses of yeasts, fungi and parasitic microorganisms. In Fields Virology D.M. Knipe, and P. M. Howley, eds. (Philadelphia, Lippincott, Williams & Wilkins), pp. 629–658.

Wickner, R.B., Gardner, N.A., Bussey, H., Fujimura, T., and Esteban, R. (2003). RNA viruses and killer genetics of *Saccharomyces*. In The Mycota, Genetics and Biotechnology, U. Kuck, ed. (Berlin, Springer), pp. 219–237.

Dissecting the Assembly Pathway of Bacterial dsRNA Viruses: Infectious Nucleocapsids Produced by Self-Assembly

7

Minna M. Poranen, Roman Tuma, and Dennis H. Bamford

Abstract

The virions of double-stranded RNA bacteriophages of the family *Cystoviridae* possess a triple-layered structure surrounding the tri-segmented genome. Genome-containing double-shelled nucleocapsids are formed by a self-assembly process that can be accomplished *in vitro* using purified proteins and RNA constituents. The assembly process involves formation of empty particles, which package single-stranded genomic precursor RNA molecules. Subsequently, the precursor RNA molecules are replicated into the double-stranded form within the particle. Finally, the genome-containing particles acquire the nucleocapsid surface shell. Two viral enzymes, the packaging NTPase P4 and the polymerase P2, are part of the assembly process. The assembly of the inner protein capsid involves tetramerization of the major capsid protein P1 and subsequent interaction with the enzymatic components (P2 and P4). The *in vitro* reconstituted ϕ6 nucleocapsids can penetrate the host cell plasma membrane and initiate a productive infection cycle.

Introduction

The virion of bacterial double-stranded RNA (dsRNA) viruses is comprised of a multi-functional macromolecular structure with three distinct layers surrounding the segmented dsRNA genome. The two outer virion layers interact with the bacterial cell envelope for host recognition, membrane fusion, peptidoglycan degradation, and virion-induced endocytosis (for details see Chapter 12; Poranen and Bamford, 2007). The core particle, which is delivered into the cell cytoplasm, constitutes the RNA synthesis machinery that simultaneously transcribes the three enclosed dsRNA genome segments. Later in infection, the protein components of the core self-assemble and selectively encapsidate and replicate the viral single-stranded RNA (ssRNA) genomic precursors. How are these viral particles assembled and their multiple functions controlled? In this review we address these issues by focusing on two related bacterial dsRNA viruses, ϕ6 and ϕ8, for which *in vitro* reconstitution systems have provided novel information on the sequence of molecular interactions during assembly.

Bacteriophage ϕ6 is the type species of the family *Cystoviridae* (http://www.ncbi.nlm.nih.gov/ICTV/). For a long time ϕ6 was the only known dsRNA virus with a prokaryotic host and the only member of the *Cystoviridae* and its single genus *Cystovirus* (Vidaver *et al.*, 1973). Subsequently, eight additional bacterial dsRNA viruses (ϕ7 to ϕ14) were isolated and characterized (Mindich *et al.*, 1999). These viruses, which share a conserved genome organization and an overall virion structure, are grouped into subgroups based on their relationship to ϕ6. Phage ϕ8 is the most distantly related, with sequence homology between ϕ6 and ϕ8 limited to the conserved polymerase (P2) and packaging nucleotide triphosphatase (NTPase, P4) motifs (Hoogstraten *et al.*, 2000).

Cystoviruses differ from other known bacterial viruses both structurally and functionally, and may share a common evolutionary origin with eukaryotic dsRNA viruses of the family *Reoviridae* (Bamford, 2003). This similarity is most evident in the innermost protein capsid that intimately surrounds the genome but less evident in the outer virion layers that mediate host interactions. The structural similarity between different dsRNA viruses suggests that conclusions reached for the assembly pathways of cystoviruses might apply to related eukaryotic dsRNA viruses.

Genome

The ϕ6 genome comprises three dsRNA segments encoding 14 open reading frames, two of which are overlapping. Twelve viral proteins are structural while two are non-structural (P12 and P14). The three dsRNA segments are named according to their size: large (6374 bp, ϕ6), medium (4063 bp, ϕ6), and small (2948 bp, ϕ6) (Gottlieb *et al.*, 1988a; McGraw *et al.*, 1986; Mindich *et al.*, 1988). Each virion contains only one copy of each segment (Day and Mindich, 1980). During ϕ6 infection viral proteins are produced from polycistronic, full-length mRNA molecules. Each segment contains genes belonging to a distinct functional group together with non-coding areas that flank the coding regions (Gottlieb *et al.*, 1988a; McGraw *et al.*, 1986; Mindich *et al.*, 1988). The non-coding regions contain signals for genome encapsidation, replication, and transcription (Frilander *et al.*, 1995; Gottlieb *et al.*, 1994; Mindich *et al.*, 1992). The production of viral proteins is controlled at the translational level (McGraw *et al.*, 1986; Mindich *et al.*, 1988) so that the relative amounts of proteins produced during viral infection equals that found in mature virions.

Virion structure

Bacteriophage ϕ6 has a triple-layered architecture (Figure 7.1). The genome is enclosed within the nuclocapsid (NC), an icosahedral capsid built from two protein layers (P1 and P8) (Butcher *et al.*, 1997; Huiskonen *et al.*, 2006). The NC is enveloped by a lipid bilayer (Van Etten *et al.*, 1976), a feature unique to cystoviruses among both bacteriophages and dsRNA viruses. Particles that lack the membrane and the P8 layer are called cores or polymerase complexes.

Polymerase complex

The viral core is composed of four protein species: P1, P2, P4, and P7 (Figure 7.1) (Bamford and Mindich, 1980; Mindich and Davidoff-Abelson, 1980). P1 is the major component of the polymerase complex and forms the structural skeleton of the particle (Ktistakis and Lang, 1987; Olkkonen and Bamford, 1987). Protein P1 is arranged as 60 asymmetric dimers (120 copies) on a T = 1 icosahedral lattice (Butcher *et al.*, 1997; Huiskonen *et al.*, 2006). This capsid organization, which exhibits two chemically identi-

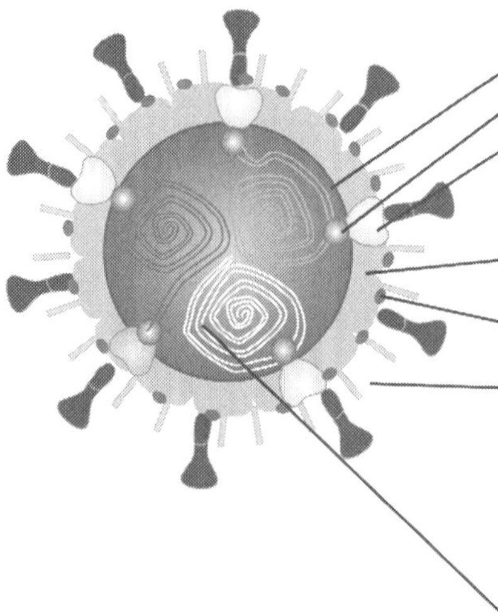

Figure 7.1 Architecture of ϕ6 virion and functions of the proteins. Four proteins (P1, P2, P4, and P7) form the polymerase complex, which is surrounded by the nucleocapsid surface shell composed of protein P8. The turret-like extrusions of protein P4 at the fivefold vertices of the polymerase complex extend to the P8 layer. The viral envelope contains four integral membrane proteins (P6, P9, P10, and P13) and the spike protein (P3). Three double-stranded RNA genome segments (S, M, and L) are enclosed within the polymerase complex.

cal subunits in different conformations within the icosahedral asymmetric unit, has been observed only in the cores of dsRNA viruses of the *Reoviridae* and *Totiviridae* families (Grimes *et al.*, 1998; Lawton *et al.*, 1997; Naitow *et al.*, 2002; Reinisch *et al.*, 2000) and is described as a 'T2' arrangement to distinguish it from the other T = 1 virion architectures (Grimes *et al.*, 1998). Within the polymerase complex the P1 framework is rather flexible and undergoes substantial but cooperative conformational changes during maturation (Figure 7.2D, F, and G) (Butcher *et al.*, 1997; Huiskonen *et al.*, 2006). These allosteric properties most likely result in the precise control of polymerase complex functions during genome encapsidation and replication (Poranen *et al.*, 2004). In addition to its structural role, P1 is also involved in the specific binding and recognition of the viral ssRNA genomic precursors (Qiao *et al.*, 2003; Qiao *et al.*, 2003). Furthermore, P1 of ϕ8 probably mediates host plasma membrane penetration during virion entry (Sun *et al.*, 2003).

Protein P2 contains the consensus sequences of RNA-dependent RNA polymerases (Koonin *et al.*, 1989). Particles lacking P2 (P1P4P7 particles) can package ssRNA but are unable to perform genome replication and transcription (Juuti and Bamford, 1995). The P2 polymerase can be purified as a monomer, is active *in vitro*, and can initiate complementary strand synthesis on the ssRNA template by a primer-independent mechanism (Makeyev and Bamford, 2000a). Molecular details of the isolated P2 are described in Chapter 14 (Koivunen *et al.*, 2007). The ϕ6 P2 (Butcher *et al.*, 2001) shares common structural features with the polymerases of hepatitis C virus (Ago *et al.*, 1999), and also with human immunodeficiency virus (Huang *et al.*, 1998), and reovirus (*Orthoreovirus, Reoviridae*)

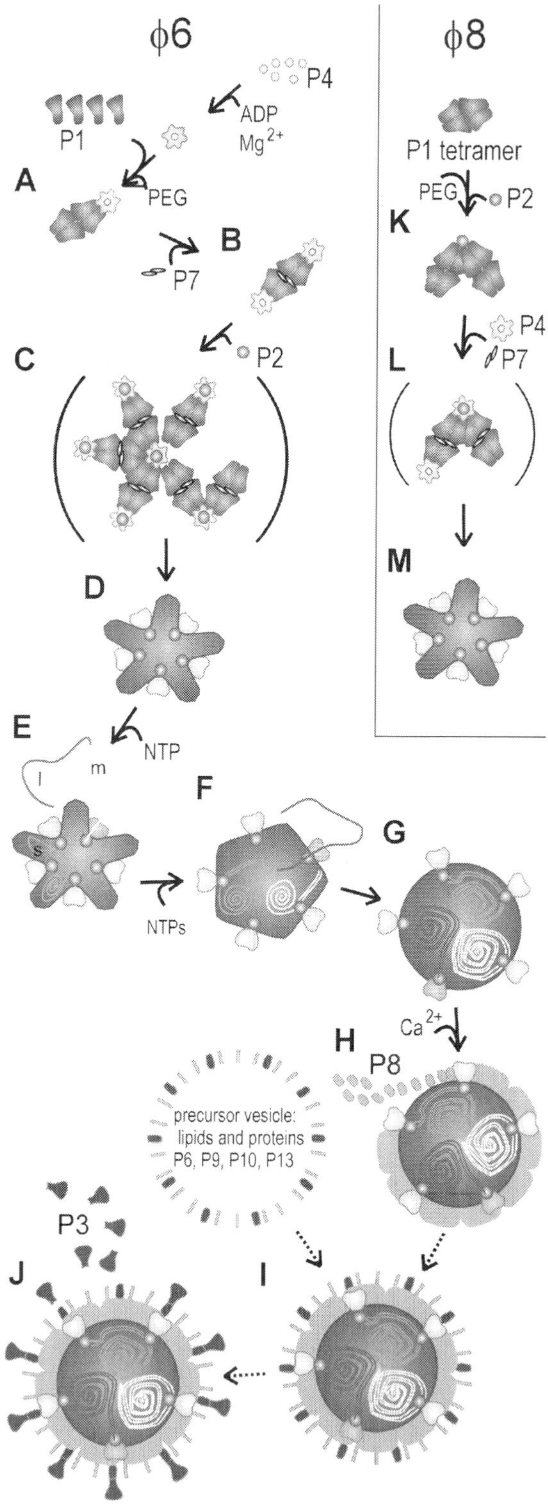
φ6
φ8
P4
ADP
Mg2+
P1
P1 tetramer
A
PEG
PEG
P2
K
B
P7
P4
P7
L
C
P2
M
D
E
NTP
l
m
s
F
G
NTPs
Ca2+
H
P8
precursor vesicle:
lipids and proteins
P6, P9, P10, P13
P3
J
I

(Tao *et al.*, 2002). Like the RNA replication devices of rotavirus (*Rotavirus, Reoviridae*), bluetongue virus (*Orbivirus, Reoviridae*), and reovirus (*Orthoreovirus, Reoviridae*) (Gouet *et al.*, 1999; Prasad *et al.*, 1996; Zhang *et al.*, 2003), the polymerase (P2) of ϕ6 localizes within the core near the fivefold vertices (Figure 7.2) (Ikonen *et al.*, 2003). In ϕ8 P2 is also involved in assembly initiation (Figure 7.2K) (see below; Kainov *et al.*, 2003a).

Protein P4 plays key roles in particle assembly, genome packaging, and transcription. The P4 hexamers reside at the icosahedral fivefold vertices and protrude from the core like turrets (Figure 7.1) (Butcher *et al.*, 1997; de Haas *et al.*, 1999). In one of the twelve icosahedral vertices, the P4 hexamer is more stably incorporated into the particle than the others, thus identifying a special vertex (Pirttimaa *et al.*, 2002). Morphologically similar turrets are found in the cores of several other dsRNA viruses. For example, the RNA-capping pentamers (λ2 and VP1, respectively) of orthoreoviruses and aquareoviruses (*Reoviridae*) form turrets in equivalent locations (Nason *et al.*, 2000; Reinisch *et al.*, 2000).

Protein P4 of ϕ6, as well as that of ϕ8, contains sequence motifs of NTP-binding proteins (Walker-motifs) and hexameric helicases (Gottlieb *et al.*, 1992; Kainov *et al.*, 2003b). Purified P4 self-assembles into ring-like hexamers that have a distinct channel at the sixfold axis (de Haas *et al.*, 1999; Mancini *et al.*, 2004). The P4 hexamers of ϕ8, ϕ12, and ϕ13 are stable in the presence of divalent cations (Mg^{2+} and Ca^{2+}) (Kainov *et al.*, 2003b), while P4 hexamers of ϕ6 also require ADP (Figure 7.2) (Juuti *et al.*, 1998). ϕ6 P4 is an active NTPase only in the hexameric form and can hydrolyze all nucleotide triphosphates (rNTP, dNTP, and ddNTP). The NTPase activities of ϕ6, ϕ12, and ϕ13 P4 are

---

Figure 7.2 (opposite) Assembly pathway for cystoviruses. (A–J) The complete assembly pathway for bacteriophage ϕ6. (K-M) Assembly pathway for ϕ8 procapsids (PCs). Solid arrows indicate reactions for which *in vitro* systems have been established while dotted arrows indicate processes that can be studied only within a cell. For *in vitro* reactions the minimal chemical conditions that activate subsequent processes are included beside the arrow. The four polymerase complex proteins of ϕ6 and ϕ8 (P1, P2, P4, and P7) self-assemble into empty procapsids (A–D and K–M). The ϕ6 PC assembly is nucleated by hexameric protein P4 (A). The hexamerization of P4 is dependent on Mg^{2+} ions and ADP. Two early assembly intermediates have been inferred for ϕ6 PC: one composed of about four P1 molecules with one P4 hexamer (A) and another composed of a P1 tetramer connecting two P4 hexamers and stabilized by a P7 dimer (B). After the formation of the nucleation complex, the assembly reaction proceeds rapidly and incorporates P2 into the particles. A hypothetical assembly intermediate, depicting how the symmetry mismatch forms, is shown in parenthesis (C). Three structures have been described for the polymerase complex: the empty compressed particle (D), the expanded dodecahedral particle (F), and the mature round particle (G). The empty particle packages the single-stranded genomic precursors s, m, and l (E). The putative special vertex functions as the ssRNA-packaging device (darker grey). The s segment is packaged first, followed by m and then l. The packaging of l activates minus-strand synthesis and the particle expands to its dodecahedral shape (F). The three segments are replicated (F) by the viral polymerase (P2), and the particle obtains its mature rounded appearance (G). Protein P8 assembles around the genome-filled polymerase complex to form nucleocapsids (H), which can infect host cell spheroplasts. Within the cell the nucleocapsids are enveloped by virus-specific membrane vesicles that contain P6, P9, P10, and P13 (I). Finally, the spike protein P3 assembles onto the enveloped virion (J). (K–M) For ϕ8 the P1 protein forms stable tetramers. The nucleation complex contains two to three P1 tetramers together with a P2 monomer (K). Proteins P4 and P7 stabilize the nucleation complex (L) for assembly of the empty compressed particle (M).

weakly stimulated by ssRNA (Juuti *et al.*, 1998; Kainov *et al.*, 2004; Kainov *et al.*, 2003b; Paatero *et al.*, 1995). In contrast, the activity of ϕ8 P4 is tightly coupled to ssRNA binding, and the purified hexameric protein also possesses RNA helicase activity (Kainov *et al.*, 2003b). The NTPase activity of P4 drives the packaging of ssRNA genome segments into the preassembled empty polymerase complexes (Frilander and Bamford, 1995; Gottlieb *et al.*, 1990; Pirttimaa *et al.*, 2002). The central channel of the P4 hexamer could provide a path for ssRNA molecules into the capsid. Recently, the structure of the P4 hexamer from ϕ12 has been solved (Mancini *et al.*, 2004), and a mechanistic model of ATP-driven RNA translocation has been proposed (Lisal *et al.*, 2004; Mancini *et al.*, 2004). The molecular basis of RNA translocation is described in Chapter 15 (Mancini and Tuma, 2007).

The smallest protein within the polymerase complex is P7. The exact location of P7 in the particle is not known, but a position near the twofold axes, at the P1 interface, has been proposed (Figure 7.2) (Juuti and Bamford, 1997; Poranen *et al.*, 2001). Purified P7 forms elongated dimers (Juuti and Bamford, 1997). The N-terminal portion of ϕ6 P7 forms the core of the dimer, while the C-termini constitute flexible protease-sensitive arms (S.J. Butcher, personal communication). The flexible part of the protein is not crucial for the stable incorporation of P7 into the polymerase complex but is essential for the viability of the virus (M.M. Poranen and S.J. Butcher, unpublished). P7-null particles have poor packaging and transcription efficiencies (Juuti and Bamford, 1997). P7 is also an assembly cofactor (see below; Poranen *et al.*, 2001). These diverse roles of P7 suggest that it might be a particle-stabilizing (glue) protein that acts as a hinge between P1 capsomers and facilitates structural transitions.

Nucleocapsid surface shell

In the ϕ6 nucleocapsid (NC), the core is surrounded by a second protein layer called the NC surface shell (Figure 7.1). This shell is formed by the viral capsid protein P8 (Bamford and Mindich, 1980; Van Etten *et al.*, 1976). The P8 protein of ϕ8 does not form a distinct shell around the core; rather, it is a minor component of the membrane (Hoogstraten *et al.*, 2000). The P8 shell of ϕ6 is arranged on a T = 13 icosahedral lattice (Butcher *et al.*, 1997; Huiskonen *et al.*, 2006), creating a mismatch between the core and the P8 layer. The same mismatch is also found in other dsRNA viruses like bluetongue virus, reovirus, and rotavirus (Mertens, 2004). The P4 turrets extend from the underlying core and protrude through the P8 layer at the fivefold vertices of NC (Figure 7.1) (Butcher *et al.*, 1997).

The double-shelled NC of ϕ6 can penetrate the host cell plasma membrane and deliver the transcriptionally active core into the host cytoplasm (Ojala *et al.*, 1990). Thus, NC particles can infect host cells in which the outer membrane and the peptidoglycan layers have been removed (spheroplasts).

Envelope

The outermost layer of the virion is a lipid–protein envelope (Figure 7.1) (Van Etten *et al.*, 1976). Four virally encoded integral membrane proteins (P6, P9, P10, and P13) are inserted into the envelope (Gottlieb *et al.*, 1988a; Sinclair *et al.*, 1975; Van Etten *et al.*, 1976). Lipids are derived from the host plasma membrane (Laurinavičius *et al.*, 2004). Host-attachment spikes (P3) are anchored to the membrane surface via protein P6 (Figure 7.1) (Stitt and Mindich, 1983a).

Virion assembly

The ϕ6 assembly pathway has been delineated using analysis of infected cells, isolation and characterization of virus mutants, and disruption and reconstitution assays. Virion assembly consists of the following steps: (1) procapsid (PC) formation, (2) genome encapsidation and replication, (3) nucleocapsid shell assembly, and (4) envelopment of the nucleocapsid. The first three assembly steps can also be studied *in vitro* using purified components or assembly intermediates (Frilander and Bamford, 1995; Gottlieb *et al.*, 1990; Olkkonen *et al.*, 1990; Olkkonen *et al.*, 1991; Poranen *et al.*, 2001). The assembly pathway of ϕ6 is summarized in Figure 7.2.

Procapsid formation

The first stable assembly intermediate detected in ϕ6-infected cells is an empty polymerase complex, the procapsid (Bamford and Mindich, 1980). Although the PC has the same protein composition as virion cores, it is devoid of viral RNA and differs morphologically from virion-derived dsRNA-filled cores. Aside from being smaller (~46 nm versus ~50 nm), the PC has a dodecahedral shape with inward-oriented 'cups' in place of fivefold facets rather than the round appearance of cores (Butcher *et al.*, 1997; Huiskonen *et al.*, 2006). No assembly intermediates preceding the PC have been isolated from infected cells.

When expressed in host cells or in heterologous organisms, the polymerase complex proteins of ϕ6 and ϕ8 self-assemble to form PC-like particles (Gottlieb *et al.*, 1988b; Sun *et al.*, 2003). These particles are stable in the absence of the outer virion layers and genomic RNA and can be readily isolated for structural and enzymatic analyses. Using heterologous expression systems it is also possible to produce incomplete PC particles lacking one of the proteins P2, P4 and P7 or both P2 and P7 (Benevides *et al.*, 2002; de Haas *et al.*, 1999; Gottlieb *et al.*, 1988b; Juuti and Bamford, 1995; Paatero *et al.*, 1998). However, these incomplete particles exhibit compromised stability and yield. Over-expression of P1 alone leads to assembly of unstable, partially unfolded and aggregated particles (Benevides *et al.*, 2002; Gottlieb *et al.*, 1988b; Paatero *et al.*, 1998). Incomplete PC particles of ϕ8 composed of P1 and P7 can also be produced (Sun *et al.*, 2003). The stable incorporation of proteins P2, P4, and P7 into different incomplete particles indicates that proteins P2, P4, and P7 can associate with the P1 skeleton independently of each other.

Assembly induced conformational changes in polymerase complex proteins have been documented by Raman spectroscopy (Benevides *et al.*, 2002). These studies have revealed structural changes in P4 and P7 proteins upon association with the P1 shell. The flexible C-termini of P7 and P4 most likely become ordered. The same C-terminal region of P4 has been shown to be essential for P4 incorporation onto the PC (Paatero *et al.*, 1998). The assembly induced conformational switching in the P4 structure may allow accommodation of the symmetry mismatch between the fivefold vertex of the P1 shell and the P4 hexamer (Benevides *et al.*, 2002; de Haas *et al.*, 1999).

Additional details of ϕ6 and ϕ8 polymerase complex morphogenesis are based on *in vitro* assembly assays (Kainov *et al.*, 2003a; Poranen *et al.*, 2001), which, for the first time, have revealed PC assembly intermediates (Figure 7.2A, B, and K). However, the assembly of ϕ6 and ϕ8 exhibit certain differences. Mainly, the major capsid protein P1 of ϕ8 forms stable tetramers that assemble more readily than the monomeric P1 of ϕ6 (Figure 7.2).

The formation of ϕ6 PC-like structures *in vitro* is absolutely dependent on the P4 packaging NTPase (in addition to P1), suggesting that the P4 hexamer is critical for the nucleation of ϕ6 PC assembly (Figure 7.2A) (Poranen *et al.*, 2001). Interestingly, the packaging portal of dsDNA bacteriophages is also probably involved in the nucleation of procapsid assembly (Valpuesta and Carrascosa, 1994). Thus, the special vertex is incorporated early in assembly, perhaps during the nucleation. How the symmetry mismatch is built into the ϕ6 PC is currently unknown; however, the P4 hexamer that nucleates PC assembly could serve as the special vertex.

ϕ8 P1 shells can be formed in the absence of any other proteins if very high P1 concentrations are used (Kainov *et al.*, 2003a). These particles are not very stable, and the P2 protein is necessary for nucleation at lower P1 concentrations (Figure 7.2K). Similarly, one rotavirus assembly model suggests that the viral polymerase-capping complex (VP1-VP3) may nucleate core assembly (Patton and Spencer, 2000). While P4 is not absolutely essential for ϕ8 PC nucleation, it significantly stabilizes assembly intermediates and prevents the formation of aberrant structures (Kainov *et al.*, 2003a). This suggests that the P4 protein also plays a crucial role during ϕ8 assembly (Figure 7.2 K-M).

In the ϕ6 assembly system, P2 does not induce any detectable differences in assembly kinetic or particle yield (Poranen *et al.*, 2001). However, P2 is always efficiently incorporated into the particles suggesting that P2 interacts with a specific assembly intermediate (Figure 7.2C). Such interactions, occurring after the rate-limiting nucleation step, are difficult to capture.

Kinetic analyses of ϕ6 PC assembly have also revealed a role for P7 in the process. The assembly rate increases with rising P7 concentration, suggesting that P7 acts as an assembly cofactor (Poranen *et al.*, 2001). Meanwhile in ϕ8 assembly, P7 has only a small effect on the assembly rate (Kainov *et al.*, 2003a).

Nucleation complex

The kinetics of ϕ6 and ϕ8 PC assembly display a distinct lag phase typical of nucleation-limited assembly (Prevelige *et al.*, 1993). Thus, kinetic analyses have revealed the approximate composition of transient assembly intermediates occurring early in PC assembly. Analyses of ϕ6 kinetics have identified a complex containing a P1 tetramer connecting two P4 hexamers as the rate-limiting, nucleation step (Figure 7.2B) (Poranen *et al.*, 2001). Such an intermediate has been observed under 'natural conditions' when all four polymerase complex components are present in the assembly reaction. When P7 is not included, a complex of one P1 tetramer with only one P4 hexamer has been detected (Figure 7.2A). Thus, the presence of P7 seems to alter ϕ6 PC assembly so that the formation of the P1 tetramer–P4 hexamer occurs more readily and the addition of a second P4 hexamer becomes rate limiting (Poranen *et al.*, 2001). Using time-resolved dynamic light scattering, ϕ6 assembly intermediates with hydrodynamic radii of 10–12 nm have been observed during the *in vitro* assembly of P1 and P4. These intermediates approximate the nucleation complex detected by kinetic methods (Figure 7.2A).

In the ϕ8 system P1 forms stable pre-assembled tetrameric building blocks, and the nucleation complex results from the interaction of several P1 tetramers with a P2 monomer (Figure 7.2K) (Kainov *et al.*, 2003a). ϕ8 complexes with hydrodynamic radii of 8–10 nm have been observed, corresponding to a nucleation complex composed of 2–3 P1 tetramers together with a P2 monomer and a P4 hexamer (Kainov *et al.*, 2003a).

Tetrameric P1 appears to be an obligatory intermediate for both ϕ6 and ϕ8 (Figure 7.2). The P1 tetramer of ϕ8 is more stable while that of ϕ6 occurs only transiently during the assembly process (Figure 7.2A and B) (Kainov *et al.*, 2003a; Poranen *et al.*, 2001). The tetramer forms a bridge between two neighbouring fivefold vertices and thus facilitates propagation of the structure. In the ϕ6 system it has been possible to identify a nucleation complex containing two P4 hexamers (Figure 7.2B), suggesting that assembly proceeds via a twofold (not fivefold)-centred assembly intermediate (Poranen *et al.*, 2001). For bluetongue virus an assembly pathway involving decamers of VP3(T2) protein has been suggested based on structural analysis of the end product, the core particle (Grimes *et al.*, 1998). Such a fivefold-centred assembly intermediate might also occur during the assembly of cystoviruses (Figure 7.2C and L), but the tetramerization of the coat protein appears to occur before the formation of these putative intermediates.

Genome encapsidation

Genome encapsidation in cystoviruses occurs in two steps (Figure 7.2E and F) (for a review see Poranen *et al.*, 2004). First, the single-stranded genome segments, which also serve as templates for protein synthesis, are packaged into the preformed empty polymerase complexes (Figure 7.2E). Next, the complementary negative strands are synthesized to form the mature dsRNA genome segments within the particle (Figure 7.2F). Proteins P2 and P4 are the key viral enzymes involved in these processes (Juuti *et al.*, 1998; Makeyev and Bamford, 2000a).

Protein P4 acts as the packaging motor and hydrolyzes NTP to catalyze the translocation of ssRNA into the particle (Frilander and Bamford, 1995; Gottlieb *et al.*, 1991; Pirttimaa *et al.*, 2002). Protein P1 selectively recognizes the genome segments using specific packaging signals at the 5′ ends of the ssRNA segments followed by unidirectional translocation of the segments from the 5′ end (Frilander *et al.*, 1992; Gottlieb *et al.*, 1994; Pirttimaa and Bamford, 2000; Qiao *et al.*, 2003; Qiao *et al.*, 2003; Qiao *et al.*, 1995a,b). The three genome segments are orderly packaged into the particle so that segment s is preferentially packaged first, followed by m and then by l (Frilander and Bamford, 1995; Qiao *et al.*, 1995a). The packaging reaction is described in detail elsewhere in this book (Mindich, 2007).

The polymerase P2 performs both genome replication and transcription. Although isolated P2 protein is relatively non-specific, the replication of the three genome segments within the polymerase complex particle is dependent on specific tRNA-like RNA secondary structures at the 3′ ends of the genome segments (Frilander *et al.*, 1992; Makeyev and Bamford, 2000a). These structures might allow correct localization of the 3′ ends of the packaged ssRNAs within the capsid (Makeyev and Bamford, 2000a). The initiation of RNA replication is a checkpoint for correct packaging and does not occur until the packaging signal of segment l (Figure 7.2F) (the last segment to be packaged) has been recognized by the particle (Frilander *et al.*, 1995; Poranen and Bamford, 1999).

In the polymerase complex particle there are two levels of asymmetry (Figure 7.2). First, the location of the hexameric P4 and monomeric P2 at each of the fivefold vertices of the P1 shell creates a symmetry mismatch at every vertex (de Haas *et al.*, 1999; Ikonen *et al.*, 2003). Second, there is asymmetry in the particle due to the different physical properties of the P4-P1 interaction in one of the twelve vertices (Pirttimaa *et al.*, 2002). These asymmetries are vital for the polymerase complex functions during genome encapsidation and replication. The asymmetry in the particle defines the special vertex that functions as an ssRNA packaging device during genome packaging (Figure 7.2E). The P4 hexamers at the other vertices cannot substitute for the P4 hexamer at the special vertex during genome packaging (Pirttimaa *et al.*, 2002). The symmetry mismatch between the P4 hexamer and the P1 shell in every vertex probably allows flexibility in P1-P4 interactions, which in turn is crucial for RNA movements across the shell during packaging and transcription. Similar mismatches in symmetry between capsid components is commonly found in the packaging vertices of dsDNA bacteriophages and also in the packaging portal of herpes viruses (Newcomb *et al.*, 2001; Orlova *et al.*, 2003; Simpson *et al.*, 2000; Valpuesta and Carrascosa, 1994).

Not only is P4 transiently required for genome encapsidation, but it is also involved in the transcription process (Pirttimaa *et al.*, 2002). Thus, P4 is needed early in the infection, prior viral protein synthesis, when the viral core particles delivered into host cytoplasm start to produce viral mRNA molecules. The special vertex may not be involved in transcription; rather, the P4 hexamers located at the other vertices take part in mRNA production (Pirttimaa *et al.*, 2002). Kainov *et al.* (2004) have shown recently that during transcription P4 of ϕ12 acts as a passive conduit for RNA transport across the P1 shell and the NTPase activity of P4 is not needed.

As discussed above, turret-like structures similar to those formed by the P4 of ϕ6 are also found at the fivefold vertices of the reovirus core particles. The core turrets of cystoviruses and reoviruses both enable RNA exit during transcription (Luongo *et al.*, 2002; Pirttimaa *et al.*, 2002). The core turrets of reovirus are involved in the catalysis of mRNA 5′-capping reaction (Mao and Joklik, 1991). Such activity is not needed in bacterial cells. The core turrets of reoviruses are pentamers of λ2 subunits (Reinisch *et al.*, 2000); thus, there is no symmetry mismatch at the fivefold vertices of reovirus core. Unlike the P4 hexamers of cystoviruses, the turrets of reoviruses are not directly involved in the active translocation of ssRNA molecules into the capsid. This could indicate that the symmetry mismatch in cystoviral core particles is mainly required for the packaging activity while it is non-essential for releasing the transcripts. The virions of bluetongue virus and rotavirus do not contain turret-like structures on the core particles (Grimes *et al.*, 1998; Prasad *et al.*, 1996) and the associated functions are most likely located in the particle interior.

Polymerase complex maturation

During genome encapsidation the polymerase complex undergoes substantial structural rearrangements and obtains its mature rounded appearance. Three conformations for the polymerase complex have been described (Butcher *et al.*, 1997): (1) the empty particle with fivefold vertices oriented inwards (Figure 7.2D), (2) an expanded dodecahedral particle (Figure 7.2F), and 3) the mature rounded dsRNA-filled core particle (Figure 7.2G). We assume that these three forms present the three different functional forms of the polymerase complex. The empty particle is active in genome packaging (Frilander and Bamford,

1995; Gottlieb *et al.*, 1991), packaging the three segments in order of size: s, m, and l (Frilander and Bamford, 1995; Qiao *et al.*, 1995a,b). Expansion to the dodecahedral shape (conformation 2) must occur during the packaging of the l segment to accommodate the three ssRNA genome segments within the particle (Poranen *et al.*, 2004). The packaging of the l segment also enables initiation of RNA replication (Frilander *et al.*, 1995; Poranen and Bamford, 1999). We have suggested earlier that the expansion of the particle and the initiation of RNA synthesis are coupled (Poranen *et al.*, 2004). In other words, the dodecahedral form of the polymerase complex is the RNA replication device (Figure 7.2F). The structural change in the P1 shell could, for example, enable the entry of the RNA 3′ termini into the template tunnel of the P2 polymerase (Butcher *et al.*, 2001; Makeyev and Bamford, 2000a) in order for the minus strand synthesis to commence. The third, rounded mature conformation of the polymerase complex (Figure 7.2G) is active in transcription (Ojala and Bamford, 1995). Transcription involves both proteins P2 and P4 (Kainov *et al.*, 2004; Makeyev and Bamford, 2000b; Pirttimaa *et al.*, 2002). P1 expansion to the transcriptionally active mature form could establish a new type of interaction between these two proteins that allows the initiation of transcription and the release of newly produced mRNA molecules.

Nucleocapsid shell assembly

The NC surface protein P8 can be purified from virions in a trimeric, highly α-helical form (Poranen *et al.*, 2001; Tuma *et al.*, 1999). Purified P8 reassembles onto both the virion-derived and the *in vitro* reconstituted, RNA-filled polymerase complexes (Figure 7.2H) (Olkkonen *et al.*, 1990; Olkkonen *et al.*, 1991; Poranen *et al.*, 2001). P8 can also self-assemble in the absence of other viral proteins into a polydisperse mixture of aberrant, shell-like, structures (M. Poranen and D. H. Bamford, unpublished observations). The self-assembly products of P8 are not topologically closed and appear as open spirals in electron micrographs (M. Poranen and D. H. Bamford, unpublished observations), indicating that the assembly of closed P8 shells is dependent on the core template. Protein VP7 from African horse sickness virus is also a trimer that forms a T = 13 shell around the virion core. In the absence of interactions with the underlying (T = 1) layer, VP7 self-assembles into flat hexagonal arrays (Burroughs *et al.*, 1994). Thus, interactions between the two inner virion layers (T = 1 and T = 13) of these dsRNA viruses seem to direct the correct assembly of the T = 13 shell. The assembly of the ϕ6 P8 shell onto the core template can occur at lower concentrations of P8 protein than the assembly without the template. This probably ensures that P8 preferentially assembles onto the cores (Olkkonen *et al.*, 1991). P8 trimers do not undergo detectable conformational changes during assembly (Tuma *et al.*, 1999), suggesting that P8 subunits assemble in a quasi-equivalent manner.

The assembly of the P8 shell onto the *in vitro* assembled and packaged polymerase complexes yields *in vitro* assembled NCs (Poranen *et al.*, 2001). These particles can infect host cell spheroplasts (i.e. cells with the outer membrane and peptidoglycan layers partially removed) and initiate a productive infection cycle (Ojala *et al.*, 1990; Poranen *et al.*, 2001). This demonstrates that the *in vitro* assembly products of ϕ6 are fully functional. Thus, the formation of bacteriophage ϕ6 NC does not require any phage- or host-encoded assembly factors and represents an example of self-assembly.

Envelopment of the nucleocapsid

While NC is being assembled, the viral envelope is being independently generated using a parallel assembly pathway (Figure 7.2). The envelope assembly occurs within the cell cytoplasm (Bamford *et al.*, 1976) with lipids derived from the host cell plasma membrane (Laurinavičius *et al.*, 2004). The phage membrane contains significantly more phosphatidylglycerol and less phosphatidylethanolamine than the host membrane (Laurinavičius *et al.*, 2004). Membrane vesicles containing viral proteins P6, P9, P10, and P13 have been detected in pulse-labelled ϕ6-infected cells. The orientation of protein in these vesicles is the same as in the virion (Stitt and Mindich, 1983a,b).

The non-structural protein P12 and the major viral membrane protein P9 are the minimal protein components needed for the formation of ϕ6-specific membrane vesicles and their localization onto the NC (Figure 7.2I) (Johnson and Mindich, 1994; Mindich and Lehman, 1983; Stitt and Mindich, 1983b). In the case of ϕ8, the membrane formation is also dependent on the integral membrane protein P10 and the spike-forming proteins P3a and P3b (Hoogstraten *et al.*, 2000).

The membrane of ϕ6 assembles and covers the NC in the absence of the spikes. The spike P3, which is anchored to the envelope via protein P6, is produced as a soluble protein and is the last component to be assembled onto the virions (Figure 7.2J) (Mindich *et al.*, 1979; Stitt and Mindich, 1983a).

Conditions for self-assembly

The specific biochemical requirements for each assembly step have arisen from optimization of the *in vitro* assembly reactions (Frilander and Bamford, 1995; Juuti *et al.*, 1998; Olkkonen *et al.*, 1991; Poranen *et al.*, 2001; van Dijk *et al.*, 1995). The ϕ6 NC is among the most complicated biological macromolecules for which self-assembly from purified components has been demonstrated to lead to a fully active product. The assembly of ϕ6 and ϕ8 PCs can be induced *in vitro* simply by adding polyethylene glycol to a mixture of the subunit proteins (Figure 7.2) (Kainov *et al.*, 2003a; Poranen *et al.*, 2001). In the case of ϕ6, particle formation is dependent on hexamerization of the P4 protein, which can be induced by adding any NTP or NDP and divalent cations (Juuti *et al.*, 1998). Genome packaging into the empty capsid is an energy-dependent process. The energy can be provided in the form of any NTP, dNTP, or ddNTP (Figure 7.2 D–E) (Frilander and Bamford, 1995). In addition, the optimal *in vitro* packaging conditions contain NH_4Ac and $MgCl_2$ (Frilander and Bamford, 1995). To initiate RNA synthesis, the rest of the NTPs must be included in the reaction mixture (Figure 7.2E–F). Low concentrations of NTPs support production of only minus-strands; transcription can be initiated by increasing the concentration of purine nucleotides (van Dijk *et al.*, 1995). Purified protein P8 assembles onto dsRNA-containing particles if calcium ions are added (Figure 7.2G–H) (Olkkonen *et al.*, 1991). Ca^{2+} ions can also halt transcription (van Dijk *et al.*, 1995), suggesting that cellular Ca^{2+} levels increase during ϕ6 infection. The detailed conditions for envelope formation and assembly onto the NC are currently unknown as there is no *in vitro* system to study these processes. Viral non-structural protein P12 (Mindich and Lehman, 1983) and most likely some cellular proteins are involved.

Efficiency of *in vitro* assembly

The structure and the functionality of the assembly products and the efficiency of assembly are important criteria when any *in vitro* assembly system is evaluated. These factors are also essential when the relevance of the *in vitro* results is interpreted. The functionality can be analyzed by different biochemical assays or, in the case of viral particles, by infectivity. Comparison to equivalent complexes formed *in vivo* measures the efficiency of *in vitro* assembly.

Examples of well-studied *in vitro* assembly systems include the bacterial ribosome (Nierhaus, 1991) and the tobacco mosaic virus (TMV) (Butler, 1999). The *Escherichia coli* ribosomal 30S subunit can be reconstituted from purified recombinant proteins (Culver and Noller, 2000). Approximately 45% of the input 16S rRNA and 10% of the input proteins are incorporated into the 30S subunits *in vitro*. The 30S subunits formed *in vitro* can then associate with the 50S ribosomal subunits and perform poly(U)-directed polyphenylanaline synthesis. The activity of the *in vitro* assembled complexes is approximately 34% of ribosomes that contain natural 30S subunits isolated from cells (Culver and Noller, 2000).

Fraenkel-Conrat and Williams (1955) were able to reconstitute infectious TMV virions from the TMV coat protein and the viral single-stranded genomic RNA. TMV particles do not have enzymatic activities, and the only measure for functionality of the assembled product is infectivity. Approximately 10% of the *in vitro* assembled particles appeared normal in electron microscopy, and 3% of the correct-sized TMV rods had infectivity similar to that of the control TMV particles. The infectivity of the viral RNA as such was about 0.1% of the control TMV.

In the ϕ6 system it is possible to produce and isolate different assembly intermediates produced within cells (PCs, cores, and NCs) and compare the *in vitro* assembly products to *in vivo* counterparts. The assembly pathway of ϕ6 includes several steps. The efficiency of these steps inside an infected cell is not known, but some estimates have been done for *in vitro* assembly reactions. Approximately 50% of proteins in the *in vitro* polymerase complex assembly reaction of ϕ6 and ϕ8 form particles (Kainov *et al.*, 2003a; Poranen *et al.*, 2001). The protein composition pattern of the *in vitro* assembled particles is the same as found in virions. Also based on their sedimentation properties and morphology, the *in vitro* assembled structures are indistinguishable from the recombinant particles purified from bacteria (Kainov *et al.*, 2003a; Poranen *et al.*, 2001). The PCs formed *in vitro* are enzymatically active and can package and replicate ssRNA with the same efficiency as the recombinant PCs purified from bacteria. Approximately 5% of the empty PCs can encapsidate all three genome segments (Frilander and Bamford, 1995; Poranen *et al.*, 2001). Some 5–10% of the genome-containing polymerase complexes become infectious following the P8 shell assembly reaction (Olkkonen *et al.*, 1991). The infectivity of the NCs demonstrates that the *in vitro* assembled polymerase complexes of ϕ6 can initiate a new round of productive infection within a host cell (Poranen *et al.*, 2001), thus confirming the biological functionality of the *in vitro* assembly product.

Acknowledgments

This work was supported by the Academy of Finland ['Finnish Centre of Excellence Program 2006–2011, grants 206926 (RT), 1213992 and 1213467 (DHB)].

References

Ago, H., Adachi, T., Yoshida, A., Yamamoto, M., Habuka, N., Yatsunami, K., and Miyano, M. (1999). Crystal structure of the RNA-dependent RNA polymerase of hepatitis C virus. Structure Fold. Des. *7*, 1417–26.

Bamford, D.H. (2003). Do viruses form lineages across different domains of life? Res. Microbiol. *154*, 231–6.

Bamford, D.H., and Mindich, L. (1980). Electron microscopy of cells infected with nonsense mutants of bacteriophage ϕ6. Virology *107*, 222–8.

Bamford, D.H., Palva, E.T., and Lounatmaa, K. (1976). Ultrastructure and life cycle of the lipid-containing bacteriophage ϕ6. J. Gen. Virol. *32*, 249–59.

Benevides, J.M., Juuti, J.T., Tuma, R., Bamford, D.H., and Thomas, G.J., Jr. (2002). Characterization of subunit-specific interactions in a double-stranded RNA virus, Raman difference spectroscopy of the ϕ6 procapsid. Biochemistry *41*, 11946–53.

Burroughs, J.N., O'Hara, R.S., Smale, C.J., Hamblin, C., Walton, A., Armstrong, R., and Mertens, P.P. (1994). Purification and properties of virus particles, infectious subviral particles, cores and VP7 crystals of African horsesickness virus serotype 9. J. Gen. Virol. *75*, 1849–57.

Butcher, S.J., Dokland, T., Ojala, P.M., Bamford, D.H., and Fuller, S.D. (1997). Intermediates in the assembly pathway of the double-stranded RNA virus ϕ6. EMBO J. *16*, 4477–87.

Butcher, S.J., Grimes, J.M., Makeyev, E.V., Bamford, D.H., and Stuart, D.I. (2001). A mechanism for initiating RNA-dependent RNA polymerization. Nature *410*, 235–40.

Butler, P.J. (1999). Self-assembly of tobacco mosaic virus, the role of an intermediate aggregate in generating both specificity and speed. Philos. Trans. R. Soc. Lond. B Biol. Sci. *354*, 537–50.

Culver, G.M., and Noller, H.F. (2000). *In vitro* reconstitution of 30S ribosomal subunits using complete set of recombinant proteins. Methods Enzymol *318*, 446–60.

Day, L.A., and Mindich, L. (1980). The molecular weight of bacteriophage ϕ6 and its nucleocapsid. Virology *103*, 376–85.

de Haas, F., Paatero, A.O., Mindich, L., Bamford, D.H., and Fuller, S.D. (1999). A symmetry mismatch at the site of RNA packaging in the polymerase complex of dsRNA bacteriophage ϕ6. J. Mol. Biol. *294*, 357–72.

Fraenkel-Conrat, H., and Williams, R.C. (1955). Reconstitution of active tobacco mosaic virus from its inactive protein and nucleic acid components. Proc. Natl Acad. Sci. USA *41*, 690–8.

Frilander, M., and Bamford, D.H. (1995). *In vitro* packaging of the single-stranded RNA genomic precursors of the segmented double-stranded RNA bacteriophage ϕ6, the three segments modulate each other's packaging efficiency. J. Mol. Biol. *246*, 418–28.

Frilander, M., Gottlieb, P., Strassman, J., Bamford, D.H., and Mindich, L. (1992). Dependence of minus-strand synthesis on complete genomic packaging in the double-stranded RNA bacteriophage ϕ6. J. Virol. *66*, 5013–7.

Frilander, M., Poranen, M., and Bamford, D.H. (1995). The large genome segment of dsRNA bacteriophage ϕ6 is the key regulator in the *in vitro* minus and plus strand synthesis. RNA *1*, 510–8.

Gottlieb, P., Metzger, S., Romantschuk, M., Carton, J., Strassman, J., Bamford, D.H., Kalkkinen, N., and Mindich, L. 1988a. Nucleotide sequence of the middle dsRNA segment of bacteriophage ϕ6, placement of the genes of membrane-associated proteins. Virology *163*, 183–90.

Gottlieb, P., Qiao, X., Strassman, J., Frilander, M., and Mindich, L. (1994). Identification of the packaging regions within the genomic RNA segments of bacteriophage ϕ6. Virology *200*, 42–7.

Gottlieb, P., Strassman, J., Bamford, D.H., and Mindich, L. 1988b. Production of a polyhedral particle in *Escherichia coli* from a cDNA copy of the large genomic segment of bacteriophage ϕ6. J. Virol. *62*, 181–7.

Gottlieb, P., Strassman, J., Frucht, A., Qiao, X.Y., and Mindich, L. (1991). *In vitro* packaging of the bacteriophage ϕ6 ssRNA genomic precursors. Virology *181*, 589–94.

Gottlieb, P., Strassman, J., and Mindich, L. (1992). Protein P4 of the bacteriophage ϕ6 procapsid has a nucleoside triphosphate-binding site with associated nucleoside triphosphate phosphohydrolase activity. J. Virol. *66*, 6220–2.

Gottlieb, P., Strassman, J., Qiao, X.Y., Frucht, A., and Mindich, L. (1990). *In vitro* replication, packaging, and transcription of the segmented double-stranded RNA genome of bacteriophage ϕ6, studies with procapsids assembled from plasmid-encoded proteins. J. Bacteriol. *172*, 5774–82.

Gouet, P., Diprose, J.M., Grimes, J.M., Malby, R., Burroughs, J.N., Zientara, S., Stuart, D.I., and Mertens, P.P. (1999). The highly ordered double-stranded RNA genome of bluetongue virus revealed by crystallography. Cell 97, 481–90.

Grimes, J.M., Burroughs, J.N., Gouet, P., Diprose, J.M., Malby, R., Zientara, S., Mertens, P.P., and Stuart, D.I. (1998). The atomic structure of the bluetongue virus core. Nature 395, 470–8.

Hoogstraten, D., Qiao, X., Sun, Y., Hu, A., Onodera, S., and Mindich, L. (2000). Characterization of ϕ8, a bacteriophage containing three double-stranded RNA genomic segments and distantly related to ϕ6. Virology 272, 218–24.

Huang, H., Chopra, R., Verdine, G.L., and Harrison, S.C. (1998). Structure of a covalently trapped catalytic complex of HIV-1 reverse transcriptase, implications for drug resistance. Science 282, 1669–75.

Huiskonen, J.T., de Haas, F., Bubeck, D., Bamford, D.H., Fuller, S.D., and Butcher, S.J. (2006). Structure of the bacteriophage ϕ6 nucleocapsid suggests a mechanism for sequential RNA packaging. Structure 14, 1039–48.

Ikonen, T., Kainov, D.E., Serimaa, R.E., Timmins, P., and Tuma, R. (2003). Localization of replication machinery within the polymerase complex of dsRNA bacteriophage ϕ6. J. Appl. Cryst. 36, 525-529.

Johnson, M.D. 3rd, and Mindich, L. (1994). Plasmid-directed assembly of the lipid-containing membrane of bacteriophage ϕ6. J. Bacteriol. 176, 4124–32.

Juuti, J.T., and Bamford, D.H. (1997). Protein P7 of phage ϕ6 RNA polymerase complex, acquiring of RNA packaging activity by *in vitro* assembly of the purified protein onto deficient particles. J. Mol. Biol. 266, 891–900.

Juuti, J.T., and Bamford, D.H. (1995). RNA binding, packaging and polymerase activities of the different incomplete polymerase complex particles of dsRNA bacteriophage ϕ6. J. Mol. Biol. 249, 545–54.

Juuti, J.T., Bamford, D.H., Tuma, R., and Thomas, G.J., Jr. (1998). Structure and NTPase activity of the RNA-translocating protein (P4) of bacteriophage ϕ6. J. Mol. Biol. 279, 347–59.

Kainov, D.E., Butcher, S.J., Bamford, D.H., and Tuma, R. 2003a. Conserved intermediates on the assembly pathway of double-stranded RNA bacteriophages. J. Mol. Biol. 328, 791–804.

Kainov, D.E., Lísal, J., Bamford, D.H., and Tuma, R. (2004). Packaging motor from double-stranded RNA bacteriophage ϕ12 acts as an obligatory passive conduit during transcription. Nucleic Acids Res. 32, 3515–21.

Kainov, D.E., Pirttimaa, M., Tuma, R., Butcher, S.J., Thomas, G.J., Jr., Bamford, D.H., and Makeyev, E.V. 2003b. RNA packaging device of double-stranded RNA bacteriophages, possibly as simple as hexamer of P4 protein. J. Biol. Chem. 278, 48084–91.

Koivunen, M.R.L., Sarin, L.P., and Bamford, D.H. (2007). Structure-function insights into the RNA-dependent RNA polymerase of the dsRNA bacteriophage ϕ6. In: Segmented double-stranded RNA viruses: structure and molecular biology, J.T. Patton, ed. (Caister Academic Press, Norfolk), pp. 239–58.

Koonin, E.V., Gorbalenya, A.E., and Chumakov, K.M. (1989). Tentative identification of RNA-dependent RNA polymerases of dsRNA viruses and their relationship to positive strand RNA viral polymerases. FEBS Lett. 252, 42–6.

Ktistakis, N.T., and Lang, D. (1987). The dodecahedral framework of the bacteriophage ϕ6 nucleocapsid is composed of protein P1. J. Virol. 61, 2621–3.

Laurinavičius, S., Käkelä, R., Bamford, D.H., and Somerharju, P. (2004). The origin of phospholipids of the enveloped bacteriophage ϕ6. Virology 326, 182–90.

Lawton, J.A., Estes, M.K., and Prasad, B.V. (1997). Three-dimensional visualization of mRNA release from actively transcribing rotavirus particles. Nat. Struct Biol. 4, 118–21.

Lisal, J., Kainov, D.E., Bamford, D.H., Thomas, G.J., Jr., and Tuma, R. (2004). Enzymatic mechanism of RNA translocation in double-stranded RNA bacteriophages. J. Biol. Chem. 279, 1343–50.

Luongo, C.L., Zhang, X., Walker, S.B., Chen, Y., Broering, T.J., Farsetta, D.L., Bowman, V.D., Baker, T.S., and Nibert, M.L. (2002). Loss of activities for mRNA synthesis accompanies loss of λ2 spikes from reovirus cores, an effect of λ2 on λ1 shell structure. Virology 296, 24–38.

Makeyev, E.V., and Bamford, D.H. 2000a. Replicase activity of purified recombinant protein P2 of double-stranded RNA bacteriophage ϕ6. EMBO J. 19, 124–33.

Makeyev, E.V., and Bamford, D.H. 2000b. The polymerase subunit of a dsRNA virus plays a central role in the regulation of viral RNA metabolism. EMBO J. 19, 6275–84.

Mancini, E.J., and Tuma, R. (2007). Structure and function of P4, a dsRNA virus packaging motor. In: Segmented double-stranded RNA viruses: structure and molecular biology, J.T. Patton, ed. (Caister Academic Press, Norfolk), pp. 259–74.

Mancini, E.J., Kainov, D.E., Grimes, J.M., Tuma, R., Bamford, D.H., and Stuart, D.I. (2004). Atomic snapshots of an RNA packaging motor reveal conformational changes linking ATP hydrolysis to RNA translocation. Cell *118*, 743–55.

Mao, Z.X., and Joklik, W.K. (1991). Isolation and enzymatic characterization of protein λ2, the reovirus guanylyltransferase. Virology *185*, 377–86.

McGraw, T., Mindich, L., and Frangione, B. (1986). Nucleotide sequence of the small double-stranded RNA segment of bacteriophage ϕ6, novel mechanism of natural translational control. J. Virol. *58*, 142–51.

Mertens, P. (2004). The dsRNA viruses. Virus Res. *101*, 3–13.

Mindich, L. (2007). Genomic RNA packaging and replication in the *Cystoviridae*. In: Segmented double-stranded RNA viruses: structure and molecular biology, J.T. Patton, ed. (Caister Academic Press, Norfolk), pp. 351–70.

Mindich, L., and Davidoff-Abelson, R. (1980). The characterization of a 120 S particle formed during ϕ6 infection. Virology *103*, 386–91.

Mindich, L., and Lehman, J. (1983). Characterization of ϕ6 mutants that are temperature sensitive in the morphogenetic protein P12. Virology *127*, 438–45.

Mindich, L., Lehman, J., and Huang, R. (1979). Temperature-dependent compositional changes in the envelope of ϕ6. Virology *97*, 171–6.

Mindich, L., Nemhauser, I., Gottlieb, P., Romantschuk, M., Carton, J., Frucht, S., Strassman, J., Bamford, D.H., and Kalkkinen, N. (1988). Nucleotide sequence of the large double-stranded RNA segment of bacteriophage ϕ6, genes specifying the viral replicase and transcriptase. J. Virol. *62*, 1180–5.

Mindich, L., Qiao, X., Onodera, S., Gottlieb, P., and Strassman, J. (1992). Heterologous recombination in the double-stranded RNA bacteriophage ϕ6. J. Virol. *66*, 2605–10.

Mindich, L., Qiao, X., Qiao, J., Onodera, S., Romantschuk, M., and Hoogstraten, D. (1999). Isolation of additional bacteriophages with genomes of segmented double-stranded RNA. J. Bacteriol. *181*, 4505–8.

Naitow, H., Tang, J., Canady, M., Wickner, R.B., and Johnson, J.E. (2002). L-A virus at 3.4 Å resolution reveals particle architecture and mRNA decapping mechanism. Nat. Struct Biol. *9*, 725–8.

Nason, E.L., Samal, S.K., and Venkataram Prasad, B.V. (2000). Trypsin-induced structural transformation in aquareovirus. J. Virol. *74*, 6546–55.

Newcomb, W.W., Juhas, R.M., Thomsen, D.R., Homa, F.L., Burch, A.D., Weller, S.K., and Brown, J.C. (2001). The UL6 gene product forms the portal for entry of DNA into the herpes simplex virus capsid. J. Virol. *75*, 10923–32.

Nierhaus, K.H. (1991). The assembly of prokaryotic ribosomes. Biochimie *73*, 739–55.

Ojala, P.M., and Bamford, D.H. (1995). *In vitro* transcription of the double-stranded RNA bacteriophage ϕ6 is influenced by purine NTPs and calcium. Virology *207*, 400–8.

Ojala, P.M., Romantschuk, M., and Bamford, D.H. (1990). Purified ϕ6 nucleocapsids are capable of productive infection of host cells with partially disrupted outer membranes. Virology *178*, 364–72.

Olkkonen, V.M., and Bamford, D.H. (1987). The nucleocapsid of the lipid-containing double-stranded RNA bacteriophage ϕ6 contains a protein skeleton consisting of a single polypeptide species. J. Virol. *61*, 2362–7.

Olkkonen, V.M., Gottlieb, P., Strassman, J., Qiao, X.Y., Bamford, D.H., and Mindich, L. (1990). *In vitro* assembly of infectious nucleocapsids of bacteriophage ϕ6, formation of a recombinant double-stranded RNA virus. Proc. Natl Acad. Sci. USA *87*, 9173–7.

Olkkonen, V.M., Ojala, P.M., and Bamford, D.H. (1991). Generation of infectious nucleocapsids by *in vitro* assembly of the shell protein on to the polymerase complex of the dsRNA bacteriophage ϕ6. J. Mol. Biol. *218*, 569–81.

Orlova, E.V., Gowen, B., Droge, A., Stiege, A., Weise, F., Lurz, R., van Heel, M., and Tavares, P. (2003). Structure of a viral DNA gatekeeper at 10 Å resolution by cryo-electron microscopy. EMBO J. *22*, 1255–62.

Paatero, A.O., Mindich, L., and Bamford, D.H. (1998). Mutational analysis of the role of nucleoside triphosphatase P4 in the assembly of the RNA polymerase complex of bacteriophage ϕ6. J. Virol. *72*, 10058–65.

Paatero, A.O., Syvaoja, J.E., and Bamford, D.H. (1995). Double-stranded RNA bacteriophage ϕ6 protein P4 is an unspecific nucleoside triphosphatase activated by calcium ions. J. Virol. *69*, 6729–34.

Patton, J.T., and Spencer, E. (2000). Genome replication and packaging of segmented double-stranded RNA viruses. Virology *277*, 217–25.

Pirttimaa, M.J., and Bamford, D.H. (2000). RNA secondary structures of the bacteriophage ϕ6 packaging regions. RNA *6*, 880–9.

Pirttimaa, M.J., Paatero, A.O., Frilander, M.J., and Bamford, D.H. (2002). Nonspecific nucleoside triphosphatase P4 of double-stranded RNA bacteriophage ϕ6 is required for single-stranded RNA packaging and transcription. J. Virol. *76*, 10122–7.

Poranen, M.M., and Bamford, D.H. (1999). Packaging and replication regulation revealed by chimeric genome segments of double-stranded RNA bacteriophage ϕ6. RNA *5*, 446–54.

Poranen, M.M., and Bamford, D.H. (2007). Entry of a segmented dsRNA virus into the bacterial cell. In: Segmented double-stranded RNA viruses: structure and molecular biology, J.T. Patton, ed. (Caister Academic Press, Norfolk), pp. 215–26.

Poranen, M.M., Paatero, A.O., Tuma, R., and Bamford, D.H. (2001). Self-assembly of a viral molecular machine from purified protein and RNA constituents. Mol. Cell *7*, 845–54.

Poranen, M.M., Pirttimaa, M.J., and Bamford, D.H. (2004). Encapsidation of the segmented double-stranded RNA genome of bacteriophage ϕ6. In, Viral genome packaging machines, genetics, structure, and mechanism. C.E. Catalano, ed. Landes Bioscience, Georgetown. p. 117–34.

Prasad, B.V., Rothnagel, R., Zeng, C.Q., Jakana, J., Lawton, J.A., Chiu, W., and Estes, M.K. (1996). Visualization of ordered genomic RNA and localization of transcriptional complexes in rotavirus. Nature *382*, 471–3.

Prevelige, P.E., Jr., Thomas, D., and King, J. (1993). Nucleation and growth phases in the polymerization of coat and scaffolding subunits into icosahedral procapsid shells. Biophys J. *64*, 824–35.

Qiao, J., Qiao, X., Sun, Y., and Mindich, L. (2003). Isolation and analysis of mutants of double-stranded-RNA bacteriophage ϕ6 with altered packaging specificity. J. Bacteriol. *185*, 4572–7.

Qiao, X., Casini, G., Qiao, J., and Mindich, L. 1995a. *In vitro* packaging of individual genomic segments of bacteriophage ϕ6 RNA, serial dependence relationships. J. Virol. *69*, 2926–31.

Qiao, X., Qiao, J., and Mindich, L. (2003). Analysis of specific binding involved in genomic packaging of the double-stranded-RNA bacteriophage ϕ6. J. Bacteriol. *185*, 6409–14.

Qiao, X., Qiao, J., and Mindich, L. (1995b). Interference with bacteriophage ϕ6 genomic RNA packaging by hairpin structures. J. Virol. *69*, 5502–5.

Reinisch, K.M., Nibert, M.L., and Harrison, S.C. (2000). Structure of the reovirus core at 3.6 Å resolution. Nature *404*, 960–7.

Simpson, A.A., Tao, Y., Leiman, P.G., Badasso, M.O., He, Y., Jardine, P.J., Olson, N.H., Morais, M.C., Grimes, S., Anderson, D.L., Baker, T.S., and Rossmann, M.G. (2000). Structure of the bacteriophage ϕ29 DNA packaging motor. Nature *408*, 745–50.

Sinclair, J.F., Tzagoloff, A., Levine, D., and Mindich, L. (1975). Proteins of bacteriophage ϕ6. J. Virol. *16*, 685–95.

Stitt, B.L., and Mindich, L. 1983a. The structure of bacteriophage ϕ6, protease digestion of ϕ6 virions. Virology *127*, 459–62.

Stitt, B.L., and Mindich, L. 1983b. Morphogenesis of bacteriophage ϕ6, a presumptive viral membrane precursor. Virology *127*, 446–58.

Sun, Y., Qiao, X., Qiao, J., Onodera, S., and Mindich, L. (2003). Unique properties of the inner core of bacteriophage ϕ8, a virus with a segmented dsRNA genome. Virology *308*, 354–61.

Tao, Y., Farsetta, D.L., Nibert, M.L., and Harrison, S.C. (2002). RNA synthesis in a cage–structural studies of reovirus polymerase λ3. Cell *111*, 733–45.

Tuma, R., Bamford, J.K., Bamford, D.H., and Thomas, G.J., Jr. (1999). Assembly dynamics of the nucleocapsid shell subunit (P8) of bacteriophage ϕ6. Biochemistry *38*, 15025–33.

Valpuesta, J.M., and Carrascosa, J.L. (1994). Structure of viral connectors and their function in bacteriophage assembly and DNA packaging. Q Rev. Biophys *27*, 107–55.

van Dijk, A.A., Frilander, M., and Bamford, D.H. (1995). Differentiation between minus- and plus-strand synthesis, polymerase activity of dsRNA bacteriophage ϕ6 in an *in vitro* packaging and replication system. Virology *211*, 320–3.

Van Etten, J.V., Lane, L., Gonzalez, C., Partridge, J., and Vidaver, A. (1976). Comparative properties of bacteriophage ϕ6 and ϕ6 nucleocapsid. J. Virol. *18*, 652–8.

Vidaver, A.K., Koski, R.K., and Van Etten, J.L. (1973). Bacteriophage ϕ6, a lipid-containing virus of *Pseudomonas phaseolicola*. J. Virol. *11*, 799–805.
Zhang, X., Walker, S.B., Chipman, P.R., Nibert, M.L., and Baker, T.S. (2003). Reovirus polymerase λ3 localized by cryo-electron microscopy of virions at a resolution of 7.6 Å. Nat. Struct Biol. *10*, 1011–8.

Infectious Bursal Disease Virus: A Segmented Double-Stranded RNA Virus With a T = 13 Capsid that Lacks a T = 1 Core

8

José R. Castón, José F. Rodríguez, and José L. Carrascosa

Abstract

Infectious Bursal Disease Virus (IBDV) is the best-characterized member of the family *Birnaviridae*. These viruses have bipartite dsRNA genomes enclosed in single-layered icosahedral capsids with T = 13*l* geometry. IBDV shares functional strategies and structural features with many other icosahedral dsRNA viruses, except that it lacks the T = 1 (or pseudo T = 2) core common to the *Reoviridae*, *Cystoviridae*, and *Totiviridae*. Interestingly, the IBDV capsid protein exhibits structural domains homologous to those of the capsid proteins of some positive ssRNA viruses, such as the nodaviruses and tetraviruses. The T = 13 shell of the IBDV capsid is formed by trimers of VP2, a protein generated by removal of the C-terminal domain from its precursor, pVP2. The trimming of pVP2 is performed on immature particles as part of the maturation process. The other major structural protein, VP3, is a multifunctional component that influences the inherent structural polymorphism of pVP2. The virus-encoded RNA-dependent RNA polymerase, VP1, is incorporated into the capsid through its association with VP3. VP3 also interacts extensively with the viral dsRNA genome, an organization that probably contributes to virion stability. The role of pVP2 and VP3 specific sequences in capsid assembly has been studied using heterologous expression systems. Such analyses have shown that the formation of the T = 13 capsid involves the interaction of an amphiphilic α-helix located at the C-terminal domain of pVP2 with a specific sequence at the C-terminus of VP3. X-ray crystallography has provided additional insight into the structure and interaction of VP2 trimers, the basic building block of the IBDV particle.

Introduction

Double-stranded (ds) RNA viruses are a ubiquitous group of viruses that infect practically all organisms, from the mammalian reoviruses to the bacteriophage ϕ6, including fungal viruses. One of the most extended features amongst dsRNA viruses is the presence of a T = 1 core that surrounds the dsRNA genome. This core remains undisturbed even upon entry of the virus into the cell, thus protecting the dsRNA genome from host defence mechanisms. The presence of RNA-dependent RNA polymerase within the core also allows the structure to act as a machine that drives and organizes RNA synthesis. Specifically, during transcription, RNA polymerases in the core can direct the synthesis of viral mRNAs from the dsRNA genome, via a process in which the structural integrity of the core is retained. During the replication cycle, RNA polymerases that are associated with new-formed cores direct minus-strand synthesis using viral mRNAs as templates, producing dsRNA genome segments. The structure of the core is unique, representing an icosahedron with T = 1 architecture built from 60 asymmetric dimers of a single protein (alternatively referred to as a 'T = 2' organization).

So far, the most profound exception to the rule that all dsRNA viruses contain a T = 1 core is illustrated by the *Birnaviridae*, a family that includes Infectious Bursal Disease Virus (IBDV), a prototypic member belonging to the genus *Avibirnavirus*. Instead of a T = 1 protein shell, the bipartite dsRNA genome of IBDV is surrounded by a T = 13 shell of protein. Another notable exception to this rule is provided by *Penicillium chrysogenum virus* (PcV) that has an authentic T = 1 capsid, formed from 60 equivalent capsid monomers (Castón *et al.*, 2003). Interestingly, sequence analysis suggests that the PcV capsid monomer consists of two halves, that show some similarity with each other, as though the monomer has been formed by gene duplication. If this is the case, then the PcV capsid monomer may be pictured as representing two covalently linked capsid subunits, thereby making the PcV capsid a T = 1 structure formed by 60 covalently linked dimers.

Pathogenesis

IBDV infects birds and is an important pathogen of chickens and other domestic fowl of economic importance. It is transmitted without intermediate vectors and destroys the bursa of Fabricius of young birds often causing death. Periodic outbreaks of highly virulent strains have stimulated research in the development of improved vaccination approaches for controlling and preventing disease.

Segmented dsRNA genome

The IBDV genome is composed of two dsRNA segments, A and B (Dobos *et al.*, 1979) (Figure 8.1). Transcription of the viral dsRNA is carried out by semi-conservative strand displacement. Segment A (3.2 kbp) contains two partially overlapping open reading frames (ORFs). The largest ORF encodes a 110-kDa polyprotein that undergoes auto-cleavage by the viral protease VP4, yielding pVP2 (54.4 kDa), VP3 (29 kDa), and VP4 (26.7 kDa). The C-terminal 71 residues of pVP2 (pVP2 C-terminal domain, pVP2CTD), are removed during further processing, rendering the mature VP2 (47 kDa). VP4 is a non-canonical Ser-Lys Lon protease (Birghan *et al.*, 2000), and its atomic structure has been recently determined (Feldman *et al.*, 2006). The smaller ORF of segment A codes for VP5 (15 kDa), a non-essential polypeptide, that is involved in virus dissemination (Lombardo *et al.*, 2000) and that inhibits apoptosis at the early stage of virus infection (Liu and Vakharia, 2006). Segment B (2.8 kbp) codes for VP1 (97.8 kDa), the viral RNA-dependent RNA polymerase (von Einem *et al.*, 2004).

Capsid structure

IBDV infectious particles assemble and accumulate in the cell cytoplasm, and consist of two major proteins (VP2 and VP3) and one (VP1) or two (VP4) minor proteins. VP2 represents the main marker for virulence and pathogenesis, and also induces neutralizing antibodies (Brandt *et al.*, 2001). The virions have a single-shelled capsid with an external

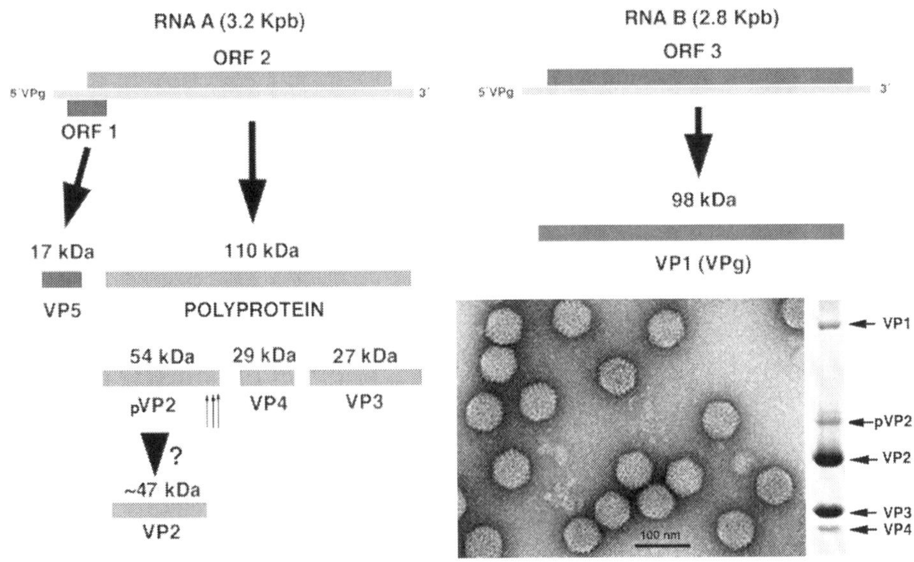

Figure 8.1 Genome organization and expression of infectious bursal disease virus (IBDV). Shown here is an electron micrograph of negatively stained virions of IBDV, and its structural polypeptides in purified virus particles.

diameter of 65–70 nm that is formed from 780 quasi-equivalent VP2 subunits, organized with T = 13*l* architecture (Figure 8.2) (Böttcher *et al.*, 1997; Castón *et al.*, 2001; Coulibaly *et al.*, 2005). The VP2 subunits are arranged as 260 trimers in the capsid, producing 12 pentamers and 120 hexamers, and from a geometrical perspective, exist in five distinct conformations (a–e). Whether protein VP4 is a true structural component of the virion is, as yet, unclear (Granzow *et al.*, 1997). In IBDV-infected cells, VP4 forms tubular structures (type II tubules) with a diameter of about 25 nm.

The architecture of the IBDV capsid is very similar to the outer capsids of many other dsRNA viruses in that they share T = 13*l* organization. However, with the exception of IBDV and the other birnaviruses, all dsRNA viruses share the presence of an internal core with a characteristic T = 1 architecture made of protein dimers. Such cores are even found in dsRNA fungal viruses, such as LA and P4 (Castón *et al.*, 1997) and Hv190SV (Caston *et al.*, 2006), as well as cypoviruses, insect viruses of the *Reoviridae* family (Hill *et al.*, 1999) all which lack an external T = 13 capsid. Until recently, the T = 1 core, as a functional unit involved with genome transcription and replication, was considered to be an essential component of dsRNA viruses. However, the existence of the birnaviruses clearly establish that a segmented dsRNA genome can be replicated without the use of a T = 1 core. This novelty of the birnaviruses has prompted a number of analyses not only aimed at better understanding the architecture of these dsRNA viruses but also at better understanding how a segmented dsRNA genome can be replicated and transcribed in the absence of a T = 1 nucleocapsid core.

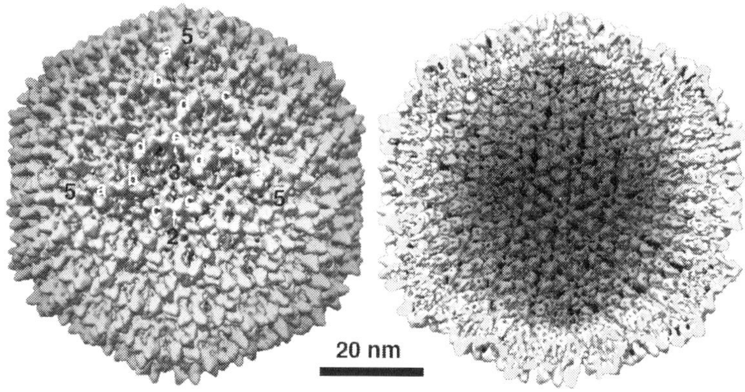

Figure 8.2 Three-dimensional structure of IBDV. The surface-shaded representation, viewed along a three-fold axis of icosahedral symmetry, shows a T = 13l structure built by five classes of VP2 trimers on the outer surface, labelled a to e (left). A model with the front half of the protein shell removed is shown (right). (See also Plate 8.2.)

Capsid assembly

One of the most successful approaches to understanding the role of IBDV proteins in particle assembly has been the use of heterologous expression systems capable of supporting the formation of virus-like particles (VLPs) (Chevalier *et al.*, 2002; Da Costa *et al.*, 2002; Fernandez-Arias *et al.*, 1998; Lombardo *et al.*, 1999; Martínez-Torrecuadrada *et al.*, 2000). In the following paragraphs, we will briefly review the findings obtained with such expression systems.

Expression of the IBDV polyprotein in mammalian cells using an inducible recombinant vaccinia virus resulted in the correct processing of the polyprotein and the accumulation of abundant levels of VLPs, composed of variable amounts of pVP2 and VP2, in the cytoplasm. The size and shape of these VLPs are similar to those of authentic IBDV particles (Fernandez-Arias *et al.*, 1998; Lombardo *et al.*, 1999), demonstrating the feasibility and authenticity of this expression system (Figure 8.3). Smaller capsids were also observed, representing a T = 7 arrangement of equivalent trimeric capsomers. Unlike the T = 13 capsid, with its five classes of trimeric capsomers, the T = 7 icosahedron displayed only three different classes.

The expression of the IBDV polyprotein in insect cells using the baculovirus expression system also results in the processing and accumulation of pVP2, VP3 and VP4. However, the assemblies produced with this expression system were structurally varied and, in general, the formation of IBDV-like particles was rather inefficient. The use of a FastBac transfer vector to generate a baculovirus encoding the polyprotein led to the production of rigid tubular structures (type I tubules), with diameters of ~50 nm, in insect cells. These tubules, made up of pVP2, exhibited an equivalent hexagonal arrangement of trimer-clustered units, similar to the type I tubules observed in IBDV virions (Martínez-Torrecuadrada *et al.*, 2000). A comparable result has been obtained with other baculovirus-based constructs (Chevalier *et al.*, 2002). Overexpression from yet another baculovirus led to the assembly of VLPs, flexible tubules, and intermediate assembly products formed

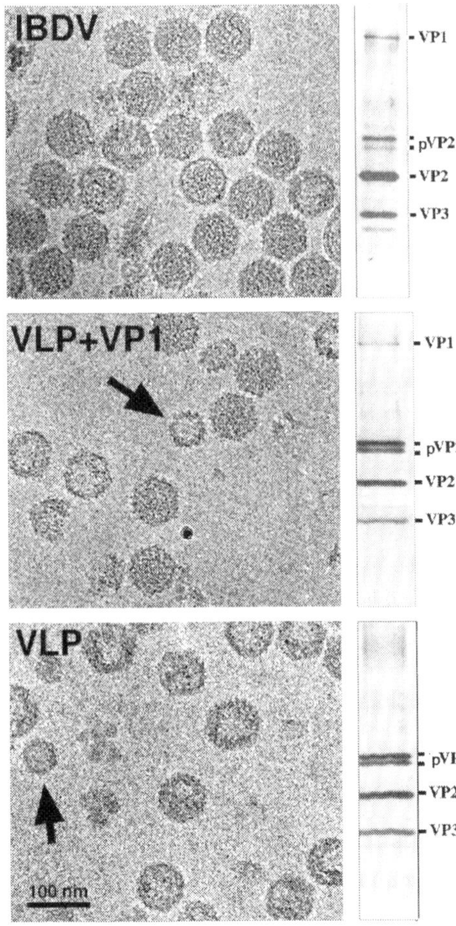

Figure 8.3 Cryoelectron microscopy of purified IBDV virions (IBDV), virus-like particles (VLPs) obtained by co-expression of the IBDV polyprotein and VP1 (VLP+VP1), and VLPs obtained by expression of the IBDV polyprotein (VLP). A small fraction of the VLPs are smaller and follows icosahedral symmetry (arrows). Their structural polypeptides are shown on the right.

by icosahedral caps elongated in tubules. All these structures consisted of pVP2 and variable amounts of VP2 and VP3 (Martínez-Torrecuadrada *et al.*, 2000) (Figure 8.4). Remarkably, deletion or replacement of the last C-terminal residue of VP3 stimulated the production of T = 13 VLPs (see below). In comparison with the vaccinia-based expression system (Lombardo *et al.*, 1999), our results with the baculovirus expression system supports the idea that insect cells have important cellular factor(s) that interfere negatively with VLP assembly (e.g. cellular proteins and/or host environment). This host-dependent effect must be taken into account when analyzing the assembly pathway of the IBDV particle using heterologous expression systems (Lee *et al.*, 2006b).

Expression of the mature-form of VP2 using different baculovirus-based vectors results in the formation of isometric particles of ~26 nm diameter (Castón *et al.*, 2001;

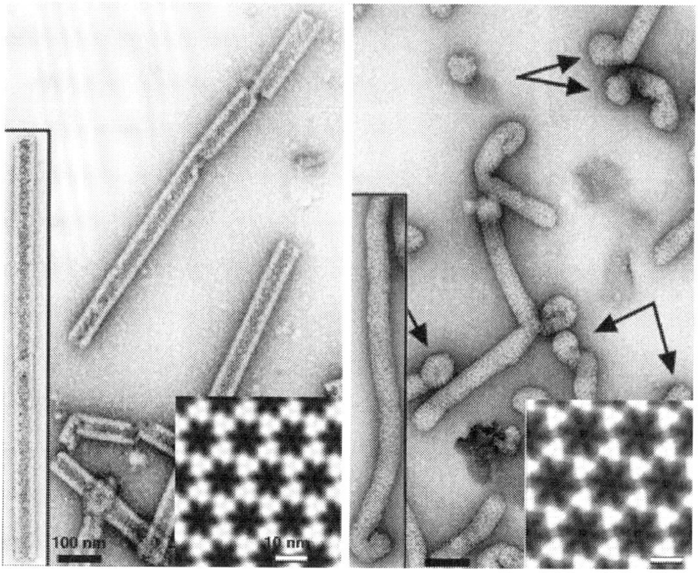

Figure 8.4 Electron microscopy of IBDV-related assemblies obtained with different recombinant baculoviruses. Structural analysis of IBDV type I tubules (left) and long capped flexible tubular structures (right) reveal (p)VP2 trimers arranged only as hexamers, but similar to VP2 trimers in the virion capsid. Caps at the end of the tubes are indicated (arrows).

Coulibaly *et al.*, 2005; Lee *et al.*, 2003). Examination of these subviral particles (SVP) showed that they were dodecameric T = 1 structures, composed of 20 trimers of VP2 (Figure 8.5) (Castón *et al.*, 2001). This result indicates that VP2 trimers alone have the intrinsic information required for their assembly into pentameric units, but not hexamer units. This suggests that VP2 requires additional information (or interactions) to form the hexameric components of icosahedral particles with a T > 1. The atomic structure of the T = 1 SVP has been solved by X-ray crystallography (Coulibaly *et al.*, 2005; Garriga *et al.*, 2006; Lee *et al.*, 2006a). The VP2 subunit is folded into three domains named projection (P), shell (S), and base (B). Domains S and P are β barrels with a jelly roll topology, oriented such that their β-strands are oriented tangentially and radially, respectively, to the spherical particle. The B domain consists of α-helices formed from extensions of the N- and C-termini of VP2, facing the inner surface of the capsid (Figure 8.5). Together, domains S and B exhibit remarkable structural similarity to the capsid proteins of nodavirus and tetravirus, two positive-sense ssRNA viruses (Coulibaly *et al.*, 2005). Besides lateral interactions between the VP2 trimers, SVPs are stabilized by Ca^{2+} ions located at the three-fold icosahedral axes that act as a glue in linking together the three subunits of each trimer. Further stabilization is accomplished by the long C-terminal helix (α4) of the VP2 subunit that is projected away from the core of the protein, oriented towards the three-fold axis of the neighbouring VP2 trimer (Garriga *et al.*, 2006).

pVP2 processing

In IBDV-infected cells, VP2 is derived from the precursor pVP2. The post-translational maturation of pVP2 (512 amino acids) into VP2 (441 amino acids) requires two cleavage

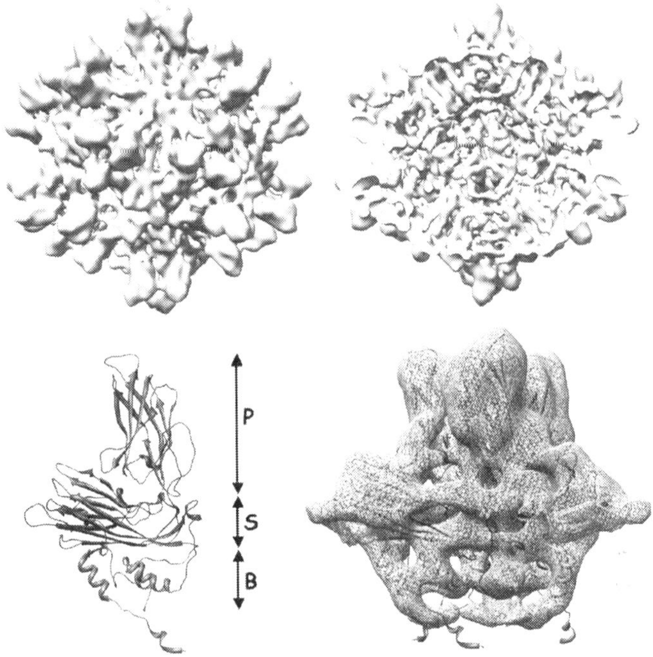

Figure 8.5 Three-dimensional structure of the VP2 subviral particle. Shown are surface-shaded representations of the of the outer (top, left) and inner (top, right) surfaces of VP2 capsids viewed along a twofold axis of icosahedral symmetry. The VP2 atomic model is shown (bottom, left), and the secondary structure elements are colour-coded by domain (red, P; blue, S; green, B). A VP2 trimer atomic model fitted into the corresponding density in the cryo-EM map is shown on the bottom right. (See also Plate 8.5.)

events, the first mediated by VP4 and the second by an unknown protease. The first event involves the cleavage of three Ala–Ala linkages located within the pVP2 CTD (positions 487, 494 and 501) (Sánchez and Rodríguez, 1999). Afterwards, the intermediate form of pVP2 is cleaved by an unknown mechanism between residues 441–442, yielding the mature VP2. The post-translational proteolysis of pVP2 seems to be designed to avoid the intrinsic tendency of mature VP2 to mistakenly assemble into T = 1 SVPs. Regardless, the maturation of pVP2 to VP2 is a process that is linked to the assembly of the T = 13 capsid (Chevalier *et al.*, 2002), and the small polypeptides generated during the trimming of pVP2 remain incorporated in the progeny viral particles (Da Costa *et al.*, 2002). It has been suggested that these short peptides might play a role in the life cycle of IBDV (Chevalier *et al.*, 2005). This post-assembly pVP2 maturation pathway has also been demonstrated for Infectious Pancreatic Necrosis Virus (IPNV), another member of the *Birnaviridae* family (Villanueva *et al.*, 2004).

The difference in the assembly behaviours of VP2 and pVP2 suggests that information affecting inter-subunit interactions is either contained in pVP2 CTD or is induced by the interaction of pVP2 CTD with additional viral components. The expression of the precursor pVP2 in the absence of any other viral protein, using a baculovirus-based system, has shed light about its specific role in capsid formation. pVP2 alone is organized in low ordered tubular structures consistent with a hexagonal lattice involving the presence of hexameric arrangement of trimers. On the other hand, it has been shown that the pVP2 expression using another baculovirus construction (Chevalier *et al.*, 2002) assembled not only tubes, but also isometric particles with diameters of around 30 nm. Nevertheless, the heterogeneous nature of these particles suggests that they may be remnants of degraded tubes. It is tempting to suggest that pVP2 has information to assemble into trimers that can arrange mainly into hexameric arrays, but they lack sufficient information to produce the complex interaction pattern required for building a T = 13 capsid. Taken together, these results clearly suggest that (i) the pVP2CTD plays an important role in determining different conformations of VP2, and (ii) IBDV capsid formation requires the interaction of VP2 with other viral components.

Effectively, VP3 also participates in the inherent ability of pVP2 to acquire different conformations through the pVP2 C-terminal end (Oña *et al.*, 2004). The peptide 443-GFKDIIRAIR-452 of pVP2CTD, which is absent in the mature protein VP2, forms an amphipathic α-helix. VP2 containing this α-helix is able to assemble into the T = 13 capsid, but only when expressed as a chimeric protein with an N-terminal His tag (Saugar *et al.*, 2005). In this configuration, the amphipathic α-helix acts as a temporally bound molecular switch, and the His tag partially emulates the VP3 C-terminal region and acts as a molecular triggering factor. We hypothesize that electrostatic interactions between these two morphogenic elements, which can be disengaged in the baculovirus expression system, are transmitted to VP2 allowing the protein to acquire the competent conformations for capsid assembly. The last five residues in the VP3 C-terminal end align with a similar acidic segment of the His tag, and there are no other similar segments in the VP3 sequence. Indirect studies have confirmed the importance and uniqueness of the VP3 C-terminal end (Chevalier *et al.*, 2004; Gonzalez *et al.*, 2005; Maraver *et al.*, 2003b). In conclusion, VP3, whose function is successfully restored by the His tag, is a canonical scaffolding protein, rather than an intrinsic component of the capsid (Coulibaly *et al.*, 2005; Saugar *et al.*, 2005).

VP1

VP3 not only interacts with pVP2 but also with the viral RNA-dependent RNA polymerase VP1 and the dsRNA genome (Maraver *et al.*, 2003a; Maraver *et al.*, 2003b; Tacken *et al.*, 2002). In fact, the interaction of VP3 with VP1 is responsible for the encapsidation of the polymerase into virus-like particles (VLPs) (Lombardo *et al.*, 1999). The region of VP3 responsible for interactions with VP1 has been mapped by deletion mutagenesis to a highly charged 16 amino-acid domain located near the C-terminus of VP3. This domain is sufficient to direct interactions with VP1 *in vitro* and to interfere with the virus production *in vivo* (Maraver *et al.*, 2003a). On the other hand, the domain of VP3 that directs the dsRNA binding activity of the protein has been mapped by deletion mutagenesis to a 69 amino-acid region located near the N-terminus. Based on *in vitro* assays, the dsRNA binding activity is sequence independent, a characteristic that is consistent with a general role for VP3 in stabilizing the genomic RNA in virus particles (Kochan *et al.*, 2003). In IPNV, VP3 is intimately associated with the viral dsRNA segments, producing filamentous ribonucleoprotein structures (Hjalmarsson *et al.*, 1999).

VP3

The key role for VP3 in IBDV capsid assembly has been supported by analyses using baculovirus-expression vectors (Maraver *et al.*, 2003b). VP3 expressed in insect cells, either alone or in context of the IBDV polyprotein, was found to undergo proteolytic processing, yielding a VP3 species with a C-terminal deletion of 13 amino acids. However, co-expression of the protein with VP1 partially prevented the trimming of VP3, probably by interacting with a region of VP3 that overlaps the proteolytic cleavage site (Maraver *et al.*, 2003a). Furthermore, the co-expression of VP1 and VP3 induces the formation of viroplasm-like aggregates similar to those characteristically found in IBDV-infected cells, suggesting the recovery of efficient assembly of VLPs. This was corroborated by the finding that under those expression conditions VLP production is higher than in the absence of VP1 (Maraver *et al.*, 2003b). An alternative approach, by fusion of heterologous proteins to the C-terminal end of the polyprotein, prevents the degradation of the VP3 C-terminal region, and yields VLPs indistinguishable from infectious IBDV virions (Chevalier *et al.*, 2004). The role of VP3 as a scaffold to control the capsid assembly is also consistent with the finding that VP3 is able to self-assemble, as many bacteriophage scaffolding proteins do (Tacken *et al.*, 2000). The use of deletion mutants has allowed mapping of the VP3 oligomerization domain between residues 224 to 247, near the C-terminal end of the protein (Maraver *et al.*, 2003b). The overlapping of this region with the 16 amino acid region of VP3 involved in the interaction of VP1 suggests the existence of a complex set of interactions underlying the assembly of IBDV. It is intriguing that VP3 stoichiometry inside the virion remains constant at ~600 copies. A plausible reason may be its temporal stabilizing role, carried out by a non-icosahedral but regular scaffold structure designed to stabilize an unstable capsid made up of 780 (p) VP2 copies, i.e. a procapsid-like assembly.

Summary

The results obtained so far using heterologous expression of the structural IBDV components reveal that assembly of the capsid demands a number of sequential events involving the pVP2 and VP3 polypeptides (Figure 8.6). Additionally, virus formation requires the proper interaction of VP1 and the RNA with the capsid through interactions with specific regions of VP3. Protein pVP2 is the initial capsid building unit, and it has intrinsic information to assemble the trimers that are the characteristic feature of the IBDV shell. The transient interaction of pVP2 (or a shorter pVP2 variant) with VP3 produces complexes that acquire intrinsic flexibility to accommodate the quasi-equivalent environments required to build a $T = 13$ capsid. If the VP3 C-terminal or the pVP2 amphipathic a helix segments are removed, trimer formation is preserved but they are no longer capable of building simultaneously the viral pentameric and hexameric interactions, and only aberrant assemblies are produced (tubes and $T = 1$ capsids). The integrity of both morphogenic elements restores the ability of the VP3/pVP2 trimers to yield icosahedral capsids with $T > 1$. The maturation of pVP2 to VP2 takes place on existing capsids, suggesting that this step confers an irreversible character to the assembly pathway, in a way similar to what happens in many other viral morphogenetic pathways. The sequential proteolysis of pVP2 produces minor peptides that remain in the capsid and could play a role in the viability of the virus.

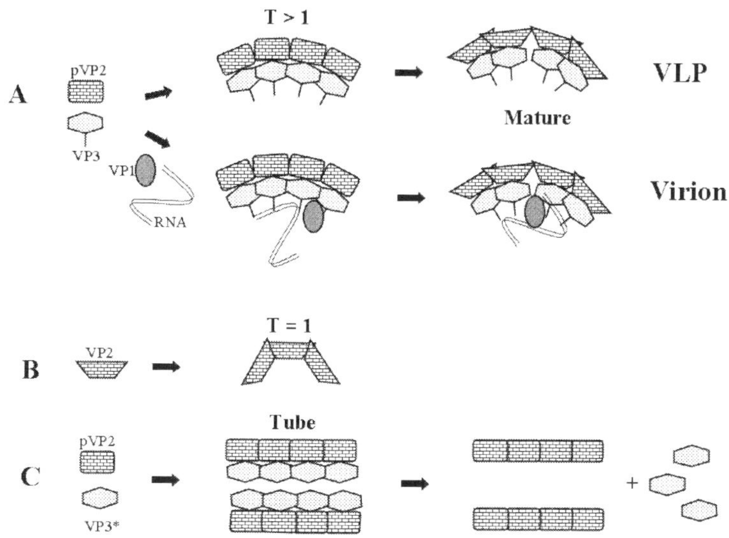

Figure 8.6 A working IBDV capsid assembly pathway. The details of the assembly pathway are described in the main text.

The interaction of VP3 with the polymerase VP1 and dsRNA points to a key role for this protein in providing the appropriate organization of the IBDV capsid to support transcription of the viral genome. As IBDV lacks the T = 2 nucleocapsid characteristic of the other dsRNA viruses, it is clear that the T = 13 capsid has to supply some of the functions carried out by the T = 2 complex. Thus, at the fivefold axis there exist mostly electronegative channels wide enough (diameter, ~20 Å) for the extrusion of ssRNA, as found in similar locations of the rotavirus middle shell. The extended interaction of VP3 with the dsRNA segments must play an important role not only in the stabilization and packaging of the genome, but also in positioning these segments for the interaction with the polymerase complex. In addition, VP2 is presently being considered as an intermediate between the outer shell proteins of the dsRNA T = 13 reoviruses and those of the positive sense ssRNA T = 3 nodaviruses and T = 4 tetraviruses, and emerging parallels in RNA replication are being also considered (Ahlquist, 2005; 2006). Further structural analyses of the IBDV life cycle are under way to get a more detailed picture of the functional relationships among these viral components.

Acknowledgements

This work was partly supported by grants BFU2005–06987, AGL2003–07189 from the DGI of the Ministerio de Ciencia y Tecnología, and 07B/0041/2002 from the Comunidad Autonoma de Madrid. JRC holds a contract from the 'Ramón y Cajal' program (MEC).

References

Ahlquist, P. (2005) Virus evolution: fitting lifestyles to a T. Curr. Biol. *15*, R465-R467.

Ahlquist, P. (2006) Parallels among positive-strand RNA viruses, reverse-transcribing viruses and double-stranded RNA viruses. Nat. Rev. Microbiol. *4*, 371–382.

Birghan, C., Mundt, E. and Gorbalenya, A.E. (2000) A non-canonical Lon proteinase lacking the ATPase domain employs the Ser-Lys catalytic dyad to exercise broad control over the life cycle of a double-stranded RNA virus. EMBO. J. *19*, 114–123.

Böttcher, B., Kiselev, N.A., Stel'Mashchuk, V.Y., Perevozchikova, N.A., Borisov, A.V. and Crowther, R.A. (1997) Three-dimensional structure of infectious bursal disease virus determined by electron cryomicroscopy. J. Virol. *71*, 325–330.

Brandt, M., Yao, K., Liu, M., Heckert, R.A. and Vakharia, V.N. (2001) Molecular determinants of virulence, cell tropism, and pathogenic phenotype of infectious bursal disease virus. J. Virol. *75*, 11974–11982.

Castón, J.R., Ghabrial, S.A., Jiang, D., Rivas, G., Alfonso, C., Roca, R., Luque, D. and Carrascosa, J.L. (2003) Three-dimensional structure of Penicillium chrysogenum virus: a double-stranded RNA virus with a genuine T = 1 capsid. J. Mol. Biol. *331*, 417–431.

Caston, J.R., Luque, D., Trus, B.L., Rivas, G., Alfonso, C., Gonzalez, J.M., Carrascosa, J.L., Annamalai, P. and Ghabrial, S.A. (2006) Three-dimensional structure and stoichiometry of Helmintosporium victoriae190S totivirus. Virology *347*, 323–332.

Castón, J.R., Martínez-Torrecuadrada, J.L., Maraver, A., Lombardo, E., Rodríguez, J.F., Casal, J.I. and Carrascosa, J.L. (2001) C terminus of infectious bursal disease virus major capsid protein VP2 is involved in definition of the T number for capsid assembly. J. Virol. *75*, 10815–10828.

Castón, J.R., Trus, B.L., Booy, F.P., Wickner, R.B., Wall, J.S. and Steven, A.C. (1997) Structure of L-A virus: a specialized compartment for the transcription and replication of double-stranded RNA. J. Cell. Biol. *138*, 975–985.

Chevalier, C., Galloux, M., Pous, J., Henry, C., Denis, J., Da Costa, B., Navaza, J., Lepault, J. and Delmas, B. (2005) Structural peptides of a nonenveloped virus are involved in assembly and membrane translocation. J. Virol. *79*, 12253–12263.

Chevalier, C., Lepault, J., Da Costa, B. and Delmas, B. (2004) The last C-terminal residue of VP3, glutamic acid 257, controls capsid assembly of infectious bursal disease virus. J. Virol. *78*, 3296–3303.

Chevalier, C., Lepault, J., Erk, I., Da Costa, B. and Delmas, B. (2002) The maturation process of pVP2 requires assembly of infectious bursal disease virus capsids. J. Virol. *76*, 2384–2392.

Coulibaly, F., Chevalier, C., Gutsche, I., Pous, J., Navaza, J., Bressanelli, S., Delmas, B. and Rey, F.A. (2005) The birnavirus crystal structure reveals structural relationships among icosahedral viruses. Cell *120*, 761–772.

Da Costa, B., Chevalier, C., Henry, C., Huet, J.C., Petit, S., Lepault, J., Boot, H. and Delmas, B. (2002) The capsid of infectious bursal disease virus contains several small peptides arising from the maturation process of pVP2. J. Virol. *76*, 2393–2402.

Dobos, P., Hill, B.J., Hallett, R., Kells, D.T., Becht, H. and Teninges, D. (1979) Biophysical and biochemical characterization of five animal viruses with bisegmented double-stranded RNA genomes. J. Virol. *32*, 593–605.

Feldman, A.R., Lee, J., Delmas, B. and Paetzel, M. (2006) Crystal structure of a novel viral protease with a serine/lysine catalytic dyad mechanism. J. Mol. Biol. *358*, 1378–1389.

Fernandez-Arias, A., Risco, C., Martinez, S., Albar, J.P. and Rodriguez, J.F. (1998) Expression of ORF A1 of infectious bursal disease virus results in the formation of virus-like particles. J. Gen. Virol. 79 *(Pt 5)*, 1047–1054.

Garriga, D., Querol-Audí, J., Abaitua, F., Saugar, I., Pous, J., Verdaguer, N., Castón, J.R. and Rodríguez, J.F. (2006) The 2.6-Angstrom structure of infectious bursal disease virus-derived T = 1 particles reveals new stabiling elements of the virus capsid. J. Virol. *80*, 6895–6905.

Gonzalez, D., Rodriguez, J.F. and Abaitua, F. (2005) Intracellular interference of infectious bursal disease virus. J. Virol. *79*, 14437–14441.

Granzow, H., Birghan, C., Mettenleiter, T.C., Beyer, J., Kollner, B. and Mundt, E. (1997) A second form of infectious bursal disease virus-associated tubule contains VP4. J. Virol. *71*, 8879–8885.

Hill, C.L., Booth, T.F., Prasad, B.V., Grimes, J.M., Mertens, P.P., Sutton, G.C. and Stuart, D.I. (1999) The structure of a cypovirus and the functional organization of dsRNA viruses. Nat. Struct. Biol. *6*, 565–568.

Hjalmarsson, A., Carlemalm, E. and Everitt, E. (1999) Infectious pancreatic necrosis virus: identification of a VP3-containing ribonucleoprotein core structure and evidence for O-linked glycosylation of the capsid protein VP2. J. Virol. *73*, 3484–3490.

Kochan, G., González, D. and Rodríguez, J.F. (2003) Characterization of the RNA-binding activity of VP3, a major structural protein of Infectious bursal disease virus. Arch. Virol. *148*, 723–744.

Lee, C.C., Ko, T.P., Chou, C.C., Yoshimura, M., Doong, S.R., Wang, M.Y. and Wang, A.H. (2006a) Crystal structure of infectious bursal disease virus VP2 subviral particle at 2.6A resolution: Implications in virion assembly and immunogenicity. J. Struct. Biol. *155*, 74–86.

Lee, C.C., Ko, T.P., Lee, M.S., Chou, C.C., Lai, S.Y., Wang, A.H. and Wang, M.Y. (2003) Purification, crystallization and preliminary X-ray analysis of immunogenic virus-like particles formed by infectious bursal disease virus (IBDV) structural protein VP2. Acta Crystallogr. D Biol. Crystallogr. *59*, 1234–1237.

Lee, M.S., Doong, S.R., Lai, S.Y., Ho, J.Y. and Wang, M.Y. (2006b) Processing of infectious bursal disease virus (IBDV) polyprotein and self-assembly of IBDV-like particles in Hi-5 cells. Biotechnol. Prog. *22*, 763–769.

Liu, M. and Vakharia, V.N. (2006) Nonstructural protein of infectious bursal disease virus inhibits apoptosis at the early stage of virus infection. J. Virol. *80*, 3369–3377.

Lombardo, E., Maraver, A., Castón, J.R., Rivera, J., Fernández-Arias, A., Serrano, A., Carrascosa, J.L. and Rodríguez, J.F. (1999) VP*1*, the putative RNA-dependent RNA polymerase of infectious bursal disease virus, forms complexes with the capsid protein VP3, leading to efficient encapsidation into virus-like particles. J. Virol. *73*, 6973–6983.

Lombardo, E., Maraver, A., Espinosa, I., Fernandez-Arias, A. and Rodriguez, J.F. (2000) VP5, the nonstructural polypeptide of infectious bursal disease virus, accumulates within the host plasma membrane and induces cell lysis. Virology *277*, 345–357.

Maraver, A., Clemente, R., Rodríguez, J.F. and Lombardo, E. (2003a) Identification and molecular characterization of the RNA polymerase-binding motif of infectious bursal disease virus inner capsid protein VP3. J. Virol. *77*, 2459–2468.

Maraver, A., Oña, A., Abaitua, F., Gonzalez, D., Clemente, R., Ruiz-Díaz, J.A., Castón, J.R., Pazos, F. and Rodríguez, J.F. (2003b) The oligomerization domain of VP3, the scaffolding protein of infectious bursal disease virus, plays a critical role in capsid assembly. J. Virol. *77*, 6438–6449.

Martínez-Torrecuadrada, J.L., Castón, J.R., Castro, M., Carrascosa, J.L., Rodríguez, J.F. and Casal, J.I. (2000) Different architectures in the assembly of infectious bursal disease virus capsid proteins expressed in insect cells. Virology *278*, 322–331.

Oña, A., Luque, D., Abaitua, F., Maraver, A., Castón, J.R. and Rodríguez, J.F. (2004) The C-terminal domain of the pVP2 precursor is essential for the interaction between VP2 and VP3, the capsid polypeptides of infectious bursal disease virus. Virology *322*, 135–142.

Sánchez, A.B. and Rodríguez, J.F. (1999) Proteolytic processing in infectious bursal disease virus: identification of the polyprotein cleavage sites by site-directed mutagenesis. Virology *262*, 190–199.

Saugar, I., Luque, D., Ona, A., Rodriguez, J.F., Carrascosa, J.L., Trus, B.L. and Caston, J.R. (2005) Structural polymorphism of the major capsid protein of a double-stranded RNA virus: an amphipathic alpha helix as a molecular switch. Structure. *13*, 1007–1017.

Tacken, M.G., Peeters, B.P., Thomas, A.A., Rottier, P.J. and Boot, H.J. (2002) Infectious bursal disease virus capsid protein VP3 interacts both with VP*1*, the RNA-dependent RNA polymerase, and with viral double-stranded RNA. J. Virol. *76*, 11301–11311.

Tacken, M.G., Rottier, P.J., Gielkens, A.L. and Peeters, B.P. (2000) Interactions in vivo between the proteins of infectious bursal disease virus: capsid protein VP3 interacts with the RNA-dependent RNA polymerase, VP1. J. Gen. Virol. *81*, 209–218.

Villanueva, R.A., Galaz, J.L., Valdes, J.A., Jashes, M.M. and Sandino, A.M. (2004) Genome assembly and particle maturation of the birnavirus infectious pancreatic necrosis virus. J. Virol. *78*, 13829–13838.

von Einem, U.I., Gorbalenya, A.E., Schirrmeier, H., Behrens, S.E., Letzel, T. and Mundt, E. (2004) VP1 of infectious bursal disease virus is an RNA-dependent RNA polymerase. J. Gen. Virol. *85*, 2221–2229.

Part II

Protein Structure and Function

Structural Basis of Mammalian Orthoreovirus Cell Attachment

9

Pierre Schelling,* Jacquelyn A. Campbell,* Thilo Stehle, and Terence S. Dermody

Abstract

Mammalian orthoreoviruses are non-enveloped viruses that display strain-specific disease phenotypes following infection of newborn mice. In several cases, these differences in disease expression segregate genetically with the viral attachment protein, $\sigma1$. The homotrimeric $\sigma1$ protein is an elongated fibre with an N-terminal tail topped with a C-terminal head. A domain in the tail of serotype 3 $\sigma1$ binds to α-linked sialic acid, whereas the head domain binds to junctional adhesion molecule-A (JAM-A). The crystal structure of a C-terminal region of $\sigma1$ has significantly advanced our understanding of several important properties of this protein, such as its multimeric state, its conformational dynamics, and its interaction with receptors sialic acid and JAM-A. The analysis of repetitive motifs in the crystallized fragment also has served to approximate key parameters of the full-length protein, such as its overall structure and flexibility. Furthermore, numerous structural and functional relationships between $\sigma1$ and the adenovirus attachment protein, fibre, have suggested an evolutionary link in the receptor-binding strategies of these two viruses. The existence of such a link is also supported by structural and functional similarities shared by the $\sigma1$ receptor JAM-A and the coxsackievirus and adenovirus receptor (CAR). Both JAM-A and CAR contain two immunoglobulin-like domains, form homodimers at regions of cell–cell contact, and contain sequences in their cytoplasmic tails that anchor the proteins to the actin cytoskeleton. It thus appears that the strategy for attachment and entry of reovirus is less similar to other members of the *Reoviridae* family and more similar to that of adenovirus.

Introduction

Mammalian orthoreoviruses (called reoviruses here) have served as highly tractable models for studies of viral pathogenesis. There are three major reovirus serotypes, which can be differentiated on the basis of haemagglutination profiles (Rosen, 1960) and neutralization assays (Virgin *et al.*, 1988; Virgin *et al.*, 1991). Each of the serotypes is represented by a prototype strain: type 1 Lang (T1L), type 2 Jones (T2J), and type 3 Dearing (T3D). Strains of all three serotypes are capable of infecting most mammals, including humans, but reovirus-induced disease is restricted to the very young (Tyler, 2001). The pathogenesis

*These authors contributed equally to this manuscript.

of reovirus infection has been studied most extensively using newborn mice, in which serotype-specific patterns of disease have been identified. The best characterized of these models is reovirus pathogenesis in the murine central nervous system (CNS).

Following oral inoculation of newborn mice, both type 1 and type 3 reoviruses undergo primary replication in intestinal epithelial cells and Peyer's patches. However, after primary replication there is a marked dichotomy in both the route of spread and tropism of these serotypes. Type 1 reovirus spreads hematogenously to ependymal cells, where it causes subacute hydrocephalus (Weiner *et al.*, 1977; Weiner *et al.*, 1980; Tyler *et al.*, 1986). In sharp contrast, type 3 reovirus spreads by neural routes to the CNS, where it infects neurons and causes acute, lethal encephalitis (Weiner *et al.*, 1977; Weiner *et al.*, 1980; Tyler *et al.*, 1986; Morrison *et al.*, 1991). This difference in spread and tissue tropism has been mapped using type 1 and type 3 reovirus reassortant viruses to the S1 gene (Weiner *et al.*, 1977; Weiner *et al.*, 1980; Tyler *et al.*, 1986), which encodes the viral attachment protein, σ1 (Weiner *et al.*, 1980; Lee *et al.*, 1981). This link between σ1 and the route of dissemination in the host and cell tropism within the CNS has made its structure and function the focus of considerable interest.

Structure of σ1

General Properties of σ1

Reovirus attachment protein σ1 is a fibre-like molecule that is anchored into the virion core at the 12 vertices of the icosahedral particle (Furlong *et al.*, 1988). Viral attachment to the cell surface by σ1 leads to internalization of the virus by receptor-mediated endocytosis, which is likely clathrin-dependent (Borsa *et al.*, 1981; Sturzenbecker *et al.*, 1987; Rubin *et al.*, 1992a; Baer *et al.*, 1999; Ehrlich *et al.*, 2004). Within endosomes, virions undergo acid-dependent, proteolytic disassembly to form infectious subvirion particles (ISVPs) (Chang and Zweerink, 1971; Silverstein *et al.*, 1972; Sturzenbecker *et al.*, 1987; Baer and Dermody, 1997). ISVPs penetrate endosomal membranes and release transcriptionally active cores into the cytoplasm (Borsa *et al.*, 1979; Lucia-Jandris *et al.*, 1993; Tosteson *et al.*, 1993; Hooper and Fields, 1996a, 1996b; Chandran *et al.*, 2001; Chandran and Nibert, 2003). The cores do not contain σ1 protein (Dryden *et al.*, 1993; Reinisch *et al.*, 2000), and the viral RNA egresses through the same channels at the 12 vertices that were formerly occupied by σ1 (Dryden *et al.*, 1993; Reinisch *et al.*, 2000). It has been proposed that σ1 undergoes conformational changes during viral disassembly and that these changes facilitate key steps in the cell entry process (Furlong *et al.*, 1988; Dryden *et al.*, 1993; Nibert *et al.*, 1995; Chappell *et al.*, 2002).

Insights into σ1 structure from biochemical studies and sequence analysis

The first experimental insights into σ1 structure came from composite transmission electron micrographs of σ1 molecules purified from virions (Fraser *et al.*, 1990). These images confirmed earlier predictions of a fibre-like structure of σ1 (Bassel-Duby *et al.*, 1985) and clearly revealed two distinct morphological domains: a long N-terminal fibrous tail region that inserts into the virion capsid and a C-terminal globular head that projects away from the virion surface. The σ1 protein was shown to exhibit substantial flexibility at three discrete regions within the fibrous tail: a region close to the N-terminus, a region at the mid-point of the tail, and a region just below the globular head.

Additional clues about key structural features of σ1 were revealed by increasingly sophisticated analyses of its amino acid sequence. The N-terminal 30 residues are highly hydrophobic and likely serve to anchor the protein into the capsid. Additional sequences also may aid in this function as the N-terminal 121 residues of σ1 sequence are required for insertion of fusion proteins into reovirus particles (Mah *et al.*, 1990). Following the hydrophobic N-terminus, the σ1 sequence contains an extended region characterized by heptad repeats, with non-polar residues at the first and fourth positions of each repeating unit. Such repeats are characteristic of α-helical coiled coils (Crick, 1953) in which amphipathic helices are wrapped around each other to form dimers, trimers, or tetramers. Accordingly, the N-terminal portion of σ1 was predicted to form an oligomeric α-helical coiled-coil structure (Bassel-Duby *et al.*, 1985). The actual oligomeric state of σ1 was the subject of some controversy, as dimeric (Lee *et al.*, 1981; Yeung *et al.*, 1987), trimeric (Leone *et al.*, 1991; Strong *et al.*, 1991), and tetrameric (Bassel-Duby *et al.*, 1987; Fraser *et al.*, 1990) structures were proposed. More complex sequence analyses and correlation of sequence data with the observed morphology eventually led to a plausible model for the domain architecture of σ1 (Nibert *et al.*, 1990). This model postulated three structurally distinct regions of σ1, each accounting for about one third of its sequence. The N-terminal third was predicted to form an α-helical coiled-coil structure, the second third a fibre-like region composed primarily of β-sheet structure, and the final third a globular head domain (Nibert *et al.*, 1990). While this model served to explain several properties of σ1, a high-resolution structure was required to accurately describe its features and correlate them with functional data.

Crystal structure of the C-terminal region of σ1

Structural analyses of σ1 have been complicated by the flexible, fibre-like nature of the protein, its poor solubility, and difficulties in producing significant quantities of soluble and functional recombinant protein. While some groups have isolated functional σ1 protein from whole virions (Nagata *et al.*, 1987; Yeung *et al.*, 1987), others have relied on recombinant expression in either bacteria, eukaryotic cells, or *in vitro* using cell lysates (Masri *et al.*, 1986; Pelletier *et al.*, 1987; Duncan *et al.*, 1991; Bisaillon *et al.*, 1999). Although the amounts of soluble and functional σ1 recovered by isolation from virions and recombinant expression systems were sufficient for biochemical experiments, they were inadequate for extensive structural studies of σ1.

A breakthrough came with the development of an internally truncated version of σ1 derived from strain T3D. The engineered protein lacked a portion of the predicted N-terminal coiled coil while retaining the hydrophobic sequence at the N-terminus and the C-terminal two thirds of the protein (Chappell *et al.*, 2000). This truncated form of σ1 protein was expressed in insect cells using a baculovirus vector, and substantial amounts of folded and relatively stable protein were purified. Cleavage of the purified protein with trypsin resulted in a stable and soluble fragment (residues 246–455) that formed functional oligomers (Barton *et al.*, 2001b). This C-terminal fragment led to the successful crystallization of T3D σ1 (Chappell *et al.*, 2002).

The crystal structure of the C-terminal σ1 fragment revealed an elongated, symmetric homotrimer, thus firmly establishing the oligomeric state of the protein as trimeric. Moreover, the structural analysis provided a wealth of information about the architecture of σ1 and potential roles of its structural features in viral attachment. The crystallized σ1 fragment can be divided into two major regions: a slender N-terminal tail formed by residues 246 to 309 and a compact head that contains residues 310 to 455 (Figure 9.1A). The short tail consists of three triple β-spiral repeats, a rare trimerization motif that has so far been seen in only two other proteins, the adenovirus fibre (Mitraki *et al.*, 2002) and avian reovirus σC (Guardado Calvo *et al.*, 2005). The β-spiral is characterized by short, vertically oriented β-strands that wind around each other and bury a substantial number of hydrophobic residues in the centre. The head domain is composed of eight antiparallel β-strands that form a compact β-barrel structure. A noticeable kink is seen between the trimer axes of the head and the tail regions, and a superposition of the three polypeptide chains reveals that their overall conformations differ in a linker region between β-spiral repeats 2 and 3 (Figure 9.1B). This linker segment is formed by an extended polypeptide chain that does not possess any secondary structure and has no contacts with other residues in the protein. Since it is capable of assuming different conformations in each molecule, this segment defines a hinge region that confers substantial flexibility onto the σ1 protein. The presence of this hinge between the head and tail regions suggests that σ1 is a dynamic molecule poised to undergo conformational rearrangements, perhaps in

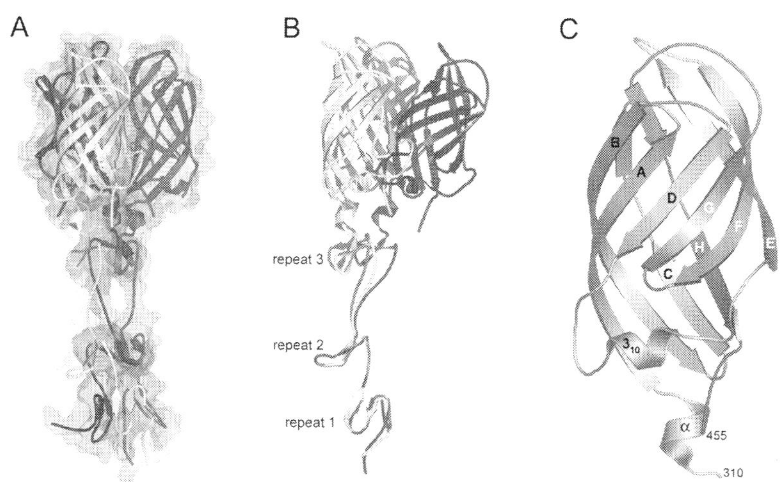

Figure 9.1 Structure of the C-terminal region of σ1. (A) Ribbon drawing and surface-contoured shape of the crystallized σ1 trimer, with the three monomers shown in shades of grey. Each monomer consists of a head domain formed by a compact β-barrel and a fibrous tail with three β-spiral repeats. (B) Superposition of the three σ1 monomers present in the crystal lattice. The superposition, which is based on the two β-spiral repeats at the N-terminus, reveals flexibility in the linker region (residues 291–294) between β-spiral repeats 2 and 3. (C) Enlarged view of the σ1 head domain. Eight β-strands (A-H) fold into a compact, cylindrical β-sheet. With the exception of the long loop connecting β-strands D and E, the loops are extremely short.

response to viral attachment or disassembly. Major features of σ1 structure are described in more detail in the following sections.

The β-spiral

The three β-spiral repeats in the crystallized portion of the σ1 tail are characterized by hydrophobic amino acids and conserved β-turn residues at specific locations. This repetitive motif folds into two anti-parallel β-strands connected by a four residue β-turn that contains a conserved glycine or proline, while a solvent-exposed variable loop connects one repeat to the next (van Raaij *et al.*, 1999). Although only three β-spiral repeats are included in the crystallized protein, the presence of characteristic residues at defined positions provides a framework for the prediction of such repeats elsewhere in the σ1 tail domain. Accordingly, full-length σ1 may contain as many as eight consecutive β-spiral repeats that span residues 170 to 309 (Chappell *et al.*, 2002). An interesting and likely significant feature of the σ1 spiral is a four-residue insertion (residues 291–294) between the second and third β-spiral repeats of the crystallized fragment. This insertion confers substantial flexibility onto the molecule just below the head domain, allowing the head to move with respect to the tail (Figure 9.1B). Flexibility at a similar location was noticed previously in electron micrographs of full-length protein isolated from virions (Fraser *et al.*, 1990). An alignment of the T1L, T2J, and T3D σ1 sequences indicates that this hinge is the only flexibility module within the eight putative triple β-spiral repeats in σ1 of all three major reovirus serotypes. Interestingly, the hinge region is predicted to be longer in the T1L and T2J σ1 proteins, suggesting that these proteins have even more flexibility at this region than does T3D σ1 (Chappell *et al.*, 2002).

The β-barrel

The head domain (residues 310–455) features eight antiparallel β-strands that fold into a compact β-barrel (50 Å long, 35 Å wide, and 25 Å thick) (Chappell *et al.*, 2002). The barrel has a slightly oval cross section, with each strand having two neighbours. The domain fold is best described as consisting of two consecutive Greek-key motifs (β-strands DABC and HEFG) (Figure 9.1C). The loops connecting the β-strands are exceedingly short, giving the barrel a rather compact and smooth appearance. A notable exception is the long loop connecting β-strands D and E (DE-loop). This loop spans the base of the barrel and features a short 3_{10} helix (Figure 9.1C).

Intersubunit contacts

While the three monomers in the σ1 trimer appear to be tightly associated, a closer inspection of the intersubunit contacts reveals that contacts within the short tail account for 60% (~1100 Å^2) of the buried surface area (Chappell *et al.*, 2002). As with most protein intersubunit interfaces, these contacts also are mediated primarily by hydrophobic residues that are conserved among the three major reovirus serotypes. In contrast, residues involved in intersubunit contacts in the head domain (~650 Å^2) are far less conserved and relatively more hydrophilic in nature. A water-filled cavity in the centre of the trimer also interrupts the interchain contacts. Thus, the tail forms intersubunit contacts that are primarily responsible for trimer stability, whereas the head contacts are less likely to contribute to stability of the trimer.

A cluster of aspartic acids

The head trimer interface contains a cluster of six conserved and buried aspartic acids, two (Asp345 and Asp346) from each monomer, at the base of the β-barrel (Figure 9.2A). All three Asp346 side chains engage in salt bridge formation with the guanidinium groups of the neighbouring Arg314 (Figure 9.2B). However, the three Asp345 side chains do not form salt bridges. They are closely opposed in a hydrophobic environment that excludes solvent. The carboxylate groups are contacted primarily by the aromatic rings of tyrosine residues from above and below. Most importantly, there is no counterion present in the electron density of the crystal structure that would neutralize the three negative charges produced by the Asp345 side chains. Given their unusual environment, it is likely that all three Asp345 side chains are protonated (and therefore uncharged). Protonated Asp345 side chains would be able to engage in favourable interactions with each other via hydrogen bond formation, as indicated in Figure 9.2B. Such interactions would likely help to stabilize the trimeric structure. Deprotonation of the Asp345 side chains, on the other hand, would result in a highly destabilizing accumulation of three negative charges at the trimer interface. In fact, molecular dynamic simulations have shown that deprotonation of Asp345 would lead to a partial separation of the β-barrels (Cavalli *et al.*, 2004). Since aspartic acid side chains are typically deprotonated at neutral pH, their unusual protonation state and surroundings have potentially important implications for σ1 function.

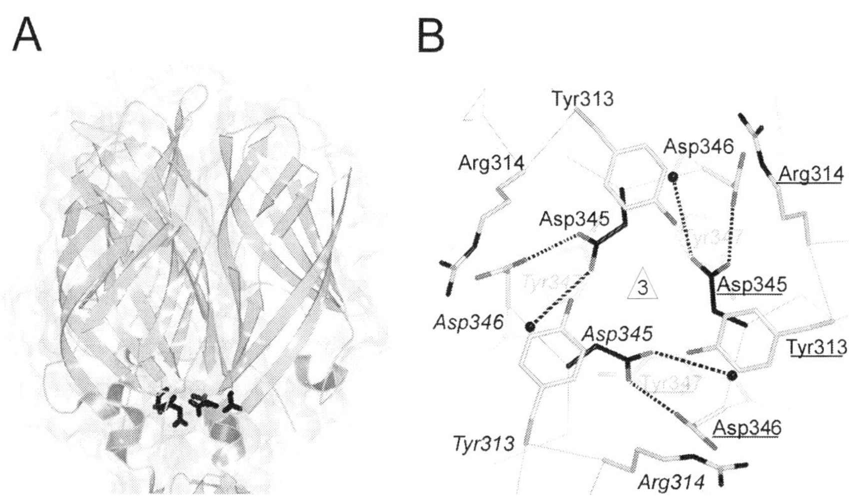

Figure 9.2 The aspartic acid cluster at the centre of the σ1 head. (A) Ribbon drawing and surface-contoured shape of the σ1 head showing the location of the aspartic acid cluster at the trimer interface. The side chains of residues Asp345 and Asp346 are depicted as black sticks. (B) View along the trimer axis, centred at the aspartic acid cluster at the base of the head. Residues Tyr313, Arg314, and Tyr347 engage in contacts with aspartic acids Asp345 and Asp346. The side chains of Asp345 are likely protonated to avoid an accumulation of negative charge at the interface. Hydrogen bonds involving protonated Asp345 are represented with dashed lines. The residues within the same monomer are indicated by regular, italic, or underlined labels.

Model of full-length σ1

Significant information about σ1 structure and conformation has been obtained using sequence analysis (Bassel-Duby *et al.*, 1985; Nibert *et al.*, 1990), viral reassortant studies (Chappell *et al.*, 1997), physico-chemical measurements (Leone *et al.*, 1992; Leone *et al.*, 1996), expression and folding studies (Strong *et al.*, 1991), and electron microscopy (Furlong *et al.*, 1988; Fraser *et al.*, 1990; Dryden *et al.*, 1993). Although electron microscopy has revealed the existence of a more retracted state of σ1 in intact virions (Furlong *et al.*, 1988; Dryden *et al.*, 1993), most data have been collected from studies in which σ1 exists in an extended conformation. Little information is available about the more retracted conformer of σ1. Indeed, when isolated from virions or produced using prokaryotic, eukaryotic, or *in vitro* expression systems, σ1 appears as an elongated molecule with clear head-and-tail morphology, leading to a largely accepted schematic model of the extended conformation of σ1. Using information provided by the crystal structure of the C-terminal half of σ1, along with existing sequence data, we developed a model for the structure of full-length, extended σ1 (Figure 9.3). The N-terminal third of σ1 sequence likely forms an α-helical coiled-coil structure (Bassel-Duby *et al.*, 1985; Duncan *et al.*, 1990; Nibert *et al.*, 1990); with the knowledge that σ1 forms trimers, this region can now be modelled as a triple α-helical coiled coil similar to that seen in the influenzavirus haemagglutinin (Weis *et al.*, 1988). The remainder of the molecule likely consists of eight consecutive β-spiral repeats followed by the β-barrel of the head domain (Chappell *et al.*, 2002). Thus, our working hypothesis for the structure of full-length σ1 consists of an N-terminal coiled-coil segment (residues 30–165) followed by eight β-spiral repeats (residues 170–309) and a globular head domain (residues 310–455). The N-terminal ~30 residues do not display the heptad repeat pattern characteristic of α-helical coiled coils. They are not expected to form an α-helix and have therefore not been included in the model. It is likely that these residues assume a unique structure that helps to anchor σ1 into the viral capsid.

The proposed full-length model fits well with the experimental and structural data collected thus far. Firstly, the overall dimensions of the proposed model agree with those observed by electron microscopy. We note that the β-spiral structure is somewhat thicker than the α-helical coiled coil, consistent with a similar difference in thickness in the rotary shadowing images. Secondly, the model explains the three observed regions of flexibility of

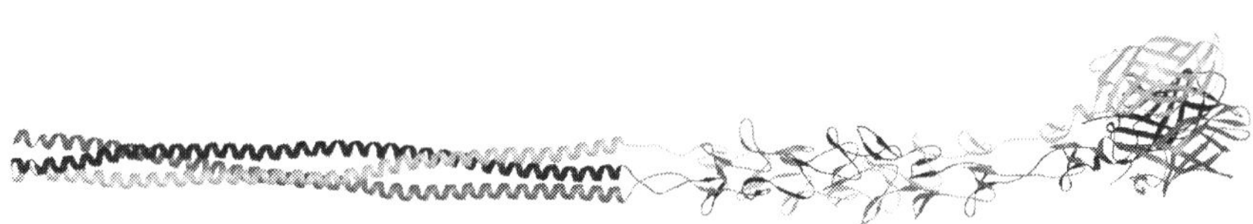

Figure 9.3 Model of full-length σ1. The model was generated based on the assumption that the N-terminal third of the protein (residues 30–165) forms a triple α-helical coiled coil (Bassel-Duby *et al.*, 1985). This section was modelled by extending an existing triple coiled-coil structure taken from the influenza virus haemagglutinin (Weis *et al.*, 1988). The middle section of the protein contains eight β-spiral repeats. The N-terminal five repeats are not included in the crystal structure and were modelled as described (Chappell *et al.*, 2002). The final third of the sequence folds into the globular β-barrel. Each of the three monomers is depicted in a different shade of grey.

σ1 at its base, midpoint, and just below the head (Fraser *et al.*, 1990). Flexibility near the base may be a result of a structural transition between the N-terminal 30 residues and the α-helical coiled coil. The σ1 midpoint corresponds to the junction between the coiled-coil and β-spiral regions, and it seems plausible that this junction introduces flexibility into the molecule. Flexibility below the σ1 head, which was also seen in rotary-shadowing images, corresponds to an insertion in the β-spiral around residue 290. Of note, other trimeric fibre-like proteins also are constructed from different trimerization motifs and exhibit flexibility at junctions between these motifs. For example, the mannose-binding lectin and related proteins contain a collagen helix followed by an α-helical coiled coil and a globular head region (Sheriff *et al.*, 1994; Epstein *et al.*, 1996). Flexibility between these segments is thought to be important for productive engagement of ligands.

Interactions of σ1 with carbohydrate receptors

Serotype-specific binding to carbohydrate receptors is mediated by σ1

Early studies examining the capacity of reovirus strains to agglutinate erythrocytes isolated from different species of animals documented a serotype-specific difference in haemagglutination that was genetically mapped using reassortant viruses to the σ1-encoding S1 gene (Weiner *et al.*, 1978; Pacitti and Gentsch, 1987). Haemagglutination produced by type 3 reovirus is inhibited by pretreatment of erythrocytes with neuraminidase (Armstrong *et al.*, 1984), suggesting that sialic acid is the erythrocyte structure bound by type 3 reovirus. Type 3 reovirus binding to some types of cells also is reduced by neuraminidase treatment, indicating that cell-surface sialic acid-containing glycoconjugates contribute to productive type 3 reovirus attachment (Gentsch and Pacitti, 1985). Both T1L and T3D reoviruses can infect L929 cells, a murine fibroblast cell line commonly used to propagate reovirus. However, only sialic acid-binding type 3 strains can infect murine erythroleukemia (MEL) cells (Rubin *et al.*, 1992b; Chappell *et al.*, 1997). Serial passage of non-sialic-acid-binding type 3 strains in MEL cells results in selection of viral variants that have acquired the capacity to bind sialic acid (Chappell *et al.*, 1997). These variants have substitutions at only three positions (Asn198, Arg202, and Pro204). In each case, the substituted residues are present in all sialic acid-binding type 3 strains sequenced to date (Dermody *et al.*, 1990; Tyler *et al.*, 2004).

Location of the σ1 sialic acid binding site

Given the close proximity of Asn198, Arg202, and Pro204 in primary sequence, it is possible that these residues contribute to a binding site for sialic acid in the tail region of σ1 (Chappell *et al.*, 1997; Chappell *et al.*, 2000). Due to the regularity of the triple β-spiral motif, the location of Asn198, Arg202, and Pro204 can be modelled with reasonable accuracy (see previous section and Chappell *et al.*, 2002). In this model, these residues form part of a solvent-exposed patch on one side of the tail (Figure 9.4A). Interestingly, they appear to cluster in a shallow groove, with Arg202 at the centre. This groove is of sufficient size to accommodate a sialic acid moiety, perhaps with Arg202 mediating contacts with the negatively charged sialic acid carboxylate group. Other viruses also use sialic acid-based carbohydrates as receptors, and in several cases (e.g. influenzavirus, polyomavirus, and rotavirus), structural information is available about the interaction between the virus and

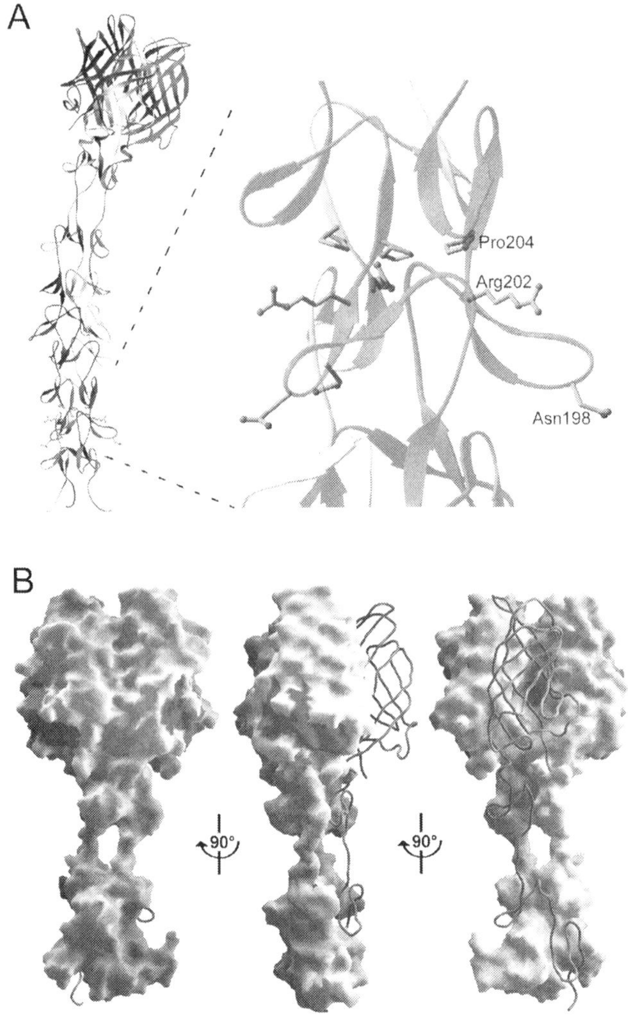

Figure 9.4 Putative receptor-binding sites of σ1. (A) Model of the complete β-spiral segment including the putative sialic acid binding site of σ1. Based on sequence pattern analysis of aligned σ1 sequences, σ1 likely contains eight β-spiral repeats that extend from residue 167 to residue 308 at the base of the head domain (Chappell *et al.*, 2002). The model shown was generated by extending the existing three-repeat structure present in the crystal structure. The close-up view shows the approximate location of amino acids Asn198, Arg202, and Pro204, which have been implicated in the interaction of T3D σ1 with sialic acid (Chappell *et al.*, 1997). According to this model, these residues would likely protrude between the second and third β-spiral repeats and form a putative binding site for sialic acid. (B) Conservation of σ1 residues predicted to interact with JAM-A. Using a sequence alignment of σ1 serotypes that can bind to JAM-A, a single extended cluster of conserved residues (shown in dark grey) is located at the base of the σ1 head domain (Chappell *et al.*, 2002). Many of the conserved residues within this cluster lie within the loop and short 3_{10} helix between β-strands D and E (DE-loop). In this representation, two σ1 monomers are surface-contoured, whereas the third is depicted as a grey ribbon.

sialic acid-based ligands (Weis *et al.*, 1988; Stehle *et al.*, 1994; Eisen *et al.*, 1997; Stehle and Harrison, 1997; Dormitzer *et al.*, 2002). Although the architecture of the binding sites differs in each case, all three viral attachment proteins recognize their ligands via shallow grooves similar in size to the one depicted for σ1. Furthermore, arginine residues at the centre of the polyomavirus VP1 and rotavirus VP4 grooves form key contacts with the ligand (Stehle *et al.*, 1994; Stehle and Harrison, 1997; Dormitzer *et al.*, 2002). In the case of polyomavirus VP1, this arginine contacts the sialic acid carboxylate group, whereas in the case of rotavirus VP4, it interacts with the glycerol side chain adjacent to the carboxylate.

The T1L strain of reovirus also recognizes carbohydrate receptors, albeit via a different region of the β-spiral (Chappell *et al.*, 2000). A recent study suggests that glycoconjugates containing sialic acid in a specific linkage, α2–3, are important for adherence of reovirus T1L to the apical surfaces of intestinal microfold (M) cells (Helander *et al.*, 2003). Thus, σ1 appears to employ the β-spiral not just as a trimerization motif but also as a receptor-binding scaffold with different binding sites. In comparison to other trimerization motifs such as α-helical coiled coils, the β-spiral is more versatile since its surface-exposed loops can be varied to accommodate different types of ligands. As reviewed elsewhere (Weigele *et al.*, 2003), β-strand-derived motifs such as triple β-spirals and triple β-helices are seen in divergent viral attachment proteins, and the multimeric arrangement of highly organized β-strands, with a hydrophobic core and connected by variable loops, may provide a stable matrix for such proteins that still enables them to interact with ligands.

Interactions of σ1 with JAM-A

Identification of JAM-A as a receptor for reovirus

Many proteins have been proposed to function as reovirus receptors, including the beta-adrenergic receptor (Co *et al.*, 1985), epidermal growth factor receptor (Tang *et al.*, 1993), and a p65/p95 complex (Saragovi *et al.*, 1995). However, to date, only junctional adhesion molecule-A (JAM-A, previously called JAM or JAM1) has been shown to directly interact with reovirus attachment protein σ1 and confer reovirus infection to non-permissive cells (Barton *et al.*, 2001b; Prota *et al.*, 2003; Campbell *et al.*, 2005). JAM-A was identified as a reovirus receptor using an expression-cloning approach in which a non-sialic-acid-binding reovirus strain was employed as an affinity ligand. COS-7 cells were transfected with cDNAs from an NT2 human teratoma cell library and screened for the capacity to support reovirus binding. Using this approach, four plasmids were identified, each of which encoded JAM-A. Three lines of evidence support the contention that JAM-A is a receptor for reovirus. First, JAM-A-specific monoclonal antibodies (mAbs) block reovirus binding and infection. Second, transfection of non-permissive cells with JAM-A-encoding cDNA confers efficient reovirus growth. Third, and most importantly, the head domain of strain T3D σ1 interacts directly with the JAM-A ectodomain, as assessed by surface plasmon resonance (Barton *et al.*, 2001b).

JAM-A was first identified in a screen of mAbs raised against murine endothelial cells in an attempt to characterize proteins localized to endothelial tight junctions (Martin-Padura *et al.*, 1998). The human (h) homolog of JAM-A is a 299-amino-acid, type I transmembrane protein that is a member of the immunoglobulin (Ig) superfamily (Ozaki *et al.*, 1999). JAM-A is expressed in many tissues, with the highest levels being found in the kidney, liver, and lung (Williams *et al.*, 1999). The JAM-A cytoplasmic tail has a classical type II PDZ binding motif and is phosphorylated by protein kinase C upon platelet activation (Wang *et al.*, 1995; Ebnet *et al.*, 2000). Several adaptor proteins have been identified as possible binding partners of the JAM-A cytoplasmic tail, including zonula occludens-1 (ZO-1), afadin (also called AF-6), multiple PDZ domain-containing protein-1, and calcium/calmodulin-dependent serine protein kinase (Ebnet *et al.*, 2000; Martinez-Estrada *et al.*, 2001; Hamazaki *et al.*, 2002). JAM-A is proposed to form homodimers located at tight junctions of polarized cells and may promote leukocyte migration and tight junction formation through homotypic interactions (Martin-Padura *et al.*, 1998; Bazzoni *et al.*, 2000a; Bazzoni *et al.*, 2000b).

JAM-A belongs to a group of related proteins termed the junctional adhesion molecule family. Two JAM-A homologues, JAM-B and JAM-C, respectively share 38% and 33% amino acid identity with JAM-A and have overlapping but also distinct tissue distribution in comparison to JAM-A. While all JAM family members are expressed by endothelial cells, JAM-B is predominantly expressed in the heart, and JAM-C is mainly expressed in the brain and kidney (Cunningham *et al.*, 2000; Aurrand-Lions *et al.*, 2001). Unlike JAM-B, which mediates homotypic interactions, JAM-C forms heterodimers with JAM-B (Arrate *et al.*, 2001). Each of the JAM family members has been proposed to be involved in leukocyte transmigration (Ostermann *et al.*, 2002), but only JAM-A serves as a receptor for reovirus (Prota *et al.*, 2003; Campbell *et al.*, 2005).

A potential JAM-A-binding site on σ1

The crystal structure of σ1 revealed a putative binding site for JAM-A at the lower edge of the σ1 head domain (Chappell *et al.*, 2002) (Figure 9.4B). Since all three reovirus serotypes can bind to JAM-A (Barton *et al.*, 2001b; Campbell *et al.*, 2005), it seems plausible that σ1 interacts with JAM-A through residues that are conserved in these serotypes. While most conserved residues are scattered about the molecule, examination of the σ1 surface reveals a single, extended patch of conserved residues at the lower edge of each β-barrel of the σ1 trimer. This predicted JAM-A-binding surface is formed primarily by a stretch of residues found mostly in the vicinity of the long DE-loop. Although it is easily accessible to ligand, this region is somewhat recessed into the protein surface and surrounded by protruding, non-conserved residues on all three edges of the trimer. Only residues from a single monomer contribute to the putative JAM-A-binding region and its borders, and the regions are not involved in σ1 intersubunit contacts. Thus, the location of conserved residues within the trimer suggests that each σ1 monomer can independently bind a JAM-A molecule.

Structure of JAM-A

The crystal structure of the hJAM-A ectodomain, expressed as a GST-fusion protein in bacteria, has been solved (Prota *et al.*, 2003). The molecule folds into two concatenated Ig-like domains, termed D1 (membrane distal) and D2 (membrane proximal) (Figure 9.5). Both D1 and D2 are formed from two antiparallel β-sheets that fold into variable type (D1) and intermediate type (D2) Ig-like domains, respectively (Harpaz and Chothia, 1994). The homodimeric interface is composed of four salt bridges and additional hydrophobic contacts involving residues on the GFCC′ faces of the two opposing D1 domains.

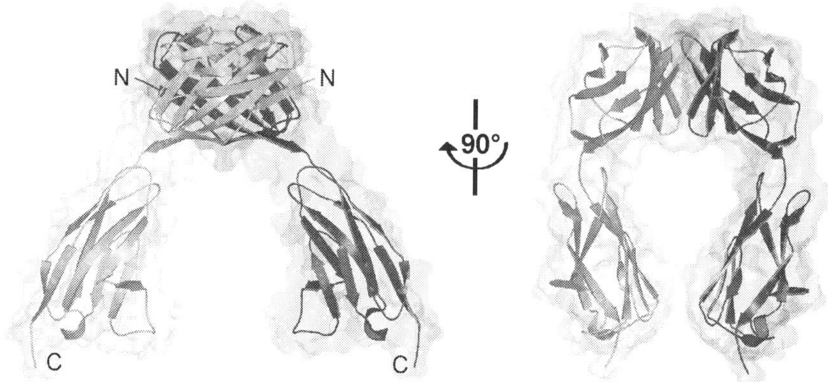

Figure 9.5 Structure of the hJAM-A ectodomain. Ribbon drawing and surface-contoured shape of the hJAM-A dimer (Prota *et al.*, 2003), with one monomer shown in black and the other in grey. Two orthogonal views are displayed.

The presence of multiple ionic interactions at a protein–protein interface is unusual, given that such interfaces are typically more hydrophobic in nature (Jones and Thornton, 1996). The ionic character of the JAM-A-JAM-A interface likely results in both pH- and ionic strength-dependency of dimer stability, as demonstrated experimentally for murine (m) JAM-A (Bazzoni *et al.*, 2000a). Additionally, although the contacting residues involved in the dimer interface of both hJAM-A and mJAM-A are identical, small differences occur in the arrangement of both homodimers in their respective crystal structures. These differences suggest that small movements of one monomer with respect to the other can occur (Prota *et al.*, 2003). As a result, the JAM-A dimer interface appears capable of rearrangement or dissociation. Such a dynamic nature of the JAM-A homodimer may be important to allow JAM-A to function in the regulation of tight junctions, and it may also play a role in viral recognition.

Implications for JAM-A binding to reovirus

Structural data for a σ1-JAM-A complex are not yet available. However, two lines of experimental evidence suggest that σ1 binds to a monomeric version of JAM-A, perhaps by engaging residues that also mediate JAM-A homodimer formation. First, chemical crosslinking of the JAM-A dimer diminishes the capacity of reovirus to bind JAM-A *in vitro* and on cells and negates the competitive effects of soluble JAM-A on reovirus attachment (Forrest *et al.*, 2003). These observations suggest that the virus recognizes a monomeric species of JAM-A and cannot interact with a covalently crosslinked dimer. Second, residues critical for reovirus binding are in the vicinity of the JAM-A dimer interface. These residues were identified by introducing mutations in several solvent-accessible residues covering most of the JAM-A D1 surface. Residues selected for this analysis are conserved in the human and murine homologues of JAM-A, both of which function as reovirus receptors (Barton *et al.*, 2001b; Prota *et al.*, 2003). Assaying the mutant constructs for the capacity to bind reovirus *in vitro* identified residues Ser57 and Tyr75 as especially important for efficient reovirus attachment (Forrest *et al.*, 2003). Substitutions of Ser57 or Tyr75 with lysine or alanine, respectively, dramatically reduced virus binding, whereas similar substitutions of other JAM-A residues had little or no effect. The location of Ser57 and Tyr75 is shown in Figure 9.6. Ser57 is located close to Tyr75, at the edge of the JAM-A dimer interface. Tyr75 is part of the dimer interface and forms a hydrogen bond with Glu114 (Prota *et al.*, 2003). Thus, Tyr75 would not be accessible to ligand in the context of a dimeric JAM-A structure.

The notion that σ1 engages JAM-A via residues at the JAM-A dimer interface is furthermore supported by structural similarities between the JAM-A dimer interface and a surface-exposed region of σ1. The JAM-A dimer interface is composed of four β-strands (G, F, C, and C') that form a large, concave surface. Strikingly, the size and curvature of this surface closely resemble a similarly concave area of σ1 that is also composed of four β-strands (B, A, D, and G) (Figure 9.7). The superimposition of these two concave β-sheets on each other yields low r.m.s. deviations (Stehle and Dermody, 2004). It is noteworthy that the cluster of conserved σ1 residues previously suggested to be part of the JAM-A binding site (see previous section, (Chappell *et al.*, 2002), and Figure 9.4B) is located at the lower edge of this BADG surface. Therefore, it is possible that σ1-JAM-A complex formation may occur by σ1 substituting for one of the two monomers in the JAM-A homodimer. In such a scenario, the BADG surface of σ1 would engage the GFCC' groove of a monomeric JAM-A molecule in a manner reminiscent of that seen in the JAM-A dimer.

Reovirus binding and tight junctions

JAM-A is localized to tight junctions found in endothelial and epithelial barriers where it engages in both homotypic and heterotypic interactions. JAM-A is also expressed by several types of non-polarized cells, including dendritic cells (Rescigno *et al.*, 2001a). Tight junctions play important roles in the establishment of distinct apical and basolateral regions in polarized cells and form a barrier to the transport of water and solutes (Balda and Matter, 2000; Zahraoui *et al.*, 2000). Tight junctions are composed of complex networks of interacting fibrils that seal the paracellular space. Occludin and members of the claudin

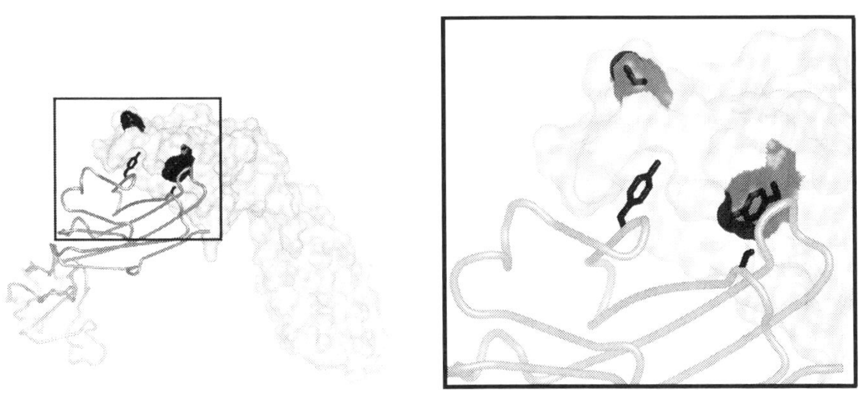

Figure 9.6 Residues at the hJAM-A dimer interface have been implicated in binding to σ1. Mutation of residues Ser57 and Tyr75 at the hJAM-A dimer interface dramatically reduces σ1 binding (Forrest *et al.*, 2003). In this representation, one monomer of JAM-A is shown as a ribbon, while the other is surface-contoured. The side chains of residues Ser57 and Tyr75 are shown in black. An expanded view of the JAM-A dimer interface is shown on the right.

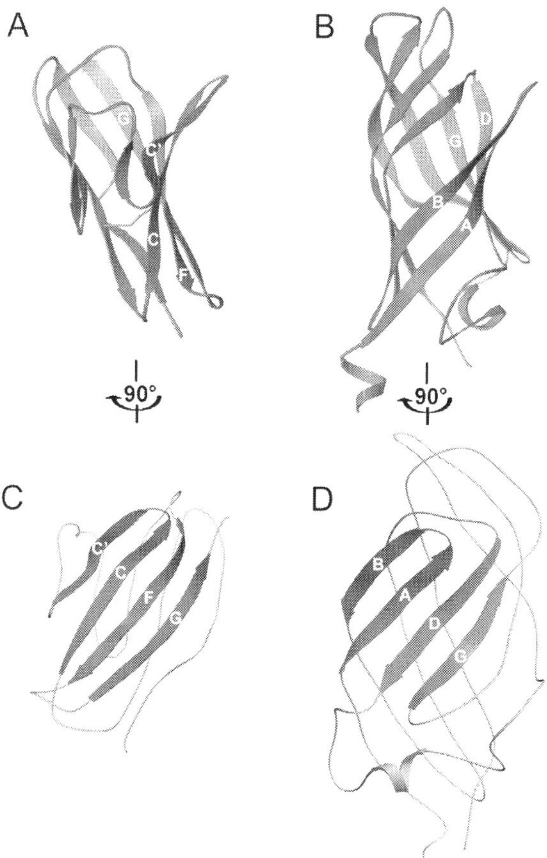

Figure 9.7 Four β-strands at the surface of σ1 mimic four β-strands at the hJAM-A dimer interface. Ribbon drawings of β-strands GFCC′ of hJAM-A (panels A and C) and β-strands BADG of σ1 (panels B and D) reveal a striking similarity. In both molecules, these four-stranded β-sheets are concave with a similar curvature. The GFCC′ sheet of JAM-A forms the JAM-A-JAM-A dimer interface, whereas the BADG sheet of σ1 lies just above the putative JAM-A binding site. The structural similarity suggests that σ1 may engage JAM-A by binding to its dimer interface in a manner similar to that seen in the JAM-A-JAM-A homodimer. Two orthogonal views are shown.

family are concentrated in the fibrils and make important contributions to tight junction barrier function (Tsukita *et al.*, 2001). These proteins are tethered to the actin cytoskeleton by interactions with numerous cytoplasmic proteins, including ZO-1 (Cordenonsi *et al.*, 1999; Wittchen *et al.*, 1999).

How might reovirus engage JAM-A in intestinal epithelia? We think it unlikely that reovirus contacts JAM-A in mature tight junctions in enterocytes at the villous tips. JAM-A is sequestered at these sites, perhaps explaining why mature enterocytes are not readily infected by reovirus (Rubin *et al.*, 1985; Wolf *et al.*, 1987; Bass *et al.*, 1988). Instead, it is possible that reovirus engages JAM-A on M cells to initiate infection in the gut. This interaction would promote transcytosis of the virus to underlying Peyer's patches and epithelial cells in the villous crypts, a scenario that is consistent with current data (Wolf *et al.*, 1981; Wolf *et al.*, 1983; Rubin *et al.*, 1985; Bass *et al.*, 1988; Morrison *et al.*, 1991; Amerongen *et al.*, 1994; Helander *et al.*, 2003). Alternatively, reovirus may bind JAM-A expressed on dendritic cells at the luminal surface, which are known to shuttle bacteria across epithelial monolayers by opening tight junctions and sampling the intestinal lumen (Rescigno *et al.*, 2001b). This interaction could lead to productive infection of these cells and resultant systemic dissemination as a consequence of dendritic cell migration.

The capacity of reovirus to use JAM-A as a receptor also has implications for reovirus infection and pathogenesis in the CNS. JAM-A-specific mAbs promote tight junction breakdown of the endothelium lining CNS blood vessels, resulting in enhanced disease in response to viral or bacterial infections (Lechner *et al.*, 2000). If reovirus-JAM-A interactions lead to a similar destabilization of tight junctions in CNS endothelium, this effect might promote breakdown of the blood-brain barrier, permitting cerebral oedema and neural inflammation, conditions associated with reovirus-induced encephalitis (Tyler and Fields, 1988).

Relationship of attachment to entry

Role of sialic acid in cell attachment and pathogenesis

The role of sialic acid in reovirus attachment and infection has been studied using reovirus strains that differ only in the capacity to bind this carbohydrate. To generate these strains, the S1 gene segments of a MEL-adapted strain, T3C44-MA, and its non-sialic-acid-binding parental strain, T3C44, were introduced into the genetic background of T1L by reassortment (Barton *et al.*, 2001a). The resultant viruses, termed T3SA+ and T3SA-, respectively, differ by a proline-to-leucine substitution at amino acid 204 in σ1. Studies using T3SA+ and T3SA- suggest that reovirus uses an adhesion-strengthening mechanism of attachment (Barton *et al.*, 2001a). For sialic acid-binding reovirus strains, the initial interaction between the viral particle and the host cell is likely to be mediated by cell-surface sialic acid. This interaction may involve multivalent virion-sialic acid interactions and has an apparent avidity of $\sim 10^{-9}$ M as measured on a biosensor surface (Barton *et al.*, 2001a). By virtue of its rapid association rate, this interaction adheres the virion to the cell surface, thereby facilitating its subsequent interaction with JAM-A. Using surface plasmon resonance, the K_D of the purified head domain of σ1 for the extracellular domain of hJAM-A was calculated to be 6×10^{-8} M (Barton *et al.*, 2001b). Such a two-step adhesion process is conceptually similar to that postulated for other viruses such as HIV (Dalgleish *et al.*, 1984; Klatzmann *et al.*, 1984) and herpes simplex virus (Spear *et al.*, 1992; Krummenacher *et al.*, 1999).

In addition to attaching the virus to the cell surface, the capacity of type 3 reovirus to bind sialic acid influences tissue tropism and disease phenotypes. Both T3SA+ and T3SA- grow well in the intestine following peroral inoculation of newborn mice and disseminate to distant sites (Barton *et al.*, 2003). However, T3SA+ produces higher titres in the brain and liver at early timepoints after inoculation. These observations, which correlate well with findings made using cultured cells (Barton *et al.*, 2001a), indicate that engagement of sialic acid-containing receptors allows T3SA+ to spread faster *in vivo*. Additionally, newborn mice infected with T3SA+ but not T3SA– develop jaundice, steatorrhoea, and oily fur (Barton *et al.*, 2003). These findings are consistent with previous studies of reovirus infections of mice and correlate with injury to bile duct epithelium (Papadimitriou, 1968; Phillips *et al.*, 1969; Parashar *et al.*, 1992; Wilson *et al.*, 1994). Concordantly, liver tissue from mice infected with T3SA+ demonstrates intense inflammation focused at intrahepatic bile ducts and high levels of T3SA+ antigen in bile duct epithelial cells (Barton *et al.*, 2003). Thus, utilization of sialic acid as a coreceptor targets reovirus to bile duct epithelial cells. Interestingly, the disease produced by T3SA+ is similar in some respects to biliary atresia in human infants, which has been linked to reovirus in approximately 50% of cases in one series (Tyler *et al.*, 1998).

Conformational changes of σ1

During viral disassembly, proteolytic cleavage of reovirus outer-capsid protein σ3, which lies in close proximity to σ1 on the virion surface (Nibert *et al.*, 1995; Nason *et al.*, 2001), coincides with a change in σ1 from a retracted to an extended conformation (Furlong *et al.*, 1988; Dryden *et al.*, 1993; Nibert *et al.*, 1995). The fact that σ1 undergoes dramatic structural changes upon disassembly of virions to ISVPs suggests that the two conformers of σ1 facilitate distinctly different functions. In general, viral attachment proteins must meet two absolute requirements: (i) evasion of host neutralizing immune responses by avoiding display of antigenic epitopes and (ii) efficient attachment to target cells. Viruses have developed different strategies to satisfy these requirements. Some viruses (such as rhinoviruses) hide their receptor-binding epitopes in recessed 'canyons' that are inaccessible to antibodies (Rossmann, 1989; Rossmann *et al.*, 2000), while others (such as HIV) cover their receptor-binding epitopes with poorly immunogenic glycans and also undergo dramatic conformational changes upon receptor engagement (Kwong *et al.*, 1998). In the case of mammalian reoviruses, different structural forms of σ1 may allow the virus to evade host defence mechanisms while also enabling it to bind receptors. In the retracted state, the C-terminal receptor-binding epitopes may be partially shielded on the virion surface, likely by contacts with other outer-capsid proteins. In the extended state, the σ1 receptor-binding epitopes protrude farther from the virion surface and may be better positioned to bind or crosslink receptors. The crystal structure of the C-terminal half of σ1, and the model of full-length σ1 derived from it (Figure 9.3), offer some clues about what the two states might look like and how a structural transition between them might occur.

The extended state of σ1

The crystallized σ1 trimer features an unusual cluster of aspartic acids at the head trimer interface (Figure 9.2B). Molecular dynamics simulations indicate that the three Asp345 side chains must be protonated (and therefore uncharged) in the crystal structure since their deprotonation would destabilize the trimer and lead to a partial separation of the head domains (Cavalli *et al.*, 2004). Thus, formation of the trimer seen in the crystal structure is likely facilitated by conditions that allow Asp345 to be protonated. The low pH in the endocytic compartment would provide one possible environment where such conditions exist. Once assembled, the σ1 trimer is remarkably stable and resistant to proteases such as trypsin, even at neutral pH (P. Schelling and T. Stehle, unpublished observations). Furthermore, the Asp345 side chains are inaccessible to solvent in the crystallized fragment, and thus this form of σ1 is unlikely to undergo a conformational change that would be linked to deprotonation of Asp345 (Chappell *et al.*, 2002). Thus, the σ1 trimer seen in the crystal structure appears to be the end product of a structural transition. Given its overall appearance and stability, this form probably represents the extended state of σ1.

The retracted state of σ1

There is little experimental data to predict the structure of the retracted conformation of σ1. We think it likely that such a state would involve a partially detrimerized σ1 protein, in which the head domains (and perhaps also the β-spiral regions) are separated from each other. The σ1 head is thought to be close to the σ3 protein in intact virions (Nibert *et al.*, 1995; Nason *et al.*, 2001), and this relationship is difficult to reconcile with a fully extended structure of σ1 in which the head protrudes from the virion surface by approximately 35 nm (Furlong *et al.*, 1988). Thus, the σ1 head must be folded back onto the particle surface in the retracted state. An event such as the engagement of a receptor, or a change in pH, could destabilize the retracted conformation and favour the formation of the extended trimer seen in the crystal structure. However, given the remarkable flexibility of the σ1 protein, it is also possible that the entire trimer exists in a compacted form with a kink at the coiled-coil/spiral junction. Such an arrangement might also bring the σ1 head domain in close proximity to the virion surface and result in a less extended structure. Finally, incomplete formation of the β-spiral region may lead to a thicker, less extended structure along the vertical axis in which both the coiled-coil region and globular head domain are largely left intact. The precise nature of the retracted form of σ1 will require higher-resolution structural studies of intact virions.

Similarities in the attachment strategies of reovirus and adenovirus

Attachment proteins

One of the most remarkable aspects of reovirus attachment is that the mechanism of cell entry used by mammalian reoviruses does not appear to be shared by other *Reoviridae* family members such as the rotaviruses and orbiviruses. Although rotaviruses and orbiviruses share a number of general properties with reoviruses, neither of these viruses contains a protein that resembles σ1. For example, although the rotavirus haemagglutinin VP4 also forms spikes that protrude from the viral surface, these spikes contain globular protein and are not fibre-like in nature (Dormitzer *et al.*, 2002). Moreover, VP4 emanates from channels surrounding the 12 vertices of the icosahedron (Estes, 2001), whereas σ1 is inserted into these vertices. The viral haemagglutinin of orbiviruses, VP2, differs in similar ways from σ1 (Roy, 2001). No evidence to date suggests that either rotaviruses or orbiviruses attach to proteins at tight junctions.

Surprisingly, the attachment and entry mechanisms of mammalian reoviruses appear to be more similar to those of adenoviruses than those of other *Reoviridae* family members. Adenoviruses and reoviruses clearly belong to different families, yet as early as 1988, the head-and-tail morphology of σ1 was described as being 'reminiscent of the fibre protein of adenoviruses' (Furlong *et al.*, 1988). Indeed, strikingly similar X-ray structures of both molecules showed that the similarities between them extend to the atomic level (van Raaij *et al.*, 1999; Chappell *et al.*, 2002). As reviewed elsewhere (Stehle and Dermody, 2003), the extensive parallels between the two attachment proteins include a conserved fold of the globular head domain and a conserved β-spiral trimerization motif. Furthermore, both attachment proteins possess a similar mechanism of flexibility just below the head domain that allows the head to move with respect to the tail. Finally, in both cases there exists a unique symmetry mismatch between the trimeric attachment protein and the pentameric capsid protein into which it is inserted. The adenovirus fibre is inserted into the penton base protein, whereas the pentameric λ2 protein forms an insertion slot for σ1. Symmetry mismatches of this nature are rare. They are usually characterized by weak interactions and indicate a tendency for rearrangement. It should be noted, however, that although both are non-enveloped, icosahedral viruses, adenoviruses and reoviruses differ substantially in design and capsid composition. Most importantly, adenoviruses have a dsDNA genome, whereas reoviruses contain dsRNA. Nevertheless, the observed similarities between their attachment proteins are highly unusual and suggest a common evolutionary origin.

Receptors

Parallels between the reovirus and adenovirus attachment proteins extend to their receptors, JAM-A and CAR, respectively. Both JAM-A and CAR are integral components of tight junctions (Martin-Padura *et al.*, 1998; Cohen *et al.*, 2001), and thus they are found at similar locations. Furthermore, they both belong to the Ig superfamily and contain two extracellular domains, the N-terminal, most membrane-distal of which engages the virus. Most strikingly, both receptors form highly similar and unique homodimers via homologous surfaces of their respective N-terminal Ig domains (Prota *et al.*, 2003). The extent of the similarities between the viral attachment proteins and the corresponding receptors suggests that reovirus and adenovirus use conserved strategies for cell attachment and entry. Structural information about how adenovirus attaches to CAR has been available for several years (Bewley *et al.*, 1999). Given the extensive similarities in the structures used by reovirus and adenovirus to attach to cells, we anticipate that their respective receptor complexes also will share many properties.

Conclusion

Analysis of reovirus attachment mechanisms indicates that multiple receptors are used to effect stable virus binding to cells. However, a precise understanding of the serotype-dependent differences in tropism exhibited by reovirus in the murine CNS remains elusive. Type 1 and type 3 reoviruses vary in the types of cell-surface carbohydrates used as coreceptors (Chappell *et al.*, 2000), but all reovirus serotypes bind to JAM-A (Barton *et al.*, 2001b; Campbell *et al.*, 2005). These observations make it unlikely that JAM-A is the sole determinant of reovirus tropism. It is possible that JAM-A serves as a serotype-independent reovirus receptor at some sites within the host and other receptors, perhaps carbohydrate in nature, confer serotype-dependent tropism. Definitive assessment of the role of JAM-A in reovirus pathogenesis awaits the results of studies using JAM-A-null mice.

Differences in receptor utilization also might influence pathogenesis by virtue of activating different types of signalling pathways. Since type 1 and type 3 reoviruses differ in the capacity to induce apoptosis (Tyler *et al.*, 1995; Rodgers *et al.*, 1997; Connolly *et al.*, 2001), a property linked to receptor binding (Barton *et al.*, 2001b; Connolly *et al.*, 2001), it is possible that post-attachment signalling plays a role in the production of disease. Support for this idea comes from studies of reovirus-induced myocarditis, in which treatment of mice with inhibitors of calpain to block pro-apoptotic signalling diminishes tissue injury (DeBiasi *et al.*, 2001). Of note, intercellular junctions are sites of enhanced membrane recycling, endocytic uptake, and intracellular signalling (Zahraoui *et al.*, 2000). Therefore, it is possible that engagement of the junction-associated JAM-A protein by reovirus alters the physiologic function of this molecule and leads to a pathological outcome.

Although significant progress has been made in the functional and structural characterization of reovirus attachment protein σ1 and its receptor JAM-A, the adaptive advantage of reovirus binding to JAM-A and carbohydrate receptors remains to be established. Moreover, additional proteins appear to be required for reovirus internalization following initial engagement of JAM-A. Reovirus infection is diminished by β1 integrin-specific antibodies but not by antibodies specific for other integrin subunits (Maginnis *et al.*, 2006). Concordantly, cells deficient in β1 integrin expression have a diminished susceptibility to reovirus infection due to a post-attachment block to viral entry (Maginnis *et al.*, 2006). These data provide strong evidence that following attachment to JAM-A, β1 integrins facilitate internalization of reovirus into cells. How JAM-A and β1 integrins interact to orchestrate reovirus cell entry is not understood.

Despite all of the information gathered about σ1 structure and function, much of σ1 biology remains a mystery. Although data are now available on the structure of the C-terminal region of σ1, putative σ1 conformational changes, and interactions between σ1 and its cellular receptors, questions concerning most of the unique structural features and functional properties of σ1 have only been partially answered. Thus, there are many important avenues left to explore. Experimental approaches using high-resolution structural analyses, coupled with well-defined viral variants and mouse strains with targeted deficiencies in specific components of the cell entry apparatus, should be especially illuminating. We also find it intriguing that clues about reovirus attachment and cell entry might come from studies of unrelated viruses that have evolved similar strategies to bind and enter host cells.

Acknowledgments

We thank members of our laboratory for many useful discussions and Jim Chappell for review of the manuscript. This research was supported by Public Health Service awards T32 CA09385 (J.A.C.), R37 AI38296 (T.S.D.), and R01 GM67853 (T.S. and T.S.D.) and the Elizabeth B. Lamb Center for Pediatric Research.

References

Amerongen, H.M., Wilson, G.A., Fields, B.N., and Neutra, M.R. (1994). Proteolytic processing of reovirus is required for adherence to intestinal M cells. J. Virol. *68*, 8428–8432.

Armstrong, G.D., Paul, R.W., and Lee, P.W. (1984). Studies on reovirus receptors of L cells, virus binding characteristics and comparison with reovirus receptors of erythrocytes. Virology *138*, 37–48.

Arrate, M.P., Rodriguez, J.M., Tran, T.M., Brock, T.A., and Cunningham, S.A. (2001). Cloning of human junctional adhesion molecule 3 (JAM3) and its identification as the JAM2 counter-receptor. J. Biol. Chem. *276*, 45826–45832.

Aurrand-Lions, M., Duncan, L., Ballestrem, C., and Imhof, B.A. (2001). JAM-2, a novel immunoglobulin superfamily molecule, expressed by endothelial and lymphatic cells. J. Biol. Chem. *276*, 2733–2741.

Baer, G.S., and Dermody, T.S. (1997). Mutations in reovirus outer-capsid protein σ3 selected during persistent infections of L cells confer resistance to protease inhibitor E64. J. Virol. *71*, 4921–4928.

Baer, G.S., Ebert, D.H., Chung, C.J., Erickson, A.H., and Dermody, T.S. (1999). Mutant cells selected during persistent reovirus infection do not express mature cathepsin L and do not support reovirus disassembly. J. Virol. *73*, 9532–9543.

Balda, M.S., and Matter, K. (2000). Transmembrane proteins of tight junctions. Semin. Cell Dev. Biol. *11*, 281–289.

Barton, E.S., Connolly, J.L., Forrest, J.C., Chappell, J.D., and Dermody, T.S. (2001a). Utilization of sialic acid as a coreceptor enhances reovirus attachment by multistep adhesion strengthening. J. Biol. Chem. *276*, 2200–2211.

Barton, E.S., Forrest, J.C., Connolly, J.L., Chappell, J.D., Liu, Y., Schnell, F.J., Nusrat, A., Parkos, C.A., and Dermody, T.S. (2001b). Junction adhesion molecule is a receptor for reovirus. Cell *104*, 441–451.

Barton, E.S., Youree, B.E., Ebert, D.H., Forrest, J.C., Connolly, J.L., Valyi-Nagy, T., Washington, K., Wetzel, J.D., and Dermody, T.S. (2003). Utilization of sialic acid as a coreceptor is required for reovirus-induced biliary disease. J. Clin. Invest. *111*, 1823–1833.

Bass, D.M., Trier, J.S., Dambrauskas, R., and Wolf, J.L. (1988). Reovirus type 1 infection of small intestinal epithelium in suckling mice and its effect on M cells. Lab. Invest. *58*, 226–235.

Bassel-Duby, R., Jayasuriya, A., Chatterjee, D., Sonenberg, N., Maizel, J.V.Jr., and Fields, B.N. (1985). Sequence of reovirus haemagglutinin predicts a coiled-coil structure. Nature *315*, 421–423.

Bassel-Duby, R., Nibert, M., Homcy, C., Fields, B., and Sawutz, D. (1987). Evidence that the sigma 1 protein of reovirus serotype 3 is a multimer. J. Virol. *61*, 1834–1841.

Bazzoni, G., Martinez-Estrada, O.M., Mueller, F., Nelboeck, P., Schmid, G., Bartfai, T., Dejana, E., and Brockhaus, M. 2000a. Homophilic interaction of junctional adhesion molecule. J. Biol. Chem. *275*, 30970–30976.

Bazzoni, G., Martinez-Estrada, O.M., Orsenigo, F., Cordenonsi, M., Citi, S., and Dejana, E. 2000b. Interaction of junctional adhesion molecule with the tight junction components ZO-1, cingulin, and occludin. J. Biol. Chem. *275*, 20520–20526.

Bewley, M.C., Springer, K., Zhang, Y.B., Freimuth, P., and Flanagan, J.M. (1999). Structural analysis of the mechanism of adenovirus binding to its human cellular receptor, CAR. Science *286*, 1579–1583.

Bisaillon, M., Senechal, S., Bernier, L., and Lemay, G. (1999). A glycosyl hydrolase activity of mammalian reovirus sigma1 protein can contribute to viral infection through a mucus layer. J. Mol. Biol. *286*, 759–773.

Borsa, J., Morash, B.D., Sargent, M.D., Copps, T.P., Lievaart, P.A., and Szekely, J.G. (1979). Two modes of entry of reovirus particles into L cells. J. Gen. Virol. *45*, 161–170.

Borsa, J., Sargent, M.D., Lievaart, P.A., and Copps, T.P. (1981). Reovirus, evidence for a second step in the intracellular uncoating and transcriptase activation process. Virology *111*, 191–200.

Campbell, J.A., Schelling, P., Wetzel, J.D., Johnson, E.M., Forrest, J.C., Wilson, G.A.R., Aurrand-Lions, M., Imhof, B.A., Stehle, T., and Dermody, T.S. (2005). Junctional adhesion molecule-A serves as a receptor for prototype and field-isolate strains of mammalian reovirus. J. Virol. *79*, 7967–7978.

Cavalli, A., Prota, A.E., Stehle, T., Dermody, T.S., Recanatini, M., Folkers, G., and Scapozza, L. (2004). A molecular dynamics study of reovirus attachment protein σ1 reveals conformational changes in σ1 structure. Biophys. J. *86*, 3423–3431.

Chandran, K., and Nibert, M.L. (2003). Animal cell invasion by a large nonenveloped virus, reovirus delivers the goods. Trends Microbiol. *11*, 374–382.

Chandran, K., Zhang, X., Olson, N.H., Walker, S.B., Chappell, J.D., Dermody, T.S., Baker, T.S., and Nibert, M.L. (2001). Complete *in vitro* assembly of the reovirus outer capsid produces highly infectious particles suitable for genetic studies of the receptor-binding protein. J. Virol. *75*, 5335–5342.

Chang, C.T., and Zweerink, H.J. (1971). Fate of parental reovirus in infected Cell Virology *46*, 544–555.

Chappell, J.D., Duong, J.L., Wright, B.W., and Dermody, T.S. (2000). Identification of carbohydrate-binding domains in the attachment proteins of type 1 and type 3 reoviruses. J. Virol. *74*, 8472–8479.

Chappell, J.D., Gunn, V.L., Wetzel, J.D., Baer, G.S., and Dermody, T.S. (1997). Mutations in type 3 reovirus that determine binding to sialic acid are contained in the fibrous tail domain of viral attachment protein sigma1. J. Virol. *71*, 1834–1841.

Chappell, J.D., Prota, A.E., Dermody, T.S., and Stehle, T. (2002). Crystal structure of reovirus attachment protein sigma1 reveals evolutionary relationship to adenovirus fiber. EMBO J. *21*, 1–11.

Co, M.S., Gaulton, G.N., Tominaga, A., Homcy, C.J., Fields, B.N., and Greene, M.I. (1985). Structural similarities between the mammalian beta-adrenergic and reovirus type 3 receptors. Proc. Natl Acad. Sci. USA *82*, 5315–5318.

Cohen, C.J., Shieh, J.T., Pickles, R.J., Okegawa, T., Hsieh, J.T., and Bergelson, J.M. (2001). The coxsackievirus and adenovirus receptor is a transmembrane component of the tight junction. Proc. Natl Acad. Sci. USA *98*, 15191–15196.

Connolly, J.L., Barton, E.S., and Dermody, T.S. (2001). Reovirus binding to cell surface sialic acid potentiates virus-induced apoptosis. J. Virol. *75*, 4029–4039.

Cordenonsi, M., D'Atri, F., Hammar, E., Parry, D.A., Kendrick-Jones, J., Shore, D., and Citi, S. (1999). Cingulin contains globular and coiled-coil domains and interacts with ZO-1, ZO-2, ZO-3, and myosin. J. Cell Biol. *147*, 1569–1582.

Crick, F.H.C. (1953). The packing of alpha-helices, simple coiled-coils. Acta Crystallogr. *6*, 689–697.

Cunningham, S.A., Arrate, M.P., Rodriguez, J.M., Bjercke, R.J., Vanderslice, P., Morris, A.P., and Brock, T.A. (2000). A novel protein with homology to the junctional adhesion molecule. Characterization of leukocyte interactions. J. Biol. Chem. *275*, 34750–34756.

Dalgleish, A.G., Beverley, P.C.L., Clapham, P.R., Crawford, D.H., Greaves, M.F., and Weiss, R.A. (1984). The CD4 (T4) antigen is an essential component of the receptor for the AIDS retrovirus. Nature *312*, 763–767.

DeBiasi, R., Edelstein, C., Sherry, B., and Tyler, K. (2001). Calpain inhibition protects against virus-induced apoptotic myocardial injury. J. Virol. *75*, 351–361.

Dermody, T.S., Nibert, M.L., Bassel-Duby, R., and Fields, B.N. (1990). Sequence diversity in S1 genes and S1 translation products of 11 serotype 3 reovirus strains. J. Virol. *64*, 4842–4850.

Dormitzer, P.R., Sun, Z.Y., Wagner, G., and Harrison, S.C. (2002). The rhesus rotavirus VP4 sialic acid binding domain has a galectin fold with a novel carbohydrate binding site. EMBO J. *21*, 885–897.

Dryden, K.A., Wang, G., Yeager, M., Nibert, M.L., Coombs, K.M., Furlong, D.B., Fields, B.N., and Baker, T.S. (1993). Early steps in reovirus infection are associated with dramatic changes in supramolecular structure and protein conformation, analysis of virions and subviral particles by cryoelectron microscopy and image reconstruction. J. Cell Biol. *122*, 1023–1041.

Duncan, R., Horne, D., Cashdollar, L.W., Joklik, W.K., and Lee, P.W. (1990). Identification of conserved domains in the cell attachment proteins of the three serotypes of reovirus. Virology *174*, 399–409.

Duncan, R., Horne, D., Strong, J.E., Leone, G., Pon, R.T., Yeung, M.C., and Lee, P.W. (1991). Conformational and functional analysis of the C-terminal globular head of the reovirus cell attachment protein. Virology *182*, 810–819.

Ebnet, K., Schulz, C.U., Meyer Zu Brickwedde, M.K., Pendl, G.G., and Vestweber, D. (2000). Junctional adhesion molecule interacts with the PDZ domain-containing proteins AF-6 and ZO-1. J. Biol. Chem. *275*, 27979–27988.

Ehrlich, M., Boll, W., Van Oijen, A., Hariharan, R., Chandran, K., Nibert, M.L., and Kirchhausen, T. (2004). Endocytosis by random initiation and stabilization of clathrin-coated pits. Cell *118*, 591–605.

Eisen, M.B., Sabesan, S., Skehel, J.J., and Wiley, D.C. (1997). Binding of the influenza A virus to cell-surface receptors, structures of five hemagglutinin-sialyloligosaccharide complexes determined by X- ray crystallography. Virology *232*, 19–31.

Epstein, J., Eichbaum, Q., Sheriff, S., and Ezekowitz, R.A. (1996). The collectins in innate immunity. Curr. Opin. Immunol. *8*, 29–35.

Estes, M.K. (2001). Rotaviruses and their replication. In Fields Virology ed. D.M. Knipe and P.M. Howley. Lippincott-Raven, Philadelphia. p. 1747–1785.

Forrest, J.C., Campbell, J.A., Schelling, P., Stehle, T., and Dermody, T.S. (2003). Structure-function analysis of reovirus binding to junctional adhesion molecule 1, implications for the mechanism of reovirus attachment. J. Biol. Chem. *278*, 48434–48444.

Fraser, R.D., Furlong, D.B., Trus, B.L., Nibert, M.L., Fields, B.N., and Steven, A.C. (1990). Molecular structure of the cell-attachment protein of reovirus, correlation of computer-processed electron micrographs with sequence-based predictions. J. Virol. *64*, 2990–3000.

Furlong, D.B., Nibert, M.L., and Fields, B.N. (1988). Sigma 1 protein of mammalian reoviruses extends from the surfaces of viral particles. J. Virol. *62*, 246–256.

Gentsch, J.R., and Pacitti, A.F. (1985). Effect of neuraminidase treatment of cells and effect of soluble glycoproteins on type 3 reovirus attachment to murine L cells. J. Virol. *56*, 356–364.

Guardado Calvo, P., Fox, G.C., Hermo Parrado, X.L., Llamas-Saiz, A.L., Costas, C., Martínez-Costas, J., Benavente, J., and van Raaij, M.J. (2005). Structure of the carboxy-terminal receptor-binding domain of avian reovirus fibre sigmaC. J. Mol. Biol. *354*, 137–149.

Hamazaki, Y., Itoh, M., Sasaki, H., Furuse, M., and Tsukita, S. (2002). Multi-PDZ domain protein 1 (MUPP1) is concentrated at tight junctions through its possible interaction with claudin-1 and junctional adhesion molecule. J. Biol. Chem. *277*, 455–461.

Harpaz, Y., and Chothia, C. (1994). Many of the immunoglobulin superfamily domains in cell adhesion molecules and surface receptors belong to a new structural set which is close to that containing variable domains. J. Mol. Biol. *238*, 528–539.

Helander, A., Silvey, K.J., Mantis, N.J., Hutchings, A.B., Chandran, K., Lucas, W.T., Nibert, M.L., and Neutra, M.R. (2003). The viral sigma1 protein and glycoconjugates containing alpha2–3-linked sialic acid are involved in type 1 reovirus adherence to M cell apical surfaces. J. Virol. *77*, 7964–7977.

Hooper, J.W., and Fields, B.N. 1996a. Monoclonal antibodies to reovirus σ1 and μ1 proteins inhibit chromium release from mouse L cells. J. Virol. *70*, 672–677.

Hooper, J.W., and Fields, B.N. 1996b. Role of the μ1 protein in reovirus stability and capacity to cause chromium release from host cells. J. Virol. *70*, 459–467.

Jones, S., and Thornton, J.M. (1996). Principles of protein–protein interactions. Proc. Natl Acad. Sci. USA *93*, 13–20.

Klatzmann, D., Champagne, E., Chamaret, S., Gruest, J., Guetard, D., Hercend, T., Gluckman, J.C., and Montagnier, L. (1984). T-lymphocyte T4 molecule behaves as the receptor for human retrovirus LAV. Nature *312*, 767–768.

Krummenacher, C., Rux, A.H., Whitbeck, J.C., Ponce-de-Leon, M., Lou, H., Baribaud, I., Hou, W., Zou, C., Geraghty, R.J., Spear, P.G., Eisenberg, R.J., and Cohen, G.H. (1999). The first immunoglobulin-like domain of HveC is sufficient to bind herpes simplex virus gD with full affinity, while the third domain is involved in oligomerization of HveC. J. Virol. *73*, 8127–8137.

Kwong, P.D., Wyatt, R., Robinson, J., Sweet, R.W., Sodroski, J., and Hendrickson, W.A. (1998). Structure of an HIV gp120 envelope glycoprotein in complex with the CD4 receptor and a neutralizing human antibody. Nature *393*, 648–659.

Lechner, F., Sahrbacher, U., Suter, T., Frei, K., Brockhaus, M., Koedel, U., and Fontana, A. (2000). Antibodies to the junctional adhesion molecule cause disruption of endothelial cells and do not prevent leukocyte influx into the meninges after viral or bacterial infection. J. Infect. Dis. *182*, 978–982.

Lee, P.W., Hayes, E.C., and Joklik, W.K. (1981). Protein sigma 1 is the reovirus cell attachment protein. Virology *108*, 156–163.

Leone, G., Coffey, M.C., Gilmore, R., Duncan, R., Maybaum, L., and Lee, P.W. (1996). C-terminal trimerization, but not N-terminal trimerization, of the reovirus cell attachment protein Is a post-translational and Hsp70/ATP- dependent process. J. Biol. Chem. *271*, 8466–8471.

Leone, G., Duncan, R., Mah, D.C., Price, A., Cashdollar, L.W., and Lee, P.W. (1991). The N-terminal heptad repeat region of reovirus cell attachment protein sigma 1 is responsible for sigma 1 oligomer stability and possesses intrinsic oligomerization function. Virology *182*, 336–345.

Leone, G., Maybaum, L., and Lee, P.W. (1992). The reovirus cell attachment protein possesses two independently active trimerization domains, basis of dominant negative effects. Cell *71*, 479–488.

Lucia-Jandris, P., Hooper, J.W., and Fields, B.N. (1993). Reovirus M2 gene is associated with chromium release from mouse L cells. J. Virol. *67*, 5339–5345.

Maginnis, M.S., Forrest, J.C., Kopecky-Bromberg, S.A., Dickeson, S.K., Santoro, S.A., Zutter, M.M., Nemerow, G.R., Bergelson, J.M., and Dermody. T.S. (2006). β1 integrin mediates internalization of mammalian reovirus. J. Virol. *80*, 2760–2770.

Mah, D.C., Leone, G., Jankowski, J.M., and Lee, P.W. (1990). The N-terminal quarter of reovirus cell attachment protein sigma 1 possesses intrinsic virion-anchoring function. Virology *179*, 95–103.

Martinez-Estrada, O.M., Villa, A., Breviario, F., Orsenigo, F., Dejana, E., and Bazzoni, G. (2001). Association of junctional adhesion molecule with calcium/calmodulin-dependent serine protein kinase (CASK/LIN-2) in human epithelial caco-2 cells. J. Biol. Chem. *276*, 9291–9296.

Martin-Padura, I., Lostaglio, S., Schneemann, M., Williams, L., Romano, M., Fruscella, P., Panzeri, C., Stoppacciaro, A., Ruco, L., Villa, A., Simmons, D., and Dejana, E. (1998). Junctional adhesion molecule, a novel member of the immunoglobulin superfamily that distributes at intercellular junctions and modulates monocyte transmigration. J. Cell Biol. *142*, 117–127.

Masri, S.A., Nagata, L., Mah, D.C., and Lee, P.W. (1986). Functional expression in Escherichia coli of cloned reovirus S1 gene encoding the viral cell attachment protein sigma 1. Virology *149*, 83–90.
Mitraki, A., Miller, S., and van Raaij, M.J. (2002). Review, conformation and folding of novel Beta-structural elements in viral fiber proteins, the triple Beta-spiral and triple Beta-helix. J. Struct. Biol. *137*, 236–247.
Morrison, L.A., Sidman, R.L., and Fields, B.N. (1991). Direct spread of reovirus from the intestinal lumen to the central nervous system through vagal autonomic nerve fibers. Proc. Natl Acad. Sci. USA *88*, 3852–3856.
Nagata, L., Masri, S.A., Pon, R.T., and Lee, P.W. (1987). Analysis of functional domains on reovirus cell attachment protein sigma 1 using cloned S1 gene deletion mutants. Virology *160*, 162–168.
Nason, E.L., Wetzel, J.D., Mukherjee, S.K., Barton, E.S., Prasad, B.V., and Dermody, T.S. (2001). A monoclonal antibody specific for reovirus outer-capsid protein σ3 inhibits σ1-mediated hemagglutination by steric hindrance. J. Virol. *75*, 6625–6634.
Nibert, M.L., Chappell, J.D., and Dermody, T.S. (1995). Infectious subvirion particles of reovirus type 3 Dearing exhibit a loss in infectivity and contain a cleaved sigma 1 protein. J. Virol. *69*, 5057–5067.
Nibert, M.L., Dermody, T.S., and Fields, B.N. (1990). Structure of the reovirus cell-attachment protein, a model for the domain organization of sigma 1. J. Virol. *64*, 2976–2989.
Ostermann, G., Weber, K.S., Zernecke, A., Schroder, A., and Weber, C. (2002). JAM-1 is a ligand of the beta2 integrin LFA-1 involved in transendothelial migration of leukocytes. Nat. Immunol. *14*, 14.
Ozaki, H., Ishii, K., Horiuchi, H., Arai, H., Kawamoto, T., Okawa, K., Iwamatsu, A., and Kita, T. (1999). Cutting edge, combined treatment of TNF-alpha and IFN-gamma causes redistribution of junctional adhesion molecule in human endothelial cells. J. Immunol. *163*, 553–557.
Pacitti, A., and Gentsch, J.R. (1987). Inhibition of reovirus type 3 binding to host cells by sialylated glycoproteins is mediated through the viral attachment protein. J. Virol. *61*, 1407–1415.
Papadimitriou, J.M. (1968). The biliary tract in acute murine reovirus 3 infection. Am.J. Pathol. *52*, 595–601.
Parashar, K., Tarlow, M.J., and McCrae, M.A. (1992). Experimental reovirus type 3-induced murine biliary tract disease. J. Pediatr. Surg. *27*, 843–847.
Pelletier, J., Nicholson, R., Bassel-Duby, R., Fields, B.N., and Sonenberg, N. (1987). Expression of reovirus type 3 (Dearing) sigma 1 and sigma s polypeptides in Escherichia coli. J. Gen. Virol. *68*, 135–145.
Phillips, P.A., Keast, D., Papadimitriou, J.M., Walters, M.N., and Stanley, N.F. (1969). Chronic obstructive jaundice induced by reovirus type 3 in weanling mice. Pathology. *1*, 193–203.
Prota, A.E., Campbell, J.A., Schelling, P., Forrest, J.C., Watson, M.J., Peters, T.R., Aurrand-Lions, M., Imhof, B.A., Dermody, T.S., and Stehle, T. (2003). Crystal structure of human junctional adhesion molecule *1*, implications for reovirus binding. Proc. Natl Acad. Sci. USA *100*, 5366–5371.
Reinisch, K.M., Nibert, M.L., and Harrison, S.C. (2000). Structure of the reovirus core at 3.6 A resolution. Nature *404*, 960–967.
Rescigno, M., Rotta, G., Valzasina, B., and Ricciardi-Castagnoli, P. (2001a). Dendritic cells shuttle microbes across gut epithelial monolayers. Immunobiology. *204*, 572–581.
Rescigno, M., Urbano, M., Valzasina, B., Francolini, M., Rotta, G., Bonasio, R., Granucci, F., Kraehenbuhl, J.P., and Ricciardi-Castagnoli, P. (2001b). Dendritic cells express tight junction proteins and penetrate gut epithelial monolayers to sample bacteria. Nat. Immunol. *2*, 361–367.
Rodgers, S.E., Barton, E.S., Oberhaus, S.M., Pike, B., Gibson, C.A., Tyler, K.L., and Dermody, T.S. (1997). Reovirus-induced apoptosis of MDCK cells is not linked to viral yield and is blocked by Bcl-2. J. Virol. *71*, 2540–2546.
Rosen, L. (1960). Serologic grouping of reovirus by hemagglutination-inhibition. Am. J. Hyg. *71*, 242–249.
Rossmann, M.G. (1989). The canyon hypothesis. Hiding the host cell receptor attachment site on a viral surface from immune surveillance. J. Biol. Chem. *264*, 14587–14590.
Rossmann, M.G., Bella, J., Kolatkar, P.R., He, Y., Wimmer, E., Kuhn, R.J., and Baker, T.S. (2000). Cell recognition and entry by rhino- and enteroviruses. Virology *269*, 239–247.
Roy, P. (2001). Orbiviruses. Lippincott Williams & Wilkins, Philadelphia.
Rubin, D.H., Kornstein, M.J., and Anderson, A.G. (1985). Reovirus type 1 intestinal infection, a novel replicative cycle with ileal disease. J. Virol. *53*, 391–398.

Rubin, D.H., Weiner, D.B., Dworkin, C., Greene, M.I., Maul, G.G., and Williams, W.V. (1992a). Receptor utilization by reovirus type 3, distinct binding sites on thymoma and fibroblast cell lines result in differential compartmentalization of virions. Microb. Pathog. *12,* 351–365.
Rubin, D.H., Wetzel, J.D., Williams, W.V., Cohen, J.A., Dworkin, C., and Dermody, T.S. (1992b). Binding of type 3 reovirus by a domain of the sigma 1 protein important for hemagglutination leads to infection of murine erythroleukemia cells. J. Clin. Invest. *90,* 2536–2542.
Saragovi, H.U., Bhandoola, A., Lemercier, M.M., Akbar, G.K., and Greene, M.I. (1995). A receptor that subserves reovirus binding can inhibit lymphocyte proliferation triggered by mitogenic signals. DNA Cell Biol. *14,* 653–664.
Sheriff, S., Chang, C.Y., and Ezekowitz, R.A. (1994). Human mannose-binding protein carbohydrate recognition domain trimerizes through a triple alpha-helical coiled-coil. Nat. Struct. Biol. *1,* 789–794.
Silverstein, S.C., Astell, C., Levin, D.H., Schonberg, M., and Acs, G. (1972). The mechanism of reovirus uncoating and gene activation *in vivo.* Virology *47,* 797–806.
Spear, P.G., Shieh, M.T., Herold, B.C., WuDunn, D., and Koshy, T.I. (1992). Heparan sulfate glycosaminoglycans as primary cell surface receptors for herpes simplex virus. Adv. Exp. Med. Biol. *313,* 341–353.
Stehle, T., and Dermody, T.S. (2004). Structural similarities in the cellular receptors used by adenovirus and reovirus. Viral Immunol. *17,* 129–143.
Stehle, T., and Dermody, T.S. (2003). Structural evidence for common functions and ancestry of the reovirus and adenovirus attachment proteins. Rev. Med. Virol. *13,* 123–132.
Stehle, T., and Harrison, S.C. (1997). High-resolution structure of a polyomavirus VP1-oligosaccharide complex, implications for assembly and receptor binding. EMBO J. *16,* 5139–5148.
Stehle, T., Yan, Y., Benjamin, T.L., and Harrison, S.C. (1994). Structure of murine polyomavirus complexed with an oligosaccharide receptor fragment. Nature *369,* 160–163.
Strong, J.E., Leone, G., Duncan, R., Sharma, R.K., and Lee, P.W. (1991). Biochemical and biophysical characterization of the reovirus cell attachment protein sigma 1, evidence that it is a homotrimer. Virology *184,* 23–32.
Sturzenbecker, L.J., Nibert, M.L., Furlong, D.B., and Fields, B.N. (1987). Intracellular digestion of reovirus particles requires a low pH and is an essential step in the viral infectious cycle. J. Virol. *61,* 2351–2361.
Tang, D., Strong, J.E., and Lee, P.W. (1993). Recognition of the epidermal growth factor receptor by reovirus. Virology *197,* 412–414.
Tosteson, M.T., Nibert, M.L., and Fields, B.N. (1993). Ion channels induced in lipid bilayers by subvirion particles of the nonenveloped mammalian reoviruses. Proc. Natl Acad. Sci. USA *90,* 10549–10552.
Tsukita, S., Furuse, M., and Itoh, M. (2001). Multifunctional strands in tight junctions. Nat. Rev. Mol. Cell Biol. *2,* 285–293.
Tyler, K.L. (2001). Mammalian reoviruses. In Fields Virology ed. D.M. Knipe and P.M. Howley. Lippincott-Raven Press, Philadelphia. pp. 1729–1745.
Tyler, K.L., Barton, E.S., Ibach, M.L., Robinson, C., Valyi-Nagy, T., Campbell, J.A., Clarke, P., O'Donnell, S.M., Wetzel, J.D., and Dermody, T.S. (2004). Isolation and molecular characterization of a novel type 3 reovirus from a child with meningitis. J. Infect. Dis. *189,* 1664–1675.
Tyler, K.L., and Fields, B.N. (1988). *Reoviridae,* the reoviruses. In Laboratory Diagnosis of Infectious Disease, Principles and Practice (vol. II). Viral, Rickettsial and Chlamydial Diseases. ed. F.A. Murphy. Springer-Verlag, New York. p. 353–374.
Tyler, K.L., McPhee, D.A., and Fields, B.N. (1986). Distinct pathways of viral spread in the host determined by reovirus S1 gene segment. Science *233,* 770–774.
Tyler, K.L., Sokol, R.J., Oberhaus, S.M., Le, M., Karrer, F.M., Narkewicz, M.R., Tyson, R.W., Murphy, J.R., Low, R., and Brown, W.R. (1998). Detection of reovirus RNA in hepatobiliary tissues from patients with extrahepatic biliary atresia and choledochal cysts. Hepatology. *27,* 1475–1482.
Tyler, K.L., Squier, M.K., Rodgers, S.E., Schneider, B.E., Oberhaus, S.M., Grdina, T.A., Cohen, J.J., and Dermody, T.S. (1995). Differences in the capacity of reovirus strains to induce apoptosis are determined by the viral attachment protein sigma 1. J. Virol. *69,* 6972–6979.
van Raaij, M.J., Mitraki, A., Lavigne, G., and Cusack, S. (1999). A triple beta-spiral in the adenovirus fibre shaft reveals a new structural motif for a fibrous protein. Nature *401,* 935–938.
Virgin, H.W., IV, Mann, M.A., Fields, B.N., and Tyler, K.L. (1991). Monoclonal antibodies to reovirus reveal structure/function relationships between capsid proteins and genetics of susceptibility to antibody action. J. Virol. *65,* 6772–6781.

Virgin, H.W., Bassel-Duby, R., Fields, B.N., and Tyler, K.L. (1988). Antibody protects against lethal infection with the neurally spreading reovirus type 3 (Dearing). J. Virol. *62*, 4594–4604.

Wang, F., Naik, U.P., Ehrlich, Y.H., Osada, S., Ohno, S., and Kornecki, E. (1995). Stimulatory antibody-induced activation and selective translocation of protein kinase C isoenzymes in human platelets. Biochem. J. *311*, 401–406.

Weigele, P.R., Scanlon, E., and King, J. (2003). Homotrimeric, beta-stranded viral adhesins and tail proteins. J. Bacteriol. *185*, 4022–4030.

Weiner, H.L., Drayna, D., Averill, D.R., Jr., and Fields, B.N. (1977). Molecular basis of reovirus virulence, role of the S1 gene. Proc. Natl Acad. Sci. USA *74*, 5744–5748.

Weiner, H.L., Powers, M.L., and Fields, B.N. (1980). Absolute linkage of virulence and central nervous system cell tropism of reoviruses to viral hemagglutinin. J. Infect. Dis. *141*, 609–616.

Weiner, H.L., Ramig, R.F., Mustoe, T.A., and Fields, B.N. (1978). Identification of the gene coding for the hemagglutinin of reovirus. Virology *86*, 581–584.

Weis, W., Brown, J.H., Cusack, S., Paulson, J.C., Skehel, J.J., and Wiley, D.C. (1988). Structure of the influenza virus haemagglutinin complexed with its receptor, sialic acid. Nature *333*, 426–431.

Williams, L.A., Martin-Padura, I., Dejana, E., Hogg, N., and Simmons, D.L. (1999). Identification and characterisation of human junctional adhesion molecule (JAM). Mol. Immunol. *36*, 1175–1188.

Wilson, G.A., Morrison, L.A., and Fields, B.N. (1994). Association of the reovirus S1 gene with serotype 3-induced biliary atresia in mice. J. Virol. *68*, 6458–6465.

Wittchen, E.S., Haskins, J., and Stevenson, B.R. (1999). Protein interactions at the tight junction. Actin has multiple binding partners, and ZO-1 forms independent complexes with ZO-2 and ZO-3. J. Biol. Chem. *274*, 35179–35185.

Wolf, J.L., Dambrauskas, R., Sharpe, A.H., and Trier, J.S. (1987). Adherence to and penetration of the intestinal epithelium by reovirus type 1 in neonatal mice. Gastroenterology. *92*, 82–91.

Wolf, J.L., Kauffman, R.S., Finberg, R., Dambrauskas, R., Fields, B.N., and Trier, J.S. (1983). Determinants of reovirus interaction with the intestinal M cells and absorptive cells of murine intestine. Gastroenterology. *85*, 291–300.

Wolf, J.L., Rubin, D.H., Finberg, R., Kaufman, R.S., Sharpe, A.H., Trier, J.S., and Fields, B.N. (1981). Intestinal M cells, a pathway of entry of reovirus into the host. Science *212*, 471–472.

Yeung, M.C., Gill, M.J., Alibhai, S.S., Shahrabadi, M.S., and Lee, P.W. (1987). Purification and characterization of the reovirus cell attachment protein sigma 1. Virology *156*, 377–385.

Zahraoui, A., Louvard, D., and Galli, T. (2000). Tight junction, a platform for trafficking and signaling protein complexes. J. Cell Biol. *151*, F31–36.

Structure and Functions of the Orthoreovirus σ3 Protein

10

Leslie A. Schiff

Abstract

Protein σ3 serves a number of distinct roles in the orthoreovirus life cycle. First, as a major structural protein that forms the outermost layer of the reovirus particle, it imparts significant environmental stability to virions. Virion σ3 also plays a critical role as a determinant of cell entry, as it must be degraded from particles during their activation for membrane penetration. In addition, σ3 is one of the most abundant proteins synthesized in reovirus-infected cells. Early in infection σ3 is thought to serve a regulatory function through its capacity to bind dsRNA and interfere with dsRNA-activated innate immune pathways. Both the structural and regulatory functions are influenced by the capacity of σ3 to interact with its structural partner, μ1. High-resolution structures have been determined for both the 365-residue σ3 protein and σ3 in complex with its capsid partner μ1. This structural data has provided a framework around which new and existing biochemical and biological information can be analyzed to provide new insight into the molecular details of the reovirus life cycle.

Introduction

Six hundred subunits of σ3 form the outermost layer of the reovirus virion (Coombs, 1998; Dryden *et al.*, 1993). As a structural protein, it is involved in the processes of viral assembly and cell entry, and σ3 is a major determinant of the environmental stability of reovirus particles (Drayna and Fields, 1982; Spendlove *et al.*, 1970). Given the relatively limited coding capacity of the orthoreoviruses, it is not surprising that some virion proteins also serve regulatory roles within the infected cell. One of the most abundant proteins in the reovirus-infected cell, σ3 is believed to serve one or more regulatory functions through its capacity to bind dsRNA and thereby impact dsRNA-activated interferon-induced antiviral pathways (Yue and Shatkin, 1997). Both its structural and regulatory functions are influenced by the capacity of σ3 to interact with its structural partner, outer capsid protein μ1. This chapter will review the relationships between σ3 structure and its functions in viral entry, assembly, particle stability and avoidance of interferon-induced antiviral pathways.

σ3 protein structure

Domain structure

The structure of the 365-residue σ3 protein was determined at 1.8 Å resolution, following expression in insect cells from a baculovirus construct (Olland *et al.*, 2001). The crystallographic asymmetric unit consisted of two σ3 subunits, associated as a dimer. Each σ3 monomer can be described as an elongated, finger-shaped molecule with two lobes organized around a long central α helix. Although the aminoterminus of σ3 is located in the small lobe and carboxyterminus in the larger lobe, the two lobes of σ3 do not represent distinct amino- and carboxyterminal domains. Rather, the protein chain passes back and forth between the lobes several times. The finding that the virion-reactive monoclonal antibody 4F2 recognizes a determinant within the large lobe suggested that this lobe is oriented towards the virion surface. This conclusion was supported when the structure of the σ3-μ1 heterohexamer was determined (Liemann *et al.*, 2002). Notably, the large lobe contains several extended loops that provide preferred sites for proteolytic cleavages during viral disassembly and entry into cells (Jané-Valbuena *et al.*, 2002). Structural features related to viral entry will be discussed further. σ3 is acetylated at its amino-terminus (Jané-Valbuena *et al.*, 2002; Liemann *et al.*, 2002; Mendez *et al.*, 2003); the importance of this modification for the protein's structure and/or function is not yet known.

Zinc-binding

Within the virion and within infected cells, σ3 contains a tightly bound zinc ion (Schiff *et al.*, 1988). Although it coordinates zinc through a CCHC zinc-binding motif similar to the retroviral RNA-binding zinc knuckle, zinc binding does not influence the nucleic acid-binding capacity of σ3 (Shepard *et al.*, 1996). Instead, zinc appears to be important for the intracellular stability of σ3 and its capacity to interact with μ1 (Mabrouk and Lemay, 1994a; Shepard *et al.*, 1996). The latter function fits well with the location of σ3's zinc-binding motif in the small lobe, which packs against μ1 and anchors σ3 to the virion (Liemann *et al.*, 2002; Olland *et al.*, 2001).

Conservation of sequence and structure

Comparative sequence analysis revealed that protein σ3 is more conserved than the minor capsid protein σ1, which determines serotype, but is more variable than μ1, which underlies σ3 and forms the T = 13*l* outer capsid shell (Centonze *et al.*, 1995). These differences likely reflect the interactions each of these proteins must make with other structural proteins and the degree to which they are subject to immune selection. Among 16 T1 and T3 reovirus field isolates whose S4 genes were sequenced, there was 96.2% average amino acid conservation in the σ3 coding sequence (Atwater *et al.*, 1986; Giantini *et al.*, 1984; Kedl *et al.*, 1995; Seliger *et al.*, 1992). The σ3 from strains Jones and Dearing are the most dissimilar, with only 89.6% sequence identity. Interestingly, analysis of the σ3 protein structure revealed that most variable residues map to one surface of σ3 that is composed largely of loops. By contrast, the conserved face of σ3 is largely helical and this face was proposed to mediate crucial contacts with other virion proteins, particularly μ1 (Olland *et al.*, 2001). This prediction is supported by the analysis of the crystal structure the σ3–μ1 complex (Liemann *et al.*, 2002). The particle association of σ3 and the nature of the σ3/μ1 interface will be discussed briefly below.

Particle association of σ3

Appearance of σ3 on the virion

Cryo-electron microscopy and image reconstruction reveal that within the orthoreovirus outer capsid, σ3 subunits project 34 Å above the outer extent of μ1 (Metcalf *et al.*, 1991) so that the virion surface has rings of six petal-like knobs surrounding the sixfold axes. These knobs represent the large lobe of σ3. Stable protease-treated particles (ISVPs, or intermediate subvirion particles) can be prepared that lack σ3 and possess a cleaved form of μ1 (Borsa *et al.*, 1973; Silverstein *et al.*, 1972). Difference maps derived from virion and ISVP image reconstructions reveal minimal differences in μ1 or other underlying virion structures. This has led to the suggestion that σ3 simply decorates the T = 13 icosahedral lattice of the outer shell (Dryden *et al.*, 1993; Jané-Valbuena *et al.*, 1999). However, as discussed further below, this description would be inaccurate if taken to imply a lack of intimate contacts between σ3 and its underlying capsid partner, μ1. In addition to its interactions with μ1, σ3 makes contacts on the virion surface with the vertex-associated protein λ2 (Dryden *et al.*, 1993; Liemann *et al.*, 2002).

Association of σ3 with μ1 during virion assembly

During virion morphogenesis, cytoplasmic hetero-oligomers of σ3 and μ1 are believed to assemble onto the surface of progeny core particles. In support of this model, studies with reovirus mutants demonstrate that temperature-sensitive lesions in either the σ3 or μ1-encoding gene segment can lead to assembly defects at the non-permissive temperature (Danis *et al.*, 1992; Fields, 1971; Morgan and Zweerink, 1974; Shing and Coombs, 1996); under these conditions progeny core particles accumulate in infected cells. The fact that some mutants with temperature-sensitive lesions in the μ1-encoding M2 gene can be phenotypically suppressed by compensatory changes in the σ3-encoding S4 gene (McPhillips and Ramig, 1984) is further evidence of the importance of σ3–μ1 interactions for virion morphogenesis. In light of the available crystal structure of the σ3–μ1 complex, analysis of such pseudorevertants might provide some insight into the process of virion morphogenesis. The M2 mutants tsA 201 and tsA 438 each have two amino acid substitutions relative to the parental Dearing sequence: Ala to Val substitutions at position 305, and Pro to Ser substitutions at position 315 (Jayasuriya, 1991). These residues lie at the juxtaposition of domains III and IV of μ1, in a region that makes contact with σ3 in the heterohexamer (Liemann *et al.*, 2002). Sequence analysis of two S4 gene pseudorevertants (RtsA 201 101 and RtsA 201 121) reveals that either a Gly to Ser substitution at position 47 in σ3 or a Glu to Ala substitution at position 217 enable outer capsids containing the tsA 201 mutant μ1 protein to assemble at non-permissive temperatures. Residue 47 in σ3 makes contacts with residues in the large lobe of the other σ3 subunit in σ3 dimers as well as with μ1 sequences in the σ3–μ1 complex. Thus, changes at this position might be predicted to impact the stability of cytoplasmic σ3 dimers and/or the competing σ3–μ1 interaction. In contrast, residue 217 in σ3 sits on the variable face of the large lobe of σ3 that does not apparently interact with either other σ3 molecules or μ1. Clearly, a larger panel of similar pseudorevertants will need to be analyzed to understand the molecular mechanisms by which changes in σ3 can functionally compensate for temperature-sensitive lesions in μ1.

Several other aminoterminal residues within σ3 have been identified as playing important roles in σ3–μ1 interactions and virion assembly. These determinants have been revealed either by sequencing the S4 genes of tsG mutants with assembly defects or after biochemical/immunological analysis of mutant proteins expressed by *in vitro* transcription and translation. Mutations at the zinc coordinating residues C51 and C54 abolish σ3–μ1 interaction (Shepard *et al.*, 1996), presumably by destabilizing the small lobe of σ3. Mutations at positions 16, 64, or 66 and 67 also either reduce or abolish μ1-binding (Bergeron *et al.*, 1998; Mabrouk and Lemay, 1994b). The crystal structure of the σ3–μ1 complex reveals that aspects of μ1 anchor σ3 residues 69–71 to the virion and that σ3 residues 16, 64 and 67 do, in fact, make close contacts with μ1.

Recently, σ3–μ1 complexes expressed from baculoviruses have been used to stoichiometrically recoat purified core particles. By a variety of biochemical, physiochemical and structural criteria, including the conformation of the vertex-associated λ2 turrets, the resultant particles resemble native virions (Chandran *et al.*, 1999). The congruence between the atomic structure of σ3/μ1 complexes and lower resolution images reconstructed from cryo-electron microscopy suggests that σ3/μ1 complexes do not undergo significant structural rearrangements during virion morphogenesis (Liemann *et al.*, 2002). Further use of the recoating approach with mutant σ3 and μ1 molecules will be of great utility in dissecting the determinants of reovirus outer capsid assembly.

Structure of σ3/μ1 complexes

Complexes of σ3 and μ1 were isolated after co-expression in insect cells infected with recombinant baculovirus (Chandran *et al.*, 1999). Crystals were obtained and the structure of $\mu1_3\sigma3_3$ heterohexamers determined and refined to a resolution of 2.8 Å (Liemann *et al.*, 2002). Analysis of the overall molecular architecture of the σ3–μ1 complex confirmed many of the predictions made from the σ3 structure and its fit within 18 Å reconstruction from electron cryomicroscopy of a type 1 Lang virion. As discussed above, the small lobe anchors σ3 to μ1 and thus is important for particle assembly. Interestingly, each σ3 monomer contacts two μ1 subunits within each trimer (Liemann *et al.*, 2002) suggesting that σ3 serves to lock the twisting μ1 head domains in place. This model is consistent with the observation that co-expression of σ3 and μ1 is required for proper μ1 folding and trimerization (Chandran *et al.*, 1999).

Relationship between σ3 subunits and σ3 in the μ1/σ3 heterohexamer

Comparing the structure of σ3 in complex with μ1 with that of a σ3 monomer reveals few differences. The only secondary structure element found in the σ3-μ1 complex that is not also evident in the σ3 dimer is a small β sheet that includes σ3 residues 69–71 (Liemann *et al.*, 2002). As mentioned above, these residues are involved in anchoring σ3 to μ1. Biochemical studies indicated that free σ3 is significantly more protease sensitive than μ1-bound σ3 (Shepard *et al.*, 1995). However comparisons between the structure of free σ3 and σ3 complexed with μ1 do not support a model in which μ1-binding induces major tertiary structural changes in σ3. This suggests that it is disruption of the σ3 dimer interface that heightens the sensitivity of σ3 to proteases. Because the conserved face of σ3 is also part of the dimer contact, μ1 binding to σ3 and σ3 dimerization were postulated to be mutually exclusive (Olland *et al.*, 2001). In fact, the crystal structure of the σ3–μ1 complex revealed significant overlap between σ3 residues that contact μ1 and those that stabilize the dimer (Liemann *et al.*, 2002). In infected cells, therefore, formation of the

$\mu 1_3 \sigma 3_3$ complex probably competes with σ3 dimerization. The emerging model suggests that μ1 binding is a regulatory switch, competing with σ3 dimer formation and shifting usage of σ3 away from dsRNA binding towards capsid assembly when μ1 concentration is sufficiently high (Liemann *et al.*, 2002; Olland *et al.*, 2001; Schmechel *et al.*, 1997; Tillotson and Shatkin, 1992).

Structure and function of σ3 in viral entry

Removal of σ3 from infecting virions

Genetic and biochemical data indicate that σ3 is a major determinant of virion thermostability (Jané-Valbuena *et al.*, 1999). The 600 copies of σ3 that decorate the virion surface serve to protect and shield the underlying penetration protein, μ1. σ3's removal, either in the intestinal lumen or in an endosome, is the first step in a disassembly pathway that activates reovirus particles for membrane penetration. Like many animal viruses, reovirus infection involves protease-mediated activation for cell entry. When proteolysis of σ3 is blocked, reovirus infection cannot progress [for examples see (Baer *et al.*, 1999; Golden *et al.*, 2002; Sturzenbecker *et al.*, 1987)]. Structural analysis of the σ3–μ1 heterohexamer complex suggests that proteolysis is required to remove σ3 from infected virions because of the tight insertion of σ3 between adjacent μ1 subunits (Liemann *et al.*, 2002).

L292 mouse fibroblasts have been used extensively to analyze early events in reovirus infection. The presence of σ3 on virions is not believed to hinder cell attachment or consequent endocytic uptake into these cells. By contrast, in the intestinal tract, proteolytic degradation of σ3 may be required to promote particle binding to intestinal M cells (Amerongen *et al.*, 1994). Recent work in cell culture reveals that particles that lack σ3 (ISVPs) have a much broader host range than do intact virions; many cells that otherwise support reovirus infection cannot efficiently mediate σ3 removal (Golden *et al.*, 2002). The viral and cellular determinants that impact reovirus uncoating are the subject of much current interest.

The importance of σ3's protease-hypersensitive region for cell entry

As mentioned above, the exposed face of the large distal lobe of σ3 contains several extended loops that were postulated to provide sites for proteolytic cleavages during reovirus entry/disassembly (Olland *et al.*, 2001). In the enteric system, this proteolysis is mediated by the secreted pancreatic alkaline proteases chymotrypsin and trypsin (Bass *et al.*, 1990; Bodkin *et al.*, 1989). The resultant ISVP-like particles then enter cells either through receptor-mediated endocytosis or by direct membrane penetration (Borsa *et al.*, 1979; Lucia-Jandris *et al.*, 1993; Sturzenbecker *et al.*, 1987; Tosteson *et al.*, 1993). Although the σ3 sequence contains as many as 30 aromatic residues that could serve as chymotrypsin cleavage sites, the enzyme generally cleaves specifically: after Tyr 238 in the strain Lang σ3 and after Phe 210 in the strain Dearing σ3 (Jané-Valbuena *et al.*, 2002). In contrast to the extracellular uncoating characteristic of enteric infections, infection of mouse fibroblasts requires intracellular proteolysis of σ3 (Borsa *et al.*, 1979; Sturzenbecker *et al.*, 1987). In fibroblasts, the acid-dependent cysteine proteases cathepsin L and cathepsin B play critical roles in the process of σ3 removal (Ebert *et al.*, 2002; Ebert *et al.*, 2004). For infection of other cell types, distinct endo/lysosomal, secreted or cell surface proteases appear to be involved. For example, the acid-insensitive cysteine protease cathepsin S mediates σ3 degradation in the macrophage-like cell line P388D (Golden *et al.*, 2004) and neutrophil elastase, an acid-insensitive inflammatory serine protease, mediates uncoating in some monocyte cell lines (Golden and Schiff, 2005).

Many different proteases can cleave σ3 from reovirus virions. Recently, the preferred sites of proteolysis have been mapped and localized on the σ3 structure. Aminoterminal sequence analysis of σ3 peptides from protease-treated virions of strains Lang and Dearing suggests that the protease hypersensitive region extends between amino acids 210 and 242 (Jané-Valbuena *et al.*, 2002). Another recent study utilized MALDI-Qq-TOFMS to analyze σ3 digestion products after limited proteolysis with a panel of different proteases (Mendez *et al.*, 2003). Results were in good accordance with the previous report in that most of the tested proteases cleaved the strain Lang σ3 between amino acids 217 and 238. Cleavage occurred under a wide range of pH conditions, from acidic to neutral. Surprisingly, the protease hypersensitive regions on the strain Dearing and Lang σ3 molecules are found in distinct positions on the protein structure. On the strain Dearing σ3, the hypersensitive region occupies a loop near the constriction that separates the small and large lobes whereas the hypersensitive region on the strain Lang σ3 is found near the apex of the large lobe (Jané-Valbuena *et al.*, 2002; Olland *et al.*, 2001). It is not clear how or if this difference impacts the biology of reovirus cell entry. Analysis of additional reovirus strains will likely provide some insight into this question.

Carboxyterminal σ3 sequences modulate susceptibility to proteolysis

Although the sites of initial σ3 proteolysis lie within a central region of the molecule, carboxyterminal regions (between amino acids 266 and 365) modulate cleavage patterns and susceptibility to proteolysis (Jané-Valbuena *et al.*, 2002). These carboxyterminal sequences are also located in the large lobe, in proximity to the protease hypersensitive region (Olland *et al.*, 2001). Amino acid 354 appears to be a particularly important regulatory determinant. Mutant viruses in which a Tyr at this position is replaced by a His can replicate in L929 cells treated with the cysteine protease inhibitor E64 or the weak base ammonium chloride (which interferes with acid-dependent cysteine proteases like cathepsins L and B) (Wilson *et al.*, 2002). These mutant viruses appear hypersensitive to proteolysis. In *in vitro* reactions with chymotrypsin, their σ3 molecules are degraded with faster kinetics than wild-type molecules (Wetzel *et al.*, 1997). In addition, they can be digested with proteases that do not efficiently uncoat wild-type virions (Ebert *et al.*, 2001; Wilson *et al.*, 2002).

Cryo-EM analysis and image reconstruction were used to compare wild-type strain Dearing virions and virions from a mutant with the Tyr to His substitution at position 354. This analysis revealed a cleft in σ3 density in wild type virions that was filled in reconstructions of particles containing the 354 change (Wilson *et al.*, 2002). Tyr 354 in the strain Dearing σ3 molecule lies at end of region that connects an α-helix in the virion-proximal small lobe to two carboxyterminal β strands in the virion-distal large lobe. It has been suggested that the substitution of a His at this site might result in repulsion from a hydrophobic pocket that includes Tyr 223, Tyr 225, Leu 228, Tyr 238, Leu 242, Val 243 and Phe 349) (Olland *et al.*, 2001). Since Tyr 354 is among the residues joining the small and large lobes of the protein, conformational changes might be transmitted to a lower region of the protein. Structural changes in σ3 containing His 354 could allow enhanced access to residues 238–250 within the hydrophobic pocket, which form a protease-sensitive region in the strain Lang and Dearing σ3. This model is supported by the recent analysis of a larger group of disassembly mutants, most of which have amino acid changes that cluster with residue 354 in the virion distal lobe of σ3 (Clark *et al.*, 2006).

Does σ3 undergo acid pH-induced conformational changes during virion entry?

In mouse fibroblasts, agents that raise vesicular pH block infection by intact reovirus virions but not by ISVPs, which lack σ3 (Sturzenbecker *et al.*, 1987). This was the first evidence to suggest that reovirus uncoating required exposure of virions to low pH and proteases in the endocytic compartment. The acidic endocytic environment could be required to promote a requisite pH-sensitive conformational change in one or more capsid proteins, and/or it could be required to activate specific proteases involved in σ3 removal. As discussed above, specific endocytic proteases have been identified which play critical roles in reovirus uncoating in mouse fibroblasts (Baer *et al.*, 1999; Ebert *et al.*, 2002; Ebert *et al.*, 2004). These cysteine proteases, cathepsins L and B, are dependent upon acidic pH for maximal activity. Recently, we have determined that cells that express acid-insensitive proteases (either naturally or by gene transfection) can be infected in the presence of agents that raise vesicular pH (including the weak base ammonium chloride or inhibitors of the vacuolar proton ATPase) (Golden *et al.*, 2004; Golden and Schiff, 2005). Thus, available data are consistent with a model in which the requirement for acidic pH reflects the activation requirements of proteases involved in σ3 removal. However, whether there is an essential, acid-dependent step in reovirus-cell entry remains an open question. An alternative proposal is that carboxyterminal residues in σ3 act as an acid-primed latch to protect σ3 cleavage sites; in this model, the density shift observed in mutant σ3s from ammonium chloride and E64-resistant viruses would correspond to an open latch (Wilson *et al.*, 2002).

Changes in μ1 that may occur in concert with σ3 degradation during virion disassembly

Although the prevailing model for reovirus entry holds that σ3 must be removed in order to expose the membrane-interactive determinants within the underlying μ1 protein, there is some evidence to suggest that a small region of μ1 may be subject to proteolysis at the same time, or even before σ3. The study in which MALDI-Qq-TOFMS was used to follow virion digestion revealed that 30 seconds of trypsin treatment released a peptide corresponding to 8aa from the C-terminus of μ1C. This was followed by a peptide corresponding to 214–236 of σ3. Other regions of μ1C were not observed until protein σ3 was completely digested. Similar results were obtained with a second protease, Glu-C, in that a μ1-related peptide (in this case corresponding to residues 513–525) was detected prior to σ3-related peptides. These findings are consistent with the fact that μ1-specific monoclonal antibodies react with intact virions (Hayes *et al.*, 1981; Virgin *et al.*, 1994) and the suggestion from the σ3-μ1 crystal structure that an unresolved portion of μ1 constitutes an exposed blob on the virus surface (Liemann *et al.*, 2002).

In virion preparations, μ1 generally appears in a cleaved form consisting of an aminoterminal myristoylated fragment, μ1N, and a carboxyterminal fragment, μ1C. It has been appreciated for some time that the cleavage of μ1 into these two particle-associated fragments depends upon μ1's prior association with σ3 (Tillotson and Shatkin, 1992). The role of σ3 in this interaction is not as a protease (Mabrouk and Lemay, 1994b), but rather as a chaperone to promote appropriate folding. Evidence now suggests that the aminoterminal cleavage of μ1, which is essential for infectivity (Odegard *et al.*, 2004) occurs in conjunction with viral entry rather than assembly as previously surmised (Nibert *et al.*, 2005). Thus, during cell entry, proteolytic degradation of σ3 and autocatalytic cleavage of μ1 are proposed to generate a metastable reovirus particle that is primed for membrane penetration (Liemann *et al.*, 2002).

Structure and function in dsRNA-binding

Preceding sections of this chapter have focused on σ3 protein structure–function relationships as they relate to viral entry and σ3's role as a major capsid protein. However, in addition to its structural roles, the reovirus σ3 protein serves regulatory functions in infected cells as a result of its capacity to bind dsRNA and thereby influence dsRNA-activated innate immune molecules.

σ3 binds dsRNA and blocks activation of dsRNA-activated innate immune molecules

Almost thirty years ago, filter-binding assays revealed that reovirus protein σ3 is capable of binding dsRNA in a sequence-independent manner (Huismans and Joklik, 1976). More recently, gel-shift analysis was used to demonstrate that purified σ3 binds RNA duplexes with a minimal length of 32–45 base pairs (Yue and Shatkin, 1997). The dsRNA-binding activity of σ3 is not believed to play a role in virion morphogenesis. Rather, σ3 is thought to sequester viral dsRNA that would otherwise activate the interferon-induced dsRNA-activated kinase PKR and/or the dsRNA-activated 2′5′oligo-adenylate synthetase/RNase L system. The former is an eIF2α kinase that exerts its antiviral effect by inhibiting translation initiation, while the latter is an endoribonuclease system that is activated in infected cells to non-specifically degrade mRNA and rRNA (Hovanessian, 1991). In cell extracts (Imani and Jacobs, 1988), and in heterologous viral systems (Beattie *et al.*, 1994; Lloyd and Shatkin, 1992), σ3 can block the activation of PKR. Transient expression of σ3 results in translational stimulation of co-transfected reporter gene constructs (Giantini and Shatkin, 1989; Seliger *et al.*, 1992). Because the process of transfection itself is known to activate PKR, the mechanism for this translational stimulation was assumed to involve inactivation of PKR. Consistent with this, mutations in σ3 that affect its dsRNA binding capacity and co-expression of μ1, which interferes with its dsRNA-binding capacity, interfere with translational stimulation (Tillotson and Shatkin, 1992; Yue and Shatkin, 1997). Based on this wealth of biochemical and molecular data, it has long been presumed that PKR is responsible for host translational shutoff in reovirus-infected cells. However recent work from our lab, using cells from knockout mice, demonstrates that both PKR and RNase L contribute to the strong host shutoff characteristic of most reovirus infections (Smith *et al.*, 2005).

In transfected cells (Yue and Shatkin, 1996), and upon infection with some reovirus isolates (Schmechel *et al.*, 1997), σ3 localizes to the nucleus. Nuclear localization of σ3 is inhibited by mutations that destroy dsRNA binding and by co-expression with μ1

(Yue and Shatkin, 1996), and it is enhanced by mutations that increase dsRNA-binding (Bergeron *et al.*, 1998). These results suggest that affinity for dsRNA is important for the partial nuclear localization of σ3. As yet there is no biological function ascribed to nuclear σ3, however the observation that PKR may also localize to the nucleus is intriguing (Jeffrey *et al.*, 1995; Jimenez-Garcia *et al.*, 1993).

μ1-binding interferes with the regulatory function of σ3

Co-expression of μ1 with σ3 interferes with the capacity of σ3 to stimulate translation in transfected cells (Tillotson and Shatkin, 1992; Yue and Shatkin, 1997) and high levels of μ1-bound σ3 correlate with translational inhibition in infected cells (Schmechel *et al.*, 1997). These observations can now be interpreted in light of structural data which suggests that σ3-dsRNA binding and σ3–μ1 binding are mutually exclusive (Liemann *et al.*, 2002; Olland *et al.*, 2001).

Biochemical and structural analysis of σ3's dsRNA-binding function

As mentioned above, when the atomic structure of σ3 was determined, the crystallographic asymmetric unit was found to be a dimer. The dimer has the geometric structure of a parallelogram, with one of the broad surfaces having an even charge distribution, and the other having a large, positively charged patch that extends across the subunit interface. The location and extent of this patch, together with the observation that σ3 dimerizes in solution, led to the suggestion that the dimer is the RNA-binding form of σ3 (Olland *et al.*, 2001).

Analysis of mutant σ3 proteins expressed *in vitro* or in infected cells led to the identification of specific amino acid residues that positively or negatively influence σ3's capacity to bind dsRNA. dsRNA-binding was not negatively affected by mutations in the zinc-binding domain. In fact mutations that disrupt this aminoterminal domain actually increased dsRNA binding (Shepard *et al.*, 1996). Similarly, the mutation at residue 16 in the σ3 from tsG 453 interfered with σ3's capacity to interact with μ1 and increased dsRNA binding, nuclear localization and interferon-resistance (Bergeron *et al.*, 1998). Biochemical analysis of σ3 fragments (Schiff *et al.*, 1988) and analysis of recombinant, mutant σ3 proteins (Miller and Samuel, 1992) first implicated sequences in the carboxyterminus of σ3 in nucleic acid binding. Further studies identified the importance of specific residues, including Arg 236, Arg 239, Lys 291 and Lys 293 (Denzler and Jacobs, 1994; Mabrouk *et al.*, 1995; Wang *et al.*, 1996). The observation that Arg substitutions for Lys 291 or 293 restored dsRNA binding argued for the importance of the positive charge at these positions (Denzler and Jacobs, 1994).

Placing these residues in the context of the σ3 crystal structure provides some insight into the molecular basis for σ3's dsRNA-binding capacity. First, Arg 236 and Arg 239 are buried within the large lobe of σ3, making it unlikely that they mediate direct interactions with dsRNA. The effect of mutations at these positions can be explained however, because these residues make numerous contacts with neighbouring atoms, suggesting that they are involved in stabilizing the large lobe of σ3. Likewise, Lys 291 makes extensive contacts with neighbouring atoms and is buried near the groove that separates the large and small lobes of σ3. The most likely candidate for mediating direct contacts with dsRNA is Lys 293, which is part of the positively charged surface patch. Importantly this residue does not make contacts with other atoms in the structure. Interestingly, while sequence and mutational analysis led to the prediction that σ3 had a dsRNA-binding domain similar in structure to other dsRNA-binding proteins (Yue and Shatkin, 1996), structural analysis reveals that the residues implicated in dsRNA binding are spread out and do not form a discrete dsRNA-binding domain.

Summary and relationship of σ3 to proteins encoded by other dsRNA viruses

As detailed above, the reovirus σ3 protein serves several distinct functions during the viral life cycle. As a structural protein, it imparts environmental stability to the virion, and must be proteolytically removed in order for the particle to become activated for membrane penetration. Activation likely involves the autocatalytic aminoterminal cleavage of σ3's outer capsid partner, μ1. In infected cells, newly synthesized cytoplasmic σ3, probably in the form of a dimer, binds dsRNA and protects the infected cell from certain dsRNA-activated, interferon-induced antiviral pathways. Later in the infection cycle, levels of newly synthesized μ1 increase, and interactions between σ3 and μ1 shift the function of σ3 away from dsRNA-binding, towards capsid assembly. Interactions with σ3 promote proper μ1 folding and trimerization. Assembly of the outer capsid onto nascent reovirus cores requires the prior formation of σ3–μ1 complexes.

What are the similarities between σ3 and capsid proteins of other dsRNA viruses?

Other dsRNA virus families include the *Totiviridae* (one genome segment), *Birnaviridae* (two segments), *Cystoviridae* (three segments) and *Reoviridae* (ten to twelve genome segments). The simpler totiviruses and birnaviruses have a single-layered virion, the cystoviruses have a double-layered capsid and the more complex members of the *Reoviridae* (including rotaviruses, orbiviruses and reoviruses) have a triple-layered capsid structure. Even among the viruses that share a triple-layered capsid structure, structural homologues can be difficult to discern. In the well-studied rotaviruses, there is no direct structural homologue of the reovirus capsid protein σ3. The outermost capsid layer in the rotavirus virion consists of 780 copies of the 38-kDa glycoprotein VP7 and 120 copies of the 88-kDa spike protein VP4 (reviewed in Prasad and Chiu, 1994). Rotavirus is an enteric pathogen and in cell culture it is typically cultivated in the presence of trypsin (Clark *et al.*, 1981; Gillespie *et al.*, 1984). Biochemical studies reveal that trypsin cleavage stabilizes the VP4 spike and that the VP4 cleavage products VP5* and VP8* remain virion-associated (Crawford *et al.*, 2001). VP4 and its cleavage products are involved in the processes of cell attachment and penetration, and thus serve functions similar to three different reovirus capsid proteins: σ1, σ3 and μ1. *In vitro* experiments indicate that trypsin-activated rotavirus particles and recombinant VP5* are lipophilic (reviewed in Jayaram 2004). Thus, loose analogies might be made between VP5* and μ1, the orthoreovirus penetration protein that is exposed upon σ3 degradation. Like σ1, the cleavage fragments of VP4 mediate interactions with cell surface receptors. The proteolytic cleavage of VP4 enhances infectivity several-fold and facilitates viral entry into cells (Estes *et al.*, 1981). This observation parallels our own finding that proteolysis of σ3 greatly expands the cellular host range of orthoreoviruses (Golden *et al.*, 2002). Similarly, proteolytic activation of bluetongue virus is reported to impact host range. Interestingly, in this system, removal of the outer capsid layer is reported to decrease the infectivity of resultant particles for some mammalian cells, while the resultant ISVPs have increased infectivity for cells derived from the insect vector (Mertens *et al.*, 1987; Mertens *et al.*, 1996).

Perhaps the clearest structural comparisons can be drawn between the outer capsid proteins of orthoreoviruses and aquareoviruses. Although the reovirus σ3 protein and VP7 of aquareoviruses are not similar at the primary sequence level (Lupiani *et al.*, 1997), they are roughly similar in size. Cryoelectron microscopy suggests that the 600 subunits of VP7 and σ3 have a similar shape and make similar interactions with underlying trimeric structures (Nason *et al.*, 2000). VP7 is described as having a bulbous head that clamps onto the three sides of underlying VP5 trimers. This is highly reminiscent of the σ3–μ1 interactions described above (Liemann *et al.*, 2002). Aquareoviruses also appear to undergo protease-mediated activation for cell entry through degradation of VP7 and conformational rearrangements in VP5 (Nason *et al.*, 2000).

Do other dsRNA viruses express proteins with the dsRNA-binding functions of σ3?

Like mammalian reoviruses, avian reoviruses are highly resistant to the antiviral effects of interferon suggesting that they encode a dsRNA-binding protein that blocks the activation of PKR and/or the 2′5′ oligoadenylate synthetase system. Recent studies suggest that the avian reovirus protein σA binds dsRNA (Martínez-Costas *et al.*, 2000) and serves as an inhibitor PKR (Gonzalez-Lopez *et al.*, 2003). σA is a minor component of the avian reovirus core (Martínez-Costas *et al.*, 1997). Thus, σA appears to share regulatory functions, but not structural functions, with σ3. Orthoreoviruses also possess core proteins with dsRNA binding capacity (Dermody *et al.*, 1991; Schiff *et al.*, 1988), but there are no data to suggest that these proteins interfere with dsRNA-activated pathways in infected cells.

Finally, two different rotavirus proteins are reported to bind dsRNA. Again, these are not direct homologues of σ3 as they are non-structural proteins. The NSP3 gene of group C rotaviruses encodes three proteins, two of which bind dsRNA. The smallest of these, p8, may be the smallest natural peptide described which is capable of binding dsRNA. Like σ3, NSP3, when expressed by transfection, can inhibit the activation of PKR and rescue the replication of an interferon-sensitive vaccinia virus mutant (Langland *et al.*, 1994). More information on NSP3 can be found in Chapter 17. The other rotavirus protein with a similar function is the NSP5 encoded by group A rotaviruses (Vende *et al.*, 2002). Unlike σ3, NSP5 has affinity for both ssRNA and dsRNA. A function for NSP5 in avoidance of innate immune pathways has been proposed, but to date there is no direct evidence to indicate that this protein interferes with activation of dsRNA-activated antiviral pathways.

Acknowledgements

I would like to thank Max Nibert for advice and intellectual support during the preparation of this chapter. L.A.S. was supported by award AI045990 from the National Institutes of Health and additional grants from the University of Minnesota Academic Health Center and the Minnesota Medical Foundation.

References

Amerongen, H.M., Wilson, G.A., Fields, B.N., and Neutra, M.R. (1994). Proteolytic processing of reovirus is required for adherence to intestinal M cells. J. Virol. *68*, 8428–8432.

Atwater, J.A., Munemitsu, S.M., and Samuel, C.E. (1986). Biosynthesis of reovirus-specified polypeptides. Molecular complementary DNA cloning and nucleotide sequence of the reovirus serotype 1 Lang strain S4 messenger RNA which encodes the major capsid surface polypeptide s3. Biochem. Biophys. Res. Commun. *136*, 183–192.

Baer, G.S., Ebert, D.H., Chung, C.J., Erickson, A.H., and Dermody, T.S. (1999). Mutant cells selected during persistent reovirus infection do not express mature cathepsin L and do not support reovirus disassembly. J. Virol. *73*, 9532–9543.

Bass, D.M., Bodkin, D., Dambrauskas, R., Trier, J.S., Fields, B.N., and Wolf, J.L. (1990). Intraluminal proteolytic activation plays an important role in replication of type 1 reovirus in the intestines of neonatal mice. J. Virol. *64*, 1830–1833.

Beattie, E., Denzler, K.L., Tartaglia, J., Perkus, M.E., Paoletti, E., and Jacobs, B.L. (1994). Reversal of the interferon-sensitive phenotype of a vaccinia virus lacking E3L by expression of the reovirus S4 gene. J. Virol. *69*, 499–505.

Bergeron, J., Mabrouk, T., Garzon, S., and Lemay, G. (1998). Characterization of the thermosensitive ts453 reovirus mutant: increased dsRNA binding of sigma 3 protein correlates with interferon resistance. Virology *246*, 199–210.

Bodkin, D.K., Nibert, M.L., and Fields, B.N. (1989). Proteolytic digestion of reovirus in the intestinal lumens of neonatal mice. J. Virol. *63*, 4676–4681.

Borsa, J., Copps, T.P., Sargent, M.D., Long, D.G., and Chapman, J.D. (1973). New intermediate subviral particles in the *in vitro* uncoating of reovirus virions by chymotrypsin. J. Virol. *11*, 552–564.

Borsa, J., Morash, B.D., Sargent, M.D., Copps, T.P., Lievaart, P.A., and Szekely, J.G. (1979). Two modes of entry of reovirus particles into L cells. J. Gen. Virol. *45*, 161–170.

Centonze, V.E., Chen, Y., Severson, T.F., Borisy, G.G., and Nibert, M.L. (1995). Visualization of individual reovirus particles by low-temperature, high-resolution scanning microscopy. J. Struct. Biol. *115*, 215–225.

Chandran, K., Walker, S.B., Chen, Y., Contreras, C.M., Schiff, L.A., Baker, T.S., and Nibert, M.L. (1999). *In vitro* recoating of reovirus cores with baculovirus-expressed outer- capsid proteins mu1 and sigma3. J. Virol. *73*, 3941–3950.

Clark, K.M., Wetzel, J.D., Gu, Y., Ebert, D.H., McAbee, S.A., Stoneman, E.K., Baer, G.S., Zhu, Y., Wilson, G.J., Prasad, B.V., and Dermody, T.S. (2006). Reovirus variants selected for resistance to ammonium chloride have mutations in viral outer-capsid protein sigma3. J. Virol. *80*, 671–681.

Clark, S.M., Roth, J.R., Clark, M.L., Barnett, B.B., and Spendlove, R.S. (1981). Trypsin enhancement of rotavirus infectivity:mechanism of enhancement. J. Virol. *39*, 816–822.

Coombs, K.M. (1998). Stoichiometry of reovirus structural proteins in virus, ISVP, and core particles. Virology *243*, 218–228.

Crawford, S.E., Mukherjee, S.K., Estes, M.K., Lawton, J.A., Shaw, A.L., Ramig, R.F., and Prasad, B.V. (2001). Trypsin cleavage stabilizes the rotavirus VP4 spike. J. Virol. *75*, 6052–6061.

Danis, C., Garzon, S., and Lemay, G. (1992). Further characterization of the ts453 mutant of mammalian orthoreovirus serotype 3 and nucleotide sequence of the mutated S4 gene. Virology *190*, 494–498.

Denzler, K.L., and Jacobs, B.L. (1994). Site-directed mutagenic analysis of reovirus sigma 3 protein binding to dsRNA. Virology *204*, 190–199.

Dermody, T.S., Schiff, L.A., Nibert, M.L., Coombs, K.M., and Fields, B.N. (1991). The S2 gene nucleotide sequences of prototype strains of the three reovirus serotypes: characterization of reovirus core protein sigma-2. J. Virol. *65*, 5721–5731.

Drayna, D., and Fields, B.N. (1982). Genetic studies on the mechanism of chemical and physical inactivation of reovirus. J. Gen. Virol. *63*, 149–159.

Dryden, K.A., Wang, G., Yeager, M., Nibert, M.L., Coombs, K.M., Furlong, D.B., Fields, B.N., and Baker, T.S. (1993). Early steps in reovirus infection are associated with dramatic changes in supramolecular structure and protein conformation: analysis of virions and subviral particles by cryoelectron microscopy and image reconstruction. J. Cell Biol. *122*, 1023–1041.

Ebert, D.H., Deussing, J., Peters, C., and Dermody, T.S. (2002). Cathepsin L and cathepsin B mediate reovirus disassembly in murine fibroblast cells. J. Biol. Chem. *277*, 24609–24617.

Ebert, D.H., Kopecky-Bromberg, S.A., and Dermody, T.S. (2004). Cathepsin B Is Inhibited in Mutant Cells Selected during Persistent Reovirus Infection. J. Biol. Chem. *279*, 3837–3851.

Ebert, D.H., Wetzel, J.D., Brumbaugh, D.E., Chance, S.R., Stobie, L.E., Baer, G.S., and Dermody, T.S. (2001). Adaptation of reovirus to growth in the presence of protease inhibitor E64 segregates with a mutation in the carboxy terminus of viral outer-capsid protein sigma 3. J. Virol. *75*, 3197–3206.

Estes, M.K., Graham, D.Y., and Mason, B.B. (1981). Proteolytic enhancement of rotavirus infectivity: molecular mechanisms. J. Virol. *39*, 879–888.

Fields, B.N. (1971). Temperature-sensitive mutants of reovirus type 3 features of genetic recombination. Virology *46*, 142–148.

Giantini, M., Seliger, L.S., Furuichi, Y., and Shatkin, A.J. (1984). Reovirus type 3 genome segment S4: nucleotide sequence of the gene encoding a major virion surface protein. J. Virol. *52*, 984–987.

Giantini, M., and Shatkin, A.J. (1989). Stimulation of chloramphenicol acetyltransferase mRNA translation by reovirus capsid polypeptide sigma 3 in cotransfected COS cells. J. Virol. *63*, 2415–2421.

Gillespie, J., Kalica, A., Conner, M., Schiff, E., Barr, M., Holmes, D., and Frey, M. (1984). The isolation, propagation and characterization of tissue-cultured equine rotaviruses. Vet. Microbiol. *9*, 1–14.

Golden, J.W., Bahe, J.A., Lucas, W.T., Nibert, M.L., and Schiff, L.A. (2004). Cathepsin S supports acid-independent infection by some reoviruses. J. Biol. Chem. *279*, 8547–8557.

Golden, J.W., Linke, J., Schmechel, S., Thoemke, K., and Schiff, L.A. (2002). Addition of exogenous protease facilitates reovirus infection in many restrictive cells. J. Virol. *76*, 7430–7443.

Golden, J.W., and Schiff, L.A. (2005). Neutrophil elastase, an acid-independent serine protease, facilitates reovirus uncoating and infection in U937 promonocyte cells. Virol. J. *2*, 48.

Gonzalez-Lopez, C., Martínez-Costas, J., Esteban, M., and Benavente, J. (2003). Evidence that avian reovirus sigmaA protein is an inhibitor of the double-stranded RNA-dependent protein kinase. J. Gen. Virol. *84*, 1629–1639.

Hayes, E.C., Lee, P.W.K., Miller, S.E., and Joklik, W.K. (1981). The interaction of a series of hybridoma IgGs with reovirus particles. Demonstration that the core protein lambda 2 is exposed on the particle surface. Virology *108*, 147–155.

Hovanessian, A.G. (1991). Interferon-induced and double-stranded RNA-activated enzymes: a specific protein kinase and 2′,5′-oligoadenylate synthetases. J. Interferon Res. *11*, 199–205.

Huismans, H., and Joklik, W.K. (1976). Reovirus-coded polypeptides in infected cells: Isolation of two native monomeric polypeptides with affinity for single-stranded and double-stranded RNA, respectively. Virology *70*, 411–424.

Imani, F., and Jacobs, B.L. (1988). Inhibitory activity for the interferon-induced protein kinase is associated with the reovirus serotype 1 sigma 3 protein. Proc. Natl Acad. Sci. USA *85*, 7887–7891.

Jané-Valbuena, J., Breun, L.A., Schiff, L.A., and Nibert, M.L. (2002). Sites and determinants of early cleavages in the proteolytic processing pathway of reovirus surface protein sigma3. J. Virol. *76*, 5184–5197.

Jané-Valbuena, J., Nibert, M.L., Spencer, S.M., Walker, S.B., Baker, T.S., Chen, Y., Centonze, V.E., and Schiff, L.A. (1999). Reovirus virion-like particles obtained by recoating infectious subvirion particles with baculovirus-expressed sigma3 protein: an approach for analyzing sigma3 functions during virus entry. J. Virol. *73*, 2963–2973.

Jayasuriya, A.K. A. (1991) Molecular characterization of the reovirus M2 gene, Harvard University, Cambridge, Mass.

Jeffrey, I.W., Kadereit, S., Meurs, E.F., Metzger, T., Bachmann, M., Schwemmle, M., Hovanessian, A.G., and Clemens, M.J. (1995). Nuclear localization of the interferon-inducible protein kinase PKR in human cells and transfected mouse cells. Exp. Cell Res. *218*, 17–27.

Jimenez-Garcia, L.F., Green, S.R., Matthews, M.B., and Spector, D.L. (1993). Organization of the double-stranded RNA-activated protein kinase DAI and virus-associated VA RNAI in adenovirus-2-infected HeLa cells. J. Cell Sci. *106*, 11–22.

Kedl, R., Schmechel, S., and Schiff, L. (1995). Comparative sequence analysis of the reovirus S4 genes from 13 serotype 1 and serotype 3 field isolates. J. Virol. *69*, 552–559.

Langland, J.O., Pettiford, S.M., Jiang, B., and Jacobs, B.L. (1994). Products of the porcine NSP3 gene bind specifically to double-stranded RNA and inhibit activation of the interferon-induced protein kinase, PKR. J. Virol. *68*, 3821–3829.

Liemann, S., Chandran, K., Baker, T.S., Nibert, M.L., and Harrison, S.C. (2002). Structure of the reovirus membrane-penetration protein, mu 1, in a complex with is protector protein, sigma 3. Cell *108*, 283–295.

Lloyd, R.M., and Shatkin, A.J. (1992). Translational stimulation by reovirus polypeptide sigma 3: substitution for VAI RNA and inhibition of phosphorylation of the alpha subunit of eukaryotic initiation factor 2. J. Virol. *66*, 6878–6884.
Lucia-Jandris, P., Hooper, J.W., and Fields, B.N. (1993). Reovirus M2 gene is associated with chromium release from mouse L cells. J. Virol. *67*, 5339–5345.
Lupiani, B., Reddy, S.M., Subramanian, K., and Samal, S.K. (1997). Cloning, sequence analysis and expression of the major outer capsid protein gene of an aquareovirus. J. Gen. Virol. *78 (Pt 6)*, 1379–1383.
Mabrouk, T., Danis, C., and Lemay, G. (1995). Two basic motifs of reovirus sigma-3 protein are involved in double-stranded RNA binding. Biochem. Cell Biol. *73*, 137–145.
Mabrouk, T., and Lemay, G. (1994a). Mutations in a CCHC zinc-binding motif of the reovirus sigma 3 protein decrease its intracellular stability. J. Virol. *68*, 5287–5290.
Mabrouk, T., and Lemay, G. (1994b). The sequence similarity of reovirus sigma 3 protein to picornaviral proteases is unrelated to its role in mu 1 viral protein cleavage. Virology *202*, 615–620.
Martínez-Costas, J., Gonzalez-Lopez, C., Vakharia, V.N., and Benavente, J. (2000). Possible involvement of the double-stranded RNA-binding core protein sigmaA in the resistance of avian reovirus to interferon. J. Virol. *74*, 1124–1131.
Martínez-Costas, J., Grande, A., Varela, R., Garcia-Martinez, C., and Benavente, J. (1997). Protein architecture of avian reovirus S1133 and identification of the cell attachment protein. J. Virol. *71*, 59–64.
McPhillips, T.H., and Ramig, R.F. (1984). Extragenic suppression of temperature-sensitive phenotype in reovirus: mapping suppressor mutations. Virology *135*, 428–439.
Mendez, II, She, Y.M., Ens, W., and Coombs, K.M. (2003). Digestion pattern of reovirus outer capsid protein sigma 3 determined by mass spectrometry. Virology *311*, 289–304.
Mertens, P.P., Burroughs, J.N., and Anderson, J. (1987). Purification and properties of virus particles, infectious subviral particles, and cores of bluetongue virus serotypes 1 and 4. Virology *157*, 375–386.
Mertens, P.P., Burroughs, J.N., Walton, A., Wellby, M.P., Fu, H., O'Hara, R.S., Brookes, S.M., and Mellor, P.S. (1996). Enhanced infectivity of modified bluetongue virus particles for two insect cell lines and for two Culicoides vector species. Virology *217*, 582–593.
Metcalf, P., Cyrklaff, M., and Adrian, M. (1991). The three-dimensional structure of reovirus obtained by cryo-electron microscopy. EMBO J. *10*, 3129–3136.
Miller, J.E., and Samuel, C.E. (1992). Proteolytic cleavage of the reovirus sigma 3 protein results in enhanced double-stranded RNA-binding activity: identification of a repeated basic amino acid motif within the C-terminal binding region. J. Virol. *66*, 5347–5356.
Morgan, E.M., and Zweerink, H.J. (1974). Reovirus morphogenesis: core-like particles in cells infected at 39 with wild-type reovirus and temperature-sensitive mutants of groups B and G. Virology *59*, 556–565.
Nason, E.L., Samal, S.K., and Prasad, B.V.V. (2000). Trypsin-induced structural transformation in aquareovirus. J. Virol. *74*, 6546–6555.
Nibert, M.L., Odegard, A.L., Agosto, M.A., Chandran, K., and Schiff, L.A. (2005). Putative Autocleavage of Reovirus mu1 Protein in Concert with Outer-capsid Disassembly and Activation for Membrane Permeabilization. J. Mol. Biol. *345*, 461–474.
Odegard, A.L., Chandran, K., Zhang, X., Parker, J.S., Baker, T.S., and Nibert, M.L. (2004). Putative autocleavage of outer capsid protein micro1, allowing release of myristoylated peptide micro1N during particle uncoating, is critical for cell entry by reovirus. J. Virol. *78*, 8732–8745.
Olland, A.M., Jané-Valbuena, J., Schiff, L.A., Nibert, M.L., and Harrison, S.C. (2001). Structure of the reovirus outer capsid and dsRNA-binding protein sigma 3 at 1.8 A resolution. EMBO J. *20*, 979–989.
Prasad, B.V., and Chiu, W. (1994). Structure of rotavirus. Curr. Top. Microbiol. Immunol. *185*, 9–29.
Schiff, L.A., Nibert, M.L., Co, M.S., Brown, E.G., and Fields, B.N. (1988). Distinct binding sites for zinc and double-stranded RNA in the reovirus outer capsid protein sigma 3. Mol. Cell. Biol. *8*, 273–283.
Schmechel, S., Chute, M., Skinner, P., Anderson, R., and Schiff, L. (1997). Preferential translation of reovirus mRNA by a sigma 3-dependent mechanism. Virology *232*, 62–73.
Seliger, L.S., Giantini, M., and Shatkin, A.J. (1992). Translational effects and sequence comparisons of the three serotypes of the reovirus S4 gene. Virology *187*, 202–210.
Shepard, D.A., Ehnstrom, J.G., and Schiff, L.A. (1995). Association of reovirus outer capsid proteins sigma 3 and mu 1 causes a conformational change that renders sigma 3 protease sensitive. J. Virol. *69*, 8180–8184.

Shepard, D.A., Ehnstrom, J.G., Skinner, P.J., and Schiff, L.A. (1996). Mutations in the zinc-binding motif of the reovirus capsid protein sigma-3 eliminate its ability to associate with capsid protein mu-1. J. Virol. *70*, 2065–2068.

Shing, M., and Coombs, K.M. (1996). Assembly of the reovirus outer capsid requires mu 1/sigma 3 interactions which are prevented by misfolded sigma 3 protein in temperature-sensitive mutant tsG453. Virus Res. *46*, 19–29.

Silverstein, S.C., Astell, C., Levin, D.H., Schonberg, M., and Acs, G. (1972). The mechanisms of reovirus uncoating and gene activation *in vivo*. Virology *47*, 797–806.

Smith, J.A., Schmechel, S.C., Williams, B.R., Silverman, R.H., and Schiff, L.A. (2005). Involvement of the Interferon-Regulated Antiviral Proteins PKR and RNase L in Reovirus-Induced Shutoff of Cellular Translation. J. Virol. *79*, 2240–2250.

Spendlove, R.S., McClain, M.E., and Lennette, E.H. (1970). Enhancement of reovirus infectivity by extracellular removal or alteration of the virus capsid by proteolytic enzymes. J. Gen. Virol. *8*, 83–94.

Sturzenbecker, L.J., Nibert, M., Furlong, D., and Fields, B.N. (1987). Intracellular digestion of reovirus particles requires a low pH and is an essential step in the viral infectious cycle. J. Virol. *61*, 2351–2361.

Tillotson, L., and Shatkin, A.J. (1992). Reovirus polypeptide sigma 3 and N-terminal myristoylation of polypeptide mu 1 are required for site-specific cleavage to mu 1C in transfected cells. J. Virol. *66*, 2180–2186.

Tosteson, M.T., Nibert, M.L., and Fields, B.N. (1993). Ion channels induced in lipid bilayers by subvirion particles of the nonenveloped mammalian reoviruses. Proc. Natl Acad. Sci. USA *90*, 10549–10552.

Vende, P., Taraporewala, Z.F., and Patton, J.T. (2002). RNA-binding activity of the rotavirus phosphoprotein NSP5 includes affinity for double-stranded RNA. J. Virol. *76*, 5291–5299.

Virgin, H.W.t., Mann, M.A., and Tyler, K.L. (1994). Protective antibodies inhibit reovirus internalization and uncoating by intracellular proteases. J. Virol. *68*, 6719–6729.

Wang, Q., Bergeron, J., Mabrouk, T., and Lemay, G. (1996). Site-directed mutagenesis of the double-stranded RNA binding domain of bacterially-expressed sigma 3 reovirus protein. Virus Res. *41*, 141–151.

Wetzel, J.D., Wilson, G.J., Baer, G.S., Dunnigan, L.R., Wright, J.P., Tang, D.S., and Dermody, T.S. (1997). Reovirus variants selected during persistent infections of L cells contain mutations in the viral S1 and S4 genes and are altered in viral disassembly. J. Virol. *71*, 1362–1369.

Wilson, G.J., Nason, E.L., Hardy, C.S., Ebert, D.H., Wetzel, J.D., Venkataram Prasad, B.V., and Dermody, T.S. (2002). A single mutation in the carboxy terminus of reovirus outer-capsid protein sigma 3 confers enhanced kinetics of sigma 3 proteolysis, resistance to inhibitors of viral disassembly, and alterations in sigma 3 structure. J. Virol. *76*, 9832–9843.

Yue, Z., and Shatkin, A.J. (1996). Regulated, stable expression and nuclear presence of retrovirus double-stranded RNA-binding protein sigma 3 in HeLa cells. J. Virol. *70*, 3497–3501.

Yue, Z., and Shatkin, A.J. (1997). Double-stranded RNA-dependent protein kinase (PKR) is regulated by reovirus structural proteins. Virology *234*, 364–371.

Rotavirus Cell Entry

11

Philip R. Dormitzer

Abstract

Rotavirus cell entry is mediated by a series of molecular interactions and rearrangements. While we now have an initial map of the rotavirus entry pathway, this map is far from complete. Some steps in rotavirus entry, such as proteolytic priming, initial attachment, and virion uncoating, correspond to well-understood steps in the entry pathways of other viruses, allowing extrapolation from data specific to rotavirus. However, the central event in rotavirus entry, the passage of a large non-enveloped particle through a lipid bilayer, though also a necessary step in some other viral entry pathways, is not well understood for any non-enveloped virus. Structural data indicate that the rotavirus outer capsid protein, VP4, undergoes a fold-back rearrangement associated with a two- to threefold reorganization of the protruding portion of the VP4 spike. While the function of this molecular event has not been determined experimentally, its resemblance to the fusogenic rearrangements of enveloped virus fusion proteins suggests a shared element between the entry mechanisms of enveloped viruses and a non-enveloped virus.

Introduction

Because their double-stranded RNA genome segments are not templates for transcription or translation by host enzymes, members of the *Reoviridae* must translocate both their genomes and their transcriptional machinery through a host cell membrane to initiate infection. The diversity of outer layer architectures among the *Reoviridae* (Prasad *et al.*, 1988; Dryden *et al.*, 1993; Nason *et al.*, 2004) indicates that different members of the family have evolved structurally distinct membrane penetration mechanisms. It remains to be determined whether these outer layers deliver their payloads in similar or in fundamentally different ways.

For rotavirus, the subviral double-layered particle (DLP) is the transcriptionally active payload (Figure 11.1a). It is 710 Å in diameter, and its two concentric icosahedral layers, formed by VP2 (the inner capsid protein) and VP6 (the middle capsid protein), contain the polymerase (VP1), a capping enzyme (VP3), and eleven double-stranded RNA genome segments. In the intact triple-layered virion (TLP) an outer capsid layer, formed by VP7, surrounds the DLP (Prasad *et al.*, 1988; Yeager *et al.*, 1990). VP4, a spike protein, is anchored in the VP6 layer, passes through the VP7 layer, and emerges as projections from the virion surface at 60 symmetry-related positions (Shaw *et al.*, 1993; Yeager *et al.*, 1994).

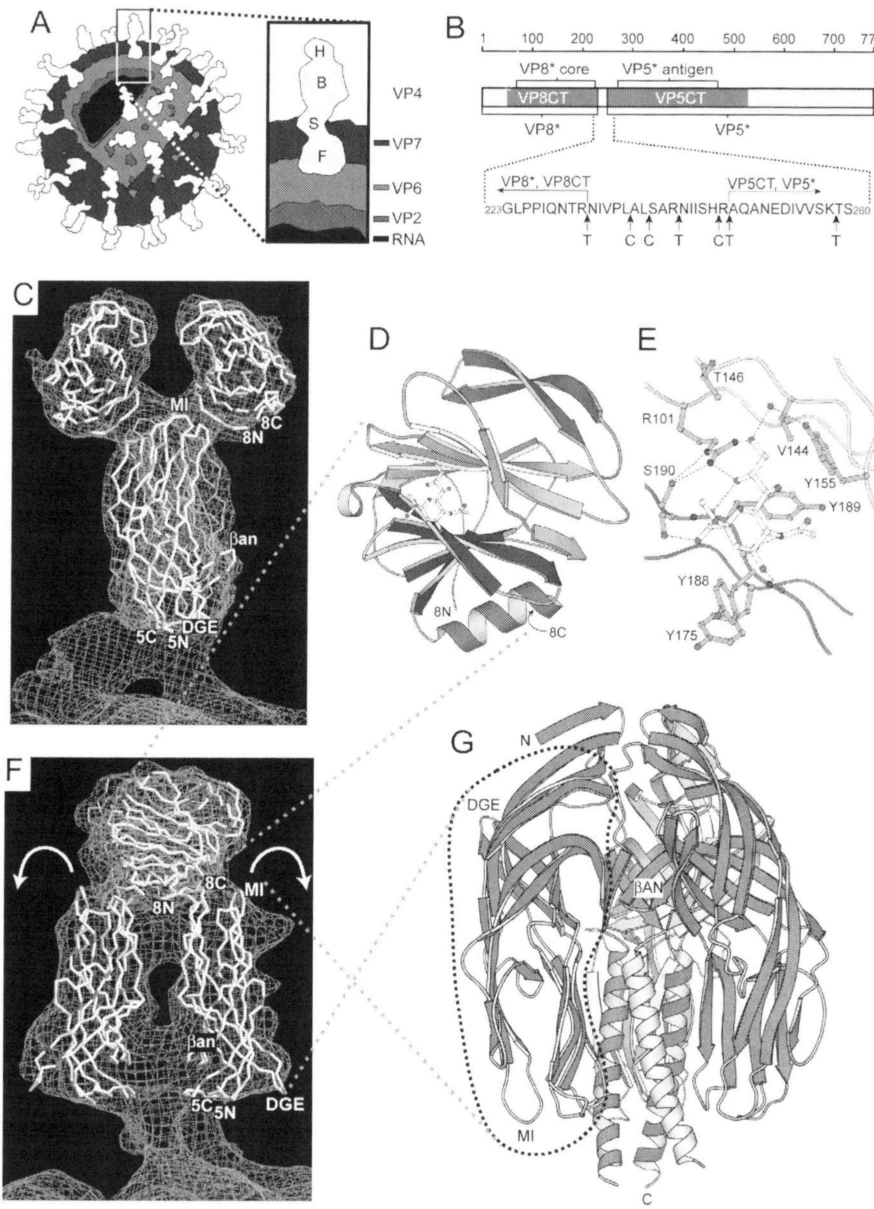

Figure 11.1 Rotavirus VP4. A. Schematic diagram of the rotavirus virion and trypsin-primed VP4 (Dormitzer *et al.*, 2004). The diagram is based on an electron cryomicroscopy-based image reconstruction (Yeager *et al.*, 1990). The proteins that make up each layer of the triple-layered icosahedral particle are indicated next to the expanded view of the spike. The subviral DP contains the RNA, VP2 and VP6 layers. The molecular envelope of the primed VP4 spike has head (H), body (B), stalk (S), and foot (F) domains. B. Linear diagram of VP4 and its fragments. The cleavage products of trypsin digestion of VP4 on the virion, VP8* (residues 1–231) and VP5* (residues 248–776), are given below the bar; the products of sequential chymotrypsin and trypsin digestion of VP4 in solution, VP8CT (residues 46–231) and VP5CT (residues 248-approximately 528), are labelled in the green bars; and the compact globular domains identified

VP4 and VP7 make up the apparatus that translocates the DLP across a membrane and into the cytoplasm. Both proteins are shed during entry (Cohen *et al.*, 1979). Because purified VP4 and VP7 are soluble and undergo entry-related conformational changes in simple *in vitro* systems (Dormitzer *et al.*, 2000, 2001), rotavirus provides a tractable system to analyze non-enveloped virus cell entry using biochemical and structural tools.

This review summarizes current knowledge of the rotavirus entry pathway. For purposes of discussion, the pathway is divided into the following steps: the priming VP4 rearrangement, attachment, internalization, membrane penetration, virion uncoating, and the second VP4 rearrangement. Actually, this scheme overstates current knowledge. The order in which membrane penetration, virion uncoating, and the second VP4 rearrangement occur has not been determined. Also, internalization and membrane penetration may or may not be the same event.

The priming VP4 rearrangement

Rotavirus is primed for efficient infectivity when trypsin cleaves VP4 into two fragments, VP8* and VP5* (Figure 11.1b; Estes *et al.*, 1981). VP8* contains the viral haemagglutinin (Fiore *et al.*, 1991). VP5* contains a sequence that is similar to the internal fusion peptides of some alphaviruses, which enter cells through a class II membrane fusion mechanism (Mackow *et al.*, 1988; Gibbons *et al.*, 2004). Both VP8* and VP5* remain associated with each other and with the virion, at least during the initial stages of attachment and entry. Western blots using antibodies raised against peptides corresponding to different potential VP4 cleavage products show that trypsin primes virions by cleaving after three closely spaced arginines (Figure 11.1b), defining a small priming region between VP8* and VP5* (Arias *et al.*, 1996). The trypsin cleavage after R247, which generates the VP5* N-terminus, is less efficient than the cleavages after R241 and R231. Only the relatively

---

by high resolution structural studies, the VP8* core (residues 65–224) and the VP5* antigen domain (residues 267–470), are given above the bar. On the expanded view of the trypsin priming region, the three sites at which trypsin cleaves VP4 on the virion (after arginines 231, 241, and 247) and the additional local site at which trypsin cleaves VP4 in solution (after lysine 258) are indicated by 'T.' The local sites at which chymotrypsin cleaves VP4 in solution (after leucines 236 and 238 and histidine 246) are indicated by 'C.' C. Cα traces of the VP8* core (PDB accession code 1KQR) and a globular domain of VP5CT (PDB accession code 1SLQ) fit to the molecular envelope of the spike from a 12 Å resolution electron cryomicroscopy-based image reconstruction of primed virions (Dormitzer *et al.*, 2004). The N- and C-termini of the VP8* core (8N and 8C), the N- and C-termini of the globular domain of VP5CT (5N and 5C), an exposed hydrophobic potential membrane interaction loop (MI), and the potential integrin binding loop (DGE). D. Ribbon diagram of the VP8* core (Dormitzer *et al.*, 2002b). The colours of the ribbons demarcate structural elements. The bound sialoside is depicted with balls and sticks, and the N- and C-termini are indicated. E. Molecular details of sialoside binding (Dormitzer *et al.*, 2002b). Hydrogen bonds are indicated by dotted lines. F. The image of panel C rotated 90° about a vertical axis. The rotation of the body domains during the twofold to threefold reorganization of the protruding regions is indicated by arrows. Dotted lines link the fit Cα traces to the corresponding ribbon diagrams. The lines linking panel F and G are crossed to indicate the domain inversion associated with the transition to the folded-back state. G. Ribbon diagram of VP5CT (Dormitzer *et al.*, 2004). Each subunit is a different colour. The globular domain that is fit to the molecular envelope of the spike is circled with a dotted line. 'βAN' marks the β-annulus. Other labels match those in panel F. (See also Plate 11.1.)

inefficient, usually final cleavage after R247 is associated with increased infectivity (Arias *et al.*, 1996). The reason that only this cleavage primes is not known. Possibly, interactions of residues between R241 and R247 with other VP4 residues prevent a priming-associated rearrangement.

Electron cryomicroscopy-based image reconstructions of rotavirus virions demonstrate rigidification of VP4 by trypsin priming (Crawford *et al.*, 2001). Before it is cleaved by trypsin, VP4 is not visible in averaged image reconstructions, but it is clearly present by biochemical assays, suggesting that it is flexible. Upon trypsin priming, VP4 forms relatively rigid, protruding spikes (Figure 11.1c and f). Hydrodynamic analysis shows that, in solution, uncleaved VP4 (strain RRV) is elongated and predominantly monomeric (Dormitzer *et al.*, 2001). Protease sensitivity analysis shows that, unlike VP4 on virions, VP4 in solution is not cleaved only, or even primarily, in the priming region, and the pattern of cleavage in the priming region is different from the pattern on virion-mounted VP4 (Figure 11.1b, Dormitzer *et al.*, 2001).

Electron cryomicroscopy-based image reconstructions of primed virions (Shaw *et al.*, 1993; Yeager *et al.*, 1994) show that the spikes have 'head', 'body', and 'stalk' regions, which protrude beyond the VP7 layer, and a 'foot' region, which is buried beneath the VP7 layer (Figure 11.1a). The spikes emerge from the side of each peripentonal channel that is furthest from an icosahedral fivefold vertex. The head and body subunits have approximate dyad symmetry (Figure 11.1 c and f). The two heads of each spike do not contact each other, but the two body subunits make proximal and distal dyad contacts, leaving a hole between the contacts on the approximate dyad axis. The angled stalk, which connects the body subunits to the foot, is asymmetrical. Electron cryomicroscopy reconstructions of rotavirus reassortants reveal subtle differences in spike conformation (Pesavento *et al.*, 2003). Some neutralization escape mutations on VP8* of a sialic acid-independent rotavirus strains are in locations that would be inaccessible to antibody on the spikes seen in reconstructions of sialic acid-dependent strains (Monnier *et al.*, 2006). These observations suggest variation in spike structure between strains and/or the presence of VP4 that can be bound by neutralizing antibodies but does not contribute to the molecular envelope of image reconstructions obtained to date.

Indeed, despite the dyad apparent in the protruding portion of each spike, it appears that each cluster of VP4 on the virion may contain three molecules. Subtracting image reconstructions of particles without spikes from image reconstructions of particles with spikes reveals the molecular envelope of the foot (Yeager *et al.*, 1994). The foot has threefold rotational symmetry, which has also been described as pseudo-hexameric symmetry. Thus, there is a symmetry mismatch between the envelope of the protruding region and of the foot. Electron cryomicroscopy image reconstructions of virions exposed to high pH show coalesced stubs of VP4 in a triskelion arrangement centred on each peri-pentonal channel, consistent with three VP4 molecules forming each cluster (Pesavento *et al.*, 2005). These findings suggest that the two cleaved VP4s furthest from the fivefold axis in each cluster interact with VP6 and/or VP7 in a manner that allows them to form dyad contacts in the region external to the VP7 shell. These interactions stabilize the conformation of these two molecules so that they form the protruding spike density in the reconstructions. In this case, the protruding part of the third molecule must remain flexible, resulting in its absence from averaged image reconstructions.

The X-ray crystal structures of two VP4 domains fit electron cryomicroscopy image reconstructions of the spikes (Figure 11.1c and f; Dormitzer *et al.*, 2002b; Dormitzer *et al.*, 2004; Yoder and Dormitzer, 2006). The RRV VP8* core (residues 65–224), which binds sialic acid, neatly fits the heads. Thus, in sialic acid-dependent rotavirus strains, cleavage-induced rigidification positions a receptor-binding domain at the tips of the spikes.

The crystal structure of a VP5* antigen domain (RRV residues 267–470) dimer (not shown) fits the spike body (Yoder and Dormitzer, 2006), although a distortion of the dimer would be required to fill the electron cryomicroscopy envelope as neatly as the two globular domains fit separately (Figure 11.1c and f). The main twofold contact of the dimer in the crystal structure corresponds to the proximal twofold contact of the spike. Based on the location of the VP8* core N-terminus at the neck between head and body, it is likely that VP4 residues 1–64, which are missing from the VP8* core structure, insert between the apices of the VP5* antigen domain, separating them, so that the domains more completely fill the spike molecular envelope while retaining their proximal dimer interface. The N-terminal residues of VP8* serve a dual function, tethering the heads to the rest of the spike and forming the distal dyad contact. In the primed spike, the VP8* heads shield the hydrophobic apices of the VP5* antigen domain, which form the spike 'shoulders.' Thus, priming produces a structure that is poised to bind the cell surface and shields a hydrophobic potential membrane interaction region beneath a tethered receptor-binding domain.

Entry pathways of both enveloped and non-enveloped viruses include a proteolytic priming step (reviewed in Skehel *et al.*, 2000 and Hogle, 2002). Membrane disruption by both types of viruses is driven by the energy released when a membrane interaction protein refolds to a lower energy state. Synthesis as an uncleaved precursor allows folding to an initial energy minimum. The cleavage of a peptide bond then releases a covalent 'safety catch', so that a lower energy state becomes accessible. An activation energy barrier prevents immediate refolding to a lower energy state after cleavage. This barrier is overcome by a triggering interaction (often with a proton or a receptor). Triggering by conditions met upon contact with the target cell or after endocytic internalization provides for productive 'firing' at the appropriate time during entry.

The best understood example, influenza haemagglutinin (HA), though the fusion protein of an enveloped virus, resembles VP4. Both are spike proteins with sialic acid-binding head domains that are N-terminal and virion-distal to membrane interaction domains. In both HA and VP4, cleavage between the domains primes the viruses. The fusogenic structural transitions that mediate influenza entry are understood in detail based on X-ray crystal structures of uncleaved, proteolytically primed, and low-pH triggered states of HA (Wilson *et al.*, 1981; Bullough *et al.*, 1994; Chen *et al.*, 1998). When influenza is primed, only 19 residues, which bracket the cleavage site in HA, shift position.

Some structural data demonstrate local priming rearrangements of other non-enveloped viruses. Crystal structures of a poliovirus assembly intermediate and of the mature poliovirus virion show that when the myristolated capsid protein, VP0, autolytically cleaves itself, the segment containing the sessile bond shifts 25 Å on the internal capsid surface (Hogle *et al.*, 1985; Basavappa *et al.*, 1994). Although VP0 cleavage does not cause a large scale capsid rearrangement, when the newly generated termini shift, previously flexible N-terminal regions of the capsid proteins form a network of interactions that stabilizes the poliovirus capsid. A similar autolytic cleavage of the myristolated reovirus membrane penetration protein, μ1, together with the proteolytic removal of a protector protein, σ3, primes reovirus for entry. In a crystal structure of the reovirus μ1/σ3 complex, in which μ1 has cleaved itself, the newly generated termini are separated by 9 Å, indicating a local conformational change (Liemann *et al.*, 2002). Electron cryomicroscopy image reconstructions of reovirus virions, reovirus cores recoated with autocleavage-defective μ1 mutants, and primed reovirus particles (ISVPs) show that the priming cleavages are associated with only modest differences in the molecular envelope attributed to μ1 (but much larger changes in σ1 and σ3; Dryden *et al.*, 1993; Odegard *et al.*, 2004). Although the flexibility of uncleaved rotavirus VP4 has, thus far, precluded a high-resolution structural understanding of its rearrangement during rotavirus priming, the rigidification of the spikes suggests that priming causes a larger conformational change in VP4 than is observed in the membrane interaction proteins of influenza virus, poliovirus, or reovirus.

Interaction with cell surface receptors

The initial attachment of rotavirus to host cells is characterized by diversity of potential receptors, variability in receptor usage between virus strains, and the lack of any identified high affinity interactions. Evidence has been presented for rotavirus interactions with host cell carbohydrates, proteins, and lipids through binding to VP8*, VP5*, and VP7. It is not known whether any cellular receptor for rotavirus has a truly central role in the entry pathway. No identified receptor interaction accounts for the wide differences in permissiveness to rotavirus entry between cell types.

Because binding to sialic acid (N-acetyl neuraminic acid) mediates haemagglutination (Bastardo *et al.*, 1980), because rotavirus strains that bind sialic acid grow more readily in culture, and because sialic acid can be selectively stripped from the cell surface by neuraminidase, this carbohydrate was the first recognized rotavirus receptor. Binding to sialic acid mediates the initial cell attachment of sialic acid-dependent rotavirus strains (Fukudome *et al.*, 1989). In strain SA11, sialic acid binding is reported to be sodium-dependent (Keljo *et al.*, 1988b). Sialic acid is the only molecule for which there is direct experimental evidence of receptor function in an infected animal: neuraminidase treatment of isolated intestinal loops from mouse pups diminishes the number of villus tip enterocytes infected by a sialoside-dependent rotavirus strain by more than 80% (Ludert *et al.*, 1996).

Most rotavirus strains do not bind sialic acid (Ciarlet *et al.*, 1999). Although a sialic acid-dependent rotavirus strain has been isolated from a human infant in a probable case of zoonotic transmission (Nakagomi *et al.*, 2000), strains known to cause rotavirus gastroenteritis in humans are universally sialic acid independent (Gentsch *et al.*, 1996; Ciarlet *et al.*, 2002b). Sialic acid dependence correlates with the polarization of entry: sialic acid-independent strains enter cultured polarized epithelial cells efficiently from either the apical or basolateral surface; sialic-acid dependent strains enter efficiently only from the apical surface, reflecting the primarily apical distribution of sialic acid (Ciarlet *et al.*, 2001).

Passage of sialic acid-dependent rotavirus strains (RRV or OSU) on neuraminidase-treated cells readily selects sialic acid-independent variants with mutations in VP4 (Mendez *et al.*, 1993; Ludert *et al.*, 1998; Lopez *et al.*, 2005). These variants actually can bind sialic acid (as indicated by neuraminidase-sensitive haemagglutination or inhibition of infection by glycophorin A), but do not require sialic acid on the cell surface for entry. Thus, the sialic acid-independent phenotype is not simply a consequence of an inability to bind sialic acid. Because the sialic acid-independent RRV variants have VP4 mutations outside the sialic acid binding domain and attach more rapidly to neuraminidase-treated cells than to untreated cells (Mendez *et al.*, 1999), their phenotype may reflect increased efficiency of subsequent entry events without first binding sialic acid. To date, sialoside triggering of a major molecular rearrangement or a new intermolecular interaction has not been demonstrated.

Nuclear magnetic resonance (NMR) analysis shows that the binding of the RRV VP8* core to sialosides is extremely weak (Kd = 1.2 mM for α(2,3)-sialyllactose) (Dormitzer *et al.*, 2002a). Therefore, attachment of sialic acid-dependent rotaviruses to cells must rely on the increased avidity provided by multivalent binding. The affinity of influenza haemagglutinin for its sialoside receptor is similarly weak (Kd = 2.7 mM for α(2,6)-sialyllactose) (Hanson *et al.*, 1992). The preference of influenza HA for sialic acid in either α(2,3) or α(2,6) linkage to galactose is an important determinant of influenza host range (Connor *et al.*, 1994). By contrast, rotavirus (strain RRV) has the same affinity for 3′ sialyllactose, 6′ sialyllactose, and a simple monomeric sialoside (Dormitzer *et al.*, 2002a). The RRV VP8* core binds N-acetyl sialosides more tightly than it binds N-glycolyl sialosides (kD = 11 mM), quashing the hypothesis that a universal preference for N-glycolyl neuraminic acid accounts for the absence of sialic acid-dependent rotavirus isolates that cause human disease – humans, unlike other mammals, do not synthesize N-glycolyl neuraminic acid (Chou *et al.*, 1998). Other rotavirus strains may preferentially bind N-glycolyl sialosides, as indicated by more potent inhibition of infection (strain OSU) by an N-glycolyl glycosphingolipid ($NeuGcGM_3$) than by its N-acetyl equivalent ($NeuAcGM_3$) and by preferential binding (strains SA11 and NCDV) to $NeuGcGM_3$ in a solid phase assay (Rolsma *et al.*, 1998; Delorme *et al.*, 2001).

The crystal structure of the RRV VP8* core in complex with sialoside reveals molecular details of the binding interaction (Figure 11.1d and e; Dormitzer *et al.*, 2002b). Bound sialoside occupies the cleft between two β-sheets. The sialoside stabilizes a network of hydrogen bonds and salt bridges that span this cleft. Residue R101 is critical to these interactions, making bidentate salt bridges to the glycerol group of the sialoside and an additional salt bridge that holds the sialoside binding cleft closed. Comparison of the crystal structure of the sialoside-bound VP8* core to an NMR solution structure of the unbound core (Dormitzer *et al.*, 2002b) suggests a binding-induced fit of VP8* to the sialoside. The VP8* core has the same fold as the members of a family of galactose-binding animal lectins, the galectins (Barondes *et al.*, 1994). Thus, the heads of the VP4 spikes may have been added when a host gene encoding a carbohydrate-binding domain inserted into a viral gene encoding a membrane interaction protein. Subsequent modification, in particular, the elaboration of a β-ribbon that neatly blocks the VP8* surface that corresponds to the galactose binding site of galectins, may have been involved in altering the carbohydrate binding site and specificity of VP8*.

The sialosides bound during rotavirus infection may be located on cell surface glycoproteins or glycolipids. In support of binding to sialosides on glycoproteins, rotavirus (strain RRV) specifically binds sialic acid in the O-linked oligosaccharide of a protein (not yet identified) that has a distribution matching the cell type tropism, tissue tropism, host age restriction, and host range restriction of rotavirus infection (Bass *et al.*, 1991). In support of entry via sialosides on glycolipids, incubating neuraminidase-treated cells with sialic acid-containing gangliosides, such as GM_1 and $NeuGcGM_3$, restores their susceptibility to sialic acid-dependent rotavirus strains SA11 and OSU (Superti *et al.*, 1991; Rolsma *et al.*, 1998). The developmental decrease in the quantity of $NeuGcGM_3$ and $NeuAcGM_3$ in neonatal pig intestinal mucosa correlates with decreasing susceptibility to rotavirus gastroenteritis (Rolsma *et al.*, 1998).

Sialic acid moieties on glycolipids have been implicated in neuraminidase-resistant rotavirus entry. Binding of a neuraminidase-insensitive rotavirus strain (UK) to $NeuAcGM_3$ but not to $NeuGcGM_3$ in a solid phase binding assay suggests that apparent sialoside independence could simply reflect binding to non-terminal (neuraminidase-resistant) sialosides on glycolipids (Delorme *et al.*, 2001). Infectivity blocking data suggest a possible interaction of two human rotavirus strains (KUN and MO) with GM_{1a}, a ganglioside with an internal sialic acid moiety (Guo *et al.*, 1999). However, both sialic acid-dependent and sialic acid-independent strains also bind glycosphingolipids that lack sialic acid – in particular, asialo-GM_1 and asialo-GM_2 (Willoughby *et al.*, 1990; Srnka *et al.*, 1992; Delorme *et al.*, 2001). As asialo-GM_1 does not appear in the neonatal mouse intestine until the end of the developmental window of high susceptibility to rotavirus, the biological significance of binding in these assays is uncertain (Willoughby *et al.*, 1990).

Non-conservative amino acid substitutions at key sialoside-binding positions in VP8* of most sialoside-independent strains and the obliteration of the sialic acid binding site in the crystal structure of a VP8* core from a sialoside-independent strain (DS-1) that causes human disease suggest that circulating sialic acid-independent strains do not bind sialic acid in any context, at least not using the same binding site as established sialoside-dependent strains (Monnier *et al.*, 2006). The presence of a leader sequence bound in a widened cleft adjacent to the obliterated SA-binding cleft in the DS-1 crystal structure suggests that VP8* of SA-independent strains may bind an alternative receptor (Monnier *et al.*, 2006).

An antibody against VP5* and a recombinant VP5* preparation block the attachment of wild-type RRV to neuraminidase-treated cells and block the attachment of sialoside-independent RRV variants to untreated cells (Zarate *et al.*, 2000b). These observations suggest that the interaction of VP5* with a cell surface molecule other than sialic acid mediates cell attachment by sialoside-independent rotavirus strains and may also occur secondarily during the entry of sialoside-dependent strains. The target for VP5*-mediated attachment may be an integrin. Integrins make up a large family of hetero-dimeric cell surface molecules. They mediate dynamic changes in cellular shape, tissue organization, and cellular adhesion by anchoring cells to the extracellular matrix, binding specific counter-receptors and plasma proteins, and carrying out bidirectional signalling across the plasma membrane (Arnaout *et al.*, 2002). Rotavirus interactions with several integrins have been proposed: VP5* with $\alpha2\beta1$ and $\alpha4\beta7$ integrins, and VP7 with $\alpha x\beta2$, $\alpha4\beta1$, and $\alpha v\beta3$ integrins.

VP5* binding to the $\alpha2$ I domain of $\alpha2\beta1$ integrin has been the most extensively investigated. Antibody blocking and competitive binding studies suggest that $\alpha2\beta1$ integrin can serve as a cell surface receptor during rotavirus infection of cells in culture (Coulson *et al.*, 1997; Zarate *et al.*, 2000a; Graham *et al.*, 2006). Transfecting cDNA encoding the

$\alpha2$ integrin chain into a poorly permissive cell line that constitutively expresses the $\beta1$ chain increases the replication of several rotavirus strains (Emsley *et al.*, 2000; Graham *et al.*, 2003). $\alpha2\beta1$ integrin anchors enterocytes to collagen and laminin in the basement membrane. Its apparent receptor activity is surprising, as it is primarily distributed on the basolateral surface of enterocytes and is more abundant in small intestinal crypts than on villi (Beaulieu, 1992), where rotavirus infects. Because rotavirus is shed apically (Jourdan *et al.*, 1997), access to $\alpha2\beta1$ integrin in the gut would presumably require disruption of tight junctions. Rotavirus infection does disrupt tight junctions (Ciarlet *et al.*, 2001), but this disruption occurs after cell entry, so it could only facilitate spread of rotavirus to neighbours of infected cells.

In a solid phase assay, the isolated $\alpha2$ I domain binds a recombinant RRV VP5* fragment but not an equivalent VP5* fragment with a mutated DGE (residues 308–310) potential integrin binding motif (Graham *et al.*, 2003). As rotavirus binding to $\alpha2\beta1$ integrin on the cell surface requires cleaved VP4 (Zarate *et al.*, 2004), the DGE motif may be inaccessible prior to priming. It is not known whether VP5* in the primed spike conformation and/or the folded-back conformation binds $\alpha2\beta1$ integrin. In the primed spike, the DGE-containing loop forms the bottom outside corner of the body (Figure 11.1 c, f; Dormitzer *et al.*, 2004). The $\alpha2$ I domain, which is at the membrane-distal end of the integrin (Emsley *et al.*, 1997), could reach the bottom of the spike body from the surface of a target cell. Modelling the $\alpha2$ I domain bound to the DGE motif on the spike shows no prohibitive steric hindrances between the bound domain and the surface of the VP7 layer (unpublished data). In the folded umbrella conformation (Figure 11.1g), the DGE motif projects into solvent at the tip of a highly exposed loop that extends from each subunit at the top of the trimer (Dormitzer *et al.*, 2004).

Binding to the $\alpha2$ I domain of $\alpha2\beta1$ integrin also mediates the attachment of echovirus I. A recombinant $\alpha2$ I domain preparation blocks attachment of echovirus I to HeLa cells with an IC_{50} of 3–30 nM (King *et al.*, 1995). While the ability of the $\alpha2$ I domain to block rotavirus binding or infectivity has not been tested, DGE-containing peptides block attachment of SA11 rotavirus to MA104 cells with an IC_{50} of approximately 100 μM (Graham *et al.*, 2003). Thus, high affinity rotavirus binding to $\alpha2\beta1$ integrins has yet to be established.

The hypothesis that DGE on VP4 is an integrin binding site was originally based on a report that DGEA is the $\alpha2\beta1$ integrin binding site on type I collagen (Staatz *et al.*, 1991; Coulson *et al.*, 1997). Although high concentrations of small, DGEA-containing, unstructured peptides block the adherence of $\alpha2\beta1$ integrin-bearing platelets to type I collagen (IC_{50} = 2.8 mM), DGEA on folded collagen is not bound by $\alpha2\beta1$ integrin (Grab *et al.*, 1996; Knight *et al.*, 1998). Rather, the recognition site on type I collagen is GFOGER ('O' for hydroxyproline), folded into a triple-helix (IC_{50} = 7 μM for inhibition of platelet adhesion; Knight *et al.*, 2000). A crystal structure of the $\alpha2$ integrin I domain in complex with triple-helical GFOGER demonstrates a ternary complex between the integrin 'MIDAS' motif, a metal ion, and the glutamic acid side chain of the peptide (Emsley *et al.*, 2000). Ternary complexes with a metal ion typically mediate integrin binding. For the $\alpha2$ I domain, magnesium or manganese ions (but not calcium at extracellular concentrations) mediate binding (Arnaout *et al.*, 2002). While the aspartic acid or glutamic acid side chains of the VP5* DGE could potentially bind the $\alpha2$ I domain in a similar manner, there is no known metal dependence of the VP5*-integrin interaction. Furthermore, mutation of key residues in the α2 integrin I domain MIDAS motif does not affect rotavirus binding (Londrigan *et al.*, 2003). Thus, the basis for α2 I domain binding to rotavirus VP5* is obscure.

Similar, though less extensive, evidence indicates that VP5* of some strains may bind α4β7 integrin (Graham *et al.*, 2005) and that VP7 of some strains may bind α4β1, αvβ3, or αxβ2 integrins (Hewish *et al.*, 2000; Graham *et al.*, 2003). The well-established αvβ3 recognition site, RGD, does not mediate the rotavirus interaction with αvβ3 integrin (Xiong *et al.*, 2001; Zarate *et al.*, 2004). Interaction with αxβ2 may be mediated by the GPR tripeptide (Coulson *et al.*, 1997). This motif is conserved in VP7 of mammalian rotavirus strains and has been proposed as an αxβ2 binding site, based the inhibition of neutrophil adherence to fibrinogen by high concentrations of synthetic peptides containing the sequence GPRP (IC_{50} = 0.38 mM) (Loike *et al.*, 1991).

The correlation between the presence of tripeptide binding motifs and apparent integrin usage by rotavirus is spotty (Graham *et al.*, 2003), and the pleotropic cellular effects of integrin signalling complicate interpretations of modest changes in viral infectivity resulting from the manipulation of integrin interactions. Blockade of rotavirus infectivity, but not attachment, by peptides or antibodies expected to interfere with integrin binding suggests that integrin interactions may function in post-attachment entry events (Guerrero *et al.*, 2000a; Zarate *et al.*, 2000a; Ciarlet *et al.*, 2002a; Graham *et al.*, 2003). Variable results have been obtained in these assays, particularly for α2β1 integrin, probably due to differences between attachment to suspended and anchored cells. A post-attachment role for integrin binding has also been observed for adenovirus. After cell attachment mediated by an interaction between the adenovirus fibre and CAR (the coxsackie and adenovirus receptor), internalization is promoted by interaction of the penton base with αvβ3 and αvβ5 integrins (Wickham *et al.*, 1993; Bergelson *et al.*, 1997). αvβ5 integrin is specifically implicated in membrane penetration by adenovirus (Wang *et al.*, 2000).

Cultured cell lines vary in susceptibility to rotavirus infection by more than 7 logs (MA104 rhesus monkey embryonic kidney vs. BHK-21 Syrian hamster kidney cells; Ciarlet *et al.*, 2002a). Transfection of rotavirus DLPs equalizes susceptibility, demonstrating that entry is the barrier to infection of resistant cell lines (Bass *et al.*, 1992; Ciarlet *et al.*, 2002a). Although differences in cell surface receptor expression could account for the differences in susceptibility, no rotavirus-receptor interactions documented to date account for the huge observed variation. Decreases in infectivity upon neuraminidase treatment of cells (for sialoside-dependent strains) or upon blocking integrin interactions (for sialoside-independent strains) are generally 10-fold or less (Fukudome *et al.*, 1989; Ludert *et al.*, 1996; Mendez *et al.*, 1996; Coulson *et al.*, 1997; Ludert *et al.*, 1998; Hewish *et al.*, 2000; Ciarlet *et al.*, 2002a; Graham *et al.*, 2004). Although recombinant expression of α2β1 integrin enhances the susceptibility of CHO cells to SA11 rotavirus by 190-fold, trypsin priming actually decreases SA11 infectivity in CHO cells approximately 60-fold, complicating the interpretation of this result (Londrigan *et al.*, 2003). In other poorly permissive cell lines, transfection of integrin-encoding cDNA enhances susceptibility approximately 10-fold (Hewish *et al.*, 2000). Extensive variation in receptor usage between strains (despite a common tropism for villus tip enterocytes *in vivo*) and poor correspondence between rotavirus tissue tropism and the distribution of integrins and sialic acid indicate that the primary determinants of rotavirus tropism have yet to be determined (Ciarlet *et al.*, 1999; Ciarlet *et al.*, 2002a; Graham *et al.*, 2003). While the identified receptors may play a role in adherence to the cell surface at different stages of the entry pathway, convincing evidence that binding to these receptors triggers conformational changes that mediate entry has yet to be obtained.

Heat shock cognate protein 70 (HSC70), a protein chaperone that binds short, hydrophobic peptides in extended conformations (Bukau *et al.*, 1998), has also been proposed as a rotavirus receptor (Guerrero *et al.*, 2000b; Guerrero *et al.*, 2002; Zarate *et al.*, 2003). In an octyl-β-glucoside extract of MA104 cells, HSC70 has been identified as a component that blocks rotavirus infectivity by 60%, even after the protein is denatured by boiling in SDS, reduction, and acetone precipitation. Incubation of cells with antibodies to HSC70 or with recombinant HSC70 itself reduces rotavirus infectivity by 40–80%, interfering with a post-attachment event. HSC70 also binds rotavirus virions via its peptide binding domain and reduces their infectivity in an ATP-dependent manner (Perez-Vargas *et al.*, 2006). HSC70 is involved in cellular processes of potential relevance to entry, such as translocating viral glycoproteins across the ER membrane and clathrin uncoating (Lambert *et al.*, 2003; Fotin *et al.*, 2004). The predominantly intracellular distribution of HSC70 and the inherent non-specificity of protein binding by chaperones make the mechanism of this molecule's direct or indirect influence on cell entry unclear.

Internalization

The fundamental question of whether productive internalization of rotavirus into cells proceeds through direct penetration of the plasma membrane, through endocytosis followed by penetration of an endocytic vesicle membrane, or by either route alternatively remains controversial. Electron microscopic images consistent with either route of entry have been obtained (Quan *et al.*, 1983; Suzuki *et al.*, 1985; Ludert *et al.*, 1987). However, with particle-to-pfu ratios ranging from less than 10 to more than 10 000 (Petrie *et al.*, 1981; Hundley *et al.*, 1985; Keljo *et al.*, 1988b; Mendez *et al.*, 1999), it is probable that for most virus preparations most observed internalization events do not lead to productive infection. Indeed, internalization, as measured by sequestration from proteases added to the medium, is as rapid in non-permissive cells as in permissive cells (Bass *et al.*, 1992). However, in permissive cell lines, internalization is associated with an eclipse phase – little infectious virus can be recovered during the two hours after internalization. Ten to one-hundredfold more infectious virus can be recovered during this time in non-permissive cell lines.

Primed and unprimed rotavirus particles enter cells with different kinetics. Trypsin-primed particles are removed from the cell surface with a half-life of 3–5 min (assayed by maintained infectivity despite the addition of neutralizing antibodies to the medium); uncleaved particles are removed with a half-life of 30–50 min (assayed by loss of an infectivity boost upon adding 0.5 μg/ml trypsin to the medium; Kaljot *et al.*, 1988). These data have been interpreted as indicating that trypsin-primed particles are internalized productively and rapidly by direct plasma membrane penetration, while uncleaved particles are internalized non-productively and slowly by endocytosis. The release of ^{51}Cr from preloaded epithelial cells upon addition of trypsin-primed, triple-layered virions provides additional data suggestive of direct plasma membrane penetration (Kaljot *et al.*, 1988).

In most experiments that investigate the route of rotavirus cell entry, compounds that inhibit intracellular processes are added to the medium of cultured cells, and effects on infection are monitored. Most broadly, incubating cells with inhibitors of energy metabolism (sodium azide or dinitrophenol) early in infection inhibits rotavirus infectivity minimally (< 20% decrease in titre), but inhibits reovirus infectivity more significantly (84–98%), arguing against a requirement for endocytosis during rotavirus entry (Kaljot *et al.*, 1988). Dansylcadaverine (an inhibitor of endocytosis-mediated receptor trafficking) and cytochalasin D (an inhibitor of actin polymerization) do not block rotavirus entry into MA104 cells (highly permissive for rotavirus) and actually enhance rotavirus entry into L cells (poorly permissive for rotavirus) at concentrations that block reovirus entry (Fukuhara *et al.*, 1987; Bass *et al.*, 1995; Cuadras *et al.*, 1997).

A recent study suggests that rotavirus may enter cells through a clathrin- and caveolin-independent endocytic pathway (Sanchez-San Martin *et al.*, 2004). Inhibiting clathrin-mediated endocytosis by treatment of MA104 cells with hypertonic sucrose or chlorpromazine or by transfection with cDNA encoding a dominant negative Eps15 mutant fails to block rotavirus infectivity. Inhibiting caveolin-mediated endocytosis by treatment with filipin or nystatin or by transfection with cDNA encoding dominant negative caveolin type 1 and 3 mutants also fails to block rotavirus infectivity. Thus, neither clathrin-mediated nor caveolin-mediated endocytosis is required for rotavirus cell entry. By contrast, transfection of cDNA encoding a dominant-negative dynamin mutant reduces rotavirus infectivity by 84%. Dynamins are lipid stretching and fission devices (Praefcke *et al.*, 2004). In a GTP-consuming process, they mediate the constriction and scission of the neck of endocytic vesicles, budding of other organelles, and actin-membrane interactions. Clathrin- and caveolin-independent endocytic pathways are dependent on dynamin activity (Conner *et al.*, 2003). As such pathways are currently defined negatively, proof of the hypothesis that rotavirus enters through a non-clathrin, non-caveolin mediated endocytic pathway awaits the development of specific markers for or inhibitors of such pathways. The potential role for dynamins in rotavirus entry must be reconciled with prior results indicating that blockade of cellular energy metabolism has little impact on entry (Kaljot *et al.*, 1988).

Mechanisms of entry following endocytosis are easier to envision than mechanisms of entry at the plasma membrane. Translocation across the plasma membrane necessitates not only local membrane disruption, but also movement of the particle through the breach before repair of the membrane by lateral diffusion of lipids. As an endocytic vesicle may be only slightly larger than a virus particle and is surrounded by cytoplasm, vesicle disruption could readily release an endocytosed particle into the cytoplasm. The small volume inside a vesicle could maintain high local concentrations of membrane-active proteins released from viral particles. Changes in the chemical environment of the vesicle interior could trigger entry events. Multiple studies using a variety of endosomal acidification inhibitors have indicated that a fall in pH is not required for productive rotavirus infection (Fukuhara *et al.*, 1987; Ludert *et al.*, 1987; Kaljot *et al.*, 1988; Keljo *et al.*, 1988a; Bass *et al.*, 1995; Cuadras *et al.*, 1997; Liprandi *et al.*, 1997; Sanchez-San Martin *et al.*, 2004). However, one report (Chemello *et al.*, 2002) does suggest that a drop in endosomal calcium concentration, driven by the electrical gradient generated by a bafilomycin-sensitive endosomal proton pump, may facilitate rotavirus infection (see 'Viral uncoating').

Membrane penetration

Numerous membrane disrupting activities of rotavirus outer capsid protein fragments have been documented. In a receptor-independent process, rotavirus particles disrupt artificial liposomes, as assayed by fluorophore dequenching (Nandi *et al.*, 1992). Cleaved particles permeabilize rotavirus-susceptible cell lines to ethidium bromide and release fluorophores from preloaded cell-derived membranes (Ruiz *et al.*, 1994, 1997). Minutes after they are added to the medium of MA104 cells in suspension, trypsin-cleaved rotavirus particles create punctate plasma membrane disruptions, visible by electron microscopy (Ruiz *et al.*, 1997). Cleaved rotavirus particles also mediate cell–cell 'fusion from without' when added to suspended MA104 cells in the presence of cholesterol (Falconer *et al.*, 1995). Each of the above activities is either inhibited by calcium, stable to calcium chelation, or dependent on the addition of a calcium chelator (EDTA or EGTA). By contrast, rotavirus infectivity is stabilized by calcium (Pando *et al.*, 2002) and eliminated by chelation-induced uncoating. Thus, these phenomena do not provide snapshots of rotavirus entry. Rather, these phenomena probably reflect the activities of outer capsid proteins that have been solubilized from rotavirus virions and cleaved by protease.

Other systems appear to provide a more direct assay of membrane disruption during entry. When added to cells, triple-layered rotavirus virions mediate the release of preloaded ^{51}Cr and the co-entry of α-sarcin (a translation inhibitor). These activities are blocked by EDTA and other inhibitors of rotavirus infection and show cell type specificity that matches susceptibility to infection (Kaljot *et al.*, 1988; Cuadras *et al.*, 1997; Liprandi *et al.*, 1997). While chromium release suggests permeabilization of the plasma membrane, α-sarcin co-entry is not specific to mode or site of entry. Enveloped viruses that enter cells through fusion of a viral envelope with either the plasma membrane or an endosomal membrane also promote α-sarcin co-entry (Carrasco, 1995).

Experiments using recombinant, non-infectious, virus-like particles (VLPs) have identified distinct membrane permeabilization activities of VP4 and VP7 fragments. Cleaved VP4 is required for VLP-mediated co-entry of α-sarcin and cell–cell fusion-from-without (Gilbert *et al.*, 1997; Liprandi *et al.*, 1997). Indeed, fusion-from-without shows sialic acid dependence matching the phenotype of the strain of origin of VP4. Mutational analysis shows that fusion-from-without requires the same cleavage (after VP4 residue R247) that primes authentic particles for infection (Gilbert *et al.*, 1998). By contrast, VP7 that has been solubilized from VP2/6/7 VLPs by calcium chelation, then cleaved by trypsin, disrupts cell-derived membrane vesicles in the absence of VP4 (Charpilienne *et al.*, 1997).

Recombinant VP5* preparations also have membrane disruption activity. An *E. coli*-expressed preparation of VP5* permeabilizes liposomes, as assayed by fluorophore release (Denisova *et al.*, 1999). Recombinant VP5* fragments also permeabilize the cell membrane of bacteria in which they are expressed to allow entry of ONPG (Dowling *et al.*, 2000) and permeabilize mammalian cells in which they expressed to allow entry of divalent cations (Golantsova *et al.*, 2004). Mutational analysis localizes membrane disruption activity to the hydrophobic sequence from residue 385 to 404 that forms part of the apex of the VP5* antigen domain (Dowling *et al.*, 2000).

Some data on the physical nature of the interactions between rotavirus protein fragments and lipids have been obtained. The recombinant VP5*-mediated release of carboxyfluorescein (0.38 kD), but not fluorescein isothiocyantate-dextran (4 kD), from liposomes suggests that this preparation makes very small membrane perforations (Dowling *et al.*, 2000). By contrast, outer capsid proteins solubilized from cleaved rotavirus particles decrease light scattering by cell-derived membrane vesicles, demonstrating a decrease in vesicle diameter (Ruiz *et al.*, 1997). This finding suggests a more global membrane disruption. Small angle X-ray diffraction indicates that addition of recombinant VP4 or rotavirus particles to sphingomyelin-containing artificial lipid bilayers changes their phase equilibrium from inverted hexagonal to lamellar, provided that cholesterol is also present (Sapin *et al.*, 2002). Rotavirus-induced fusion-from-without also requires added cholesterol (Falconer *et al.*, 1995). Depleting cholesterol in cultured cells with methyl-β-cyclodextrin inhibits rotavirus infectivity (Sanchez-San Martin *et al.*, 2004). Changes in membrane fluidity, altered interaction between rotavirus proteins and membrane lipids, or decreased association of rotavirus receptors with lipid microdomains (rafts) (Isa *et al.*, 2004) could mediate this inhibition. By contrast, liposome disruption by rotavirus is insensitive to lipid composition (Nandi *et al.*, 1992).

A variety of hypotheses about how rotavirus translocates the DLP into the cytoplasm could be constructed from this diverse set of experimental observations. In one version, a small VP5*-induced membrane breach lowers local calcium concentrations, solubilizing VP7, which is then cleaved into membrane-disrupting fragments (Charpilienne *et al.*, 1997; Dowling *et al.*, 2000). However, due to the non-specificity of many protein-membrane interactions, activities observed in model systems do not necessarily correspond to biological functions. For example, signal peptides permeabilize synthetic membrane vesicles to carboxyfluorescein and promote vesicle fusion (Nagaraj *et al.*, 1987). Yet, their biological function, mediating protein translocation across the endoplasmic reticulum (ER) membrane, reflects their ability to specifically bind a signal recognition particle and a protein-conducting channel (Rapoport *et al.*, 2004), not their ability to permeabilize or fuse membranes. Similarly, in isolation, a 40 residue peptide from the coiled-coil region of influenza HA2 inserts into phospholipid vesicles (Yu *et al.*, 1994). However, this region of HA2 does not insert into a membrane during influenza virus entry (Durrer *et al.*, 1996).

Given these limitations in extrapolating from model systems, links between the interactions of rotavirus outer capsid protein fragments with lipid bilayers and entry mechanisms need to be established experimentally. In particular, it is necessary to establish what cleavage products and conformational states of VP4 and VP7 are present during rotavirus entry and determine the membrane interactions of these entities. It is not known whether VP7 released from virions is cleaved during entry. Nor is it known whether the conformation (or conformations) of VP5* during translocation of the DLP is the same as the conformation of the membrane-active component of recombinant VP5* preparations. Indeed, the relevant activity of VP5* may occur during a transition between conformations and may not be carried out by any stable conformation. Thus, the membrane disruption activities of VP5* and VP7 that mediate rotavirus entry remain uncertain.

Virion uncoating

Rotavirus uncoating is easily studied *in vitro*. Chelating calcium from rotavirus TLPs triggers uncoating and activates transcription from the resulting DLPs (Cohen *et al.*, 1979). Uncoating in low calcium environments reflects the molecular characteristics of VP7, the outer capsid glycoprotein. VP7 forms calcium-dependent trimers; upon chelation of calcium, these trimers dissociate (Dormitzer *et al.*, 2000). Dissociation of VP7 trimers probably mediates uncoating triggered by calcium chelators. Neutralizing antibodies against rotavirus that bind VP7 are specific for the trimeric conformation, stabilize the trimers to low calcium, and block rotavirus uncoating (Dormitzer *et al.*, 1992; Dormitzer *et al.*, 1994; Ludert *et al.*, 2002).

The structural basis for the similar phenomenon of non-enveloped plant virus swelling upon calcium chelation has been determined (Robinson *et al.*, 1982). In tomato bushy stunt virus, calcium is liganded by side chain carboxylates of aspartates on either side of the interface between trimeric capsomer subunits. Upon loss of the divalent cation at elevated pH, the negatively charged carboxylates repel each other, destabilizing inter-subunit contacts and making the capsid swell. A similar chemistry could underlie the destabilization of inter-subunit contacts within VP7 capsomers. In rotavirus strains selected for stability in low calcium, VP7 prolines 75 and 279 are replaced by leucine and serine, respectively (Gajardo *et al.*, 1997). Although these residues could bind calcium directly (with their main chain carbonyls), they probably influence calcium affinity indirectly, by determining the orientation or spacing of calcium-binding residues.

Data on the role of calcium loss in mediating rotavirus uncoating during cell entry are inconsistent. The K_{Ca} for uncoating under physiologic conditions is sufficiently high that exposure to cytoplasmic calcium concentrations could trigger uncoating (Martin *et al.*, 2002). Endosomes also provide a low calcium environment, as endosomal calcium falls in association with the formation of an electrical gradient through the activity of a proton pump (Gerasimenko *et al.*, 1998). Treating cells with a calcium ionophore (A23187) prevents the majority of internalized rotavirus strain OSU particles from uncoating, as assayed by isopycnic gradient centrifugation (Ludert *et al.*, 1987). Also consistent with a role for low endosomal calcium in triggering entry, maintaining 10 mM CaEGTA (raising the extracellular calcium reservoir) and/or 500 nM bafilomycin A1 (inhibiting the endosomal proton pump) throughout infection inhibits rotavirus infectivity by approximately 10-fold and also reduces the co-entry of α-sarcin (Chemello *et al.*, 2002). These agents do not significantly inhibit infection if added 15 minutes after virus inoculation.

On the other hand, raising intracytoplasmic calcium concentrations using A23187, ionomycin (another calcium ionophore), or thapsigargin (an endoplasmic reticulum Ca^{2+}-ATPase pump inhibitor) fails to block infection by rotavirus strain RRV (Cuadras *et al.*, 1997). As the site of rotavirus entry is not known, the effect of these drugs on the calcium concentration at the site of entry cannot be determined. In fact, there is little direct evidence to indicate where uncoating occurs (at the plasma membrane or in an endosome) or when uncoating occurs (before, during, or after membrane penetration). When primed TLPs are added to cells at 37° C, VP7 remains sensitive to a mixture of EDTA and 500 μg/ml trypsin, while VP2 and VP6 lose their sensitivity; when unprimed particles are added, VP2, VP6, and VP7 all lose their protease sensitivity (Fukuhara *et al.*, 1988). This observation suggests that only the DLP of primed particles is internalized, leaving VP7 at the cell surface, and that unprimed particles are internalized whole, presumably by endocytosis. These data suggest that uncoating occurs at the plasma membrane before or during membrane penetration.

One intriguing finding suggests an alternative mechanism for uncoating during entry. Neutralizing antibodies that select neutralization escape mutations at VP4 residues 180 and 183 (in a loop of the VP8* core adjacent to its C-terminal α-helix) uncoat SA11 rotavirus particles (Zhou *et al.*, 1994). This result suggests that an antibody binding-induced conformational change, which propagates from the distal end of the VP4 spike to the VP7 layer, can trigger uncoating. If an analogous conformational change, induced by VP8* binding to a cell surface molecule, drives uncoating during entry, the calcium-dependent conformational change in VP7 may be more important in regulating outer capsid assembly in the ER lumen (a high calcium environment) than in regulating outer capsid disassembly during entry.

A second VP4 rearrangement

A crystal structure of the N-terminal half of VP5* (Figure 11.1g) indicates that, after the priming rearrangement, VP4 undergoes a second rearrangement (Dormitzer *et al.*, 2004). Because the serial proteolysis of VP4 that produced the crystallizable VP5* fragment also triggered a conformational change, the biochemical manipulations that produced the fragment affect the interpretation of the resulting structure. When digesting soluble VP4, trypsin does not cleave to create the authentic VP5* N-terminus at A248 in most molecules, instead producing an N-terminus at T259 (Figure 11.1b; Dormitzer *et al.*, 2001). In a modest proportion of molecules, digestion does produce an N-terminus at A248, despite extended incubation with trypsin. This observation suggests that some molecules assume a conformation that sequesters the K258–T259 peptide bond from protease. Chymotrypsin digestion (after L236, L238, and H246 in the priming region) reliably produces the conformation that protects the K258–T259 bond. Therefore, serial chymotrypsin and trypsin digestion of VP4 produces a relatively homogeneous, crystallizable fragment of the VP5* region. This fragment, named VP5CT (Figure 11.1b), has the authentic VP5* N-terminus at A248 and is actually larger that the most abundant fragment of VP5* produced by digestion of soluble VP4 by trypsin alone.

Size exclusion chromatography of chymotrypsin-digested soluble VP4 indicates that the conformational change that protects the K258–T259 peptide bond involves an oligomerization. The requirement for chymotrypsin predigestion to prevent trypsin cleavage of the K258–T259 bond in solution, but not on the virion, suggests that uncleaved VP4 is held in a different conformation on the virion. Alternatively, a conformational change that protects the K258–T259 peptide bond could occur more rapidly on the virion after initial tryptic cleavages in the priming region. A cooperative conformational change might be facilitated by the enforced proximity of VP4 molecules anchored in clusters.

VP5CT remains oligomeric during unboiled, reducing SDS-PAGE, indicating that the association of its subunits is very stable (Dormitzer *et al.*, 2001). Analytical ultracentrifugation did not distinguish whether the oligomer is a homogeneous trimer or a tightly associated dimer that weakly self-associates into tetramers (dimers of dimers). The VP5CT crystal structure (Figure 11.1g) resolves the ambiguity, revealing a well-ordered trimer, held together by non-covalent inter-subunit contacts, including a C-terminal α-helical triple coiled-coil and a nine-stranded β-annulus, with three strands contributed by each subunit (Dormitzer *et al.*, 2004). VP5CT resembles a folded umbrella, with the coiled-coil forming the post and a globular region (the VP5* antigen domain) forming each panel. The trimer's stability, large buried surface area (25.8% of each subunit), and well-ordered threefold contacts indicate that it models a native conformational state of this region of VP4. Thus, in addition to the rearrangement that produces the spikes on primed virions, VP4 also can rearrange to form tightly associated trimers.

Like the low-energy, post-fusion conformational states of enveloped virus fusion proteins, VP5CT requires a priming cleavage and resists disruption by SDS (Bullough *et al.*, 1994; Weissenhorn *et al.*, 1996; Mothes *et al.*, 2000). During viral entry, rearrangement from the primed state (Figure 11.1c and f) probably yields this third conformation. The rearrangement following VP4 cleavage in solution bypasses the primed state, directly producing the third conformation. This difference between the behaviour of VP4 on the virion and in solution suggests that interactions of virion-associated VP4 with VP6 and VP7 prevent the VP5* region from refolding to its lowest energy conformation after cleavage. Thus, trypsin-cleaved VP4 is trapped in a 'meta-stable' state on the virion, primed for membrane interaction during entry.

Because the body of each primed spike appears to be a dyad, the transition to the third conformational state of VP4 must be associated with a twofold to threefold reorganization of this region. If two VP4 molecules form each cluster on the virion, the reorganization must occur after uncoating, when molecules released from different clusters can mingle; if three VP4 molecules form each cluster, the reorganization could occur on the virion. As described above ('The priming rearrangement'), electron cryomicroscopy reconstructions of trypsin primed virions suggest that three molecules form each cluster and leave open the possibilities that the second VP4 rearrangement could occur either on or off of the virion (Pesavento *et al.*, 2005, Yeager *et al.*, 1994).

The two- to threefold reorganization must be associated with rotation of the body subunits (Figure 11.1f). Each VP5* antigen domain has a hydrophobic apex, formed by the tips of three loops (Dormitzer *et al.*, 2004). One of these loops resembles the Semliki Forest virus membrane fusion loop in primary sequence (Mackow *et al.*, 1988), though not in fold. In the VP5CT trimer (Figure 11.1g), the globular domains pack against the coiled-coil with their hydrophobic apices adjacent to the C-termini of VP5CT, which would attach to the foot in full length VP5*. Fitting the VP5* antigen domain dimer to an image reconstruction of the primed spike (Dormitzer *et al.*, 2004) and antibody decoration data (Tihova *et al.*, 2001) place the hydrophobic apex at the 'shoulder' of the body, pointing away from the foot. Therefore, during the transition to the third VP4 conformation, the VP5* globular domain must rotate approximately 180° about an axis roughly perpendicular to the dyad axis of the spike body, translocating its hydrophobic apex by at least 55 Å towards the foot. The rotation also translocates the DGE potential integrin binding motif from the bottom corner of the body to the top of the folded-back trimer.

As the distal dyad contact of the rotavirus spike body is formed by N-terminal residues that also non-covalently tether VP8* to VP5* (see The Priming Rearrangement), dissociation of VP8* probably accompanies the domain rotation. Because the hydrophobic apex of the VP5* globular domain is at least partially masked by VP8* on the spike (Figure 11.1c and f), dissociation of VP8* would expose a hydrophobic element for potential membrane insertion. The outer surface of the coiled-coil has negative electrostatic potential and would expose a hydrophilic surface to solvent were it not for the globular domains packed against it. Thus, the coiled-coil could 'zip-up' before the globular domains fold back. A large cavity (390 $Å^3$), centred on the threefold axis near the top of the VP5CT trimer leaves room for domain movements associated with the rotation. The crystal structure of the VP5* antigen domain dimer (not shown) indicates that, during the rearrangement, a loop that forms the central strands of the inter-subunit saddle-shaped β-sheet of the dimer shifts position to insert into the β-annulus that locks the structure in the trimeric folded umbrella conformation (Yoder and Dormitzer, 2006). In addition, hydrophobic residues that pack against each other on the twofold axis of the dimer repack around the threefold axis of the trimer.

There is no direct experimental evidence that the VP4 fold-back rearrangement actually mediates membrane disruption. However, the rearrangement closely resembles the fusogenic rearrangements of enveloped virus fusion proteins. Specifically, in both type I and type II enveloped virus fusion mechanisms, a fold-back translocation approximates a hydrophobic fusion peptide, which is inserted into a target cell membrane, and a transmembrane anchor, which traverses the viral envelope. Thus, the fold-back brings host and viral lipid bilayers into close apposition, facilitating fusion (Bullough *et al.*, 1994; Weissenhorn *et al.*, 1996; Gibbons *et al.*, 2004; Modis *et al.*, 2004). Similarly, if the VP4 fold-back rearrangement occurs prior to uncoating, the hydrophobic apex of the globular domain would be approximated to the surface of the VP7 layer. If the hydrophobic apex inserts into a target cell membrane, the fold-back could be involved in penetrating the target membrane.

The type I fusion mechanism also involves at least a local dissociation of a membrane-distal receptor binding domain (such as influenza HA1 or HIV gp120) from a membrane fusion domain (Moore *et al.*, 1990; Godley *et al.*, 1992). Similarly, the dual function of the VP8* N-terminus as the distal twofold contact of the spike and the tether of the heads to the body subunits suggests that VP8* dissociates from VP5* (at least locally) in association with the fold-back. The type II enveloped virus fusion mechanism includes a twofold to threefold oligomeric re-organization of the fusion protein (Gibbons *et al.*, 2004; Modis *et al.*, 2004). A twofold to threefold reorganization of the protruding portion of the VP4 spike also occurs in association with the fold-back rearrangement.

Protein rearrangements that accompany cell entry also mediate membrane penetration by other non-enveloped viruses, such as poliovirus and reovirus. As cDNA corresponding to the transfected poliovirus genome is infectious (Racaniello *et al.*, 1981), poliovirus need only translocate a single RNA strand through the cell membrane to initiate infection. The structures that mediate membrane penetration by picornaviruses do not resemble those that mediate membrane penetration by rotavirus. Unlike rotavirus VP4 and VP7, which are soluble and biochemically 'well-behaved' as purified proteins, the picornavirus capsid proteins are so tightly intertwined structurally and functionally that their properties in isolation probably do not reflect their properties in the assembled capsid (Hogle *et al.*, 1985; Hogle, 2002). Nevertheless, gross movements of hydrophobic structural elements are involved in poliovirus entry. Poliovirus membrane penetration is associated with plate-like movements of the capsid (Belnap *et al.*, 2000), externalization of the myristolated capsid protein VP4, and exposure of an amphipathic helix at the N-terminus of the capsid protein VP1 (Fricks *et al.*, 1990).

Like rotavirus, the other members of the *Reoviridae* also must translocate their transcriptionally active subviral particles through a membrane to initiate infection. The crystal structure of the reovirus membrane penetration protein, μ1 (Liemann *et al.*, 2002), does not resemble the crystal structures of VP4 fragments. Indeed, in overall architecture μ1 more closely resembles the middle layer proteins of rotavirus (VP6) and orbivirus (VP7) (Grimes *et al.*, 1997; Mathieu *et al.*, 2001). Reovirus μ1 does undergo a major rearrangement in association with membrane disruption, as indicated by increased protease sensitivity, altered antigenicity, and increased exposure of hydrophobic regions (Chandran *et al.*, 2002). No crystallographic or electron cryomicroscopy-based structural data on the rearranged state of μ1 have been obtained. Therefore, the degree of mechanistic similarity between the rearrangements of reovirus μ1 and rotavirus VP4 remains to be determined. The obligate release of a short, N-terminal, myristolated, autolytically cleaved fragment of μ1 (μ1N) during cell entry (Odegard *et al.*, 2004) suggests a similarity between the reovirus and picornavirus membrane penetration mechanisms.

A variety of interactions, alone or in combination, can trigger the protein rearrangements that mediate membrane penetration or membrane fusion. Examples include receptor binding (HIV gp120/41, poliovirus), membrane insertion (flavivirus E glycoprotein ectodomains), and proton binding (influenza HA, flavivirus E glycoproteins) (Skehel *et al.*, 1982; Gomez Yafal *et al.*, 1993; Jones *et al.*, 1998; Stiasny *et al.*, 2002). The proximate trigger for the second VP4 rearrangement remains obscure. A fall in pH has been effectively ruled out, but receptor binding, membrane insertion, additional protease cleavage, and virion uncoating are all candidates.

Conclusions

Investigation into rotavirus cell entry has been challenged by the lack of a reverse genetics system. The recent incorporation of a recombinant gene encoding VP4 into infectious rotavirus virions is a major step in the development of an efficient reverse genetics system that can be used to study cell entry by rotavirus (Komoto *et al.*, 2006). The development of a recoating genetics system to study reovirus entry (Chandran *et al.*, 2001) and the incorporation of outer capsid proteins into rotavirus-like particles (Crawford, *et al.*, 1994) and transcapsidated rotavirus particles (Chen and Ramig, 1993) suggest that a recoating genetics system could also be used to study rotavirus entry. Additional challenges to understanding rotavirus entry remain, including generally high physical-to-infectious particle ratios, the apparent diversity and low affinity of rotavirus-receptor interactions, and the lack of a well understood model for the central event in entry, the passage of a large non-enveloped particle through a lipid bilayer. Nevertheless, some basic elements of the rotavirus cell entry pathway are clear – trypsin cleavage primes particles; particles attach to at least one of a heterogeneous collection of receptors; the virus uncoats; and a large subviral particle is translocated across a membrane and into the cytoplasm.

Investigation into rotavirus entry is aided by the physical characteristics of rotavirus particles and the outer capsid proteins that mediate entry: rotavirus particles are homogeneous and stable, and the outer capsid proteins are designed to come off the virion. These properties allow a biochemical and structural approach to understanding some key molecular properties that underlie rotavirus cell entry. This approach has shown that rotavirus particles uncoat in low calcium and upon binding of some neutralizing antibodies; that the dissociation of VP7 trimers mediates uncoating in low calcium; that trypsin cleavage rigidifies VP4; and that VP4 undergoes a fold-back rearrangement that resembles the fusogenic rearrangements of enveloped virus fusion proteins. Detailed understanding of these relatively straightforward steps, which can be modelled in simple biochemical systems, can complement the necessarily less certain knowledge of more complicated steps, which involve more extensive interaction with cellular structures and mechanisms. As the mechanism of the central event in rotavirus entry, the passage of a large particle through a lipid bilayer, is not known for any non-enveloped virus, the tractable elements of rotavirus as an experimental system provide a resource for understanding cell entry by a large and diverse class of viruses.

References

Arias, C.F., Romero, P., Alvarez, V., and Lopez, S. (1996). Trypsin activation pathway of rotavirus infectivity. J. Virol. *70*, 5832–5839.

Arnaout, M.A., Goodman, S.L., and Xiong, J.P. (2002). Coming to grips with integrin binding to ligands. Curr. Opin. Cell. Biol. *14*, 641–651.

Barondes, S.H., Cooper, D.N., Gitt, M.A., and Leffler, H. (1994). Galectins. Structure and function of a large family of animal lectins. J. Biol. Chem. *269*, 20807–20810.

Basavappa, R., Syed, F., Flore, O., Icenogle, J.P., Filman, D.J., and Hogle, J.M. (1994). Role and mechanism of the maturation cleavage of VP0 in poliovirus assembly, structure of the empty capsid assembly intermediate at 2.9 angstrom resolution. Protein Sci. *3*, 1651–1669.

Bass, D.M., Baylor, M., Chen, C., and Upadhyayula, U. (1995). Dansylcadaverine and cytochalasin D enhance rotavirus infection of murine L cells. Virology *212*, 429–437.

Bass, D.M., Baylor, M.R., Chen, C., Mackow, E.M., Bremont, M., and Greenberg, H.B. (1992). Liposome-mediated transfection of intact viral particles reveals that plasma membrane penetration determines permissivity of tissue culture cells to rotavirus. J. Clin. Invest. *90*, 2313–2320.

Bass, D.M., Mackow, E.R., and Greenberg, H.B. (1991). Identification and partial characterization of a rhesus rotavirus binding glycoprotein on murine enterocytes. Virology *183*, 602–610.

Bastardo, J.W., and Holmes, I.H. (1980). Attachment of SA-11 rotavirus to erythrocyte receptors. Infect. Immun. *29*, 1134–1140.

Beaulieu, J.F. (1992). Differential expression of the VLA family of integrins along the crypt-villus axis in the human small intestine. J. Cell Sci. *102*, 427–436.

Belnap, D.M., Filman, D.J., Trus, B.L., Cheng, N., Booy, F.P., Conway, J.F., Curry, S., Hiremath, C.N., Tsang, S.K., Steven, A.C., and Hogle, J.M. (2000). Molecular tectonic model of virus structural transitions, the putative cell entry states of poliovirus. J. Virol. *74*, 1342–1354.

Bergelson, J.M., Cunningham, J.A., Droguett, G., Kurt-Jones, E.A., Krithivas, A., Hong, J.S., Horwitz, M.S., Crowell, R.L., and Finberg, R.W. (1997). Isolation of a common receptor for Coxsackie B viruses and adenoviruses 2 and 5. Science *275*, 1320–1323.

Bukau, B., and Horwich, A.L. (1998). The Hsp70 and Hsp60 chaperone machines. Cell *92*, 351–366.

Bullough, P.A., Hughson, F.M., Skehel, J.J., and Wiley, D.C. (1994). Structure of influenza haemagglutinin at the pH of membrane fusion. Nature *371*, 37–43.

Carrasco, L. (1995). Modification of membrane permeability by animal viruses. Adv. Virus Res. *45*, 61–112.

Chandran, K., Farsetta, D.L., and Nibert, M.L. (2002). Strategy for nonenveloped virus entry, a hydrophobic conformer of the reovirus membrane penetration protein μ1 mediates membrane disruption. J. Virol. *76*, 9920–9933.

Chandran, K., Zhang, X., Olson, N.H., Walker, S.B., Chappell, J.D., Dermody, T.S., Baker, T.S., Nilbert, M.L. (2001). Complete *in vitro* assembly of the reovirus outer capsid produces highly infectious particles suitable for genetic studies of the receptor-binding proteins. J. Virol. *75*, 5335–5342.

Charpilienne, A., Abad, M.J., Michelangeli, F., Alvarado, F., Vasseur, M., Cohen, J., and Ruiz, M.C. (1997). Solubilized and cleaved VP7, the outer glycoprotein of rotavirus, induces permeabilization of cell membrane vesicles. J. Gen. Virol. *78*, 1367–1371.

Chemello, M.E., Aristimuno, O.C., Michelangeli, F., and Ruiz, M.C. (2002). Requirement for vacuolar H^+-ATPase activity and Ca^{2+} gradient during entry of rotavirus into MA104 cells. J. Virol. *76*, 13083–13087.

Chen, J., Lee, K.H., Steinhauer, D.A., Stevens, D.J., Skehel, J.J., and Wiley, D.C. (1998). Structure of the hemagglutinin precursor cleavage site, a determinant of influenza pathogenicity and the origin of the labile conformation. Cell 95, 409–417.

Chen, D., and Ramig, R.F. (1993). Rescue of infectivity by *in vitro* transcapisdation of rotavirus single-particles. Virology *192*, 422–429.
Chou, H.H., Takematsu, H., Diaz, S., Iber, J., Nickerson, E., Wright, K.L., Muchmore, E.A., Nelson, D.L., Warren, S.T., and Varki, A. (1998). A mutation in human CMP-sialic acid hydroxylase occurred after the Homo-Pan divergence. Proc. Natl Acad. Sci. USA *95*, 11751–11756.
Ciarlet, M., Crawford, S.E., Cheng, E., Blutt, S.E., Rice, D.A., Bergelson, J.M., and Estes, M.K. 2002a. VLA-2 ($\alpha 2\beta 1$) integrin promotes rotavirus entry into cells but is not necessary for rotavirus attachment. J. Virol. *76*, 1109–1123.
Ciarlet, M., Crawford, S.E., and Estes, M.K. (2001). Differential infection of polarized epithelial cell lines by sialic acid-dependent and sialic acid-independent rotavirus strains. J. Virol. *75*, 11834–11850.
Ciarlet, M., and Estes, M.K. (1999). Human and most animal rotavirus strains do not require the presence of sialic acid on the cell surface for efficient infectivity. J. Gen. Virol. *80*, 943–948.
Ciarlet, M., Ludert, J.E., Iturriza-Gomara, M., Liprandi, F., Gray, J.J., Desselberger, U., and Estes, M.K. 2002b. The initial interaction of rotavirus strains with N-acetyl-neuraminic (sialic) acid residues on the cell surface correlates with VP4 genotype, not species of origin. J. Virol. *76*, 4087–4095.
Cohen, J., Laporte, J., Charpilienne, A., and Scherrer, R. (1979). Activation of rotavirus RNA polymerase by calcium chelation. Arch. Virol. *60*, 177–186.
Conner, S.D., and Schmid, S.L. (2003). Regulated portals of entry into the cell. Nature 422.
Connor, R.J., Kawaoka, Y., Webster, R.G., and Paulson, J.C. (1994). Receptor specificity in human, avian, and equine H2 and H3 influenza virus isolates. Virology *205*, 17–23.
Coulson, B.S., Londrigan, S.L., and Lee, D.J. (1997). Rotavirus contains integrin ligand sequences and a disintegrin-like domain that are implicated in virus entry into cells. Proc. Natl Acad. Sci. USA *94*, 5389–5394.
Crawford, S.E., Labbe, M., Cohen, J., Burroughs, M.H., Zhou, Y.J., and Estes, M.K. (1994). Characterization of virus-like particles produced by the expression of rotavirus capsid proteins in insect cells. J. Virol. *68*, 5945–5952.
Crawford, S.E., Mukherjee, S.K., Estes, M.K., Lawton, J.A., Shaw, A.L., Ramig, R.F., and Prasad, B.V. (2001). Trypsin cleavage stabilizes the rotavirus VP4 spike. J. Virol. *75*, 6052–6061.
Cuadras, M.A., Arias, C.F., and Lopez, S. (1997). Rotaviruses induce an early membrane permeabilization of MA104 cells and do not require a low intracellular Ca^{2+} concentration to initiate their replication cycle. J. Virol. *71*, 9065–9074.
Delorme, C., Brussow, H., Sidoti, J., Roche, N., Karlsson, K.A., Neeser, J.R., and Teneberg, S. (2001). Glycosphingolipid binding specificities of rotavirus, identification of a sialic acid-binding epitope. J. Virol. *75*, 2276–2287.
Denisova, E., Dowling, W., LaMonica, R., Shaw, R., Scarlata, S., Ruggeri, F., and Mackow, E.R. (1999). Rotavirus capsid protein VP5* permeabilizes membranes. J. Virol. *73*, 3147–3153.
Dormitzer, P.R., Both, G.W., and Greenberg, H.B. (1994). Presentation of neutralizing epitopes by engineered rotavirus VP7's expressed by recombinant vaccinia viruses. Virology *204*, 391–402.
Dormitzer, P.R., and Greenberg, H.B. (1992). Calcium chelation induces a conformational change in recombinant herpes simplex virus-1-expressed rotavirus VP7. Virology *189*, 828–832.
Dormitzer, P.R., Greenberg, H.B., and Harrison, S.C. (2000). Purified recombinant rotavirus VP7 forms soluble, calcium-dependent trimers. Virology *277*, 420–428.
Dormitzer, P.R., Greenberg, H.B., and Harrison, S.C. (2001). Proteolysis of monomeric recombinant rotavirus VP4 yields an oligomeric VP5* core. J. Virol. *75*, 7339–7350.
Dormitzer, P.R., Nason, E.B., Prasad, B.V., and Harrison, S.C. (2004). Structural rearrangements in the membrane penetration protein of a non-enveloped virus. Nature *430*, 1053–1058.
Dormitzer, P.R., Sun, Z.-Y.J., Blixt, O., Paulson, J.C., Wagner, G., and Harrison, S.C. 2002a. The specificty and affinity of sialic acid binding by the rhesus rotavirus VP8* core. J. Virol. *76*, 10512–10517.
Dormitzer, P.R., Sun, Z.-Y.J., Wagner, G., and Harrison, S.C. 2002b. The rhesus rotavirus VP4 sialic acid binding domain has a galectin fold with a novel carbohydrate binding site. EMBO J. *21*, 885–897.
Dowling, W., Denisova, E., LaMonica, R., and Mackow, E.R. (2000). Selective membrane permeabilization by the rotavirus VP5* protein is abrogated by mutations in an internal hydrophobic domain. J. Virol. *74*, 6368–6376.
Dryden, K.A., Wang, G., Yeager, M., Nibert, M.L., Coombs, K.M., Furlong, D.B., Fields, B.N., and Baker, T.S. (1993). Early steps in reovirus infection are associated with dramatic changes in supramolecular structure and protein conformation, analysis of virions and subviral particles by cryoelectron microscopy and image reconstruction. J. Cell Biol. *122*, 1023–1041.

Durrer, P., Galli, C., Hoenke, S., Corti, C., Gluck, R., Vorherr, T., and Brunner, J. (1996). H+-induced membrane insertion of influenza virus hemagglutinin involves the HA2 amino-terminal fusion peptide but not the coiled coil region. J. Biol. Chem. *271*, 13417–13421.
Emsley, J., King, S.L., Bergelson, J.M., and Liddington, R.C. (1997). Crystal structure of the I domain from integrin alpha2beta1. J. Biol. Chem. *272*, 28512–28517.
Emsley, J., Knight, C.G., Farndale, R.W., Barnes, M.J., and Liddington, R.C. (2000). Structural basis of collagen recognition by integrin α2β1. Cell *101*, 47–56.
Estes, M.K., Graham, D.Y., and Mason, B.B. (1981). Proteolytic enhancement of rotavirus infectivity, molecular mechanisms. J. Virol. *39*, 879–888.
Falconer, M.M., Gilbert, J.M., Roper, A.M., Greenberg, H.B., and Gavora, J.S. (1995). Rotavirus-induced fusion from without in tissue culture cells. J. Virol. *69*, 5582–5591.
Fiore, L., Greenberg, H.B., and Mackow, E.R. (1991). The VP8 fragment of VP4 is the rhesus rotavirus hemagglutinin. Virology *181*, 553–563.
Fotin, A., Cheng, Y., Grigorieff, N., Walz, T., Harrison, S.C., and Kirchhausen, T. (2004). Structure of an auxilin-bound clathrin coat and its implications for the mechanism of uncoating. Nature *432*, 649–653.
Fricks, C.E., and Hogle, J.M. (1990). Cell-induced conformational change in poliovirus, externalization of the amino terminus of VP1 is responsible for liposome binding. J. Virol. *64*, 1934–1945.
Fukudome, K., Yoshie, O., and Konno, T. (1989). Comparison of human, simian, and bovine rotaviruses for requirement of sialic acid in hemagglutination and cell adsorption. Virology *172*, 196–205.
Fukuhara, N., Yoshie, O., Kitaoka, S., and Konno, T. (1988). Role of VP3 in human rotavirus internalization after target cell attachment via VP7. J. Virol. *62*, 2209–2218.
Fukuhara, N., Yoshie, O., Kitaoka, S., Konno, T., and Ishida, N. (1987). Evidence for endocytosis-independent infection by human rotavirus. Arch. Virol. *97*, 93–99.
Gajardo, R., Vende, P., Poncet, D., and Cohen, J. (1997). Two proline residues are essential in the calcium-binding activity of rotavirus VP7 outer capsid protein. J. Virol. *71*, 2211–2216.
Gentsch, J.R., Woods, P.A., Ramachandran, M., Das, B.K., Leite, J.P., Alfieri, A., Kumar, R., Bhan, M.K., and Glass, R.I. (1996). Review of G and P typing results from a global collection of rotavirus strains, implications for vaccine development. J. Infect. Dis. *174*(Suppl. 1), S30–36.
Gerasimenko, J.V., Tepikin, A.V., Petersen, O.H., and Gerasimenko, O.V. (1998). Calcium uptake via endocytosis with rapid release from acidifying endosomes. Curr. Biol. *8*, 1335–1338.
Gibbons, D.L., Vaney, M.C., Roussel, A., Vigouroux, A., Reilly, B., Lepault, J., Kielian, M., and Rey, F.A. (2004). Conformational change and protein–protein interactions of the fusion protein of Semliki Forest virus. Nature *427*, 320–325.
Gilbert, J.M., and Greenberg, H.B. (1997). Virus-like particle-induced fusion from without in tissue culture cells, role of outer-layer proteins VP4 and VP7. J. Virol. *71*, 4555–4563.
Gilbert, J.M., and Greenberg, H.B. (1998). Cleavage of rhesus rotavirus VP4 after arginine 247 is essential for rotavirus-like particle-induced fusion from without. J. Virol. *72*, 5323–5327.
Godley, L., Pfeifer, J., Steinhauer, D., Ely, B., Shaw, G., Kaufmann, R., Suchanek, E., Pabo, C., Skehel, J.J., Wiley, D.C., and *et al.* (1992). Introduction of intersubunit disulfide bonds in the membrane-distal region of the influenza hemagglutinin abolishes membrane fusion activity. Cell *68*, 635–645.
Golantsova, N.E., Gorbunova, E.E., and Mackow, E.R. (2004). Discrete domains within the rotavirus VP5* direct peripheral membrane association and membrane permeability. J. Virol. *78*, 2037–2044.
Gomez Yafal, A., Kaplan, G., Racaniello, V.R., and Hogle, J.M. (1993). Characterization of poliovirus conformational alteration mediated by soluble cell receptors. Virology *197*, 501–505.
Grab, B., Miles, A.J., Furcht, L.T., and Fields, G.B. (1996). Promotion of fibroblast adhesion by triple-helical peptide models of type I collagen-derived sequences. J. Biol. Chem. *271*, 12234–12240.
Graham, K.L., Halasz, P., Tan, Y., Hewish, M.J., Takada, Y., Mackow, E.R., Robinson, M.K., and Coulson, B.S. (2003). Integrin-using rotaviruses bind α2β1 integrin α2 I domain via VP4 DGE sequence and recognize αXβ2 and αVβ3 by using VP7 during cell entry. J. Virol. *77*, 9969–9978.
Graham, K.L., Zeng, W., Takada, Y., Jackson, D.C., and Coulson, B.S. (2004). Effects on rotavirus cell binding and infection of monomeric and polymeric peptides containing α2β1 and αxβ2 integrin ligand sequences. J. Virol. *78*, 11786–11797.
Graham, K.L., Fleming, F.E., Halasz, P., Hewish, M.J., Nagesha, H.S., Holmes, I.H., Takada, Y., and Coulson, B.S. (2005). Rotaviruses interact with α4β7 and α4β1 integrins by binding the same integrin domains as natural ligands. J. Gen. Virol. *86*, 3397–3408.

Graham, K.L., Takada, Y., and Coulson, B.S. (2006). Rotavirus spike protein VP5* binds α2β1 integrin on the cell surface and competes with virus for cell binding and infectivity. J. Gen. Virol. *87*, 1275–1283.

Grimes, J.M., Jakana, J., Ghosh, M., Basak, A.K., Roy, P., Chiu, W., Stuart, D.I., and Prasad, B.V. (1997). An atomic model of the outer layer of the bluetongue virus core derived from X-ray crystallography and electron cryomicroscopy. Structure 5, 885–893.

Guerrero, C.A., Bouyssounade, D., Zarate, S., Isa, P., Lopez, T., Espinosa, R., Romero, P., Mendez, E., Lopez, S., and Arias, C.F. (2002). Heat shock cognate protein 70 is involved in rotavirus cell entry. J. Virol. 76, 4096–4102.

Guerrero, C.A., Mendez, E., Zarate, S., Isa, P., Lopez, S., and Arias, C.F. 2000a. Integrin αvβ3 mediates rotavirus cell entry. Proc. Natl Acad. Sci. USA 97, 14644–14649.

Guerrero, C.A., Zarate, S., Corkidi, G., Lopez, S., and Arias, C.F. 2000b. Biochemical characterization of rotavirus receptors in MA104 cells. J. Virol. *74*, 9362–9371.

Guo, C.T., Nakagomi, O., Mochizuki, M., Ishida, H., Kiso, M., Ohta, Y., Suzuki, T., Miyamoto, D., Hidari, K.I., and Suzuki, Y. (1999). Ganglioside GM(1a) on the cell surface is involved in the infection by human rotavirus KUN and MO strains. J. Biochem. (Tokyo) *126*, 683–688.

Hanson, J.E., Sauter, N.K., Skehel, J.J., and Wiley, D.C. (1992). Proton nuclear magnetic resonance studies of the binding of sialosides to intact influenza virus. Virology *189*, 525–533.

Hewish, M.J., Takada, Y., and Coulson, B.S. (2000). Integrins α2β1 and α4β1 can mediate SA11 rotavirus attachment and entry into cells. J. Virol. *74*, 228–236.

Hogle, J.M. (2002). Poliovirus cell entry, common structural themes in viral cell entry pathways. Annu. Rev. Microbiol. *56*, 677–702.

Hogle, J.M., Chow, M., and Filman, D.J. (1985). Three-dimensional structure of poliovirus at 2.9 Å resolution. Science *229*, 1358–1365.

Hundley, F., B. Biryahwaho, M. Gow, and U. Desselberger. (1985). Genome rearrangements of bovine rotavirus after serial passage at high multiplicity of infection. Virology *143*, 88–103.

Isa, P., Realpe, M., Romero, P., Lopez, S., and Arias, C.F. (2004). Rotavirus RRV associates with lipid membrane microdomains during cell entry. Virology *322*, 370–381.

Jones, P.L., Korte, T., and Blumenthal, R. (1998). Conformational changes in cell surface HIV-1 envelope glycoproteins are triggered by cooperation between cell surface CD4 and co-receptors. J. Biol. Chem. *273*, 404–409.

Jourdan, N., Maurice, M., Delautier, D., Quero, A.M., Servin, A.L., and Trugnan, G. (1997). Rotavirus is released from the apical surface of cultured human intestinal cells through nonconventional vesicular transport that bypasses the Golgi apparatus. J. Virol. *71*, 8268–8278.

Kaljot, K.T., Shaw, R.D., Rubin, D.H., and Greenberg, H.B. (1988). Infectious rotavirus enters cells by direct cell membrane penetration, not by endocytosis. J. Virol. *62*, 1136–1144.

Keljo, D.J., Kuhn, M., and Smith, A. (1988a). Acidification of endosomes is not important for the entry of rotavirus into the cell. J. Pediatr. Gastroenterol. Nutr. *7*, 257–263.

Keljo, D.J., and Smith, A.K. (1988b). Characterization of binding of simian rotavirus SA-11 to cultured epithelial cells. J. Pediatr. Gastroenterol. Nutr. *7*, 249–256.

King, S.L., Cunningham, J.A., Finberg, R.W., and Bergelson, J.M. (1995). Echovirus 1 interaction with the isolated VLA-2 I domain. J. Virol. *69*, 3237–3239.

Knight, C.G., Morton, L.F., Onley, D.J., Peachey, A.R., Messent, A.J., Smethurst, P.A., Tuckwell, D.S., Farndale, R.W., and Barnes, M.J. (1998). Identification in collagen type I of an integrin α2β1-binding site containing an essential GER sequence. J. Biol. Chem. *273*, 33287–33294.

Knight, C.G., Morton, L.F., Peachey, A.R., Tuckwell, D.S., Farndale, R.W., and Barnes, M.J. (2000). The collagen-binding A-domains of integrins α1β1 and α2β1 recognize the same specific amino acid sequence, GFOGER, in native (triple-helical) collagens. J. Biol. Chem. *275*, 35–40.

Komoto, S., Sasaki, J., and Taniguchi, K. (2006). Reverse genetics system for introduction of site-specific mutations into the double-stranded RNA genome of infectious rotavirus. PNAS *103*, 4646–4651.

Lambert, C., and Prange, R. (2003). Chaperone action in the posttranslational topological reorientation of the hepatitis B virus large envelope protein: implications for translocational regulation. Proc. Natl Acad. Sci. USA *100*, 5199–5204.

Liemann, S., Chandran, K., Baker, T.S., Nibert, M.L., and Harrison, S.C. (2002). Structure of the reovirus membrane-penetration protein, μ1, in a complex with its protector protein, σ3. Cell *108*, 283–295.

Liprandi, F., Moros, Z., Gerder, M., Ludert, J.E., Pujol, F.H., Ruiz, M.C., Michelangeli, F., Charpilienne, A., and Cohen, J. (1997). Productive penetration of rotavirus in cultured cells induces coentry of the translation inhibitor α-sarcin. Virology *237*, 430–438.
Loike, J.D., Sodeik, B., Cao, L., Leucona, S., Weitz, J.I., Detmers, P.A., Wright, S.D., and Silverstein, S.C. (1991). CD11c/CD18 on neutrophils recognizes a domain at the N terminus of the A α chain of fibrinogen. Proc. Natl Acad. Sci. USA *88*, 1044–1048.
Londrigan, S.L., Graham, K.L., Takada, Y., Halasz, P., and Coulson, B.S. (2003). Monkey rotavirus binding to α2β1 integrin requires the α2 I domain and is facilitated by the homologous β1 subunit. J. Virol. *77*, 9486–9501.
Lopez, J.A., Maldonado, A.J., Gerder, M., Abanero, J., Murgich, J., Pujol, F.H., Liprandi, F., Ludert, J.E. (2005). Characterization of neuraminidase-resistant mutants derived from rotavirus porcine strain OSU. J. Virol. *79*, 10369–10375.
Ludert, J.E., Feng, N., Yu, J.H., Broome, R.L., Hoshino, Y., and Greenberg, H.B. (1996). Genetic mapping indicates that VP4 is the rotavirus cell attachment protein *in vitro* and *in vivo*. J. Virol. *70*, 487–493.
Ludert, J.E., Mason, B.B., Angel, J., Tang, B., Hoshino, Y., Feng, N., Vo, P.T., Mackow, E.M., Ruggeri, F.M., and Greenberg, H.B. (1998). Identification of mutations in the rotavirus protein VP4 that alter sialic-acid-dependent infection. J. Gen. Virol. *79*, 725–729.
Ludert, J.E., Michelangeli, F., Gil, F., Liprandi, F., and Esparza, J. (1987). Penetration and uncoating of rotaviruses in cultured cells. Intervirology *27*, 95–101.
Ludert, J.E., Ruiz, M.C., Hidalgo, C., and Liprandi, F. (2002). Antibodies to rotavirus outer capsid glycoprotein VP7 neutralize infectivity by inhibiting virion decapsidation. J. Virol. *76*, 6643–6651.
Mackow, E.R., Shaw, R.D., Matsui, S.M., Vo, P.T., Dang, M.N., and Greenberg, H.B. (1988). The rhesus rotavirus gene encoding protein VP3: location of amino acids involved in homologous and heterologous rotavirus neutralization and identification of a putative fusion region. Proc. Natl Acad. Sci. USA *85*, 645–649.
Martin, S., Lorrot, M., El Azher, M.A., and Vasseur, M. (2002). Ionic strength- and temperature-induced K(Ca) shifts in the uncoating reaction of rotavirus strains RF and SA*11*, correlation with membrane permeabilization. J. Virol. *76*, 552–559.
Mathieu, M., Petitpas, I., Navaza, J., Lepault, J., Kohli, E., Pothier, P., Prasad, B.V., Cohen, J., and Rey, F.A. (2001). Atomic structure of the major capsid protein of rotavirus, implications for the architecture of the virion. EMBO J. *20*, 1485–1497.
Mendez, E., Arias, C.F., and Lopez, S. (1993). Binding to sialic acids is not an essential step for the entry of animal rotaviruses to epithelial cells in culture. J. Virol. *67*, 5253–5259.
Mendez, E., Arias, C.F., and Lopez, S. (1996). Interactions between the two surface proteins of rotavirus may alter the receptor-binding specificity of the virus. J. Virol. *70*, 1218–1222.
Mendez, E., Lopez, S., Cuadras, M.A., Romero, P., and Arias, C.F. (1999). Entry of rotaviruses is a multistep process. Virology *263*, 450–459.
Modis, Y., Ogata, S., Clements, D., and Harrison, S.C. (2004). Structure of the dengue virus envelope protein after membrane fusion. Nature *427*, 313–319.
Monnier, N., Higo-Moriguchi, K., Sun, Z.-Y., Prasad, B.V.V., Taniguchi, K., and Dormitzer, P.R. (2006). High-resolution molecular and antigen structure of the VP8* core of a sialic acid-independent human rotavirus strain. J. Virol. *80*, 1513–1523.
Moore, J.P., McKeating, J.A., Weiss, R.A., and Sattentau, Q.J. (1990). Dissociation of gp120 from HIV-1 virions induced by soluble CD4. Science *250*, 1139–1142.
Mothes, W., Boerger, A.L., Narayan, S., Cunningham, J.M., and Young, J.A. (2000). Retroviral entry mediated by receptor priming and low pH triggering of an envelope glycoprotein. Cell *103*, 679–689.
Nagaraj, R., Joseph, M., and Reddy, G.L. (1987). Perturbation of the lipid behavior of model membranes by synthetic signal peptides. Biochim. Biophys. Acta *903*, 465–472.
Nakagomi, T., and Nakagomi, O. (2000). Human rotavirus HCR3 possesses a genomic RNA constellation indistinguishable from that of feline and canine rotaviruses. Arch. Virol. *145*, 2403–2409.
Nandi, P., Charpilienne, A., and Cohen, J. (1992). Interaction of rotavirus particles with liposomes. J. Virol. *66*, 3363–3367.
Nason, E.L., Rothagel, R., Mukherjee, S.K., Kar, A.K., Forzan, M., Prasad, B.V., and Roy, P. (2004). Interactions between the inner and outer capsids of bluetongue virus. J. Virol. *78*, 8059–8067.
Odegard, A.L., Chandran, K., Zhang, X., Parker, J.S., Baker, T.S., and Nibert, M.L. (2004). Putative autocleavage of outer capsid protein μ1, allowing release of myristoylated peptide μ1N during particle uncoating, is critical for cell entry by reovirus. J. Virol. 78, 8732–8745.

Pando, V., Isa, P., Arias, C.F., and Lopez, S. (2002). Influence of calcium on the early steps of rotavirus infection. Virology *295*, 190–200.
Perez-Vargas, J., Romero, P., Lopez, S., and Arias, C.F. (2006). The peptide-binding and ATPase domains of recombinant hsc70 are required to interact with rotavirus and reduce its infectivity. J. Virol. *80*, 3322–3331.
Pesavento, J.B., Billingsley, A.M., Roberts, E.J., Ramig, R.F., and Prasad, B.V.V. (2003). Structures of rotavirus reassortants demonstrate correlation of altered conformation of the VP4 spike and expression of unexpected VP4-associated phenotypes. J. Virol. *77*, 3291–3296.
Pesavento, J.B., Crawford, S.E., Roberts, E., Estes, M.K., and Prasad, B.V.V. (2005). pH-induced conformational change of the rotavirus VP4 spike, implications for cell entry and antibody neutralization. J. Virol. *79*, 8572–8580.
Petrie, B.L., Graham, D.Y., and Estes, M.K. (1981). Identification of rotavirus particle types. Intervirology *16*, 20–28.
Praefcke, G.J., and McMahon, H.T. (2004). The dynamin superfamily, universal membrane tubulation and fission molecules? Nat. Rev. Mol. Cell Biol. *5*, 133–147.
Prasad, B.V., Wang, G.J., Clerx, J.P., and Chiu, W. (1988). Three-dimensional structure of rotavirus. J. Mol. Biol. *199*, 269–275.
Quan, C.M., and Doane, F.W. (1983). Ultrastructural evidence for the cellular uptake of rotavirus by endocytosis. Intervirology *20*, 223–231.
Racaniello, V.R., and Baltimore, D. (1981). Cloned poliovirus complementary DNA is infectious in mammalian cells. Science *214*, 916–919.
Rapoport, T.A., Goder, V., Heinrich, S.U., and Matlack, K.E. (2004). Membrane-protein integration and the role of the translocation channel. Trends Cell Biol. *14*, 568–575.
Robinson, I.K., and Harrison, S.C. (1982). Structure of the expanded state of tomato bushy stunt virus. Nature *297*, 563–568.
Rolsma, M.D., Kuhlenschmidt, T.B., Gelberg, H.B., and Kuhlenschmidt, M.S. (1998). Structure and function of a ganglioside receptor for porcine rotavirus. J. Virol. *72*, 9079–9091.
Ruiz, M.C., Abad, M.J., Charpilienne, A., Cohen, J., and Michelangeli, F. (1997). Cell lines susceptible to infection are permeabilized by cleaved and solubilized outer layer proteins of rotavirus. J. Gen. Virol. *78*, 2883–2893.
Ruiz, M.C., Alonso-Torre, S.R., Charpilienne, A., Vasseur, M., Michelangeli, F., Cohen, J., and Alvarado, F. (1994). Rotavirus interaction with isolated membrane vesicles. J. Virol. *68*, 4009–4016.
Sanchez-San Martin, C., Lopez, T., Arias, C.F., and Lopez, S. (2004). Characterization of rotavirus cell entry. J. Virol. *78*, 2310–2318.
Sapin, C., Colard, O., Delmas, O., Tessier, C., Breton, M., Enouf, V., Chwetzoff, S., Ouanich, J., Cohen, J., Wolf, C., and Trugnan, G. (2002). Rafts promote assembly and atypical targeting of a nonenveloped virus, rotavirus, in Caco-2 cells. J. Virol. *76*, 4591–4602.
Shaw, A.L., Rothnagel, R., Chen, D., Ramig, R.F., Chiu, W., and Prasad, B.V. (1993). Three-dimensional visualization of the rotavirus hemagglutinin structure. Cell *74*, 693–701.
Skehel, J.J., Bayley, P.M., Brown, E.B., Martin, S.R., Waterfield, M.D., White, J.M., Wilson, I.A., and Wiley, D.C. (1982). Changes in the conformation of influenza virus hemagglutinin at the pH optimum of virus-mediated membrane fusion. Proc. Natl Acad. Sci. USA *79*, 968–972.
Skehel, J.J., and Wiley, D.C. (2000). Receptor binding and membrane fusion in virus entry, the influenza hemagglutinin. Annu. Rev. Biochem. *69*, 531–569.
Srnka, C.A., Tiemeyer, M., Gilbert, J.H., Moreland, M., Schweingruber, H., de Lappe, B.W., James, P.G., Gant, T., Willoughby, R.E., Yolken, R.H., and *et al.* (1992). Cell surface ligands for rotavirus, mouse intestinal glycolipids and synthetic carbohydrate analogs. Virology *190*, 794–805.
Staatz, W.D., Fok, K.F., Zutter, M.M., Adams, S.P., Rodriguez, B.A., and Santoro, S.A. (1991). Identification of a tetrapeptide recognition sequence for the alpha 2 beta 1 integrin in collagen. J. Biol. Chem. *266*, 7363–7367.
Stiasny, K., Allison, S.L., Schalich, J., and Heinz, F.X. (2002). Membrane interactions of the tick-borne encephalitis virus fusion protein E at low pH. J. Virol. *76*, 3784–3790.
Superti, F., and Donelli, G. (1991). Gangliosides as binding sites in SA-11 rotavirus infection of LLC-MK2 cells. J. Gen. Virol. *72*, 2467–2474.
Suzuki, H., Kitaoka, S., Konno, T., Sato, T., and Ishida, N. (1985). Two modes of human rotavirus entry into MA 104 cells. Arch. Virol. *85*, 25–34.

Tihova, M., Dryden, K.A., Bellamy, A.R., Greenberg, H.B., and Yeager, M. (2001). Localization of membrane permeabilization and receptor binding sites on the VP4 hemagglutinin of rotavirus, implications for cell entry. J. Mol. Biol. *314*, 985–992.
Wang, K., Guan, T., Cheresh, D.A., and Nemerow, G.R. (2000). Regulation of adenovirus membrane penetration by the cytoplasmic tail of integrin β5. J. Virol. *74*, 2731–2739.
Weissenhorn, W., Wharton, S.A., Calder, L.J., Earl, P.L., Moss, B., Aliprandis, E., Skehel, J.J., and Wiley, D.C. (1996). The ectodomain of HIV-1 env subunit gp41 forms a soluble, alpha-helical, rod-like oligomer in the absence of gp120 and the N-terminal fusion peptide. EMBO J. *15*, 1507–1514.
Wickham, T.J., Mathias, P., Cheresh, D.A., and Nemerow, G.R. (1993). Integrins αvβ3 and αvβ5 promote adenovirus internalization but not virus attachment. Cell *73*, 309–319.
Willoughby, R.E., Yolken, R.H., and Schnaar, R.L. (1990). Rotaviruses specifically bind to the neutral glycosphingolipid asialo-GM1. J. Virol. *64*, 4830–4835.
Wilson, I.A., Skehel, J.J., and Wiley, D.C. (1981). Structure of the haemagglutinin membrane glycoprotein of influenza virus at 3 Å resolution. Nature *289*, 366–373.
Xiong, J.P., Stehle, T., Diefenbach, B., Zhang, R., Dunker, R., Scott, D.L., Joachimiak, A., Goodman, S.L., and Arnaout, M.A. (2001). Crystal structure of the extracellular segment of integrin αVβ3. Science *294*, 339–345.
Yeager, M., Berriman, J.A., Baker, T.S., and Bellamy, A.R. (1994). Three-dimensional structure of the rotavirus haemagglutinin VP4 by cryo-electron microscopy and difference map analysis. EMBO J. *13*, 1011–1018.
Yeager, M., Dryden, K.A., Olson, N.H., Greenberg, H.B., and Baker, T.S. (1990). Three-dimensional structure of rhesus rotavirus by cryoelectron microscopy and image reconstruction. J. Cell Biol. *110*, 2133–2144.
Yoder, J.D., and Dormitzer, P.R. (2006). Alternative intermolecular contacts underlie the rotavirus VP5* two-fold to three-fold rearrangement. EMBO J. *25*, 1559–1568.
Yu, Y.G., King, D.S., and Shin, Y.K. (1994). Insertion of a coiled-coil peptide from influenza virus hemagglutinin into membranes. Science *266*, 274–276.
Zarate, S., Cuadras, M.A., Espinosa, R., Romero, P., Juarez, K.O., Camacho-Nuez, M., Arias, C.F., and Lopez, S. (2003). Interaction of rotaviruses with Hsc70 during cell entry is mediated by VP5. J. Virol. *77*, 7254–7260.
Zarate, S., Espinosa, R., Romero, P., Guerrero, C.A., Arias, C.F., and Lopez, S. (2000a). Integrin α2β1 mediates the cell attachment of the rotavirus neuraminidase-resistant variant nar3. Virology *278*, 50–54.
Zarate, S., Espinosa, R., Romero, P., Mendez, E., Arias, C.F., and Lopez, S. (2000b). The VP5 domain of VP4 can mediate attachment of rotaviruses to cells. J. Virol. *74*, 593–599.
Zarate, S., Romero, P., Espinosa, R., Arias, C.F., and Lopez, S. (2004). VP7 mediates the interaction of rotaviruses with integrin αvβ3 through a novel integrin-binding site. J. Virol. *78*, 10839–10847.
Zhou, Y.J., Burns, J.W., Morita, Y., Tanaka, T., and Estes, M.K. (1994). Localization of rotavirus VP4 neutralization epitopes involved in antibody-induced conformational changes of virus structure. J. Virol. *68*, 3955–3964.

Entry of a Segmented dsRNA Virus into the Bacterial Cell

12

Minna M. Poranen and Dennis H. Bamford

Abstract

The entry pathway of bacterial dsRNA viruses of the family *Cystoviridae* is a multistep process, which does not resemble the entry mechanisms employed by other bacteriophages, rather having features common with viruses infecting animal cells. As the virion passes the multilayered cell envelope of its Gram-negative host the structural layers of the virion are removed in a stepwise manner. After adsorption to the receptor entry is initiated by fusion between the virus membrane and the host outer membrane, followed by peptidoglycan penetration and finally by endocytic-like internalization of the viral particle. Spike protein P3, fusion protein P6, lytic enzyme P5, and nucleocapsid surface shell protein P8 are exposed one by one to perform their role and to direct the viral core to its destination, the host cell cytosol. The energetic state of the host cell plays an important role during the entry of dsRNA phage ϕ6.

Introduction

Members of the *Cystoviridae* family are large enveloped viruses with a complex capsid structure. Except for the outer envelope, these bacterial dsRNA viruses have many structural and functional characteristics similar to the eukaryotic dsRNA viruses of the family *Reoviridae* (Bamford *et al.*, 2002). The prototypic organism of the *Cystoviridae* family is bacteriophage ϕ6, which for a long time was the only known bacterial dsRNA virus, and the only known enveloped bacteriophage. This unusual phage has been under intensive investigation since its isolation (Vidaver *et al.*, 1973) and most of our current knowledge of cystoviruses is based on studies with ϕ6. The overall virion structure and genome organization are conserved in all cystovirus isolates, although, the sequence similarity can be limited to a few conserved motifs in the key viral enzymes. Currently cystoviruses are classified into two groups based on their relationship to ϕ6. The close relatives include ϕ7, ϕ9, ϕ10, and ϕ11, while ϕ8, ϕ12, and ϕ13 are more distantly related (Mindich *et al.*, 1999).

Common to all dsRNA viruses is that their genome is delivered into the host cell within a protein capsid (polymerase complex) which functions as an RNA polymerization device and also protects the dsRNA genome from host cell responses. This is also true for members of the family *Cystoviridae*, which infect Gram-negative eubacterial hosts. The delivery of a large viral particle into the host cytoplasm is rare in the prokaryotic world. Typically bacteriophages channel their genome across the host cell envelope and the capsid is never internalized. This review highlights the complex entry pathway of bacterial dsRNA viruses. The entry process involves five distinct steps: (1) attachment, (2) envelope fusion, (3) peptidoglycan penetration, (4) virion induced endocytosis, and (5) release from the intracellular vesicle. The viral core particle delivered into the host cytosol initiates the production of progeny virion components by transcribing the viral dsRNA genome segments.

Virion structure

The spherical virion of ϕ6 contains three structural layers one within the other (Figure 12.1). The envelope encloses the icosahedral nucleocapsid (NC), which in turn consists of two layers; the NC surface shell (T = 13) and the core (T = 1; sixty asymmetric dimers of the coat protein P1). The tri-segmented double-stranded RNA genome is contained within the core. The diameter of the enveloped virion is ~86 nm, while the NC and core are ~60 and ~50 nm, respectively (Butcher *et al.*, 1997; Huiskonen *et al.*, 2006; Kenney *et al.*, 1992). Details of the NC structure are described elsewhere (Huiskonen *et al.*, 2006).

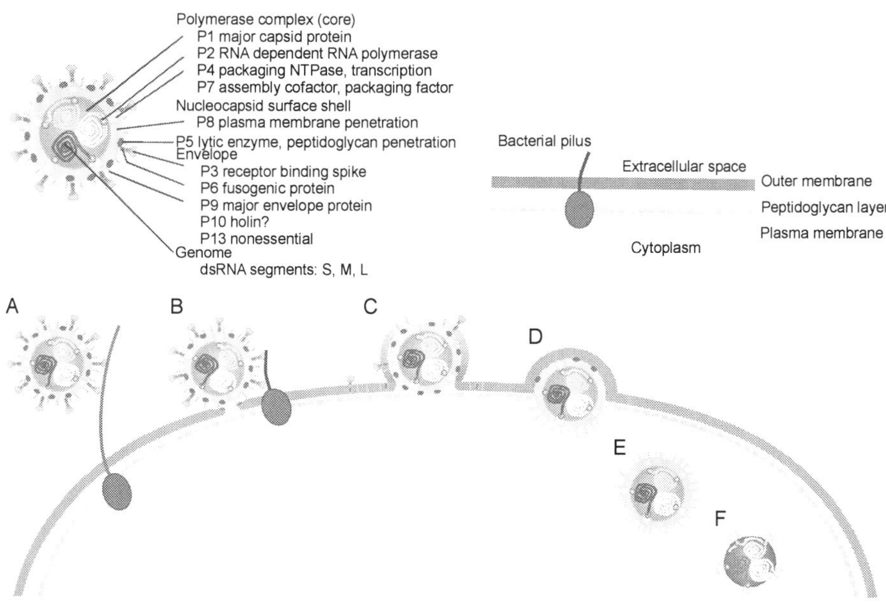

Figure 12.1 The entry pathway of bacteriophage ϕ6. The symbols for the structural components of the virion and the cell envelope are given. (A) Binding to the primary receptor, the bacterial pilus. The interaction is mediated by the spike protein P3. (B) Pilus retraction brings the bound virions into contact with the outer membrane of their host. Fusion between the viral envelope and the host outer membrane takes place. The fusion process is aided by the phage envelope protein P6. (C) NC enters into the periplasmic space. NC surface protein P5 has muralytic activity and it locally digests the peptidoglycan layer. (D) Passage of the NC through the opening in the peptidoglycan layer allows interaction with the host plasma membrane. The NC shell protein P8 interacts with the host plasma membrane. Membrane voltage-dependent changes in the P8 – plasma membrane interaction drive the formation of an NC-containing invagination and (E) a free vesicle. (E-F) Dissociation of the NC surface shell and disruption of the membrane vesicle releases the transcriptionally active core into the host cell cytoplasm.

The viral membrane is composed of five viral-encoded proteins, P3, P6, P9, P10 and P13 (Gottlieb *et al.*, 1988; Sinclair *et al.*, 1975; Van Etten *et al.*, 1976), and lipids derived from the host plasma membrane (Laurinavičius *et al.*, 2004; Sands, 1973). Four of the proteins, P6, P9, P10, and P13, span the membrane. The fifth membrane-associated protein, P3, is anchored to the P6 protein and forms the outer spikes that protrude from the membrane surface (Figure 12.1; Mindich *et al.*, 1976; Stitt and Mindich, 1983; Van Etten *et al.*, 1976). Only proteins P3 and P6 are accessible on the virion surface (Stitt and Mindich, 1983; Van Etten *et al.*, 1976). The membrane is ~6 nm thick and the surface spikes extend 2 nm from the surface. The viral envelope proteins do not follow the icosahedral symmetry of the inner protein shells (Kenney *et al.*, 1992; Jäälinoja *et al.*, 2007).

The protein shell of the NC is formed by the P8 protein organized with T = 13 icosahedral symmetry (Bamford and Mindich, 1980; Van Etten *et al.*, 1976). The regions close to the fivefold symmetry axes of the P8 lattice are occupied by hexameric P4 turrets extending from the core atop of the P8 shell (Butcher *et al.*, 1997; Huiskonen *et al.*, 2006). The stability of the P8 shell is dependent on Ca^{2+} ions (Ktistakis and Lang, 1987; Olkkonen and Bamford, 1987). In addition to P8 and P4, the core proteins P1 and P7 are accessible on the NC surface (Juuti and Bamford, 1997; Ojala *et al.*, 1993; Ojala *et al.*, 1994). A small viral protein P5 resides between the P8 layer and the membrane (Figure 12.1; Hantula and Bamford, 1988).

The viral core is composed of four proteins, P1, P2, P4 and P7 (Mindich and Davidoff-Abelson, 1980). Particles formed by these proteins are also known as polymerase complexes as they can carry out all the functions needed for the replication of the viral genome. The largest viral protein, P1, forms the icosahedral shell of the core particle (Huiskonen *et al.*, 2006; Ktistakis and Lang, 1987; Olkkonen and Bamford, 1987), while the other three proteins are involved in the enzymatic activities needed for genome replication. The structural arrangement of the ϕ6 P1 shell is similar to the arrangement of the shell protein of the cores of eukaryotic dsRNA viruses belonging to the *Reoviridae* (Grimes *et al.*, 1998; Reinisch *et al.*, 2000).

The particle to infectivity ratio for ϕ6 virions is close to one (Day and Mindich, 1980). This strongly implies that the morphological snapshots of ϕ6 viruses entering into the cell reflect the true infectious route of internalization.

Host

Members of the *Cystoviridae* family infect different species of Gram-negative bacteria. ϕ6 was initially isolated from *Pseudomonas syringae* pathovar *phaseolicola* HB10Y (Vidaver *et al.*, 1973). In addition to its original host, ϕ6 is able to infect several other *P. syringae* pathovars (Cuppels *et al.*, 1981) as well as *P. pseudoalcaligenes* cells (Mindich *et al.*, 1976). *P. syringae* is a plant pathogen. The disease symptoms caused by *P. syringae* strains are in most cases chlorotic or necrotic spots or lesions in leaves, fruits or flowers. Efficient adhesion of *P. syringae* on the plant cell surface is dependent on the pilus encoded by the bacterial chromosome (Romantschuk and Bamford, 1986; Romantschuk and Bamford, 1985). This bacterial pilus is also the primary receptor for ϕ6 (Bamford *et al.*, 1976; see below and Figure 12.1). Typically ϕ6-sensitive bacterial cells show approximately one phage-adsorbing pilus on their surface (Romantschuk and Bamford, 1985).

Other cystoviruses have been isolated from bacteria species living on the leaves of *Daucus carota sativum, Lycopersicon esculentum, Ocimum basilicum, Phaseolus vulgaris, Pisum sativum,* and *Raphanus sativum*. The common host of these viruses seem to be pseudomonads, but ϕ8 and ϕ13 can also enter into *Escherichia coli* and *Salmonella typhirium* cells and establish carrier states. ϕ8 can also productively infect some *Salmonella typhimurium* strains (Mindich *et al.*, 1999).

Host cell envelope

The cell envelope of Gram-negative bacteria represents a complex multilayered barrier for the entering bacteriophage (see Figure 12.1). The outer membrane is an atypical lipid bilayer with highly asymmetrical lipid composition: the outer leaflet consists mostly of lipopolysaccharide molecules, while the inner leaflet is composed of phospholipids. The peptidoglycan sacculus, located in the periplasm, determines the shape and size of the bacterium and allows the maintenance of high osmotic pressure inside the cell. The sieve-like peptidoglycan network restricts the passage of over 50 kDa proteins (Demchick and Koch, 1996), and thus hinders the genome delivery of many bacteriophages.

The innermost part of the bacterial envelope is the plasma membrane, which is highly energized due to the action of respiratory chain components. The transport of protons across the plasma membrane establishes a proton motive force (Δp), which is used for vital cellular processes, like ATP synthesis, transport of nutrients and flagellar motion. Also the entry of several dsDNA bacteriophages is known to be dependent on Δp (Dreiseikelmann, 1994; Khan and Macnab, 1980; Letellier *et al.*, 1999; Palmen *et al.*, 1994). The major component of proton motive force is membrane voltage ($\Delta\psi$, the transmembrane difference of electrical potential) with the other component being the pH gradient (ΔpH) (Mitchell, 1979).

Adsorption

ϕ6 adsorbs to the side of the host cell pilus via the spike protein P3 (Mindich *et al.*, 1976; Romantschuk and Bamford, 1986; Romantschuk and Bamford, 1985) which is anchored into the membrane via the integral membrane protein P6 (Figure 12.1). Phages ϕ7, ϕ9, ϕ10, and ϕ11 also infect only piliated bacterial stains (Mindich *et al.*, 1999), whereas ϕ8, ϕ12, and ϕ13 have adapted to infect pilus-less strains by attaching directly to the outermembrane lipopolysaccharides (Daugelavičius *et al.*, 2005; Mindich *et al.*, 1999). The ability to infect non-piliated cells correlates with differences in the make up of the attachment protein complex. While only a single peptide is involved (P3) in attachment of ϕ6, attachment of ϕ8, ϕ12, and ϕ13 is mediated by two or three peptides (P3a, P3b, and P3c) (Gottlieb *et al.*, 2002; Hoogstraten *et al.*, 2000; Qiao *et al.*, 2000).

How do virions that are attached at the distal end the pilus move to the host cell surface, where the outer membrane can be engaged? The pilus is a dynamic structure which can assemble and disassemble. As the pilus retracts the bound pilus-associated virions are carried towards the host cell surface (Figure 12.1; Romantschuk and Bamford, 1985). Thus the pilus allows efficient capture of the viral particles at a distance, and the retraction of the pilus directs the bound virions to the bacterial surface. Some superpiliated host strains are not sensitive to ϕ6 infection probably due to lack of pilus retraction (Romantschuk and

Bamford, 1985). The energetic state of the cell controls the pilus expression on the cell surface. Dissipation of proton motive force (Δp) causes pilus retraction and thus blocks ϕ6 adsorption to its host (Romantschuk *et al.*, 1988). Thus only energized cells are susceptible to ϕ6 infection. This ensures that only cells, which can efficiently support viral replication, are targets of the virus.

Strains of *P. syringae* are known to produce extracellular polysaccharides (Datta *et al.*, 2004) and a halo of about 150 nm corresponding to the polysaccharide capsule has been observed around *P. syringae* cells suscepticle to ϕ6 infection (Romantschuk and Bamford, 1985). When such capsulated cells were incubated with an excess of ϕ6 virions, most of the phages stayed at a distance from the cell surface and only few were seen in contact with the outer membrane. The number of contact points equalled the number of pili. When *E. coli* cells producing no extracellular polysaccharide were treated similarly, phages made readily contact with the bacterial surface (Romantschuk and Bamford, 1985). Apparently the usage of pilus as a carrier by ϕ6 and its close relatives markedly facilitates the penetration of the extracellular polysaccharide and thus increases the probability of infection.

Envelope fusion

Early electron microscopic studies indicated that the ϕ6 envelope fuses with the host outer membrane resulting in the transfer of the virion NC into the periplasmic space (Figure 12.1; Bamford and Lounatmaa, 1978; Bamford *et al.*, 1976). The fusion activity has been localized to the viral membrane protein P6. P6 has no sequence similarity with other viral membrane proteins known to have fusion activity. The fusion event leads to the mixing of phage phospholipids with the bacterial outer membrane lipids and the release of the NC into the periplasmic space without leakage of periplasmic markers to the extracellular milieu. The process is independent of divalent cations and pH (Bamford *et al.*, 1987).

Similar to fusion proteins of some animal viruses, the fusogenic activity of the P6 protein is hidden in the intact virion and becomes activated only after the virion has made a specific contact with its primary receptor, which in the case of ϕ6 is the pilus. The *in vitro* fusion activity of the ϕ6 virion is greatly enhanced when the P3 spike is removed by treatment with butylated hydroxytoluene (Bamford *et al.*, 1987). The interaction of the P3 spike protein with the pilus probably induces a conformational change in P3 which affects the P3-P6 interaction, which in turn unmasks the fusogenic region of P6, allowing fusion to take place.

Getting through the cell wall

After fusion the NCs face the rigid peptidoglycan layer (Figure 12.1). The peptidoglycan layer severely restricts the passage of large molecules and therefore the genomes of many bacteriophages encode enzymes which can degrade peptidoglycan. Generally such enzymes are needed in the end of the phage reproduction cycle to induce cell lysis. Many bacteriophages, like PRD1, T4, T7 and ϕ6 also have lytic enzymes as part of their virion components (Rydman and Bamford, 2002).

The lytic activity in ϕ6 virions has been localized to the P5 protein (Kakitani *et al.*, 1980; Mindich and Lehman, 1979). Studies of P5-mutant viruses have showed that P5 protein has a role in cell penetration during entry (Mindich and Lehman, 1979). P5 is localized between the NC and the viral envelope (Hantula and Bamford, 1988) and thus becomes exposed on the NC surface after the membrane fusion event (Figure 12.1). Simultaneous infection with multiple virions results in an overdose of lytic activity leading to premature cell lysis (Bamford *et al.*, 1987).

P5 protein has been purified and enzymatically characterized. It cleaves the peptide bond between mesodiaminopimelic acid and D-alanine (Caldentey and Bamford, 1992) and thus opens a path for the NC across the peptidoglycan (Figure 12.1). Host cells with altered cell wall structure are resistant to the P5 lysozyme, hindering virus entry (Romantschuk and Bamford, 1981). The temperature sensitivity of P5 largely explains the reduced infectivity of ϕ6 at higher temperatures (Bamford *et al.*, 1987; Mindich and Lehman, 1979).

Plasma membrane penetration

The situation faced by ϕ6 NCs at the host plasma membrane is reminiscent of that encountered by non-enveloped animal viruses. In fact, purified ϕ6 NCs can interact directly with bacterial cells from which the outer membrane and peptidoglycan layers have been removed (spheroplasts), an event able to initiate a productive infection cycle (Ojala *et al.*, 1990). This NC-spheroplast infection method has provided the means of studying in more detail the interactions between the ϕ6 NC and the host plasma membrane and the processes involved in plasma-membrane penetration (Poranen *et al.*, 1999).

The morphological analyses of synchronized infections and infections arrested with energy-depleting agents have revealed ϕ6 NCs in plasma membrane invaginations and apparent vesicles. This suggests that NCs most likely penetrate the plasma membrane by a mechanism similar to the endocytic entry process used by some animal viruses (Figure 12.1; Romantschuk *et al.*, 1988). Results supporting such an entry process for ϕ6 NCs have also been obtained by analyzing infected spheroplasts using rapid freezing and freeze substitution in addition to conventional chemical fixation techniques (Poranen *et al.*, 1999).

Since bacterial cells generally do not use endocytic mechanisms of internalization, endocytosis of the ϕ6 NC is most likely induced by the virion. P8, the T = 13 surface protein of the NC appears to play a vital role in plasma membrane penetration. This is based on the observation that uncoated NC particles, devoid of P8 (cores), are not infectious to host cell spheroplasts (Olkkonen *et al.*, 1991) and that P8-specific antibodies have a neutralizing effect on NC infectivity (Ojala *et al.*, 1990; Poranen *et al.*, 1999). Antibodies against core proteins accessible on the NC surface (P1, P4, and P7) do not affect the infectivity of NCs.

Phage ϕ8 differs from other cystoviruses. The P8 protein of ϕ8 does not form a shell around the core but instead is a minor component of the viral membrane (Hoogstraten *et al.*, 2000). Thus P8 of ϕ8 is left behind as a part of the viral membrane that undergoes fusion with the host outer membrane. The mechanism by which the ϕ8 core penetrates the plasma membrane is unknown, but apparently is not mediated by the P8, as it is for ϕ6. In fact, isolated ϕ8 cores have been shown to infect host cell spheroplasts (Sun *et al.*, 2003). By default, it is presumed that the core shell protein P1 (T = 1) mediates plasma membrane interaction during ϕ8 entry.

NC – plasma membrane interaction

The slow kinetics of ϕ6 NC-spheroplast infection suggests that the interaction between the NC and the host plasma membrane is not directed by a high-affinity receptor (Ojala *et al.*, 1990). During normal ϕ6 infection the NC will inevitably come into intimate contact with the plasma membrane (Figure 12.1) and thus there is no requirement for a specific receptor at the plasma membrane. By using the NC-spheroplast infection system, it has been possible to resolve the initial interaction of the NC with the plasma membrane into two stages: dilution-sensitive absorption and dilution-resistant binding. Both stages are perturbed by antibodies against the NC (Poranen *et al.*, 1999). In low salt conditions (NaCl or KCl), the adsorption of the particles appears to be weaker than that detected in 50 mM salt indicating that electrostatic interactions might be important in NC binding to the plasma membrane (Poranen *et al.*, 1999).

How does P8 mediate the interaction of the ϕ6 NC with the plasma membrane? The secondary structure of the P8 protein is mainly α-helical and totally devoid β-secondary structures (Bamford *et al.*, 1993; Huiskonen *et al.*, 2006). Based on time-resolved Raman spectroscopic analysis and hydrogen-isotope exchange measurements, the intersubunit interactions between the P8 components of the NC are highly dynamic suggesting transient movements in the P8 lattice and an equilibrium between a 'compact' and an 'open' states. The 'open' state could expose a buried α-helix in the lattice required for the initial association of the NC with the membrane (Tuma *et al.*, 1999). The membrane association could then stabilize the 'open' conformation of P8, leading to a tight interaction between P8 and the membrane (Figure 12.1).

Internalization of the NC

Romantschuk observed that the ϕ6 NC internalization process was stalled when cells were treated with energy depleting agents. We have analysed the energy requirements of NC entry in more detail using the NC-spheroplast infection assay and specific agents that interfere with transmembrane difference of electrical potential (Δψ), the transmembrane pH gradient (ΔpH), proton pumping and ATP stability (Poranen *et al.*, 1999).

NC entry is not affected by conditions where the cellular H^+ and K^+ concentration gradients are dissipated or when extra- or intracellular ATP content is reduced (Poranen *et al.*, 1999). Instead, ϕ6 NC internalization is strongly dependent on membrane voltage (Δψ) with a threshold value of ~110 mV. The energy in a form of Δψ is most likely required for the formation of the endocytic-like vesicle, as Δψ is not needed for NC binding to the plasma membrane.

How is the Δψ-dependency of vesicle formation explained? We suggest that Δψ allows a new type of P8–plasma membrane interaction, which then drives the formation of the vesicle. This idea is consistent with previous studies which have shown that Δψ can mediate the incorporation of transmembrane-protein α-helixes into the bacterial plasma membrane. As one example, the insertion of channel-forming colicins into the bacterial plasma membrane occurs through an electrostatic binding to the membrane surface, an event followed by the spontaneous insertion of a hydrophobic hairpin into the membrane and a Δψ-driven insertion of amphiphilic helixes (Cramer *et al.*, 1995). In a similar manner, Δψ could drive the insertion of an exposed P8 α-helix in the ϕ6 NC into the plasma membrane (as depicted in Figure 12.1).

We suggest that the first P8 subunit interacting with the plasma membrane in a $\Delta\psi$-dependent manner would nucleate the formation of the vesicle and that sequential changes in the neighbouring subunits would drive the formation of a membrane indentation and invagination leading to a NC in a tight fitting membrane vesicle (Figure 12.1). In other words, the NC could function as a template for membrane curvature.

Release from the endocytic vesicle

The final stage in φ6 entry is the release of the transcriptionally active core particle into the host cytosol. Two key questions concerning this stage are how the NC escapes from the intracellular vesicle and how the P8 shell is dissociated.

Studies with purified P8 protein have indicated that in its soluble form, P8 possesses membrane-destabilizing activity. As a result, incubation of host cell spheroplasts with P8 leads to the release of cytosolic markers from cells (P. Ojala, D.H. Bamford, unpublished). Also, unsuccessful attempts to express P8 in the absence of other viral proteins have implied that soluble P8 is harmful to the cell. These observations suggest that P8 could have a role in the disruption of the vesicle, and that uncoating of the NC is required prior to the release of the core from the vesicle.

What kind of change in the intavesicular milieu could destabilize the P8 shell so that the membrane active form would be released? Two factors are known to trigger the dissociation of the P8 shell from the NC surface: (1) a decrease in Ca^{2+} concentration (Olkkonen *et al.*, 1991) and (2) an increase in H^+ concentration (M.M. Poranen, D.H. Bamford, unpublished). In addition, a hydrophobic environment may have an additional role in P8 release as non-ionic detergents can make the Ca^{2+}-dependent uncoating process more efficient (Olkkonen *et al.*, 1991). An increase in H^+ concentration in the NC-containing intracellular vesicle will be achieved if the vesicle has captured one of the many respiratory chain complexes. The pH in the vesicle would decrease leading to the P8 release from the NC surface. Alternatively a low intravesicular Ca^{2+} concentration could reduce the stability of the P8 shell, or a combination of these processes may ensue. In turn, solubilized P8 within the vesicle could trigger membrane permeabilization and the release of the transcriptionally active core to the cytosol.

Comparison to other viral systems

The accumulated data on the φ6 entry process highlight common mechanisms operating in both prokaryotic and eukaryotic cells. Notably, processes which are rare in phage–bacterial host interactions, like membrane fusion and endocytosis (Figure 12.1), have evolved in dsRNA bacteriophages that enable the delivery of its core into the host cytoplasm.

The usage of bacterial pilus as a primary receptor is not unique for φ6 and its close relatives, but several non-related phages like the filamentous ssDNA phage M13 and the icosahedral ssRNA phage MS2 also recognize their bacterial host via a pilus. φ6, like MS2, attach on the side of the pilus while M13 virions interact specifically with the tip of the filament. The pilus efficiently translocates the bound φ6 virions to the host cell surface. Similarly, several animal viruses, such as adenovirus, herpes simplex virus and baculovirus utilize cellular filaments (microtubules, actin filaments) as transport mechanisms for virus movement within the cell (Poranen *et al.*, 2002).

The $\phi 6$ envelope fusion with the host cell outer membrane was the first protein-targeted membrane-fusion event to be described in prokaryotes (Bamford *et al.*, 1987). The fusion of the $\phi 6$ envelope with the host cell resembles the pH-independent membrane fusion occurring during entry by Paramyxoviruses and Retroviruses. Specifically, fusion by these eukaryotic viruses is induced by integral membrane proteins of the virion envelope, take place independently of pH, result in the mixing of host and viral lipid layers, and is activated after specific contact between the virus and the cell.

The membrane voltage-dependency of $\phi 6$ NC penetration through the plasma membrane is not unique. The delivery of the dsDNA genome of several bacteriophages is diminished under conditions where the transmembrane proton motive force (Δp) is dissipated (Dreiseikelmann, 1994; Letellier *et al.*, 1999). Also, it has been suggested that the entry of Semliki forest virus and poliovirus might be dependent on Δp (either $\Delta\psi$ or ΔpH), in this case at the step involving the translocation of the viral genome and other components across the membrane of the endocytic vesicle (Carrasco, 1994; Carrasco, 1995). In the case of $\phi 6$ NC entry, the transmembrane voltage potential ($\Delta\psi$) was found to be a critical factor in the formation of the endocytic vesicle. In eukaryotic cells, specific viral proteins typically drive vesicle formation, and the assembly or disassembly of these vesicles is dependent on purine NTPs (Geli and Riezman, 1998). The internalization of the $\phi 6$ NC is not affected by reductions in intra- or extracellular ATP, indicating that there are no ATP-dependent cellular factors involved in the entry process (Poranen *et al.*, 1999).

Are there any similarities in the entry mechanisms of bacterial and eukaryotic dsRNA viruses? The outer capsid layers of dsRNA viruses are responsible for delivering the transcriptionally active core of the virion into the host cell cytoplasm. While the core particle is structurally and functionally conserved among different dsRNA viruses, the outer virion layers mediating the host interactions are divergent. This probably reflects the adaptation to different host systems (Mertens, 2004). The $T = 13$ intermediate shell typical of many eukaryotic dsRNA viruses may be involved in cell entry in a manner reminiscent of the NC surface shell ($T = 13$) of $\phi 6$.

The entry mechanism of rotavirus has been under extensive investigation due to the major medical importance of this virus. The interpretation of the morphological observations of rotavirus entry has been difficult. This is probably due to the relatively low specific infectivity of rotavirus particles and the difficulty in distinguishing between particles that have entered the cell through productive versus non-productive (or less productive) routes. It has been suggested that the productive pathway of rotavirus entry involves endocytic internalization and Ca^{2+}-dependent uncoating in the vesicle. Furthermore, the solubilization of the outer capsid proteins is apparently required for membrane permeabilization and vesicle lysis (Ruiz *et al.*, 1997; Ruiz *et al.*, 2000). Thus similar strategies may be involved in the internalization of the core particles of rotavirus and the bacterial dsRNA viruses.

Acknowledgements

This work was supported by the Academy of Finland ['Finnish Centre of Excellence Program 2006–2011, grants 1213992 and 1213467 (DHB)].

References

Bamford, D.H., Burnett, R.M., and Stuart, D.I. (2002). Evolution of viral structure. Theor. Popul. Biol. *61*, 461–70.

Bamford, D.H., and Lounatmaa, K. (1978). Freeze-fracturing of *Pseudomonas phaseolicola* infected by the lipid-containing bacteriophage ϕ6. J. Gen. Virol. *39*, 161–70.

Bamford, D.H., and Mindich, L. (1980). Electron microscopy of cells infected with nonsense mutants of bacteriophage ϕ6. Virology *107*, 222–8.

Bamford, D.H., Palva, E.T., and Lounatmaa, K. (1976). Ultrastructure and life cycle of the lipid-containing bacteriophage ϕ6. J. Gen. Virol. *32*, 249–59.

Bamford, D.H., Romantschuk, M., and Somerharju, P.J. (1987). Membrane fusion in prokaryotes: bacteriophage ϕ6 membrane fuses with the *Pseudomonas syringae* outer membrane. EMBO J. *6*, 1467–73.

Bamford, J.K., Bamford, D.H., Li, T., and Thomas, G.J., Jr. (1993). Structural studies of the enveloped dsRNA bacteriophage ϕ6 of Pseudomonas syringae by Raman spectroscopy. II. Nucleocapsid structure and thermostability of the virion, nucleocapsid and polymerase complex. J. Mol. Biol. *230*, 473–82.

Butcher, S.J., Dokland, T., Ojala, P.M., Bamford, D.H., and Fuller, S.D. (1997). Intermediates in the assembly pathway of the double-stranded RNA virus ϕ6. EMBO J. *16*, 4477–87.

Caldentey, J., and Bamford, D.H. (1992). The lytic enzyme of the *Pseudomonas* phage ϕ6. Purification and biochemical characterization. Biochim. Biophys. Acta *1159*, 44–50.

Carrasco, L. (1994). Entry of animal viruses and macromolecules into cells. FEBS Lett. *350*, 151–4.

Carrasco, L. (1995). Modification of membrane permeability by animal viruses. Adv. Virus Res. *45*, 61–112.

Cramer, W.A., Heymann, J.B., Schendel, S.L., Deriy, B.N., Cohen, F.S., Elkins, P.A., and Stauffacher, C.V. (1995). Structure-function of the channel-forming colicins. Annu. Rev. Biophys. Biomol. Struct. *24*, 611–41.

Cuppels, D.A., Van Etten, J.L., Lambrecht, P., and Vidaver, A.K. (1981). Survey of phytopathogenic pseudomonads for a restriction and modification system active on the double-stranded ribonucleic acid phage ϕ6. Curr. Microb. *5*, 247–49.

Datta, S., Kumar, R.S., Ayyadurai, N., Kumar, S., and Sakthivel, N. (2004). Polysaccharides of pseudomonas pathovar strains that infect pea, tomato, and soya bean. Curr. Microbiol. *49*, 35–41.

Daugelavičius, R., Cvirkaitė, V., Gaidelytė, A., Bakienė, E., Gabrėnaitė-Verkhovskaya, R., and Bamford, D.H. (2005). Penetration of enveloped double-stranded RNA bacteriophages ϕ13 and ϕ6 into *Pseudomonas syringae* cells. J. Virol. *79*, 5017–26.

Day, L.A., and Mindich, L. (1980). The molecular weight of bacteriophage ϕ6 and its nucleocapsid. Virology *103*, 376–85.

Demchick, P., and Koch, A.L. (1996). The permeability of the wall fabric of *Escherichia coli* and *Bacillus subtilis*. J. Bacteriol. *178*, 768–73.

Dreiseikelmann, B. (1994). Translocation of DNA across bacterial membranes. Microbiol. Rev. *58*, 293–316.

Geli, M.I., and Riezman, H. (1998). Endocytic internalization in yeast and animal cells: similar and different. J. Cell Sci. *111*, 1031–7.

Gottlieb, P., Metzger, S., Romantschuk, M., Carton, J., Strassman, J., Bamford, D.H., Kalkkinen, N., and Mindich, L. (1988). Nucleotide sequence of the middle dsRNA segment of bacteriophage ϕ6: placement of the genes of membrane-associated proteins. Virology *163*, 183–90.

Gottlieb, P., Wei, H., Potgieter, C., and Toporovsky, I. (2002). Characterization of ϕ12, a bacteriophage related to ϕ6: nucleotide sequence of the small and middle double-stranded RNA. Virology *293*, 118–24.

Grimes, J.M., Burroughs, J.N., Gouet, P., Diprose, J.M., Malby, R., Zientara, S., Mertens, P.P., and Stuart, D.I. (1998). The atomic structure of the bluetongue virus core. Nature *395*, 470–8.

Hantula, J., and Bamford, D.H. (1988). Chemical crosslinking of bacteriophage ϕ6 nucleocapsid proteins. Virology *165*, 482–8.

Hoogstraten, D., Qiao, X., Sun, Y., Hu, A., Onodera, S., and Mindich, L. (2000). Characterization of ϕ8, a bacteriophage containing three double-stranded RNA genomic segments and distantly related to ϕ6. Virology *272*, 218–24.

Huiskonen, J.T., de Haas, F., Bubeck, D., Bamford, D.H., Fuller, S.D., and Butcher, S.J. (2006). Structure of the bacteriophage ϕ6 nucleocapsid suggests a mechanism for sequential RNA packaging. Structure *14*, 1039–48.

Jäälinoja, H.T., Huiskonen, J.T., and Butcher S.J. (2007). Electron cryomicroscopy comparison of the architectures of the enveloped bacteriophages ϕ6 and ϕ8. Structure *15*, 157–67.

Juuti, J.T., and Bamford, D.H. (1997). Protein P7 of phage ϕ6 RNA polymerase complex, acquiring of RNA packaging activity by *in vitro* assembly of the purified protein onto deficient particles. J. Mol. Biol. *266*, 891–900.

Kakitani, H., Iba, H., and Okada, Y. (1980). Penetration and partial uncoating of bacteriophage ϕ6 particle. Virology. *101*, 475–83.

Kenney, J.M., Hantula, J., Fuller, S.D., Mindich, L., Ojala, P.M., and Bamford, D.H. (1992). Bacteriophage ϕ6 envelope elucidated by chemical cross-linking, immunodetection, and cryoelectron microscopy. Virology *190*, 635–44.

Khan, S., and Macnab, R.M. (1980). The steady-state counterclockwise/clockwise ratio of bacterial flagellar motors is regulated by protonmotive force. J. Mol. Biol. *138*, 563–97.

Ktistakis, N.T., and Lang, D. (1987). The dodecahedral framework of the bacteriophage ϕ6 nucleocapsid is composed of protein P1. J. Virol. *61*, 2621–3.

Laurinavičius, S., Käkelä, R., Bamford, D.H., and Somerharju, P. (2004). The origin of phospholipids of the enveloped bacteriophage ϕ6. Virology *326*, 182–90.

Letellier, L., Plancon, L., Bonhivers, M., and Boulanger, P. (1999). Phage DNA transport across membranes. Res. Microbiol. *150*, 499–505.

Mertens, P. (2004). The dsRNA viruses. Virus Res. *101*, 3–13.

Mindich, L., Cohen, J., and Weisburd, M. (1976). Isolation of nonsense suppressor mutants in Pseudomonas. J. Bacteriol. *126*, 177–82.

Mindich, L., and Davidoff-Abelson, R. (1980). The characterization of a 120 S particle formed during ϕ6 infection. Virology *103*, 386–91.

Mindich, L., and Lehman, J. (1979). Cell wall lysin as a component of the bacteriophage ϕ6 virion. J. Virol. *30*, 489–96.

Mindich, L., Qiao, X., Qiao, J., Onodera, S., Romantschuk, M., and Hoogstraten, D. (1999). Isolation of additional bacteriophages with genomes of segmented double-stranded RNA. J. Bacteriol. *181*, 4505–8.

Mindich, L., Sinclair, J.F., and Cohen, J. (1976). The morphogenesis of bacteriophage ϕ6: particles formed by nonsense mutants. Virology *75*, 224–31.

Mitchell, P. (1979). The Ninth Sir Hans Lecture. Compartmentation and communication in living systems. Ligand conduction: a general catalytic principle in chemical, osmotic and chemiosmotic reaction systems. Eur. J. Biochem. *95*, 1–20.

Ojala, P.M., Juuti, J.T., and Bamford, D.H. (1993). Protein P4 of double-stranded RNA bacteriophage ϕ6 is accessible on the nucleocapsid surface: epitope mapping and orientation of the protein. J. Virol. *67*, 2879–86.

Ojala, P.M., Paatero, A.O., and Bamford, D.H. (1994). NTP binding induces conformational changes in the double-stranded RNA bacteriophage ϕ6 subviral particles. Virology *205*, 170–8.

Ojala, P.M., Romantschuk, M., and Bamford, D.H. (1990). Purified ϕ6 nucleocapsids are capable of productive infection of host cells with partially disrupted outer membranes. Virology *178*, 364–72.

Olkkonen, V.M., and Bamford, D.H. (1987). The nucleocapsid of the lipid-containing double-stranded RNA bacteriophage ϕ6 contains a protein skeleton consisting of a single polypeptide species. J. Virol. *61*, 2362–7.

Olkkonen, V.M., Ojala, P.M., and Bamford, D.H. (1991). Generation of infectious nucleocapsids by *in vitro* assembly of the shell protein on to the polymerase complex of the dsRNA bacteriophage ϕ6. J. Mol. Biol. *218*, 569–81.

Palmen, R., Driessen, A.J., and Hellingwerf, K.J. (1994). Bioenergetic aspects of the translocation of macromolecules across bacterial membranes. Biochim Biophys Acta. *1183*, 417–51.

Poranen, M.M., Daugelavičius, R., and Bamford, D.H. (2002). Common principles in viral entry. Annu. Rev. Microbiol. *56*, 521–38.

Poranen, M.M., Daugelavičius, R., Ojala, P.M., Hess, M.W., and Bamford, D.H. (1999). A novel virus-host cell membrane interaction. Membrane voltag-dependent endocytic-like entry of bacteriophage ϕ6 nucleocapsid. J. Cell Biol. *147*, 671–82.

Qiao, X., Qiao, J., Onodera, S., and Mindich, L. (2000). Characterization of ϕ13, a bacteriophage related to ϕ6 and containing three dsRNA genomic segments. Virology *275*, 218–24.
Reinisch, K.M., Nibert, M.L., and Harrison, S.C. (2000). Structure of the reovirus core at 3.6 Å resolution. Nature *404*, 960–7.
Romantschuk, M., and Bamford, D.H. (1986). The causal agent of halo blight in bean, *Pseudomonas syringae* pv. *phaseolicola*, attaches to stomata via its pili. Microb. Pathog. *1*, 139–48.
Romantschuk, M., and Bamford, D.H. (1985). Function of pili in bacteriophage ϕ6 penetration. J. Gen. Virol. *66*, 2461–9.
Romantschuk, M., and Bamford, D.H. (1981). ϕ6-resistant phage-producing mutants of *Pseudomonas phaseolicola*. J. Gen. Virol. *56*, 287–95.
Romantschuk, M., Olkkonen, V.M., and Bamford, D.H. (1988). The nucleocapsid of bacteriophage ϕ6 penetrates the host cytoplasmic membrane. EMBO J. *7*, 1821–9.
Ruiz, M.C., Abad, M.J., Charpilienne, A., Cohen, J., and Michelangeli, F. (1997). Cell lines susceptible to infection are permeabilized by cleaved and solubilized outer layer proteins of rotavirus. J. Gen. Virol. *78*, 2883–93.
Ruiz, M.C., Cohen, J., and Michelangeli, F. (2000). Role of Ca2+in the replication and pathogenesis of rotavirus and other viral infections. Cell Calcium. *28*, 137–49.
Rydman, P.S., and Bamford, D.H. (2002). Phage enzymes digest peptidoglycan to deliver DNA. AMS News *68*, 330–35.
Sands, J.A. (1973). The phospholipid composition of bacteriophage ϕ6. Biochem. Biophys Res. Commun. *55*, 111–16.
Sinclair, J.F., Tzagoloff, A., Levine, D., and Mindich, L. (1975). Proteins of bacteriophage ϕ6. J. Virol. *16*, 685–95.
Stitt, B.L., and Mindich, L. (1983). The structure of bacteriophage ϕ6: protease digestion of ϕ6 virions. Virology *127*, 459–62.
Sun, Y., Qiao, X., Qiao, J., Onodera, S., and Mindich, L. (2003). Unique properties of the inner core of bacteriophage ϕ8, a virus with a segmented dsRNA genome. Virology *308*, 354–61.
Tuma, R., Bamford, J.K., Bamford, D.H., and Thomas, G.J., Jr. (1999). Assembly dynamics of the nucleocapsid shell subunit (P8) of bacteriophage ϕ6. Biochemistry *38*, 15025–33.
Van Etten, J.V., Lane, L., Gonzalez, C., Partridge, J., and Vidaver, A. (1976). Comparative properties of bacteriophage ϕ6 and ϕ6 nucleocapsid. J. Virol. *18*, 652–8.
Vidaver, A.K., Koski, R.K., and Van Etten, J.L. (1973). Bacteriophage ϕ6: a lipid-containing virus of *Pseudomonas phaseolicola*. J. Virol. *11*, 799–805.

Crystal Structure of Reovirus Polymerase λ3

13

Yizhi Jane Tao and Stephen C. Harrison

Abstract

The RNA-dependent RNA polymerase of reovirus and those of other *Reoviridae* family members function within the confines of a protein capsid to transcribe the 10–12 tightly packed dsRNA genome segments. The crystal structures of the reovirus polymerase λ3 and its initiation and elongation complexes have been determined by X-ray crystallography. The λ3 structure shows a fingers-palm-thumb core, similar to those of other viral polymerases, surrounded by major N- and C-terminal elaborations, which create a cage-like structure with four channels leading to the catalytic site. These four channels serve as the inlet and outlet for the template, substrate, dsRNA product, and mRNA transcript during replication and transcription. A 5′ cap-binding site has been found on the surface of λ3, suggesting that λ3 may employ a template retention mechanism by which attachment of the 5′ end of the plus-strand facilitates insertion of the 3′ end of the minus-strand into the template channel.

Introduction

One of the most distinct features of the members of the *Reoviridae* is that their icosahedral virions contain a 120-subunit core structure which remains intact in the host cytosol and actively produces viral mRNAs (Nibert and Schiff, 2001). At the heart of the transcriptionally active core is the virus-encoded RNA-dependent RNA polymerase, which catalyzes both transcription (dsRNA→mRNA) and replication (mRNA→dsRNA) of the viral genome (Figure 13.1a). The reovirus polymerase λ3 is a 140-kD protein, and it is estimated that there are about 12 copies of λ3 protein in each virus particle. Because the reovirus genome consists of ten segments of dsRNA, each λ3 molecule presumably associates with one genomic RNA segment. In viral particles, the polymerase molecules are located near the fivefold vertices and occupy large volumes of the interior core as revealed by analyses of the reovirus and reocore structures by crystallography and electron microscopy reconstruction (Dryden *et al.*, 1998; Reinisch *et al.*, 2000).

Because λ3 functions inside a viral core filled with dsRNA arranged in a liquid-crystalline array, the polymerase must possess additional functionalities compared to other polymerases that can operate without spatial constraints. For example, λ3 needs to rely on mechanisms other than random diffusion to locate its dsRNA template for efficient re-initiation of genome transcription. It is also not known how λ3 maintains processivity during RNA synthesis, how the superhelicity in the dsRNA genome is relieved while the

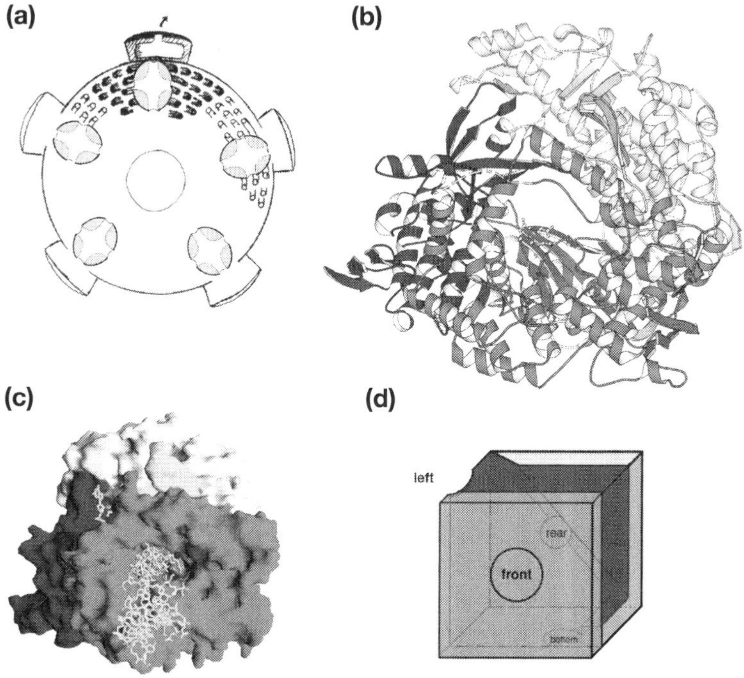

Figure 13.1 Reovirus λ3 (from Tao *et al*, 2002). (a) One λ3 molecule is positioned interiorly near each of the fivefold vertices of the core of reovirus particles. (b) A ribbon diagram of the λ3 crystal structure. The central polymerase domain is shown with its fingers in blue, palm in red and thumb in green. The N- and C-terminal domains are coloured in yellow and magenta, respectively. (c) λ3 represented by its molecular surface, coloured as in (b). A template:product duplex from HIV-RT (Huang *et al*., 1998), shown in grey, is modelled on to λ3 to illustrate the different functions of the four channels. (d) λ3 represented as a cage structure with its N-terminal domain in yellow, the central polymerase domain in blue and the C-terminal bracelet domain in magenta. NTPs and single-stranded templates move to the catalytic core through the rear and left domains, respectively. In its replicase mode, the dsRNA product leaves the polymerase through the front channel. In its transcriptase mode, the mRNA product is believed to leave the polymerase through the bottom channel, with the minus-strand template RNA leaving through the front channel. (See also Plate 13.1.)

transcription bubble moves along the template, or how the nascent transcripts fit precisely into the small openings of the protein capsid so they can be extruded from the core. The answers to the above questions should be revealed through a detailed understanding of the complex molecular interactions among the polymerase λ3, the viral genome, the major capsid protein λ1, and the other minor capsid protein μ2. The atomic structures recently determined for the λ3 apo-enzyme and functional derivatives of the polymerase reovirus shed new light on the replication and transcription processes catalyzed by the reocore (Tao *et al*., 2002) (PDB ID: 1MUK, 1N1H, 1N38, and 1N35).

Biochemical activities of λ3

The biological activities of λ3 are largely dependent on its structural context. In intact cores, λ3 functions as a transcriptase and exclusively produces mRNAs in a conservative fashion (Skehel and Joklik, 1969). Electron microscopy has shown that several mRNA molecules are extruded simultaneously from an actively transcribing core (Gillies *et al.*, 1971). The rates of RNA synthesis of the different RNA segments are inversely proportional to their lengths (Watanabe *et al.*, 1968), suggesting that initiation of transcription is not rate-limiting, and that each of the ten gene segments is transcribed independently. The virion and subvirion cannot produce full-length transcripts, but they readily synthesize short oligonucleotides when given appropriate substrates (Yamakawa *et al.*, 1982). Therefore, the polymerase λ3 is capable of initiation, but not elongation, in both virions and subvirions. It is important to note that the reocore also synthesizes predominantly short abortive transcripts, but the transcripts are occasionally extended to produce full-length products (Yamakawa *et al.*, 1981; Zarbl *et al.*, 1980). In nascent recores, λ3 functions as a replicase and performs a single round of minus-strand RNA synthesis to generate a dsRNA genome from a plus-strand RNA precursor (Morgan and Zweerink, 1975). Recombinant λ3 only has a limited poly(C)-dependent poly(G) activity (Starnes and Joklik, 1993), suggesting that complete transcriptase activities may require additional core proteins. A partial activity appears to be a general property of purified polymerase proteins of viruses in the *Reoviridae* family. For example, the purified VP1 polymerase of bluetongue virus (BTV) exhibits only replicase activity (Boyce *et al.*, 2004). Similarly, the purified VP1 polymerase of rotavirus is inactive by itself, but gains replicase activity when combined with the core shell protein VP2 (Patton *et al.*, 1997).

Overall structure of λ3

The λ3 polypeptide chain folds into a compact unit with a diameter of about 65 Å (Figure 13.1b). The structure of λ3 can be divided into three domains: a central polymerase domain, which contains 'fingers', 'palm', and 'thumb' subdomains, an N-terminal domain, which bridges the fingers and thumb on one side of the catalytic cleft, and a C-terminal 'bracelet' domain, which covers the catalytic cleft on the other side (Figure 13.1b, c, d). The polymerase domain has a 'right-hand' configuration, which closely resembles those of other RNA-dependent RNA polymerases (RdRps) (Ago *et al.*, 1999; Bressanelli *et al.*, 1999; Butcher *et al.*, 2001; Hansen *et al.*, 1997; Lesburg *et al.*, 1999). Compared to the RdRps of other plus-strand RNA viruses, λ3 has rather elaborate N- and C-terminal domains. The N-terminal domain of λ3, residues 1–380, covers one side of the active-site cleft. This domain anchors and augments the interaction between the fingers and thumb subdomains. Like the HCV and ϕ6 polymerases (Bressanelli *et al.*, 2002; Butcher *et al.*, 2001; Lesburg *et al.*, 1999), λ3 appears to have a 'closed' structure, even in its resting state. Catalysis can occur in the λ3 crystal without disturbing the structure, suggesting that the fingers and the thumb of the central polymerase domain move very little during catalysis.

The C-terminal domain of λ3, residues 891–1267, is an annular structure ('bracelet'). Its shape and orientation relative to the core polymerase recall those of the sliding clamp of processive DNA polymerases (Shamoo and Steitz, 1999; Stukenberg *et al.*, 1991), but it has an unrelated fold. Sliding clamps encircle DNA and act as tethers to ensure polymerase processivity. However, the λ3 bracelet is unlikely to open and close dynamically, like sliding clamps in DNA replication, because it folds into two tightly sealed circles. The opening of the λ3 bracelet has a diameter of about 20 Å and bears a net positive charge.

When represented by its molecular surface, λ3 resembles a cube-like cage (Figure 13.1). Access to the catalytic centre of the cube is gained through four channels, which function in the polymerase as avenues for the movement of nucleotide substrates, and RNA templates and products. The front channel, which corresponds to the bracelet opening, is the exit channel for the dsRNA product. The left channel, which is at the interface of the polymerase and bracelet domains, represents the template entry tunnel. The rear channel is situated at the interface of the polymerase and N-terminal domains, and nucleotides needed for RNA polymerization likely diffuse through this opening into the active site. The bottom channel in λ3 is a structural feature not yet observed in other polymerases. It has been proposed that viral mRNAs exit the active site through this bottom channel when the polymerase is operating in transcriptional mode (Tao *et al.*, 2002).

Initiation of RNA synthesis

Reovirus λ3 can bind non-specifically to an RNA template, but it prefers the conserved 3′ terminal sequences of reovirus genomic RNA (Tao *et al.*, 2002). Crystallographic studies have shown that a template G at position +1 provides a carbonyl for a hydrogen bond with R518 and an –NH for a hydrogen bond with the side chain of S682 (Figure 13.2). A template U at this position can also interact favourably with R518, but A and C cannot. Given the conserved 3′ terminal sequences of reovirus RNAs (plus, UCAUC-3′; minus, UAGC-3′), the preference for G or U at +1 promotes synthesis of full-length RNAs of either sense. The specific interaction between the conserved viral RNA sequences and λ3 also prevents template overshooting and ensures that the initiation of RNA synthesis occurs at the very end of its template. Of the four template bases visible in the λ3 crystal structure, the phosphate backbone follows a path similar to that taken by the template overhang in the catalytic complexes of other DNA/RNA polymerases (Cheetham and Steitz, 1999; Doublie *et al.*, 1998). The ribose and base of the templating nucleotide (position +1) stack tightly under the side chains of P530 and I528, forcing the downstream template to bend away from the catalytic pocket (Figure 13.2).

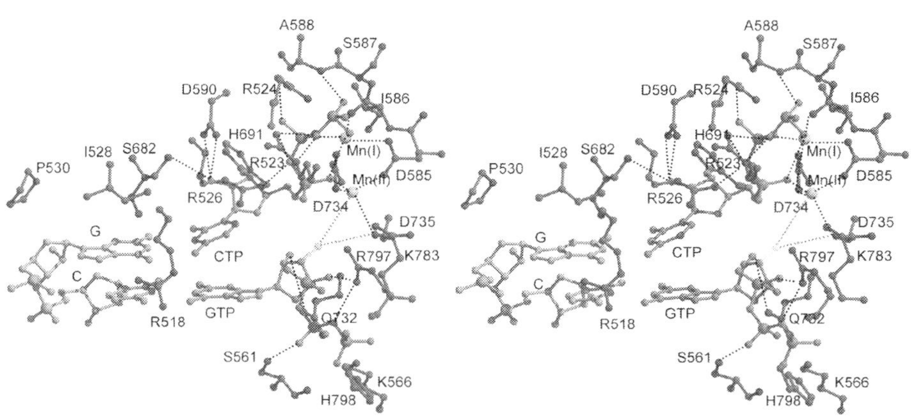

Figure 13.2 Stereo diagram of the active site in the λ3 initiation complex (from Tao *et al*, 2002). Shown are critical amino acids (stick models: blue, nitrogen; red, oxygen), 3′-end of template strand, green; 3′-deoxy rCTP at incoming nucleotide position, purple; 3′-deoxy rGTP at priming position, orange; Mn^{2+} ions, gold spheres; assigned hydrogen bonds and metal ligand interactions, dashed lines. The missing 3′-OH is shown in a lighter shade, and the related hydrogen bonds, in red. (See also Plate 13.2.)

The overall active site conformation and the positioning of the incoming and priming nucleoside triphosphates in λ3 are very similar to those of other polymerase complexes. In a λ3 initiation complex structure, the 3′-OH group of the priming rNTP is placed for in-line attack on the α phosphate of the incoming rNTP (Figure 13.2). The incoming rNTP pairs with the template base at position +1, and its triphosphate is coordinated by two divalent cations, three basic residues (R523, R524, and R526) from the NTP-binding loop at the tip of the fingers subdomain, and the main-chain –NH group of A588. The 2′-OH of the incoming rNTP forms hydrogen bonds with D590 and S682; the 2′-OH of the nucleotide at the primer position forms a hydrogen bond with Q732. These contacts presumably confer the observed specificity of λ3 for ribonucleotides. The priming rNTP stacks onto a platform made by a loop containing residues 558–565. This 'priming loop' is effectively an insertion within a strand that is found in the palms of all other polymerases, and it is unique (among known structures) to λ3.

RNA synthesis by λ3

Crystalline λ3 can catalyze phosphodiester bond formation and produce short dsRNA molecules. The polymerase contacts the RNA product along its minor groove. The loop that supports the priming rNTP retracts by about 3 Å toward the palm with respect to its position in the apo-enzyme and in the initiation complexes, and it fits snugly into the minor groove of the product duplex (Figure 13.3). If the loop were to fail to shift, it would block elongation and cause abortive initiation. Indeed, both reovirus and reocore produce large amounts of short oligonucleotides, and these abortive transcripts are predominantly 2–4 bases long (Farsetta *et al.*, 2000). Therefore, the priming loop could provide a kinetic barrier for the transition from initiation to elongation.

Another loop, formed by residues 1109–1125 from the C-terminal bracelet domain, lies in the path of the dsRNA product at the downstream end of the product exit channel. This structure might help to separate template and product in the transcriptional mode and to direct the transcript toward the bottom channel. Alternatively, separation might depend on partial constriction of the front channel, perhaps by a shift in the flexible loop between residues 957 and 964, or by binding of μ2, the second internal protein in the reovirus core. Because λ3 lies close to the fivefold pores in the λ1 shell (Dryden *et al*, 1998), mRNA molecules may pass smoothly into the turret chambers for capping and export. If μ2 is the 5′-terminal phosphatase that prepares the mRNA terminus for capping, then the leading end of a transcript must pass across it before entering the capping chamber, and the two proteins may be jointly tethered. Interestingly, based on an EM reconstruction of reovirus, it has been proposed that λ3 anchors to the inner surface of the icosahedral core shell, and that the bottom transcript channel of λ3 is aligned with a small tunnel in the capsid that is slightly skewed from the fivefold axis (Zhang *et al.*, 2003).

Cap binding site and its biological implication

λ3 possesses a cap-binding site on its molecular surface next to the template entry channel (Figure 13.4). The positively charged, methylated purine of the mRNA cap interacts with residue D1035 through two hydrogen bonds, whereas the other purine base hydrogen bonds with R851. Both bases are sandwiched between hydrophobic, largely aliphatic side chains.

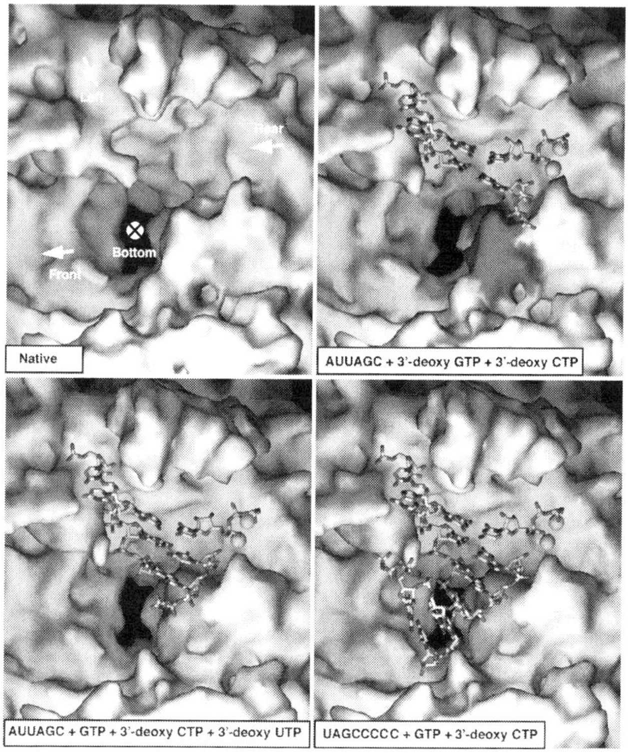

Figure 13.3 RNA synthesis by λ3 (from Tao *et al*, 2002). The active site of λ3 is shown at four progressive stages – as the native apo-enzyme (upper left), an initiation complex (upper right), after synthesis of a 2-nucleotide product (lower left) and after synthesis of a 5-nucleotide product (lower right). λ3 is rendered as a molecular surface; RNA and nucleotides, as atomic models; and metal ions, as yellow spheres. The four channels are indicated on the image of native λ3. Upon formation of the first phosphodiester bond, the loop (residues 658–665) that supports the priming nucleotide changes its conformation (red → green), thus avoiding collision with the dsRNA product. (See also Plate 13.3.)

Cap binding may be significant for two distinct stages in the replication cycle: RNA packaging into assembling cores and mRNA synthesis after viral entry into a host (Figure 13.5). In the packaging step, λ3 could recognize capped viral RNA segments by its conserved terminal elements: 5′-m^{7}G-GCUA and UCAUC-3′. Binding of the cap at one position on the polymerase surface and insertion of the 3′ end into the template channel nearby could ensure association of a single polymerase molecule with each segment of viral plus-strand RNA. We suggest that higher-order structures might then form between different RNA segments to select a full set of 10 distinct mRNAs for packaging. Interactions between the λ3 polymerase molecules and the interior of the assembling λ1 shell could promote efficient incorporation of the viral genome. Such a process would be different from the RNA packaging mechanism of bacteriophage ϕ6, in which plus-strand RNAs are inserted into pre-assembled procapsids (Mindich, 1999).

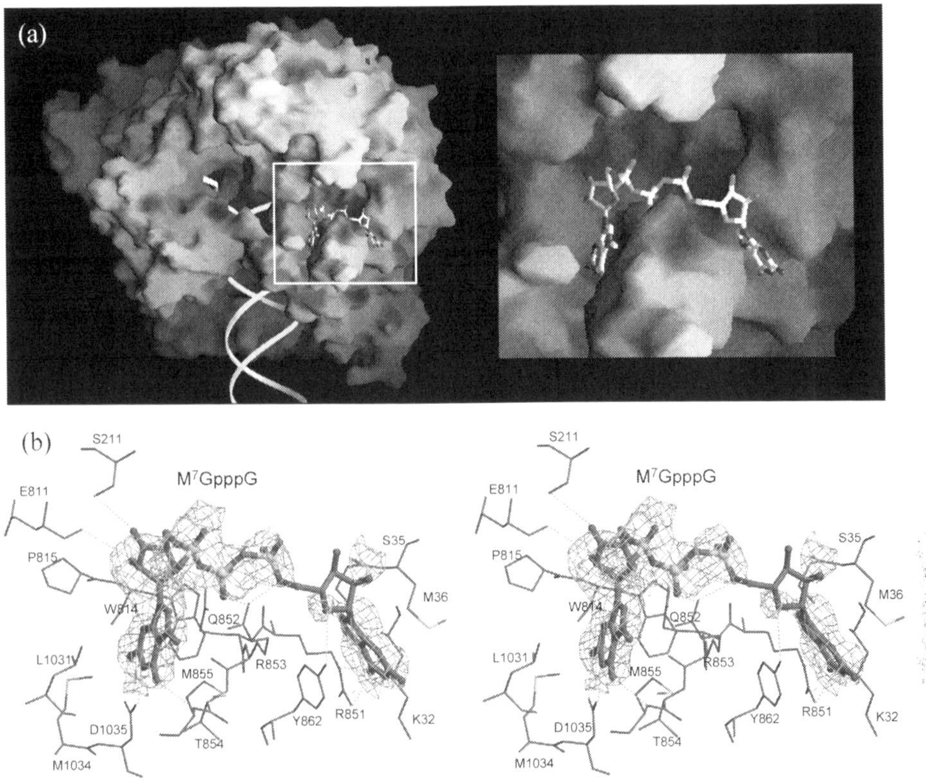

Figure 13.4 Cap binding site on λ3 (from Tao *et al*, 2002). (a) Location. λ3 is rendered as a molecular surface coloured according to electrostatic potential, with positive in blue and negative in red. On the right is an enlarged view of the pocket. The phosphate backbone of a modelled template:primer duplex is shown by green ribbons. (b) Stereo diagram of the cap-binding pocket. The view is taken in the same orientation as in (a). Hydrogen bonds are represented by pink dotted lines. Electron density corresponding to the cap analogue is shown in green. (See also Plate 13.4.)

Cap binding may also play an important role in the initiation of mRNA synthesis (Figure 13.5). Of the positive and negative strands of the reovirus dsRNA genome, only the plus-sense RNA is capped. Attachment of the 5′ end of the plus strand to the cap site may facilitate insertion of the free 3′ end of the minus strand into the template channel. When RNA synthesis approaches the 5′ end of the template, the terminus will be forced to leave the cap site and enter the template channel. As it emerges through the front channel, the cap can be recaptured by the cap-binding site, which is adjacent to the front-channel exit. Moreover, after completion of each round of mRNA synthesis, the 3′ end of the minus strand will easily reinsert and reinitiate. Were the 5′ end of the plus strand not anchored firmly to the polymerase, the 3′ end of the template would 'get lost' within the coils of dsRNA in the interior of the core, greatly decreasing the efficiency of re-initiation. Moreover, a cap-binding site prevents initiation from the wrong strand.

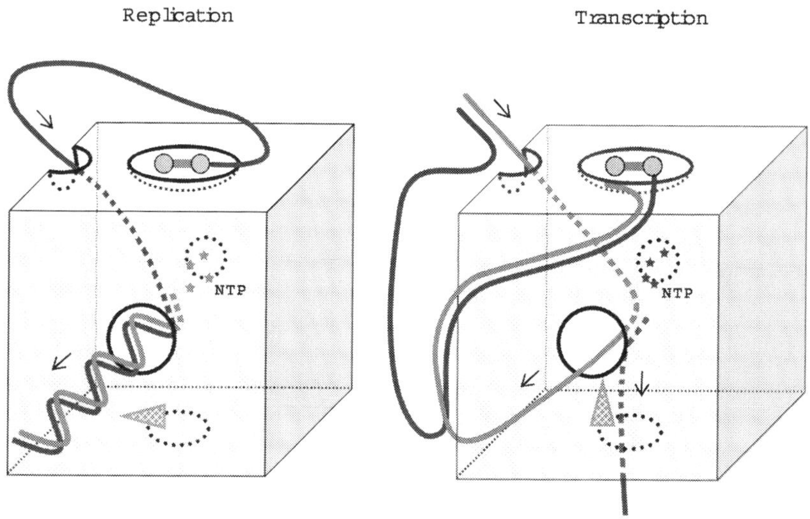

Figure 13.5 Model for spatial organization of polynucleotide synthesis. Left, replication; right, transcription. λ3 is represented by a grey cube. Darker grey lines and lighter grey lines correspond to plus- and minus-strand RNA, respectively. The saucer-like feature on the top of the λ3 cube represents the cap-binding site; the dumb bell-like object represents the 5′ cap. The template RNA exits the active site through the front channel. The nascent strand has two possible routes: either through the front channel during replication or through the bottom channel during transcription. The triangular-shaped arrow inside the cube symbolizes a presumptive separator structure that splits the nascent transcript from its template.

Similar mechanisms might be adopted by other RNA viruses in order to synthesize predominantly plus or minus-strand RNA molecules during viral replication (Ahola *et al.*, 2000). It is important to note that cap-binding activity and the enzymatic activity of λ3 are mutually independent.

Similarity with other polymerases

The architecture of λ3 polymerase has evidently evolved to ensure spatial and temporal coordination in the crowded interior of the reovirus core. The catalytic core lies at the centre of a cage with four defined channels. There is an external site for binding a 5′-m^{7}G RNA cap to tether the 5′ end of the RNA to the polymerase. λ3 also contains a C-terminal bracelet, which presumably functions as a processivity factor. Moreover, there are candidate higher order structures within the catalytic core that may function to specify different exit routes of the RNA products during replication versus transcription.

Unlike HIV reverse transcriptase and T7 DNA and RNA polymerases, the binding of λ3 to its template and substrate results in only localized rearrangements. Likewise, template translocation and nucleotide incorporation can be accomplished with little structural perturbation. The accuracy of RNA synthesis by λ3 (one error for every 10^4 bases) is comparable to the fidelity of other polymerases without proofreading activity (Echols and Goodman, 1991). Like λ3, the catalytic activity of the ϕ6 polymerase is associated with little conformational change (Salgado *et al.*, 2004).

The catalytic complex structures of λ3 have much in common with those of other RNA-dependent polymerase catalytic complexes for which structures are known (Bressanelli *et al.*, 2002; Butcher *et al.*, 2001; Doublie *et al.*, 1999; Huang *et al.*, 1998). These conserved features include the conformation and ligation of the incoming nucleotide, the conformation of the template at the −2, −1 and +1 positions, the way in which the templating base is 'jammed' against a segment of the fingers subdomain, and the way in which other interactions constrain the geometry of template-primer and template-incoming nucleotide base pairs. Stacking platforms for the priming nucleotide, in the form of the priming loop in reovirus λ3, are also found in other RNA polymerases such as those from HCV and ϕ6, although their exact structures are vastly different.

Future directions

At present the biggest question is what the exact function of μ2 is in reovirus RNA synthesis. The internal core protein μ2 (Noble and Nibert, 1997), as well as the core capsid shell protein λ1 (Bisaillon *et al.*, 1997), have been proposed to have a helicase activity. It is also possible that μ2 acts as a topoisomerase to relieve superhelicities generated in genomic dsRNA during RNA synthesis.

Moreover, there may be additional surface features of λ3 not fully revealed by the present structures that may function in reovirus replication/transcription. In particular, there may be aspects of the λ3 cage that help guide the plus-strand (non-template) side of the transcription bubble, and there is probably a docking surface for μ2. It is also not clear how the removal of the outer core capsid protein allows λ3 to produce full-length transcripts. Is it because the priming loop is able to assume the 'down' position in reocore but not in the virion? In addition, how does λ3 contribute to the overall organization of the dsRNA genome inside reocore? It has been proposed that the segmented dsRNA genome is organized in the form of distinct spindles. It is expected that multidisciplinary researches, such as those employing structural biology, mathematical modelling, reverse genetics, and molecular biology, are necessary in order to fully understand this elegantly simple replication/transcription complex.

Acknowledgement

YJT is supported by the Welch Foundation.

References

Ago, H., Adachi, T., Yoshida, A., Yamamoto, M., Habuka, N., Yatsunami, K., and Miyano, M. (1999). Crystal structure of the RNA-dependent RNA polymerase of hepatitis C virus. Structure Fold Des. 7, 1417–1426.

Ahola, T., den Boon, J.A., and Ahlquist, P. (2000). Helicase and capping enzyme active site mutations in brome mosaic virus protein 1a cause defects in template recruitment, negative-strand RNA synthesis, and viral RNA capping. J. Virol. 74, 8803–8811.

Bisaillon, M., Bergeron, J., and Lemay, G. (1997). Characterization of the nucleoside triphosphate phosphohydrolase and helicase activities of the reovirus lambda1 protein. J. Biol. Chem. 272, 18298–18303.

Boyce, M., Wehrfritz, J., Noad, R., and Roy, P. (2004). Purified recombinant bluetongue virus VP1 exhibits RNA replicase activity. J. Virol. 78, 3994–4002.

Bressanelli, S., Tomei, L., Rey, F.A., and De Francesco, R. (2002). Structural analysis of the hepatitis C virus RNA polymerase in complex with ribonucleotides. J. Virol. 76, 3482–3492.

Bressanelli, S., Tomei, L., Roussel, A., Incitti, I., Vitale, R.L., Mathieu, M., De Francesco, R., and Rey, F.A. (1999). Crystal structure of the RNA-dependent RNA polymerase of hepatitis C virus. Proc. Natl Acad. Sci. USA *96*, 13034–13039.

Butcher, S.J., Grimes, J.M., Makeyev, E.V., Bamford, D.H., and Stuart, D.I. (2001). A mechanism for initiating RNA-dependent RNA polymerization. Nature *410*, 235–240.

Cheetham, G.M., and Steitz, T.A. (1999). Structure of a transcribing T7 RNA polymerase initiation complex. Science *286*, 2305–2309.

Doublie, S., Sawaya, M.R., and Ellenberger, T. (1999). An open and closed case for all polymerases. Structure Fold Des. *7*, R31–35.

Doublie, S., Tabor, S., Long, A.M., Richardson, C.C., and Ellenberger, T. (1998). Crystal structure of a bacteriophage T7 DNA replication complex at 2.2 Å resolution. Nature *391*, 251–258.

Dryden, K.A., Farsetta, D.L., Wang, G., Keegan, J.M., Fields, B.N., Baker, T.S., and Nibert, M.L. (1998). Internal/Structures containing transcriptase-related proteins in top component particles of mammalian orthoreovirus. Virology *245*, 33–46.

Echols, H., and Goodman, M.F. (1991). Fidelity mechanisms in DNA replication. Annu. Rev. Biochem. *60*, 477–511.

Farsetta, D.L., Chandran, K., and Nibert, M.L. (2000). Transcriptional activities of reovirus RNA polymerase in recoated cores. Initiation and elongation are regulated by separate mechanisms. J. Biol. Chem. *275*, 39693–39701.

Gillies, S., Bullivant, S., and Bellamy, A.R. (1971). Viral RNA polymerases: electron microscopy of reovirus reaction cores. Science *174*, 694–696.

Hansen, J.L., Long, A.M., and Schultz, S.C. (1997). Structure of the RNA-dependent RNA polymerase of poliovirus. Structure *5*, 1109–1122.

Huang, H., Chopra, R., Verdine, G.L., and Harrison, S.C. (1998). Structure of a covalently trapped catalytic complex of HIV-1 reverse transcriptase: implications for drug resistance. Science *282*, 1669–1675.

Lesburg, C.A., Cable, M.B., Ferrari, E., Hong, Z., Mannarino, A.F., and Weber, P.C. (1999). Crystal structure of the RNA-dependent RNA polymerase from hepatitis C virus reveals a fully encircled active site. Nat. Struct. Biol. *6*, 937–943.

Mindich, L. (1999). Precise packaging of the three genomic segments of the double-stranded-RNA bacteriophage phi6. Microbiol. Mol. Biol. Rev. *63*, 149–160.

Morgan, E.M., and Zweerink, H.J. (1975). Characterization of transcriptase and replicase particles isolated from reovirus-infected cells. Virology *68*, 455–466.

Nibert, M.L., and Schiff, L.A. (2001). Reoviruses and their replication. In Fields Virology D.M. Knipe, and P. M. Howley, eds. (Lippincott-Raven), pp. 1679–1728.

Noble, S., and Nibert, M.L. (1997). Core protein mu2 is a second determinant of nucleoside triphosphatase activities by reovirus cores. J. Virol. *71*, 7728–7735.

Patton, J.T., Jones, M.T., Kalbach, A.N., He, Y.W., and Xiaobo, J. (1997). Rotavirus RNA polymerase requires the core shell protein to synthesize the double-stranded RNA genome. J. Virol. *71*, 9618–9626.

Reinisch, K.M., Nibert, M.L., and Harrison, S.C. (2000). Structure of the reovirus core at 3.6 A resolution. Nature *404*, 960–967.

Salgado, P.S., Makeyev, E.V., Butcher, S.J., Bamford, D.H., Stuart, D.I., and Grimes, J.M. (2004). The structural basis for RNA specificity and Ca2+ inhibition of an RNA-dependent RNA polymerase. Structure *12*, 307–316.

Shamoo, Y., and Steitz, T.A. (1999). Building a replisome from interacting pieces: sliding clamp complexed to a peptide from DNA polymerase and a polymerase editing complex. Cell *99*, 155–166.

Skehel, J.J., and Joklik, W.K. (1969). Studies on the *in vitro* transcription of reovirus RNA catalyzed by reovirus cores. Virology *39*, 822–831.

Starnes, M.C., and Joklik, W.K. (1993). Reovirus protein lambda 3 is a poly(C)-dependent poly(G) polymerase. Virology *193*, 356–366.

Stukenberg, P.T., Studwell-Vaughan, P.S., and O'Donnell, M. (1991). Mechanism of the sliding beta-clamp of DNA polymerase III holoenzyme. J. Biol. Chem. *266*, 11328–11334.

Tao, Y., Farsetta, D.L., Nibert, M.L., and Harrison, S.C. (2002). RNA Synthesis in a Cage-Structural Studies of Reovirus Polymerase lambda3. Cell *111*, 733–745.

Watanabe, Y., Millward, S., and Graham, A.F. (1968). Regulation of transcription of the Reovirus genome. J. Mol. Biol. *36*, 107–123.

Yamakawa, M., Furuichi, Y., Nakashima, K., LaFiandra, A.J., and Shatkin, A.J. (1981). Excess synthesis of viral mRNA 5-terminal oligonucleotides by reovirus transcriptase. J. Biol. Chem. *256*, 6507–6514.
Yamakawa, M., Furuichi, Y., and Shatkin, A.J. (1982). Reovirus transcriptase and capping enzymes are active in intact virions. Virology *118*, 157–168.
Zarbl, H., Hastings, K.E., and Millward, S. (1980). Reovirus core particles synthesize capped oligonucleotides as a result of abortive transcription. Arch. Biochem. Biophys. *202*, 348–360.
Zhang, X., Walker, S.B., Chipman, P.R., Nibert, M.L., and Baker, T.S. (2003). Reovirus polymerase lambda 3 localized by cryo-electron microscopy of virions at a resolution of 7.6 A. Nat. Struct. Biol. *10*, 1011–1018.

Structure–Function Insights into the RNA-Dependent RNA Polymerase of the dsRNA Bacteriophage ϕ6

14

Minni R. L. Koivunen, L. Peter Sarin and
Dennis H. Bamford

Abstract

RNA-dependent RNA polymerases (RdRPs) are critical components in the life cycle of double-stranded RNA (dsRNA) viruses. However, it is not fully understood how these important enzymes function during viral replication. Expression and characterization of the purified recombinant RdRP of bacteriophage ϕ6, a member of the *Cystoviridae* family, has provided the first direct demonstration of RdRP activity catalyzed by a single protein for a dsRNA virus. The recombinant ϕ6 RdRP is highly active *in vitro*, possesses RNA replication and transcription activities, and is capable of using both homologous and heterologous RNA molecules as templates. The crystal structure of the ϕ6 polymerase, solved in complex with a number of ligands, provides insights towards understanding the mechanism of primer-independent initiation of RNA-dependent RNA polymerization. Furthermore, the purified ϕ6 RdRP displays processive elongation *in vitro* and self-assembles along with polymerase complex proteins into subviral particles that are fully functional *in vitro* and *in vivo*. This chapter will review what is currently known about the structure and function of the ϕ6 RdRP with a special emphasis on the mechanism of primer-independent initiation of ϕ6 viral RNA synthesis.

Introduction

Double-stranded RNA (dsRNA) viruses are intricate molecular machines, which carry all the necessary components required for viral RNA synthesis within their capsids. Upon delivery into the infected cell, the dsRNA virion particle becomes transcriptionally active and generates single-stranded, positive-sense RNA molecules [(+)RNAs]. These (+)RNAs can serve as templates for protein synthesis or can be sequestered into newly formed subviral particles, where they function as templates for dsRNA synthesis. The particle-associated RNA-dependent RNA polymerase (RdRP) is responsible for the viral RNA metabolism and is, thus, a central component in the life cycle of a dsRNA virus. Since dsRNA viruses encode only a single protein containing conserved RdRP motifs, this protein must catalyze both transcription of genomic dsRNAs into (+)RNAs and replication of the (+)RNAs into dsRNAs. To date, RdRPs from the following dsRNA viruses have been purified and biochemically characterized to varying degrees: bacteriophages ϕ6, ϕ8, ϕ12, ϕ13 (Butcher *et al.*, 2001; Makeyev and Bamford, 2000a, b; Yang *et al.*, 2001, 2003a, b), bluetongue virus (BTV) (Boyce *et al.*, 2004), reovirus (Starnes and Joklik, 1993; Tao *et al.*, 2002), and rotavirus (Patton, 1996; Patton *et al.*, 1997; Zeng *et al.*, 1996). In this review, the structure and function of the bacteriophage ϕ6 RdRP is described and compared with the structure and function of several other viral RdRPs.

Characteristics of bacteriophage ϕ6

Bacteriophage ϕ6 is an enveloped dsRNA virus of the gram-negative pathogenic plant bacterium, *Pseudomonas syringae*, and is the type-organism of the family *Cystoviridae* (Vidaver *et al.*, 1973; http://www.ncbi.nlm.nih.gov/ICTVdb/). In the past, ϕ6 was the only known member of the *Cystoviridae* family. However, following the isolation and characterization of new bacteriophages, eight additional viruses (ϕ7 to ϕ14) have recently been added to the *Cystoviridae* family (Mindich, 1999). These differ markedly, both structurally and functionally, from other known bacterial viruses. Yet, cystoviruses share common features with eukaryotic dsRNA viruses of the family *Reoviridae*, suggesting a possible evolutionary link. So far, ϕ6 is the only dsRNA virus system where *in vitro* genome packaging, replication, and assembly using purified components is achieved.

The genome of ϕ6 consists of three dsRNA segments (S, M, and L), with one copy of each segment contained in the virion. The coding regions of the viral genome segments are flanked at the 5′ and the 3′ termini by distinct non-coding regions (Gottlieb *et al.*, 1988; McGraw *et al.*, 1986; Mindich *et al.*, 1988). These non-coding regions contain important signals for genome packaging and replication (Gottlieb *et al.*, 1994; Pirttimaa and Bamford, 2000). The viral genome is encapsidated by a proteinacous nucleocapsid (NC) and a lipid envelope. The NC is made up of two concentric protein layers: (i) NC surface shell (P8) and (ii) the polymerase complex (PC) core. The PC is composed of four protein species: the major structural protein (P1), the RdRP (P2), the hexameric NTPase (P4), and a protein cofactor (P7) (Poranen *et al.*, 2005b). P1 is the major structural component of the PC, forming the polyhedral skeleton of the particle. P4 forms turret-like protrusions on the P1 lattice, creating a symmetry mismatch at the fivefold vertices (Butcher *et al.*, 1997; de Haas *et al.*, 1999). P2 is located at the fivefold symmetry axes of the PC, whereas both the fivefold and the twofold symmetry axes has been proposed for the location of P7 (Ikonen *et al.*, 2003; Juuti and Bamford, 1997; Juuti *et al.*, 1998; Poranen *et al.*, 2001). Details on ϕ6 entry and the architecture of the lipid envelope, the NC surface P8 shell, and the PC are provided in this book (Poranen and Bamford, 2006).

Life cycle of bacteriophage ϕ6

During entry of ϕ6 into a host bacterium, the structural layers of the virion are sequentially dissociated and exposed. In the host cell cytosol, the RdRP in the PC is activated to transcribe full-length (+)RNAs using genomic dsRNAs as templates. These (+)RNAs are exported into the cytoplasm, where they serve as templates for protein synthesis (Kakitani *et al.*, 1980). The newly produced proteins encoded by the L segment (P1, P2, P4 and P7) assemble into empty PCs (Bamford and Mindich, 1980). The PC has the same protein composition as the virion core, but is devoid of RNA. The P4 NTPase of the empty PC packages the genomic precursor (+)RNAs and the RdRP converts them into dsRNA (Ewen and Revel, 1988, 1990; Kainov *et al.*, 2003; Mindich *et al.*, 1995; Poranen and Bamford, 1999). The dsRNA-filled particles then start a new round of transcription that leads to the synthesis of late proteins, encoded from the S and M genomic segments (Rimon

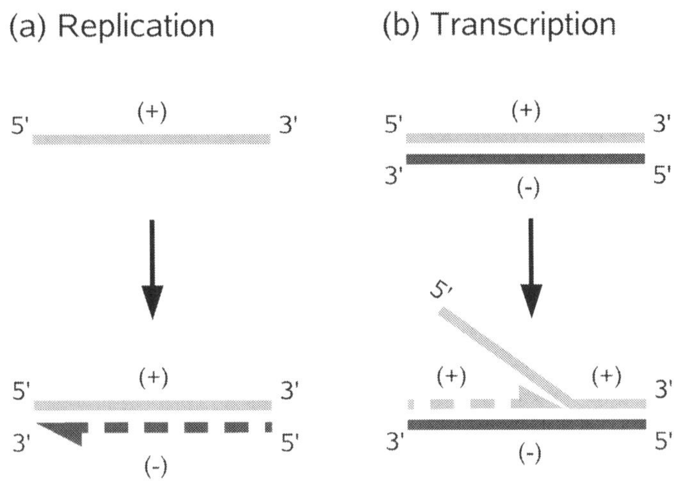

Figure 14.1 The two RNA polymerization reactions catalyzed by viral RdRPs. The polarity of the RNA strands in the figure are marked as (+) or (–), respectively. Modified from Makeyev and Grimes (2004).

and Haselkorn, 1978). Protein P8 assembles onto the dsRNA-filled particles, forming double-shelled NCs, which acquire a lipid envelope and host-attachment spike proteins to form mature virions (Bamford *et al.*, 1976; Bamford and Mindich, 1980; Mindich *et al.*, 1979; Olkkonen *et al.*, 1991). The mature virions are released from the host cell by lysis (Mindich and Lehman, 1979; Vidaver *et al.*, 1973).

Functions of the RdRP in ϕ6 life cycle

Previously, the RNA metabolism of dsRNA viruses such as reovirus, bacteriophage ϕ6, cypovirus, infectious pancreatic necrosis virus and others have been studied using purified viruses or cores containing dsRNA genomes (Cohen, 1975; Partirdge *et al.*, 1979; Shimotohno and Miura, 1973; Skehel and Joklik, 1969; Van Etten *et al.*, 1973), and isolated virus intermediates from L-A virus and rotavirus containing packaged ssRNA (Chen *et al.*, 1994; Ewen and Revel, 1990; Fujimura *et al.*, 1986; Fujimura and Wickner, 1988; Lawton *et al.*, 2000). A set of recombinant rotaviral particles has been produced using a baculovirus expression system. However, the purified recombinant RdRP protein is inactive and requires additional proteins to retain polymerase activity (Patton, 1996; Patton *et al.*, 1997; Valenzuela *et al.*, 1991; Zeng *et al.*, 1996). Furthermore, both the purified recombinant BTV VP1 and the cell lysate containing an RdRP of recombinant infectious bursal disease virus have been found to exhibit RdRP activity (Boyce *et al.*, 2004; von Einem *et al.*, 2004). Characterization of the recombinant ϕ6 RdRP by Makeyev and Bamford (2000a) was the first direct demonstration that RdRP activity is catalyzed by an individual dsRNA viral protein. The recombinant ϕ6 RdRP possesses both RNA replication and transcription activities *in vitro* (Figure 14.1). In contrast to other dsRNA viral RdRPs, the recombinant ϕ6 RdRP is a highly active, stable, soluble monomer that can be purified to near homogeneity in large quantities (Makeyev and Bamford, 2000a, b). The crystal structure of the ϕ6 RdRP has been solved in complex with a number of ligands (Figure 14.2) (Butcher *et al.*,

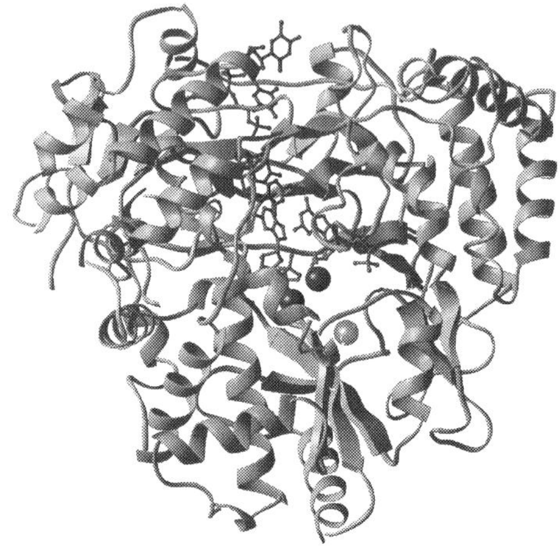

Figure 14.2 Crystal structure of the ϕ6 RdRP initiation complex; (Butcher *et al.*, 2001) Protein Data Bank accession number 1HI0). The figure shows the backbone of the ϕ6 polymerase, the oligonucleotide template, GTPs, magnesium ions, and manganese. The figure was produced using the program INSIGHT II (Molecular Simulation Inc., San Diego, CA). Modified from Laurila *et al.* (2002).

2001; Salgado *et al.*, 2004). The ϕ6 RdRP also displays processive elongation *in vitro* and purified enzyme self-assembles together with P1, P4 and P7 into viral PCs that are fully functional *in vitro* and *in vivo* (Poranen *et al.*, 2001). The robust activity of the ϕ6 RdRP in the absence of other proteins suggests that it is an ideal system to study the mechanism of RNA synthesis for RNA viruses.

ϕ6 RNA replication

During viral replication, packaged (+)RNAs are converted into dsRNAs (Figure 14.1). Both the packaged (+)RNAs and newly produced dsRNA segments are resistant to RNase degradation, suggesting that all stages of replication occur inside PC particles (Ewen and Revel, 1990; Gottlieb *et al.*, 1991). Highly purified recombinant ϕ6 RdRP efficiently replicates (+)RNA templates *in vitro* at a rate of ~120 nt/s, indicating that the presence of other PC components is not critical for replication (Makeyev and Bamford, 2000a). The ϕ6 RdRP supports end-to-end polymerization without loss of genetic information. Interestingly, the addition of manganese ions (Mn^{2+}) along with magnesium ions (Mg^{2+}) significantly stimulates the ϕ6 RdRP during *in vitro* replication (Laurila *et al.*, 2005). This enhancement of activity in the presence of Mn^{2+} is also the case for related cystovirus enzymes from bacteriophages ϕ8, ϕ12, and ϕ13 (Yang *et al.*, 2001, 2003a, b), as well as for a number of other RdRPs of RNA viruses, including those of Qβ bacteriophage, hepatitis C virus (HCV), brome mosaic virus (BMV), and poliovirus (PV) (Alaoui-Lsmaili *et al.*, 2000; Arnold *et al.*, 1999; Blumenthal, 1980; Blumenthal and Carmichael, 1979; Sun *et al.*, 1996; Zhong *et al.*, 2000b). Furthermore, the ϕ6 RdRP requires a relatively high concentration of initiatory NTPs for optimal RNA synthesis (Laurila *et al.*, 2002). These same conditions stimulate ϕ6 particle-based replication (Ojala and Bamford, 1995; van

Dijk *et al.*, 1995), suggesting that the isolated φ6 RdRP accounts, at least in most part, for the biochemical behaviour of the entire PC.

The φ6 RdRP as well as RdRPs from other cystoviruses can use heterologous RNA templates *in vitro*. However, the amount of dsRNA produced during the reaction depends on the 3′-terminal template sequence and its secondary structure. The specificity of the φ6 RdRP for the 3′-terminus was studied using a set of φ6 (+)RNAs with an internal deletions (sΔ$^+$ ssRNA) and a set of firefly luciferase mRNA variants (Makeyev and Bamford, 2001; Yang *et al.*, 2001). One of the sΔ$^+$ ssRNAs, designated Ø, contained the φ6-specific terminus ...CUCUCUCUCU-3′, whereas the other templates contained either A, C, G, or U as the 3′-terminal nucleotide. The efficiency of these templates for *in vitro* replication was found to decrease in the order C > G > A > Ø > U. A similar result was observed for the set of firefly luciferase mRNAs, where Ø ends with ...CCCAAGCUUA-3′. Luciferase mRNAs with longer 3′ termini were also tested for *in vitro* replication. The results showed that the relative efficiency of replication decreased in the order of CCC-3′ > UUUUUUUCC-3′ > CC-3′ > UC-3′ > AC-3′ ≈C-3′ > UCUCUCUCU-3′ > Ø (Makeyev and Bamford, 2000b). These results suggest that RNA templates containing one or more 3′-terminal cytosines are most efficiently replicated. Although there are differences in the template preferences between the cystovirus RdRPs of φ6, φ8 and φ13, (+)RNAs with one or more 3′-terminal cytosines are usually favoured. The observation that one to three 3′-terminal nucleotides are sufficient for stimulation of the φ6 RdRP replication reaction indicates that replication efficiency is controlled at the initiation step. Furthermore, RdRPs seem to prefer pyrimidine-rich 3′-terminal initiation sites, as a terminal C is more efficient that a terminal U. These findings correlate with template requirements of other RdRPs that also initiate efficiently on cytosine-rich 3′-termini, such as Qβ replicase and the polymerase subunits of HCV and bovine viral diarrhoea virus (BVDV) (Blumenthal and Carmichael, 1979; Kao *et al.*, 1999; Sun *et al.*, 2000; Zhong *et al.*, 2000a).

Besides the 3′-terminal sequence of the template, its secondary structure also affects the efficiency of RNA synthesis. Adding a stable hairpin-tetra loop structure to the sΔ$^+$ ssRNA decreased the replication efficiency of the template at least one order of magnitude when compared to the wild-type sΔ$^+$ ssRNA template (Laurila *et al.*, 2002). Additionally, some highly structured RNAs, such as tRNA, ribosomal RNAs of *Escherichia coli*, or genomic RNAs of bacteriophage MS2 and tobacco mosaic virus are poor templates for RdRPs (Makeyev and Grimes, 2004). These results indicate that, while the φ6 RdRP is a non-specific replicase, the efficiency of replication depends on the 3′ terminus of the template.

φ6 RNA transcription

Cystoviruses catalyze transcription from dsRNA templates using a semi-conservative mechanism (Figure 14.1). Initiation of transcription *in vivo* occurs at the 3′ termini of the negative strands of the dsRNA segments and produces full-length (+)RNA strands (Emori *et al.*, 1980; Partirdge *et al.*, 1979; Usala *et al.*, 1980). No negative strand synthesis is detected during particle-directed transcription (Pagratis and Revel, 1990). The isolated φ6 RdRP shows the same characteristics and is able to distinguish between positive and negative strand initiation signals, producing predominantly (+)RNAs *in vitro*. In the case of φ6-specific dsRNA templates, the transcription reaction is less efficient than replication *in vitro*. The lack of efficient ϕ6 RdRP-directed transcription is likely a result of inefficient initiation from dsRNA templates. Strand-displacement assays indicate that the ϕ6 RdRP needs the 3′ terminus of the template strand in a single-stranded form to initiate transcription (Makeyev and Bamford, 2000b). In contrast, ϕ6 transcription is very efficient *in vivo*, suggesting that active 'unzipping' of the dsRNA termini occurs effectively inside the PC. Interestingly, the melting temperatures of the positive strand initiation sites of all three genomic segments are notably lower than the negative strand initiation sites, suggesting that opening of the dsRNA might help regulate transcription. When assembled with P1, P4, and P7 into the PC, the ϕ6 RdRP can catalyze multiple rounds of transcription (Poranen *et al.*, 2001). However, adding purified P4 and/or P7 into the ϕ6 RdRP *in vitro* transcription reaction fails to increase the level of RNA synthesis, indicating that a correct architecture of the PC is required for efficient initiation of transcription (Makeyev and Grimes, 2004).

Nonetheless, after the initiation step ϕ6 RdRP-directed transcription is highly processive *in vitro* and the purified RdRP is capable of normal elongation, producing full-length transcripts at a rate of ~30 nt/s (Makeyev and Bamford, 2000b). A similar elongation rate has been reported for the NC system (Usala *et al.*, 1980). Hence, the ϕ6 RdRP does not seem to require additional proteins to unwind duplex RNA and displace the non-template strand during elongation, which is the case for two other viruses, PV and alfalfa mosaic virus (Cho *et al.*, 1993; de Graaff *et al.*, 1995). For dsRNA viruses that use a semi-conservative transcription mechanism, RNA helicases may be required for rewinding the dsRNA genome (Butcher *et al.*, 2001). Further insight into the role of the ϕ6 helicase protein (P4) is given by Mancini and Tuma elsewhere in this book.

Isolated cystovirus RdRPs are relatively non-specific transcriptases, capable of using heterologous templates including dsRNA from BTV, cypovirus, rotavirus, and L-A virus (Makeyev and Bamford, 2000b). However, the ϕ6 RdRP does display increased specificity when incubated together with homologous dsRNA templates. Makeyev and Bamford (2000b) showed that the S- and M-segments of ϕ6 and those from the closely related ϕ13 are transcribed more efficiently than the L segment. Furthermore, the ϕ6 RdRP favours synthesis of positive-sense, but not negative-sense, copies of these segments. These findings concur with the distribution of RNA products in particle-directed transcription (van Dijk *et al.*, 1995), suggesting that the ϕ6 RdRP controls RNA metabolism through selective initiation at transcriptional promoters (Makeyev and Bamford, 2000b). In addition to template sequence preferences, factors such as RNA secondary structure might also contribute to the selection of the correct transcriptional initiation site by the ϕ6 RdRP (Yang *et al.*, 2003a).

Structure of the ϕ6 RNA-dependent RNA polymerase

A large number of high-resolution crystal structures have been determined for viral RdRPs, in both the apo form and in complexes with substrates. The structures solved to date, including RdRPs of bacteriophage ϕ6 (Butcher *et al.*, 2001; Salgado *et al.*, 2004), HCV (Ago *et al.*, 1999; Bressanelli *et al.*, 1999, 2002; Lesburg *et al.*, 1999), BVDV (Choi *et al.*, 2004, 2006), PV (Hansen *et al.*, 1997; Thompson and Peersen, 2004), Norwalk virus (Ng *et al.*, 2004), human rhinovirus (Love *et al.*, 2004), rabbit hemorrhagic fever virus (RHDV; Ng *et al.*, 2002), reovirus (Tao *et al.*, 2002), and foot-and-mouth disease virus (FMDV; (Ferrer-Orta *et al.*, 2004), show a canonical 'right hand' structure with finger, palm and thumb sub-domains. In contrast to the 'open right hand' architecture of many other polymerases, all known viral RdRPs display a closed 'cupped right hand' conformation.

Among the viral RdRPs, the palm subdomain is the most conserved structural feature, whereas structural differences are more evident in the fingers and thumb subdomains. Very little primary amino acid sequence identity is seen between viral RdRPs, although there are clear structural homologies extending well beyond viral families. The overall structural similarity and the conservation of secondary and tertiary structure elements in the palm and thumb subdomain of polymerases of the families *Picorna-*, *Flavi-*, *Cysto-*, and *Retroviridae* suggest a possible evolutionary link, such as a common ancestor (Butcher *et al.*, 2001; Koonin and Dolja, 1993; Makeyev and Saldago, 2006).

The structure of the ϕ6 RdRP was determined by Butcher *et al.* (2001) in its apo form, in complex with a DNA template, with a bound GTP, and in an initiation-competent state. The ϕ6 RdRP displays two elaborations to the basic 'cupped right hand' architecture. The first is an N-terminal extension composed of six regions that strap the fingertips and thumb sub-domains together. The second elaboration consists of 64 residues from the C-terminus, which forms a loop that protrudes into the central cavity of the enzyme. This loop is referred to as the initiation platform and is important for primer-independent initiation (to be discussed below). Overall, the polymerase appears as a compact spherical molecule with an internally located active site and two positively charged tunnels that allow access of the RNA template (template tunnel) and NTP substrates (substrate tunnel) into the polymerase interior. The template tunnel is wide enough to accommodate single- but not double-stranded nucleic acid, and the distance from the surface of the RdRP to the active site can be spanned by a 5 nt long oligonucleotide. The plough-shaped edge of the template tunnel is adjacent to a positively charged patch on the polymerase surface. The plough area is believed to facilitate the separation of dsRNA during transcription and to guide the template strand into the tunnel while the non-coding strand slides over a patch on the polymerase surface. The substrate tunnel is wide enough to ensure an adequate diffusion of NTPs to the active site. This tunnel is positively charged and interacts with the incoming negatively charged NTPs (Butcher *et al.*, 2001; Salgado *et al.*, 2004).

Many polymerases use a two-metal catalysis of nucleotidyl transfer (Steitz, 1998), with the metal ions being coordinated by catalytic aspartates. For the ϕ6 RdRP, the catalytic aspartates are positioned at residues 324 and 453 in palm sub-domain. Interestingly, the ϕ6 RdRP apoenzyme binds metal ions at a distinct position some 6 Å away from these catalytic residues. However, in the initiation complex, both metal ions are present at the correct positions in proximity to the active site for catalysis to occur. The observation that the metal ions in the apoenzyme structure are 6 Å away from the active site suggests that the interactions between the catalytic metals and the ϕ6 RdRP must be stabilized by correctly positioned NTPs (Butcher *et al.*, 2001).

Initiation of RNA-dependent RNA synthesis

Correct initiation and adequate processivity of RNA synthesis are essential for maintaining the integrity of the viral genome. Initiation of the RNA synthesis can also function as an important regulatory step in the viral life cycle and might determine the RNA specificity of the RdRP. From a biochemical point of view, initiation from the terminal end of the genome is often the rate limiting step of RNA synthesis. A competent initiation complex consists of a polymerase, divalent cations, NTP(s) and a template with or without a primer. There exist two principally distinct mechanisms for the initiation of RNA synthesis: (i) primer-independent and (ii) primer-dependent. Most RdRPs initiate RNA synthesis primer-independently (*de novo*) (Kao *et al.*, 2001; van Dijk *et al.*, 2004). In this case, the 3′OH group of the first NTP acts as an acceptor for the second nucleotide. In contrast, primer-dependent enzymes add nucleotides to the 3′OH group of a polynucleotide or protein primer. For both mechanisms, nucleotidyl transfer is then repeated with the subsequent NTPs giving rise to an elongated RNA product. Both of these initiation systems have been widely studied using biochemical and structural methods (Appleby *et al.*, 2005; Choi *et al.*, 2004; Ferrer-Orta *et al.*, 2006b; Kao *et al.*, 2001; van Dijk *et al.*, 2004 and references therein). Based on the available information, a molecular mechanism for *de novo* initiation of ϕ6 RNA synthesis is presented below (Butcher *et al.*, 2001; Salgado *et al.*, 2004).

De novo initiation of the RNA synthesis

The defining features of *de novo* RNA synthesis are that no information of the viral genome is lost and that additional viral- or host-encoded proteins for priming are unnecessary. Nonetheless, Watson–Crick interactions between the template and two incoming NTPs (D1 and D2) might be insufficient to keep them correctly positioned at the RdRPs active site. Therefore, *de novo* initiation requires specific molecular interactions to stabilize the initiation complex. Past the initiation step, when the first phosphodiester bond of the daughter strand is formed, full-length dsRNA products are synthesized in most cases. However, abortive initiation has been reported for viral RdRPs, such as those of turnip crinkle carmovirus, reovirus), and rotavirus (Chen and Patton, 2000; Farsetta *et al.*, 2000; Nagy and Simon, 1997; Yamakawa *et al.*, 1981).

Molecular mechanism for initiating *de novo* RNA-dependent RNA polymerization

The structures of the ϕ6 RdRP complexed with NTPs and/or template suggest an atomic level model for *de novo* initiation of RNA polymerization (Butcher *et al.*, 2001; Salgado *et al.*, 2004). Particularly, the data are consistent with the following sequence of events that lead to the formation of a complete initiation complex (Figure 14.3). First, the template tunnel of the RdRP accommodates a short, single-stranded oligonucleotide which docks snugly into the specificity pocket (S pocket), well past the catalytic site of the enzyme. Next, the 3′ cytidine of the template (T1) is specifically recognized and buried into the S pocket. The T1 cytidine of the RNA template is rotated 180 degrees (compared to a T1 of a DNA template) and additional template-protein contacts are formed between the hydroxyl group of the ribose sugar of T1 and the template tunnel (Salgado *et al.*, 2004). These template-protein contacts may provide the basis for the preference of the ϕ6 RdRP for RNA rather than DNA templates. In the substrate tunnel (site I), the first nucleotide of the daughter strand (GTP) is coordinated by arginine residues within the RdRP fingertips (Figure 14.3; I–IV). Following GTP stabilization, the substrate tunnel is occupied by incoming NTPs, presumably in rapid exchange. Thereafter, the first GTP (D2) moves to the initiation platform (site P), which consists of a QYKW loop (residues 629–632).

Here, the GTP stacks against Y630 of site P, and Watson–Crick base-pairs are formed between D2 and the second terminal cytidine (T2) of the template strand (Figure 14.3; V). As a result of this specific bonding, the 3′ end of the template ratchets backwards from the S pocket (Figure 14.3; VI) allowing the second GTP (D1) to occupy site P (Figure 14.3; VII). The stacking interactions between D1 and Y630 of site P are further stabilized, causing the initiation complex to favour the formation of the first phosphodiester bond. A nucleotidyl-transferase reaction is then catalyzed by two Mg^{2+}, complexed with the incoming NTPs, and a pyrophosphate is released as a by-product. The third NTP can now enter the active site and chain elongation can proceed. When the nascent dsRNA reaches a critical length, the C-terminal domain of the RdRP is repositioned so that the extending dsRNA product can exit from the contained active site of the RdRP. This conformational rearrangement might be facilitated by contacts between the phosphate of the GTP at site P and the structural Mn^{2+} ion.

Structural studies of the active initiation complexes of the HCV and reovirus RdRPs lend further insight into the general mechanism of *de novo* RNA polymerization and support the view that the ϕ6 model could be valid for other RdRPs (O'Farrell *et al.*, 2003; Tao *et al.*, 2002; Makeyev and Grimes, 2004; van Dijk *et al.*, 2004). Importantly, structural

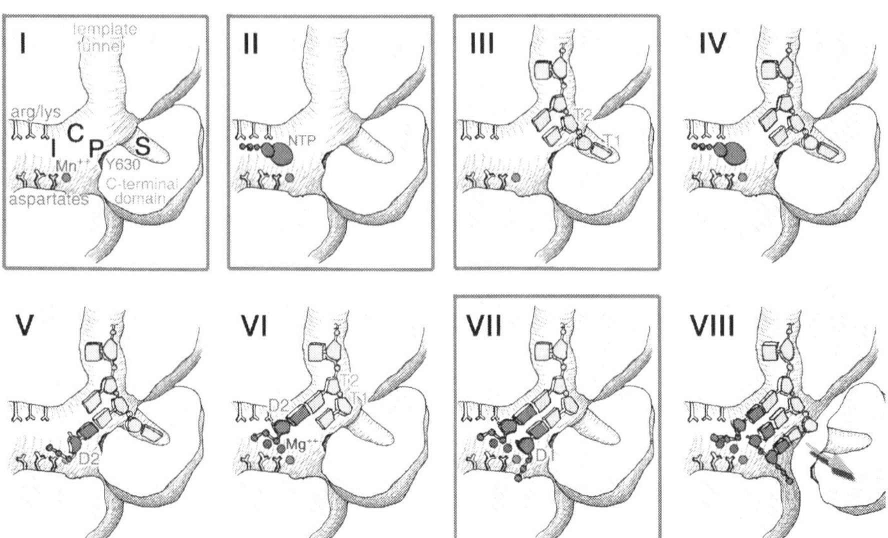

Figure 14.3 Schematic model for the *de novo* initiation mechanism of ϕ6 RdRP. (I) The apoenzyme structure is shown with bound manganese. The binding sites are marked with black letters. (II) Site I is occupied by NTP molecule. (III) The oligonucleotide template is bound in the template tunnel fitting snugly in site S. (IV) Bound template and NTP at site I. (V) GTP molecule binds site P. (VI) The template ratchets backwards, moving the T1 of the template away from the pocket S. (VII) The second GTP binds at site P. The initiation platform (P) forms stacking interactions with the incoming nucleotides and locks the initiation complex into its final form. (VIII) Catalysis has occurred, releasing pyrophosphate. Displacement of the C-terminal domain is necessary for the elongation to proceed. The boxes in the figure indicate experimental results. Adapted by permission from Macmillan Publishers Ltd: Nature, Butcher *et al.*, copyright 2001.

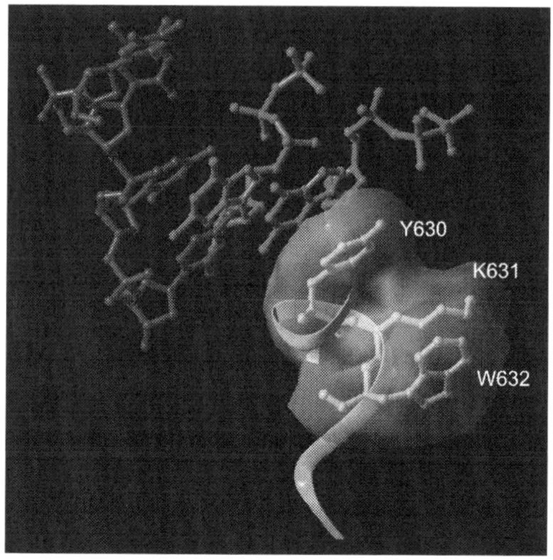

Figure 14.4 A close-up of the initiation complex of ϕ6 RdRP, showing the backbone of the ϕ6 polymerase, the oligonucleotide template, and GTPs. The residues Y630, K631, and W632 on the C-terminal initiation platform loop are labelled. The aromatic ring of Y630 stacks against one of the incoming GTP. The GTP molecules are base-paired with the template 3′end. Note the helical path of stacking interactions including the two GTPs, Y630 and W632. The figure was produced using the program INSIGHT II (Molecular Simulation Inc., San Diego, CA). Modified from Laurila *et al.* (2002).

elements equivalent to the initiation platform of the ϕ6 RdRP (Figure 14.4) have been reported for several primer-independent RdRPs. In the HCV RdRP, a β-hairpin protruding from the thumb sub-domain toward the active site at the base of the palm sub-domain (Bressanelli *et al.*, 1999; Lesburg *et al.*, 1999) along with a non-catalytic, C-terminal hydrophobic pocket, form a proposed GTP-stabilizing platform (Hong *et al.*, 2001; Leveque *et al.*, 2003; Ranjith-Kumar *et al.*, 2003b). In a similar manner, the BVDV RdRP has a 'β-thumb' region, which may be similar to the ϕ6 initiation platform (Choi *et al.*, 2004). In addition, a unique insertion in the palm sub-domain has been found in the reovirus RdRP. This insertion is thought to form a special priming loop and is proposed to stabilize the initiation complex (Tao *et al.*, 2002). These observations suggest that many RdRPs have a common stabilizing initiation platform –like element and provide evidence for a possible evolutionary link between positive-strand and dsRNA viral RdRPs.

In addition to the clear structural similarities, HCV, BVDV and ϕ6 RdRPs show notable biochemical resemblances. All three RdRPs are stimulated by divalent cations (Makeyev and Bamford, 2000a; Ranjith-Kumar *et al.*, 2002) and require elevated concentrations of the initiatory purine nucleotides for stabilization of the initiation complex (Kao *et al.*, 1999; Kao and Sun, 1996; Laurila *et al.*, 2002; Ranjith-Kumar *et al.*, 2002, 2003b; Zhong *et al.*, 2000b). Furthermore, the base-stacking interactions between the initiation platform and the incoming NTPs, which stabilize the initiation complex, could well become a paradigm for many other RdRPs (Makeyev and Grimes, 2004). Notably, many RdRPs that initiate RNA synthesis *de novo* use the priming nucleotide GTP, which is known to have high stacking propensity. Intriguingly, the presence of a structural Mn^{2+}

as a cofactor for RNA synthesis has not been appreciated until very recently. Mn^{2+} ions stimulate a number of primer-independent RdRPs including those of φ6 (Laurila *et al.*, 2005; Makeyev and Bamford, 2000a) and related cystoviral φ8, φ12 and φ13 (Yang *et al.*, 2001, 2003a, b), HCV (Alaoui-Lsmaili *et al.*, 2000; Ferrari *et al.*, 1999; Luo *et al.*, 2000; Zhong *et al.*, 2000b), BVDV (Ranjith-Kumar *et al.*, 2002), classical swine fever virus (Yi *et al.*, 2003), bacteriophage Qβ (Blumenthal, 1980; Blumenthal and Carmichael, 1979), and GB virus-B (Ranjith-Kumar *et al.*, 2003a). In the crystal structure of the φ6 RdRP, Mn^{2+} interacts with three amino acids, including one of the two catalytic aspartates of the active site (Butcher *et al.*, 2001; Salgado *et al.*, 2004). This data suggests that Mn^{2+} might be present in RdRPs under physiological conditions. In fact, Mn^{2+} has been detected also in crystal structures of HCV, RHDV, and reovirus RdRPs (Bressanelli *et al.*, 2002; O'Farrell *et al.*, 2003; Ng *et al.*, 2002; Tao *et al.*, 2002). Hence, the binding of the Mn^{2+} in close proximity to the active site of these enzymes might explain the observed stimulation of the RNA synthesis.

De novo versus primer-dependent initiation of the RNA synthesis

Compared with primer-independent RdRPs, primer-dependent RdRPs, such as FMDV, PV and RHDV RdRPs have significantly smaller thumb sub-domains, wider template tunnels, and large, exposed active sites (Arnold *et al.*, 1999; Ferrer-Orta *et al.*, 2004, 2006b; Lopez Vazquez *et al.*, 2001). The absence of initiation platforms in these RdRPs is consistent with their capability of utilizing protein primers for the initiation of the RNA synthesis (Ferrer-Orta *et al.*, 2006b; Machin *et al.*, 2001; Nayak *et al.*, 2005; Paul *et al.*, 1998). Recently, the crystal structure of the FMDV RdRP complexed with RNA, divalent cations and protein primer VPg, showed how the central cleft is wide enough to accommodate the bulky priming peptide during initiation of the RNA synthesis (Ferrer-Orta *et al.*, 2006a). In contrast, for the primer-independent φ6, HCV, and BVDV RdRPs the C-terminal extensions fill most of the active site cavity (Bressanelli *et al.*, 1999; Butcher *et al.*, 2001; Choi *et al.*, 2004; Lesburg *et al.*, 1999; Leveque *et al.*, 2003).

The role of the initiation platform of the φ6 RdRP has been studied using biochemical and structural approaches (Laurila *et al.*, 2002, 2005). The findings of these studies suggest that the protruding polypeptide loops observed in the φ6, HCV, BVDV and reovirus RdRPs are related to *de novo* initiation (Bressanelli *et al.*, 1999; Butcher *et al.*, 2001; Choi *et al.*, 2004; Lesburg *et al.*, 1999; Leveque *et al.*, 2003; Tao *et al.*, 2002). To investigate the functions of the φ6 initiation platform, either the tip of the loop (Figure 14.4) was changed (Y630 to A or F) or the whole initiation platform was minimized by changing YKW(630–632) to a smaller and flexible GSG or YKWQ(629–632) to SG. The φ6 RdRPs with mutations at Y630 were significantly less efficient in replicating (+)RNA templates than the wild-type φ6 RdRP. Interestingly, the GSG and SG mutants with smaller initiation platform loops retained their RNA synthesis activity, but only with certain (+)RNA templates. When the dsRNA polymerization products were analyzed, it was observed that these mutant RdRPs initiated RNA synthesis predominantly by back-priming. Back-priming is a primer-dependent mechanism where the 3′ end of the RNA template folds back, allowing the 3′ hydroxyl to serve as a primer for the RNA synthesis. Thus, the engineered changes in the initiation platform reverted the *de novo* initiation mode of the φ6 RdRP to the back-priming mode. Similar results have been observed for the HCV RdRP *in vitro* (Hong *et al.*, 2001; Leveque *et al.*, 2003; Ranjith-Kumar *et al.*, 2002, 2003b) further supporting the crucial role of the C-terminal, protruding loops in *de novo* RNA synthesis of RdRPs. Notably, the *de novo* initiation mode of the SG mutant was partially rescued in the presence of elevated concentration of Mn^{2+}, which is known to increase the affinity of NTPs to viral RdRPs. For structural studies, the SG initiation platform mutant was used because sufficient yields of purified protein could be produced (Laurila *et al.*, 2005). The overall structure of this mutant RdRP, even the catalytic active site, did not appear to be perturbed. In contrast, as indicated by the poorly defined density around the altered initiation platform, the whole C-terminal domain of the initiation platform was destabilized. One reason for the observed instability might be the loss of the bulky W632 side-chain, which in the wild-type ϕ6 RdRP makes bridging contacts with a region of the palm sub-domain (residues 305–308). It is possible that a truncation of the HCV RdRP C terminus of could produce a similar disruption to the initiation platform as observed for the SG mutant. This disruption might explain the preference of the HCV RdRP for the back-priming initiation mode *in vitro* (Leveque *et al.*, 2003). It is important to note that the native secondary structure at the 3′ end of ϕ6 (+)RNA templates does not support the formation of back-primed intermediates in RNA synthesis (Laurila *et al.*, 2002). Instead, the terminal sequences are folded in a tRNA-like structure, with the five 3′-proximal nucleotides in a single-stranded form (Mindich *et al.*, 1994). These termini are sufficient to span the template channel of the polymerase (Butcher *et al.*, 2001), but cannot loop back.

In conclusion, the C-terminal initiation platform of the ϕ6 RdRP has three distinct functions during the *de novo* initiation of viral RNA synthesis: (i) stabilizing the initiation complex by interacting with the initiatory nucleotides, (ii) preventing the back-priming initiation mode by physically reducing the volume of the active site, and (iii) controlling the switch from initiation to elongation (Butcher *et al.*, 2001; Laurila *et al.*, 2002, 2005). Overall, the selection between *de novo* and primer-dependent initiation seems to be accomplished by restricting access to the template-binding site (Ferrer-Orta *et al.*, 2006b).

RNA-dependent RNA polymerases of other dsRNA viruses

For dsRNA viruses, a biochemical mechanism for the *de novo* initiation of RNA synthesis for has been reported for cystoviruses and BTV (Boyce *et al.*, 2004; Makeyev and Bamford, 2000a, 2001). As described earlier, primer-independent viral RdRPs have specific structural elaborations that function to stabilize the initiation complex. For viruses belonging to the families of *Cystoviridae* and *Reoviridae*, such initiation platforms have been found in the crystal structures of the viral RdRPs. The priming loop of the reovirus RdRP and the initiation platform of the ϕ6 RdRP differ considerably from each other, both in sequence and in relative positioning in the enzyme structure (Butcher *et al.*, 2001; Tao *et al.*, 2002). This difference could indicate that the mechanism of *de novo* initiation might have arisen independently during the evolution of the different virus lineages (Makeyev and Saldago, 2006).

The *in vitro* RdRP activity of BTV, reovirus, ϕ6, and other cystoviral (ϕ8, ϕ12 and ϕ13) RdRPs is associated with a single, purified protein species (Boyce *et al.*, 2004; Makeyev and Bamford, 2000a; Starnes and Joklik, 1993; Yang *et al.*, 2001, 2003b). On the contrary, the rotavirus RdRP is completely dependent upon the inner capsid protein VP2 for its catalytic activity (Chen *et al.*, 1994; Patton, 1996; Patton *et al.*, 1997). However, it should be noted that the reovirus RdRP is capable of synthesizing only short oligonucleotides with a strong preference for template cytidines and that only low levels of activity are associated with the BTV RdRP (Boyce *et al.*, 2004; Starnes and Joklik, 1993; Tao *et al.*, 2002). These results suggest that viral- or host-encoded proteins might be necessary for efficient RNA synthesis for these viruses. In contrast, the specific activity of purified ϕ6 RdRP is high. In the ϕ6 RdRP *in vitro* replication reactions, full-length dsRNA products can be readily detected in agarose gels using ethidium bromide staining. For ϕ6 and related ϕ8, viral (+)RNAs are packaged into preassembled, empty PCs (Bamford and Mindich, 1980; Kainov *et al.*, 2003; Poranen *et al.*, 2001) and, thereafter, the (+)RNAs are converted into the dsRNA segments (Mindich, 1999; Poranen *et al.*, 2005a). In the *Reoviridae* family, ϕ6-like packaging into preformed particles, prior to the genome replication, has not been reported. On the other hand, free dsRNA has not been observed inside infected cells. Therefore, packaging, replication, and particle assembly are thought to be closely coupled processes (Antczak and Joklik, 1992; Jayaram *et al.*, 2004; Patton *et al.*, 2006; Roy and Noad, 2006). Thus, for the *Reoviridae*, it might be advantageous to maintain close interactions between the RdRP and inner capsid core proteins (Patton *et al.*, 2006; Tao *et al.*, 2002), and thereby prevent efficient synthesis of dsRNA with the RdRP alone.

Conclusions

The ϕ6 RdRP is the best characterized polymerase of any dsRNA virus. Careful studies of this enzyme have yielded valuable insights into the RNA polymerization machinery of dsRNA viruses, especially regarding the *de novo* initiation mechanism of RNA synthesis (Butcher *et al.*, 2001; Laurila *et al.*, 2002, 2005; Makeyev and Bamford, 2000a, b; Salgado *et al.*, 2004). The ϕ6 RdRP has several advantages as a model enzyme: (i) it is a stable, soluble monomer, (ii) has high specific activity, and (iii) large amounts of the ϕ6 RdRP can be efficiently produced in an *Escherichia coli* expression system. Structural analyses have provided a detailed model of the requirements for successful initiation of ϕ6 RNA synthesis (Butcher *et al.*, 2001; Salgado *et al.*, 2004). In particular, using the ϕ6 system, we now know that the C-terminal initiation platform is critical for the stabilization of the initiation complex (Butcher *et al.*, 2001; Laurila *et al.*, 2002, 2005). Still, further structural and biochemical studies are required to illuminate the intricate steps preceding and succeeding the formation of the initiation ϕ6 complex. The information gathered from these studies, will further deepen our understanding of the *de novo* initiation mechanism of RNA synthesis. Since the *de novo* initiation mode of RNA synthesis is utilized by a vast number of animal and human viral pathogens, the knowledge gained by studying the ϕ6 RdRP could help elucidate the functions of other viral RdRPs. Moreover, the non-specific nature of the ϕ6 RdRP enables the enzyme to be harnessed for biotechnological applications. Indeed, the ϕ6 polymerase is the first viral RdRP which is commercially available, opening up exiting possibilities in e.g. the field of RNA interference (Aalto et al., 2007).

References

Aalto, A.P., Sarin, L.P., van Dijk, A.A., Saarma, M., Poranen, M.M., Arumäe, U., and Bamford, D.H. (2007) Large scale production of dsRNA and siRNA pools for RNA interference utilizing bacteriophage ϕ6 RNA-dependent RNA polymerase. RNA *13*, 422–29.

Ackermann, M., and Padmanabhan, R. (2001). De novo synthesis of RNA by the dengue virus RNA-dependent RNA polymerase exhibits temperature dependence at the initiation but not elongation phase. J. Biol. Chem. *276*, 39926–37.
Ago, H., Adachi, T., Yoshida, A., Yamamoto, M., Habuka, N., Yatsunami, K., and Miyano, M. (1999). Crystal structure of the RNA-dependent RNA polymerase of hepatitis C virus. Structure Fold. Des. *7*, 1417–26.
Alaoui-Lsmaili, M.H., Hamel, M., L'Heureux, L., Nicolas, O., Bilimoria, D., Labonte, P., Mounir, S., and Rando, R.F. (2000). The hepatitis C virus NS5B RNA-dependent RNA polymerase activity and susceptibility to inhibitors is modulated by metal cations. J. Hum. Virol. *3*, 306–16.
Antczak, J.B., and Joklik, W.K. (1992). Reovirus genome segment assortment into progeny genomes studied by the use of monoclonal antibodies directed against reovirus proteins. Virology. *187*, 760–76.
Appleby, T.C., Luecke, H., Shim, J.H., Wu, J.Z., Cheney, I.W., Zhong, W., Vogeley, L., Hong, Z., and Yao, N. (2005). Crystal structure of complete rhinovirus RNA polymerase suggests front loading of protein primer. J. Virol. *79*, 277–88.
Arnold, J.J., Ghosh, S.K., and Cameron, C.E. (1999). Poliovirus RNA-dependent RNA polymerase (3D(pol)). Divalent cation modulation of primer, template, and nucleotide selection. J. Biol. Chem. *274*, 37060–9.
Bamford, D.H., and Mindich, L. (1980). Electron microscopy of cells infected with nonsense mutants of bacteriophage phi 6. Virology. *107*, 222–8.
Bamford, D.H., Palva, E.T., and Lounatmaa, K. (1976). Ultrastructure and life cycle of the lipid-containing bacteriophage phi 6. J. Gen. Virol. *32*, 249–59.
Behrens, S.E., Tomei, L., and De Francesco, R. (1996). Identification and properties of the RNA-dependent RNA polymerase of hepatitis C virus. EMBO J. *15*, 12–22.
Blumenthal, T. (1980). Q beta replicase template specificity: different templates require different GTP concentrations for initiation. Proc. Natl Acad. Sci. USA *77*, 2601–5.
Blumenthal, T., and Carmichael, G.G. (1979). RNA replication: function and structure of Qbeta-replicase. Annu. Rev. Biochem. *48*, 525–48.
Boyce, M., Wehrfritz, J., Noad, R., and Roy, P. (2004). Purified recombinant bluetongue virus VP1 exhibits RNA replicase activity. J. Virol. *78*, 3994–4002.
Bressanelli, S., Tomei, L., Rey, F.A., and De Francesco, R. (2002). Structural analysis of the hepatitis C virus RNA polymerase in complex with ribonucleotides. J. Virol. *76*, 3482–92.
Bressanelli, S., Tomei, L., Roussel, A., Incitti, I., Vitale, R.L., Mathieu, M., De Francesco, R., and Rey, F.A. (1999). Crystal structure of the RNA-dependent RNA polymerase of hepatitis C virus. Proc. Natl Acad. Sci. USA *96*, 13034–9.
Butcher, S.J., Dokland, T., Ojala, P.M., Bamford, D.H., and Fuller, S.D. (1997). Intermediates in the assembly pathway of the double-stranded RNA virus phi6. EMBO J. *16*, 4477–87.
Butcher, S.J., Grimes, J.M., Makeyev, E.V., Bamford, D.H., and Stuart, D.I. (2001). A mechanism for initiating RNA-dependent RNA polymerization. Nature *410*, 235–40.
Chen, D., and Patton, J.T. (2000). De novo synthesis of minus strand RNA by the rotavirus RNA polymerase in a cell-free system involves a novel mechanism of initiation. RNA *6*, 1455–67.
Chen, D., Zeng, C.Q., Wentz, M.J., Gorziglia, M., Estes, M.K., and Ramig, R.F. (1994). Template-dependent, *in vitro* replication of rotavirus RNA. J. Virol. *68*, 7030–9.
Cho, M.W., Richards, O.C., Dmitrieva, T.M., Agol, V., and Ehrenfeld, E. (1993). RNA duplex unwinding activity of poliovirus RNA-dependent RNA polymerase 3Dpol. J. Virol. *67*, 3010–8.
Choi, K.H., Gallei, A., Becher, P., and Rossmann, M.G. (2006). The structure of bovine viral diarrhea virus RNA-dependent RNA polymerase and its amino-terminal domain. Structure *14*, 1107–13.
Choi, K.H., Groarke, J.M., Young, D.C., Kuhn, R.J., Smith, J.L., Pevear, D.C., and Rossmann, M.G. (2004). The structure of the RNA-dependent RNA polymerase from bovine viral diarrhea virus establishes the role of GTP in *de novo* initiation. Proc. Natl Acad. Sci. USA *101*, 4425–30. Epub 2004 Mar 19.
Cohen, J. (1975). Ribonucletic acid polymerase activity in purified infectious pancreatic necrosis virus of trout. Biochem. Biophys. Res. Commun. *62*, 689–95.
de Graaff, M., Houwing, C.J., Lukacs, N., and Jaspars, E.M. (1995). RNA duplex unwinding activity of alfalfa mosaic virus RNA-dependent RNA polymerase. FEBS Lett. *371*, 219–22.
de Haas, F., Paatero, A.O., Mindich, L., Bamford, D.H., and Fuller, S.D. (1999). A symmetry mismatch at the site of RNA packaging in the polymerase complex of dsRNA bacteriophage phi6. J. Mol. Biol. *294*, 357–72.

Duijsings, D., Kormelink, R., and Goldbach, R. (2001). In vivo analysis of the TSWV cap-snatching mechanism: single base complementarity and primer length requirements. EMBO J. *20,* 2545–52.
Dykxhoorn, D.M., Novina, C.D., and Sharp, P.A. (2003). Killing the messenger: short RNAs that silence gene expression. Nat. Rev. Mol. Cell Biol. *4,* 457–67.
Emori, Y., Iba, H., and Okada, Y. (1980). Semi-conservative transcription of double-stranded RNA catalyzed by bacteriophage phi 6 RNA polymerase. J. Biochem. (Tokyo) *88,* 1569–75.
Ewen, M.E., and Revel, H.R. (1988). *In vitro* replication and transcription of the segmented double-stranded RNA bacteriophage phi 6. Virology. *165,* 489–98.
Ewen, M.E., and Revel, H.R. (1990). RNA-protein complexes responsible for replication and transcription of the double-stranded RNA bacteriophage phi 6. Virology *178,* 509–19.
Farsetta, D.L., Chandran, K., and Nibert, M.L. (2000). Transcriptional activities of reovirus RNA polymerase in recoated cores. Initiation and elongation are regulated by separate mechanisms. J. Biol. Chem. *275,* 39693–701.
Ferrari, E., Wright-Minogue, J., Fang, J.W., Baroudy, B.M., Lau, J.Y., and Hong, Z. (1999). Characterization of soluble hepatitis C virus RNA-dependent RNA polymerase expressed in Escherichia coli. J. Virol. *73,* 1649–54.
Ferrer-Orta, C., Arias, A., Agudo, R., Perez-Luque, R., Escarmis, C., Domingo, E., and Verdaguer, N. (2006a). The structure of a protein primer-polymerase complex in the initiation of genome replication. EMBO J. *25,* 880–8. Epub 2006 Feb 2.
Ferrer-Orta, C., Arias, A., Escarmis, C., and Verdaguer, N. (2006b). A comparison of viral RNA-dependent RNA polymerases. Curr. Opin. Struct. Biol. *16,* 27–34. Epub 2005 Dec 20.
Ferrer-Orta, C., Arias, A., Perez-Luque, R., Escarmis, C., Domingo, E., and Verdaguer, N. (2004). Structure of foot-and-mouth disease virus RNA-dependent RNA polymerase and its complex with a template-primer RNA. J. Biol. Chem. *279,* 47212–21. Epub 2004 Aug 03.
Fujimura, T., Esteban, R., and Wickner, R.B. (1986). *In vitro* L-A double-stranded RNA synthesis in virus-like particles from Saccharomyces cerevisiae. Proc. Natl Acad. Sci. USA *83,* 4433–7.
Fujimura, T., and Wickner, R.B. (1988). Replicase of L-A virus-like particles of Saccharomyces cerevisiae. *In vitro* conversion of exogenous L-A and M1 single-stranded RNAs to double-stranded form. J. Biol. Chem. *263,* 454–60.
Gottlieb, P., Metzger, S., Romantschuk, M., Carton, J., Strassman, J., Bamford, D.H., Kalkkinen, N., and Mindich, L. (1988). Nucleotide sequence of the middle dsRNA segment of bacteriophage phi 6: placement of the genes of membrane-associated proteins. Virology *163,* 183–90.
Gottlieb, P., Qiao, X., Strassman, J., Frilander, M., and Mindich, L. (1994). Identification of the packaging regions within the genomic RNA segments of bacteriophage phi 6. Virology *200,* 42–7.
Gottlieb, P., Strassman, J., Frucht, A., Qiao, X.Y., and Mindich, L. (1991). *In vitro* packaging of the bacteriophage phi 6 ssRNA genomic precursors. Virology *181,* 589–94.
Hansen, J.L., Long, A.M., and Schultz, S.C. (1997). Structure of the RNA-dependent RNA polymerase of poliovirus. Structure 5, 1109–22.
Hong, Z., Cameron, C.E., Walker, M.P., Castro, C., Yao, N., Lau, J.Y., and Zhong, W. (2001). A novel mechanism to ensure terminal initiation by hepatitis C virus NS5B polymerase. Virology *285,* 6–11.
Huiet, L., Feldstein, P.A., Tsai, J.H., and Falk, B.W. (1993). The maize stripe virus major noncapsid protein messenger RNA transcripts contain heterogeneous leader sequences at their 5′ termini. Virology. *197,* 808–12.
Ikonen, T., Kainov, D., Timmins, P., Serimaa, R., and Tuma, R. (2003). Locating the minor components of double-stranded RNA bacteriophage phi6 by neutron scattering. J. Appl. Cryst. *36,* 525–529.
Jayaram, H., Estes, M.K., and Prasad, B.V. (2004). Emerging themes in rotavirus cell entry, genome organization, transcription and replication. Virus Res. *101,* 67–81.
Juuti, J.T., and Bamford, D.H. (1997). Protein P7 of phage phi6 RNA polymerase complex, acquiring of RNA packaging activity by *in vitro* assembly of the purified protein onto deficient particles. J. Mol. Biol. *266,* 891–900.
Juuti, J.T., Bamford, D.H., Tuma, R., and Thomas, G.J., Jr. (1998). Structure and NTPase activity of the RNA-translocating protein (P4) of bacteriophage phi 6. J. Mol. Biol. *279,* 347–59.
Kainov, D.E., Pirttimaa, M., Tuma, R., Butcher, S.J., Thomas, G.J., Jr., Bamford, D.H., and Makeyev, E.V. (2003). RNA packaging device of double-stranded RNA bacteriophages, possibly as simple as hexamer of P4 protein. J. Biol. Chem. *278,* 48084–91. Epub 2003 Sep 8.
Kakitani, H., Iba, H., and Okada, Y. (1980). Penetration and partial uncoating of bacteriophage phi 6 particle. Virology. *101,* 475–83.

Kao, C.C., Del Vecchio, A.M., and Zhong, W. (1999). De novo initiation of RNA synthesis by a recombinant flaviviridae RNA-dependent RNA polymerase. Virology *253*, 1–7.

Kao, C.C., Singh, P., and Ecker, D.J. (2001). De novo initiation of viral RNA-dependent RNA synthesis. Virology *287*, 251–60.

Kao, C.C., and Sun, J.H. (1996). Initiation of minus-strand RNA synthesis by the brome mosaicvirus RNA-dependent RNA polymerase: use of oligoribonucleotide primers. J. Virol. *70*, 6826–30.

Koonin, E.V., and Dolja, V.V. (1993). Evolution and taxonomy of positive-strand RNA viruses: implications of comparative analysis of amino acid sequences. Crit. Rev. Biochem. Mol. Biol. *28*, 375–430.

Laurila, M.R., Makeyev, E.V., and Bamford, D.H. (2002). Bacteriophage phi 6 RNA-dependent RNA polymerase: molecular details of initiating nucleic acid synthesis without primer. J. Biol. Chem. *277*, 17117–24.

Laurila, M.R., Salgado, P.S., Stuart, D.I., Grimes, J.M., and Bamford, D.H. (2005). Back-priming mode of phi6 RNA-dependent RNA polymerase. J. Gen. Virol. *86*, 521–6.

Lawton, J.A., Estes, M.K., and Prasad, B.V. (2000). Mechanism of genome transcription in segmented dsRNA viruses. Adv. Virus Res. *55*, 185–229.

Lesburg, C.A., Cable, M.B., Ferrari, E., Hong, Z., Mannarino, A.F., and Weber, P.C. (1999). Crystal structure of the RNA-dependent RNA polymerase from hepatitis C virus reveals a fully encircled active site. Nat. Struct. Biol. *6*, 937–43.

Leveque, V.J., Johnson, R.B., Parsons, S., Ren, J., Xie, C., Zhang, F., and Wang, Q.M. (2003). Identification of a C-terminal regulatory motif in hepatitis C virus RNA-dependent RNA polymerase: structural and biochemical analysis. J. Virol. *77*, 9020–8.

Lopez Vazquez, A.L., Martin Alonso, J.M., and Parra, F. (2001). Characterisation of the RNA-dependent RNA polymerase from Rabbit hemorrhagic disease virus produced in Escherichia coli. Arch. Virol. *146*, 59–69.

Love, R.A., Maegley, K.A., Yu, X., Ferre, R.A., Lingardo, L.K., Diehl, W., Parge, H.E., Dragovich, P.S., and Fuhrman, S.A. (2004). The crystal structure of the RNA-dependent RNA polymerase from human rhinovirus: a dual function target for common cold antiviral therapy. Structure (Camb) *12*, 1533–44.

Luo, G., Hamatake, R.K., Mathis, D.M., Racela, J., Rigat, K.L., Lemm, J., and Colonno, R.J. (2000). De novo initiation of RNA synthesis by the RNA-dependent RNA polymerase (NS5B) of hepatitis C virus. J. Virol. *74*, 851–63.

Machin, A., Martin Alonso, J.M., and Parra, F. (2001). Identification of the amino acid residue involved in rabbit hemorrhagic disease virus VPg uridylylation. J. Biol. Chem. *276*, 27787–92. Epub 2001 May 21.

Makeyev, E.V., and Bamford, D.H. (2000a). Replicase activity of purified recombinant protein P2 of double-stranded RNA bacteriophage phi6. EMBO J. *19*, 124–33.

Makeyev, E.V., and Bamford, D.H. (2000b). The polymerase subunit of a dsRNA virus plays a central role in the regulation of viral RNA metabolism. EMBO J. *19*, 6275–84.

Makeyev, E.V., and Bamford, D.H. (2001). Primer-independent RNA sequencing with bacteriophage phi6 RNA polymerase and chain terminators. RNA *7*, 774–81.

Makeyev, E.V., and Grimes, J.M. (2004). RNA-dependent RNA polymerases of dsRNA bacteriophages. Virus. Res. *101*, 45–55.

Makeyev, E.V., and Saldago, P. (2006). Structure function, and evolution of viral and cell-encoded RNA-dependent RNA polymerases. In Recent Advances in RNA Virus Replication, K.L. Hefferon, ed. (Kerala, India: Research Signpost, T.C. 37/*661*(2), Fort Post Office, Trivandrum-695023, Kerala, India).

McGraw, T., Mindich, L., and Frangione, B. (1986). Nucleotide sequence of the small double-stranded RNA segment of bacteriophage phi 6: novel mechanism of natural translational control. J. Virol. *58*, 142–51.

Meister, G., and Tuschl, T. (2004). Mechanisms of gene silencing by double-stranded RNA. Nature. *431*, 343–9.

Mertens, P. (2004). The dsRNA viruses. Virus Res. *101*, 3–13.

Mindich, L. (1999). Precise packaging of the three genomic segments of the double-stranded-RNA bacteriophage phi6. Microbiol. Mol. Biol. Rev. *63*, 149–60.

Mindich, L., and Lehman, J. (1979a). Cell wall lysin as a component of the bacteriophage phi 6 virion. J. Virol. *30*, 489–96.

Mindich, L., Lehman, J., and Huang, R. (1979b). Temperature-dependent compositional changes in the envelope of phi 6. Virology. *97*, 171–6.

Mindich, L., Nemhauser, I., Gottlieb, P., Romantschuk, M., Carton, J., Frucht, S., Strassman, J., Bamford, D.H., and Kalkkinen, N. (1988). Nucleotide sequence of the large double-stranded RNA segment of bacteriophage phi 6: genes specifying the viral replicase and transcriptase. J. Virol. *62*, 1180–5.

Mindich, L., Qiao, X., Onodera, S., Gottlieb, P., and Frilander, M. (1994). RNA structural requirements for stability and minus-strand synthesis in the dsRNA bacteriophage phi 6. Virology *202*, 258–63.

Mindich, L., Qiao, X., and Qiao, J. (1995). Packaging of multiple copies of reduced-size genomic segments by bacteriophage phi 6. Virology *212*, 213–7.

Myers, J.W., Jones, J.T., Meyer, T., and Ferrell, J.E., Jr. (2003). Recombinant Dicer efficiently converts large dsRNAs into siRNAs suitable for gene silencing. Nat. Biotechnol. *21*, 324–8. Epub 2003 Feb 18.

Nagy, P.D., and Simon, A.E. (1997). New insights into the mechanisms of RNA recombination. Virology *235*, 1–9.

Nayak, A., Goodfellow, I.G., and Belsham, G.J. (2005). Factors required for the Uridylylation of the foot-and-mouth disease virus 3B1, 3B2, and 3B3 peptides by the RNA-dependent RNA polymerase (3Dpol) *in vitro*. J. Virol. *79*, 7698–706.

Ng, K.K., Cherney, M.M., Vazquez, A.L., Machin, A., Alonso, J.M., Parra, F., and James, M.N. (2002). Crystal structures of active and inactive conformations of a caliciviral RNA-dependent RNA polymerase. J. Biol. Chem. *277*, 1381–7. Epub 2001 Oct 24.

Ng, K.K., Pendas-Franco, N., Rojo, J., Boga, J.A., Machin, A., Alonso, J.M., and Parra, F. (2004). Crystal structure of norwalk virus polymerase reveals the carboxyl terminus in the active site cleft. J. Biol. Chem. *279*, 16638–45. Epub 2004 Feb 05.

O'Farrell, D., Trowbridge, R., Rowlands, D., and Jager, J. (2003). Substrate complexes of hepatitis C virus RNA polymerase (HC-J4): structural evidence for nucleotide import and de-novo initiation. J. Mol. Biol. *326*, 1025–35.

Ojala, P.M., and Bamford, D.H. (1995). *In vitro* transcription of the double-stranded RNA bacteriophage phi 6 is influenced by purine NTPs and calcium. Virology *207*, 400–8.

Olkkonen, V.M., Ojala, P.M., and Bamford, D.H. (1991). Generation of infectious nucleocapsids by *in vitro* assembly of the shell protein on to the polymerase complex of the dsRNA bacteriophage phi 6. J. Mol. Biol. *218*, 569–81.

Pagratis, N., and Revel, H.R. (1990). Detection of bacteriophage phi6 minus-strand RNA and novel mRNA isoconformers synthesized *in vivo* and *in vitro*, by strand-separating agarose gels. Virology *177*, 273–280.

Parker, W.B. (2005). Metabolism and antiviral activity of ribavirin. Virus Res. *107*, 165–71.

Partridge, J.E., Van Etten, J.L., Burbank, D.E., and Vidaver, A.K. (1979). RNA polymerase activity associated with bacteriophage phi 6 nucleocapsid. J. Gen. Virol. *43*, 299–307.

Patton, J.T. (1996). Rotavirus VP1 alone specifically binds to the 3′ end of viral mRNA, but the interaction is not sufficient to initiate minus-strand synthesis. J. Virol. *70*, 7940–7.

Patton, J.T., Jones, M.T., Kalbach, A.N., He, Y.W., and Xiaobo, J. (1997). Rotavirus RNA polymerase requires the core shell protein to synthesize the double-stranded RNA genome. J. Virol. *71*, 9618–26.

Patton, J.T., Silvestri, L.S., Tortorici, M.A., Vasquez-Del Carpio, R., and Taraporewala, Z.F. (2006). Rotavirus genome replication and morphogenesis: role of the viroplasm. Curr. Top. Microbiol. Immunol. *309*, 169–87.

Paul, A.V., van Boom, J.H., Filippov, D., and Wimmer, E. (1998). Protein-primed RNA synthesis by purified poliovirus RNA polymerase. Nature *393*, 280–4.

Pirttimaa, M.J., and Bamford, D.H. (2000). RNA secondary structures of the bacteriophage phi6 packaging regions. RNA *6*, 880–9.

Poranen, M.M., and Bamford, D.H. (1999). Packaging and replication regulation revealed by chimeric genome segments of double-stranded RNA bacteriophage phi 6. RNA *5*, 1–9.

Poranen, M.M., Paatero, A.O., Tuma, R., and Bamford, D.H. (2001). Self-assembly of a viral molecular machine from purified protein and RNA constituents. Mol. Cell *7*, 845–54.

Poranen, M.M., Pirttimaa, M.J., and Bamford, D.H. (2005a). Encapsidation of the Segmented Double-Stranded RNA Genome of Bacteriophage PHI 6. In Viral Genome Packaging Machines: Genetics, Structure and Mechanism, C.Catalano, ed. (Georgetown, Texas: Landes Bioscience/Eurekah.com, 810 South Church Street, Georgetown, Texas, U.S.A. 78626), pp. 153.

Poranen, M.M., Tuma, R., and Bamford, D.H. (2005b). Assembly of double-stranded RNA bacteriophages. Adv. Virus Res. *64*, 15–43.

Ranjith-Kumar, C.T., Gajewski, J., Gutshall, L., Maley, D., Sarisky, R.T., and Kao, C.C. (2001). Terminal nucleotidyl transferase activity of recombinant Flaviviridae RNA-dependent RNA polymerases: implication for viral RNA synthesis. J. Virol. *75*, 8615–23.

Ranjith-Kumar, C.T., Gutshall, L., Sarisky, R.T., and Kao, C.C. (2003b). Multiple interactions within the hepatitis C virus RNA polymerase repress primer-dependent RNA synthesis. J. Mol. Biol. *330*, 675–85.

Ranjith-Kumar, C.T., Kim, Y.C., Gutshall, L., Silverman, C., Khandekar, S., Sarisky, R.T., and Kao, C.C. (2002). Mechanism of *de novo* initiation by the hepatitis C virus RNA-dependent RNA polymerase: role of divalent metals. J. Virol. *76*, 12513–25.

Ranjith-Kumar, C.T., Santos, J.L., Gutshall, L.L., Johnston, V.K., Lin-Goerke, J., Kim, M.J., Porter, D.J., Maley, D., Greenwood, C., Earnshaw, D.L., Baker, A., Gu, B., Silverman, C., Sarisky, R.T., and Kao, C. (2003a). Enzymatic activities of the GB virus-B RNA-dependent RNA polymerase. Virology *312*, 270–80.

Rao, P., Yuan, W., and Krug, R.M. (2003). Crucial role of CA cleavage sites in the cap-snatching mechanism for initiating viral mRNA synthesis. EMBO J. *22*, 1188–98.

Rimon, A., and Haselkorn, R. (1978). Transcription and replication of bacteriophage phi6 RNA. Virology *89*, 206–17.

Roy, P., and Noad, R. (2006). Bluetongue virus assembly and morphogenesis. Curr. Top. Microbiol. Immunol. *309*, 87–116.

Salgado, P.S., Makeyev, E.V., Butcher, S.J., Bamford, D.H., Stuart, D.I., and Grimes, J.M. (2004). The structural basis for RNA specificity and Ca2+ inhibition of an RNA-dependent RNA polymerase. Structure (Camb) *12*, 307–16.

Shimotohno, K., and Miura, K. (1973). Transcription of double-stranded RNA in cytoplasmic polyhedrosis virus *in vitro*. Virology *53*, 283–6.

Simons, J.F., and Pettersson, R.F. (1991). Host-derived 5′ ends and overlapping complementary 3′ ends of the two mRNAs transcribed from the ambisense S segment of Uukuniemi virus. J. Virol. *65*, 4741–8.

Skehel, J.J., and Joklik, W.K. (1969). Studies on the *in vitro* transcription of reovirus RNA catalyzed by reovirus cores. Virology *39*, 822–31.

Starnes, M.C., and Joklik, W.K. (1993). Reovirus protein lambda 3 is a poly(C)-dependent poly(G) polymerase. Virology *193*, 356–66.

Steitz, T.A. (1998). A mechanism for all polymerases. Nature *391*, 231–2.

Sun, J.H., Adkins, S., Faurote, G., and Kao, C.C. (1996). Initiation of (–)-strand RNA synthesis catalyzed by the BMV RNA-dependent RNA polymerase: synthesis of oligonucleotides. Virology *226*, 1–12.

Sun, X.L., Johnson, R.B., Hockman, M.A., and Wang, Q.M. (2000). De novo RNA synthesis catalyzed by HCV RNA-dependent RNA polymerase. Biochem. Biophys. Res. Commun. *268*, 798–803.

Tao, Y., Farsetta, D.L., Nibert, M.L., and Harrison, S.C. (2002). RNA synthesis in a cage--structural studies of reovirus polymerase lambda3. Cell *111*, 733–45.

Thompson, A.A., and Peersen, O.B. (2004). Structural basis for proteolysis-dependent activation of the poliovirus RNA-dependent RNA polymerase. EMBO J. *23*, 3462–71. Epub 2004 Aug 12.

Usala, S.J., Brownstein, B.H., and Haselkorn, R. (1980). Displacement of parental RNA strands during *in vitro* transcription by bacteriophage phi 6 nucleocapsids. Cell *19*, 855–62.

Valenzuela, S., Pizarro, J., Sandino, A.M., Vasquez, M., Fernandez, J., Hernandez, O., Patton, J., and Spencer, E. (1991). Photoaffinity labeling of rotavirus VP1 with 8-azido-ATP: identification of the viral RNA polymerase. J. Virol. *65*, 3964–7.

van Dijk, A.A., Frilander, M., and Bamford, D.H. (1995). Differentiation between minus- and plus-strand synthesis: polymerase activity of dsRNA bacteriophage phi 6 in an *in vitro* packaging and replication system. Virology *211*, 320–3.

van Dijk, A.A., Makeyev, E.V., and Bamford, D.H. (2004). Initiation of viral RNA-dependent RNA polymerization. J. Gen. Virol. *85*, 1077–93.

Van Etten, J.L., Vidaver, A.K., Koski, R.K., and Semancik, J.S. (1973). RNA polymerase activity associated with bacteriophage phi 6. J. Virol. *12*, 464–71.

Vidaver, A.K., Koski, R.K., and Van Etten, J.L. (1973). Bacteriophage phi6: a lipid-containing virus of *Pseudomonas phaseolicola*. J. Virol. *11*, 799–805.

von Einem, U.I., Gorbalenya, A.E., Schirrmeier, H., Behrens, S.E., Letzel, T., and Mundt, E. (2004). VP1 of infectious bursal disease virus is an RNA-dependent RNA polymerase. J. Gen. Virol. *85*, 2221–9.

Yamakawa, M., Furuichi, Y., Nakashima, K., LaFiandra, A.J., and Shatkin, A.J. (1981). Excess synthesis of viral mRNA 5-terminal oligonucleotides by reovirus transcriptase. J. Biol. Chem. *256*, 6507–14.
Yang, H., Gottlieb, P., Wei, H., Bamford, D.H., and Makeyev, E.V. (2003b). Temperature requirements for initiation of RNA-dependent RNA polymerization. Virology *314*, 706–15.
Yang, H., Makeyev, E.V., and Bamford, D.H. (2001). Comparison of polymerase subunits from double-stranded RNA bacteriophages. J. Virol. *75*, 11088–95.
Yang, H., Makeyev, E.V., Butcher, S.J., Gaidelyte, A., and Bamford, D.H. (2003a). Two distinct mechanisms ensure transcriptional polarity in double-stranded RNA bacteriophages. J. Virol. *77*, 1195–203.
Yi, G.H., Zhang, C.Y., Cao, S., Wu, H.X., and Wang, Y. (2003). De novo RNA synthesis by a recombinant classical swine fever virus RNA-dependent RNA polymerase. Eur. J. Biochem. *270*, 4952–61.
Zeng, C.Q., Wentz, M.J., Cohen, J., Estes, M.K., and Ramig, R.F. (1996). Characterization and replicase activity of double-layered and single-layered rotavirus-like particles expressed from baculovirus recombinants. J. Virol. *70*, 2736–42.
Zhong, W., Ferrari, E., Lesburg, C.A., Maag, D., Ghosh, S.K., Cameron, C.E., Lau, J.Y., and Hong, Z. (2000a). Template/primer requirements and single nucleotide incorporation by hepatitis C virus nonstructural protein 5B polymerase. J. Virol. *74*, 9134–43.
Zhong, W., Gutshall, L.L., and Del Vecchio, A.M. (1998). Identification and characterization of an RNA-dependent RNA polymerase activity within the nonstructural protein 5B region of bovine viral diarrhea virus. J. Virol. *72*, 9365–9.
Zhong, W., Uss, A.S., Ferrari, E., Lau, J.Y., and Hong, Z. (2000b). De novo initiation of RNA synthesis by hepatitis C virus nonstructural protein 5B polymerase. J. Virol. *74*, 2017–22.

Structure and Function of P4, a dsRNA Virus Packaging Motor

15

Erika J. Mancini and Roman Tuma

Abstract

This review focuses on the biochemical, biophysical, and structural characterization of P4, the protein that forms the packaging motor of the *Cystoviridae* family of dsRNA bacteriophages. P4 exists as a hexameric ring that is positioned atop the fivefold vertices of the procapsid shell. The ATPase activity of P4 is coupled to translocation of the viral genomic precursors into the confines of a preformed procapsid. Recently determined structures of P4 from bacteriophage ϕ12 in complex with nucleotide diphosphates and triphosphates together with the essential ion Mg^{2+} have captured snapshots of the packaging motor in different states along the catalytic pathway. Surprisingly, the structure of P4 shows a close similarity to RecA-like hexameric motors including the ubiquitous family of hexameric helicases. A model for the mechanism of action of P4 and other hexameric molecular motors has been proposed, which is the first model based on high-resolution structures captured along the catalytic pathway. The model was confirmed by biochemical data. While genomic packaging by P4 in the *Cystoviridae* family of dsRNA bacteriophages may share some features with dsDNA packaging by terminases of tailed bacteriophages, the implications of the present work for the mechanism of packaging in other dsRNA viruses are not obvious, but cannot be excluded.

Introduction

Genome packaging is an essential step in the life cycle of all viruses. Different strategies ranging from co-assembly of the coat protein with the nucleic acid to active packaging into pre-assembled capsids have been demonstrated. The dsRNA bacteriophages of the *Cystoviridae* family (ϕ6 to ϕ14) utilize the latter mechanism. The recent structure of a cystovirus packaging motor provided the first insights into the mechanism of active packaging in atomic detail (Mancini *et al.*, 2004a). In this chapter we review the structural and biochemical aspects of this packaging motor. The overall life cycle, assembly, replication and other aspects of packaging in cystoviruses were reviewed recently (Mindich 2004; Poranen and Tuma 2004). Because a description of the *Cystoviridae* system is presented in detail elsewhere within this volume (see chapters by Poranen *et al.*, and Mindich), we offer only a brief description of cystoviruses to provide the necessary starting point for understanding the packaging motor action.

Cystoviruses assemble via an empty precursor capsid intermediate (procapsid or empty polymerase complex) into which the three segments of viral single-stranded (ss)RNA are sequentially packaged (Mindich 1999; Mindich 2004). Specific *pac* sites, which are located close to the 5′ end of each RNA precursor (Gottlieb *et al.*, 1994; Pirttimaa and Bamford 2000), bind to their cognate binding sites on the surface of the procapsid protein P1 (Mindich 1999; Mindich 2004; Qiao *et al.*, 1995; Qiao *et al.*, 1997; Sun *et al.*, 2004). Packaging of the ssRNAs is performed by the hexameric motor protein P4 at the expense of ATP hydrolysis (Gottlieb *et al.*, 1991; Juuti and Bamford 1995; Paatero *et al.*, 1998; Paatero *et al.*, 1995). The hexameric packaging motor, which serves as the portal for RNA entry, resides at the fivefold vertices of the procapsid (Figure 15.1) (de Haas *et al.*, 1999; Huiskonen *et al.*, 2007) and is linked to the procapsid via its C-terminus (Lisal *et al.*, 2006; Paatero *et al.*, 1998). Two biochemical classes of P4 hexamers were detected for bacteriophage ϕ6: one being essential for packaging and the other important for transcription (Pirttimaa *et al.*, 2002). However, ATPase activity is only necessary for packaging and P4 seems to act as a passive conduit for RNA extrusion during transcription (Kainov *et al.*, 2004). Apart from RNA transport, the P4 hexamer also plays an important role in assembly as reviewed previously (Poranen and Tuma, 2004) and elsewhere in this volume (see chapter by Poranen *et al.*).

In this review, we focus on the biochemical and structural properties of P4 proteins and present the recently proposed mechanism of coupling between ATP hydrolysis and

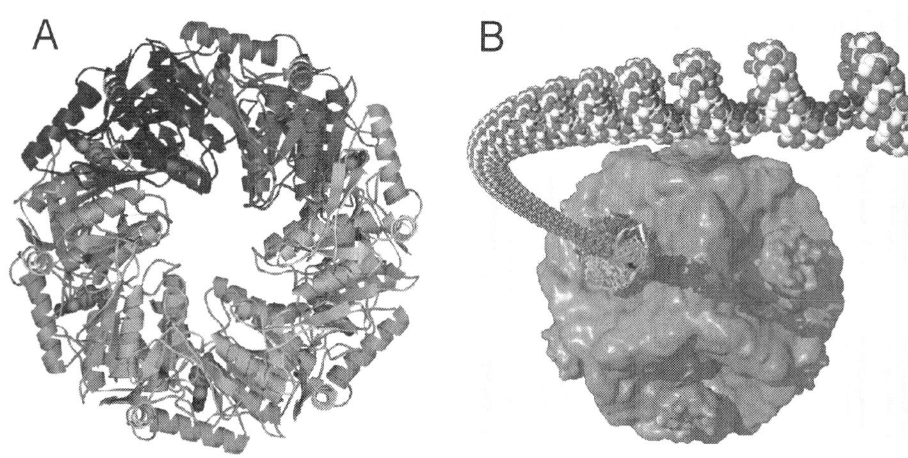

Figure 15.1 (A) Ribbon structure of P4 hexamer. In this representation different colours are used to designate distinct subunits. ADP, binding at the interface between monomers, is shown in a ball-and-stick representation. (B) The P4 hexameric motor is shown in the act of packaging ssRNA into the empty capsid. A 20 Å icosahedral reconstruction of the ϕ6 procapsid (proteins P1, P2, P4 and P7) (de Haas *et al.*, 1999) is shown in a cyan isosurface representation. Density for one of the P4 turrets has been removed and substituted with a ribbon representation of the crystal structure of the hexameric P4. A computer modelled A-form ssRNA is shown in a cpk model (atom colours are: carbon, white; oxygen, red; nitrogen, blue; phosphorus, yellow). (See also Plate 15.1.)

translocation. We also delineate the mechanistic implications of these studies for related proteins belonging to the hexameric helicase family.

Sequence similarity between P4 proteins from *Cystoviridae*

Because of low sequence identity between different cystoviruses, their genes are classified on the basis of genome organization and function rather than genetic similarity. Indeed, despite their similar biological function, the P4 proteins from *Cystoviridae* show little sequence identity other than in the loosely conserved RecA motifs (H1, H1a, H2, H3, H4) and fall into the twilight zone of protein sequence similarity (i.e. proteins with sequence identity between 20 and 25%) (Figure 15.2B). Indeed, it is likely that the P4 proteins originated from a common ancestor and in the course of evolution have changed significantly due to point mutations, insertions and deletions. The implication is that P4 proteins of the *Cystoviridae* share common structural features and biological function despite the low sequence similarity.

Biochemical properties of P4

P4 proteins from various cystoviruses (ϕ6, ϕ8, ϕ12, ϕ13) were successfully expressed in *E. coli* and purified to homogeneity in the hexameric form (Juuti *et al.*, 1998; Kainov *et al.*, 2003; Mancini *et al.*, 2004b). The hexameric form is stabilized by divalent cations and nucleotide diphosphates or triphosphates (Juuti *et al.*, 1998; Mancini *et al.*, 2004b).

All purified P4s exhibit nucleotide triphosphatase (NTPase) activity that is stimulated by ssRNA but not dsRNA or DNA (Juuti *et al.*, 1998; Kainov *et al.*, 2004; Kainov *et al.*, 2003). ϕ8 P4 exhibits the strongest coupling between RNA binding and hydrolysis while ϕ13 is only weakly coupled (Kainov *et al.*, 2003; Lisal *et al.*, 2004). The stimulated turnovers (k_{cat}) are between 1 to 6 s^{-1} depending on the enzyme and the RNA substrate (Kainov *et al.*, 2003; Lisal *et al.*, 2004; Lisal and Tuma 2005). NTPase activity also depends on the length of the ssRNA, which indicates that NTP hydrolysis is coupled to translocation along the substrate (Lisal *et al.*, 2004; Lisal and Tuma 2005). The average processivity, estimated from this dependence, is almost 2000 bases for the ϕ8 hexamer (Lisal *et al.*, 2004) but only 6–7 bases for ϕ12 P4 (Lisal and Tuma 2005). This difference most likely reflects the higher affinity of the ϕ8 enzyme for ssRNA (dissociation constant $K_d = 2\ \mu M$) with respect to the ϕ12 P4 ($K_d = 0.36$ mM). The higher translocation processivity and tight RNA binding of the ϕ8 enzyme also result in RNA-specific helicase activity (Kainov *et al.*, 2003). High affinity for nucleic acids and helicase activity was also observed for ϕ13 P4 (Kainov *et al.*, 2003).

The nucleotide specificity of the P4 enzymes is relatively low. For example, ϕ6 P4 can hydrolyze purine, pyrimidine, ribo-, deoxy- or dideoxy- nucleoside triphosphates with slight preference for GTP and UTP (Jenkins *et al.*, 1999; Juuti *et al.*, 1998; Kainov *et al.*, 2003). Only ϕ12 P4 exhibits strict purine specificity (Kainov *et al.*, 2004). The P4 affinity for nucleotides (K_d) ranges from 30 to 100 μM (Juuti *et al.*, 1998; Lisal *et al.*, 2004; Lisal and Tuma 2005). In all studied cases, the apparent Michaelis constant (K_m) was higher than K_d suggesting that the ATPase mechanism involves more than one binding site.

The catalytic cycles of the P4 proteins of the bacteriophages ϕ8 and ϕ12 were studied in greater detail (Lisal *et al.*, 2004; Lisal and Tuma 2005). ATP hydrolysis of both motors exhibits apparent cooperativity involving 2 to 3 subunits. The cooperativity is a

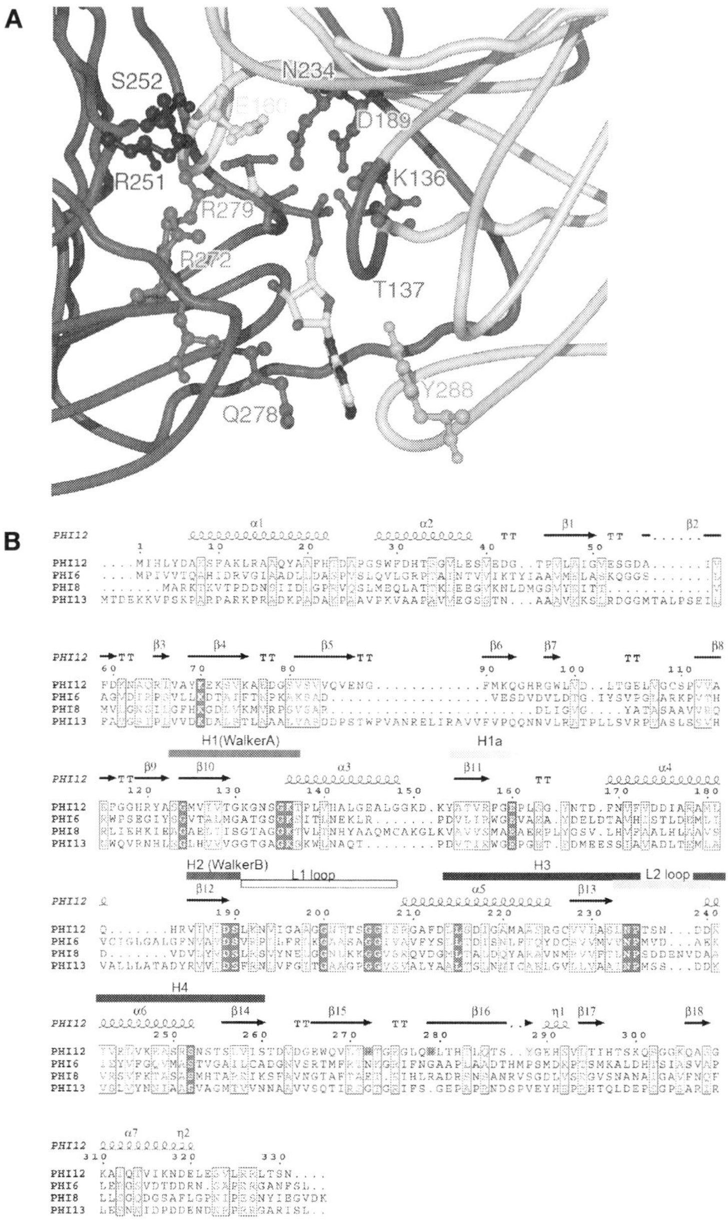

Figure 15.2 Structure and sequence of the conserved helicase motifs in P4 proteins. (A) Detail of the nucleotide binding site using the same colour coding for the conserved motifs as in the bottom panel. AMPcPP is shown in a ball-and-stick representation binding at the interface between two subunits (grey and white). (B) Sequence alignment and assignment of the conserved motifs for *Cystoviridae* P4 proteins (ϕ6, ϕ8, ϕ12 and ϕ13). The conserved motifs are called H1 (red), H1a (yellow), H2 (green), H3 (blue) and H4 (purple) in analogy to the conserved motifs in hexameric helicases. The two loops protruding in the centre of the P4 channel are the disordered loops L1 (transparent) and L2 (cyan). The putative arginine fingers Arg272 and Arg279 are shown in magenta and the adenine ring binding Tyr288 is shown in light green. (See also Plate 15.2.)

result of sequential hydrolysis of ATP bound to several subunits with similar affinity. Thus, the apparent cooperativity is not a result of cooperative binding but rather reflects coordination of hydrolysis within the hexamer. This distinguishes these motors from the structurally related F1-ATPase(Menz *et al.*, 2001). ATP binding to the second site is coupled to a conformational change in the protein and the bound RNA in ϕ8 P4 (Lisal *et al.*, 2004) while no such conformational changes were detected for ϕ12 P4 (Lisal and Tuma 2005). While the hydrolysis step seems to be limiting for both enzymes, ADP and inorganic phosphate release are fast. Another difference between the two motors is that ϕ12 P4 has non-cooperative basal ATPase activity in the absence of RNA. RNA induces apparent hydrolytic cooperativity and plays a central role in the coordination of sequential hydrolysis. RNA oligonucleotides, as short as five units, are capable of stimulating the ATPase activity of ϕ12 P4. In the context of the ϕ12 P4 structure, this suggests that two to three neighbouring subunits may interact with RNA during translocation. The proposed sequential mechanism of hydrolysis is in accord with the structure-based model for coordinated hydrolysis (see below).

Although isolated P4 hexamers alone are able to translocate RNA, their NTPase activity is regulated within the core (polymerase complex containing dsRNA). However, the degree of regulation varies between bacteriophages. While P4 hexamers associated with ϕ6 core or the nucleocapsid retain partial activity (Jenkins *et al.*, 1999; Ojala *et al.*, 1993) the P4 within the ϕ12 core is completely inactivated (Kainov *et al.*, 2004). The mechanism of such regulation is most likely achieved via core induced conformational changes within the P4 structure (Benevides *et al.*, 2002). Recent results obtained by hydrogen-deuterium exchange, which probes local dynamics of the protein backbone, provided further insight into the possible regulatory mechanism (Lisal *et al.*, 2006). P4 association with the core protects primarily the C-terminal region, indicating it as the main site of interaction of P4 with the P1 shell (Paatero *et al.*, 1998). Regions in the proximity of the active site also exhibit induced protection while structural studies provide evidence for direct coupling between the conformation of the C-terminal region and that of the catalytic core (see below).

Structural studies on P4

Recently, X-ray structures of ϕ12 P4 for all the key points along the catalytic pathway, including apo, substrate bound (AMPcPP-Mg^{2+}) and product bound (ADP-Mg^{2+}) states, have been solved (Mancini *et al.*, 2004a). These studies have provided the first mechanistic insight into how P4 couples ATP hydrolysis to RNA translocation during genome packaging. In presence and in absence of bound nucleotides, P4 ϕ12 crystallizes as a hexameric ring. The P4 monomer belongs to the α/β fold family of proteins and is composed of three domains (Figure 15.1A). The N-terminal domain features two orthogonal helices (the first of which projects from one monomer to the adjacent one) that are thought to play a role in stabilizing the hexamer. Apart from these helices, the rest of the N-terminus is predominantly made of β-strands and exhibits a fold that is observed in the three-dimensional structural family of solenoids. In particular, despite any noticeable sequence homology, the P4 N-terminus has a fold similar to the choline-binding domain of LytA, the major pneumococcal autolysin protein (Fernâandez-Tornero *et al.*, 2001). LytA contains six β-hairpins that stack in 120° successive counter-clockwise rotations to form a left-handed spiral staircase. P4 structure exhibits a similar staircase but with only 2 steps. However, the significance of such similarity is not clear at present.

The central domain of P4 has a Rossmann-type nucleotide binding fold containing an eight-stranded β-sheet flanked by five helices. Despite a very low sequence identity, the central domain of P4 has a RecA-like fold comprising the five conserved structural motifs (Figure 15.2A). They include the Walker A motif (or P-loop: Lys136 and Thr137, red), the Walker B motif (Asp189, yellow), the catalytic motif Glu160 (dark-green) and the sensor motif Asn234 (blue), all belonging to the same monomer. Finally the fifth motif, the arginine fingers Arg272 and Arg279 (both in magenta), are contributed from the adjacent monomer (Figure 15.2B). Similarly to RepA, the adenine base of the nucleotide is found sandwiched between two residues contributed from adjacent monomers, Tyr288 (light-green) *in cis* and Glu278 (grey) *in trans*. When looking at the P4 hexamer ring, there are two loops, L1 and L2, protruding towards the centre of the channel, named after the equivalent loops in RecA (Figure 15.3). The L1 loop (transparent) is almost entirely disordered and could not be modelled owing to the poorly defined electron density. L2 loop (cyan) is well resolved in our structure and it has been shown to be essential for RNA binding by site-directed mutagenesis (Kainov *et al*., submitted). The equivalent loop in RecA is also involved in nucleic acid binding (Story and Steitz 1992; Story *et al*., 1992). As explained in the following section, the L2 loop undergoes conformational changes dictated by the nucleotide binding state of the protein.

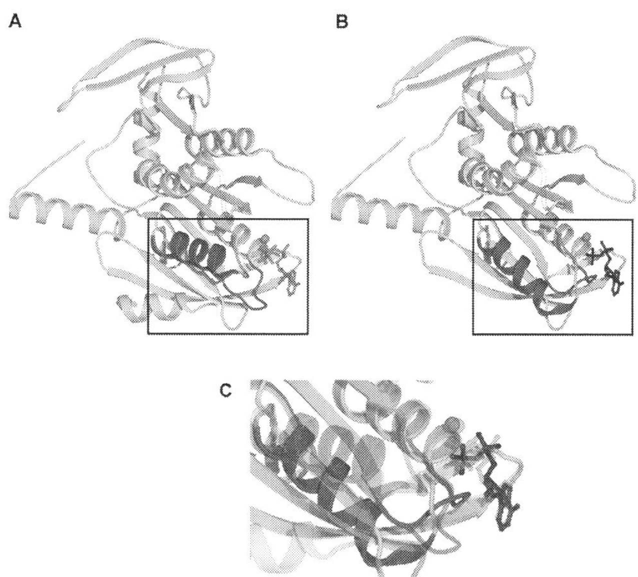

Figure 15.3 Conformational changes induced by ATP-hydrolysis. (A) Conformation of the P-loop (red), L2 loop (cyan) and helix α6 (magenta) in the AMPcPP+Mg^{2+} structure. (B) Conformation of the P-loop, L2 loop and helix α6 in the ADP+Mg^{2+} structure. (C) Superimposition of the two conformations. (See also Plate 15.3.)

The C-terminus is a rather flexible domain. Residues 199–206 and 321–331 could not be modelled and helix α7 rearranges during the conformational changes associated with ATP hydrolysis cycle and RNA translocation. It is known that the C-terminus is dispensable for catalytic activity and hexamerization but is required for interaction with the procapsid shell (Lisal *et al.*, 2006). The observed rearrangement of α7 suggests a way of transmitting regulatory signals from the core (P1 shell) to the P4 catalytic domain. For example, the central channel could be obstructed and ATPase activity diminished after successful packaging thereby avoiding passive leakage of the RNA from the core. After genome replication the channel could partially open to allow for the exit of ssRNA transcripts while the ATPase activity may be diminished by an induced conformational change in the ring (Huiskonen *et al.*, 2007). Interaction with the P1 shell stabilizes the hexamer and is thought to control ring opening (Lisal *et al.*, 2006) that has been shown to mediate RNA loading into the central channel (Lisal *et al.*, 2005). Ring opening constitutes a common mechanism of nucleic acid loading among hexameric helicases (Ahnert *et al.*, 2000).

Mechanism of coupling between RNA translocation and ATP hydrolysis

In the crystal P4 forms almost perfect sixfold symmetric rings with two loops, L1 and L2 decorating the internal channel (Figure 15.3). A lysine K241, which is conserved among cystoviruses, is located within the RNA binding loop L2 and projects into the centre of the channel assuming an ideal position to engage ssRNA. This residue is also essential for coupling RNA binding and ATPase activity (Kainov *et al.*, submitted). The C-terminus of the L2 loop is tethered to the long helix α6 (purple). The helix, in turn, links the central channel to the nucleotide-binding site of the next subunit. The substrate (AMPcPP-Mg^{2+}) and product (ADP-Mg^{2+}) bound conformations of P4 are compared in Figure 15.3A and B. Upon hydrolysis of the triphosphate the P-loop (red) collapses into the nucleotide binding site, while Arg279 is inserted into the catalytic pocket of the neighbouring subunit making it catalytically competent. The Arg279 residue was shown essential for hydrolysis (Kainov *et al.*, submitted). The collapse of the P-loop causes a deformation of the central Rossmann-fold β-sheet which in turn changes the conformation of the L2 loop. This controlled conformational change is regulated via the conserved sensory motif residue Asn234 (Kainov *et al.*, submitted). This substantial conformational change in the L2 loop extends from residue Asn234 all through the helix α6. This change involves large alterations in backbone dihedral angles of the loop. In the extreme the Lys241 experiences the largest displacement (~12 Å) as it swings from one side of the L2 loop to the other. Assuming that the Lys241 side chain makes a direct contact with RNA backbone, translocation might be driven directly via a mechanical levering action of the swinging arm. Thus, it can be argued that the AMPcPP-Mg^{2+} (Figure 15.3A) and ADP-Mg^{2+} (Figure 15.3B) structures represent the starting and the ending states of the catalytic cycle, respectively. This is consistent with biochemical analysis which links the ATP hydrolysis step with the RNA translocation. We propose that the local rearrangement of the P-loop upon ATP hydrolysis is amplified by the large swing of the α6 helix. In our working model for translocation sequential conformational changes result in the rocking motion of the L2 loops. This rocking motion coupled to ATP hydrolysis is envisioned to propagate around the P4 ring and translocate ssRNA through the central channel.

On the basis of these crystal structures a modified version of the binding change model of ATP hydrolysis has been put forward (Figure 15.4A and B). Binding and hydrolysis are sequential with five of the sites being occupied at any one time with ATP (orange and red subunits) and one with ADP+P_i (blue subunit). The ATP bound subunit engages RNA (yellow) inside the channel via the Lys241 side chain. The down conformation of helix α6 in the ADP-bound subunit activates the RNA-bound neighbouring subunit by promoting the insertion of the arginine finger (Arg279) into its active site. Hydrolysis in the RNA bound subunit triggers P-loop collapse, which is amplified via the motion of helix α6 to translocate the L2-bound RNA through the channel. Assuming that the ssRNA is in an approximately A form geometry, the translocation step would bring the RNA backbone to the vicinity of the counter clockwise ATP-bound subunit. The swing of α6 helix and L2 loop most likely triggers disengagement of RNA because ADP release is fast and RNA-independent. The cycle repeats and the conformational changes, which follow the hydrolysis wave, propagate from one subunit to another in a counter clockwise manner (Figure 15.4A). As a result ssRNA is passed from one subunit to another by motions that spiral across the inner surface of the channel. These motions draw the ssRNA down through the ring by about two bases per ATP molecule hydrolyzed (Figure 15.4B). The same stoichiometry has been recently estimated for ϕ8 P4 using a single molecule translocation assay (Lisal and Tuma, unpublished). Similar stoichiometry has been demonstrated also for the dsDNA packaging motor of bacteriophage ϕ29 (Chemla *et al.*, 2005) and bacteriophage T7 gp4 hexameric helicase (Liao *et al.*, 2005).

Structural similarity with hexameric helicases, F1-ATPase and AAA proteins

There is a strong structural similarity between the nucleotide binding domain of ϕ12 P4 and the equivalent domain of a number of molecular motors involved in various aspects of nucleic acid metabolism (Figure 15.5 and Table 15.1). RecA is the structural prototype for this family of enzymes that includes hexameric helicases, like bacteriophage T7 gp4 and

Figure 15.4 (opposite) Mechanism of ssRNA translocation. (A) Six cycles of ATP hydrolysis are shown, in a schematic illustrating the nucleotide binding state of each subunit during successive cycles. Orange subunits contain bound ATP (T), blue have ADP+Pi (D), and red are hydrolysis competent (or activated) ATP containing sites. RNA binding loops are schematically shown in green (up position) and blue (down position). This is a sequential binding and hydrolysis model where all six bound ATP sites are utilized at different points during sequential cycles. ATP hydrolysis triggers movement of the L2 loop (blue) and the bound RNA backbone (blue dot) down the channel while another backbone phosphate is interacting with the L2 loop of a neighbour (red asterisk) and makes this subunit competent for hydrolysis. These conformational changes propagate as a ripple going counter-clockwise around the ring and do not require a rotation of the ring itself or of the RNA. The notation of the subunits is maintained in panel B. (B) Representation of the hexameric ring structure such that the subunits are laid out flat in two dimensions. The RNA binding loops are indicated in green, the RNA is coloured in yellow with the phosphates bases are represented by small balls. For simplicity, only the first three cycles of ATP hydrolysis are shown. As the hydrolysis wave proceeds from right to left, RNA is drawn inside the viral capsid (in grey) by about 2 subunits per hydrolyzed ATP. (See also Plate 15.4.)

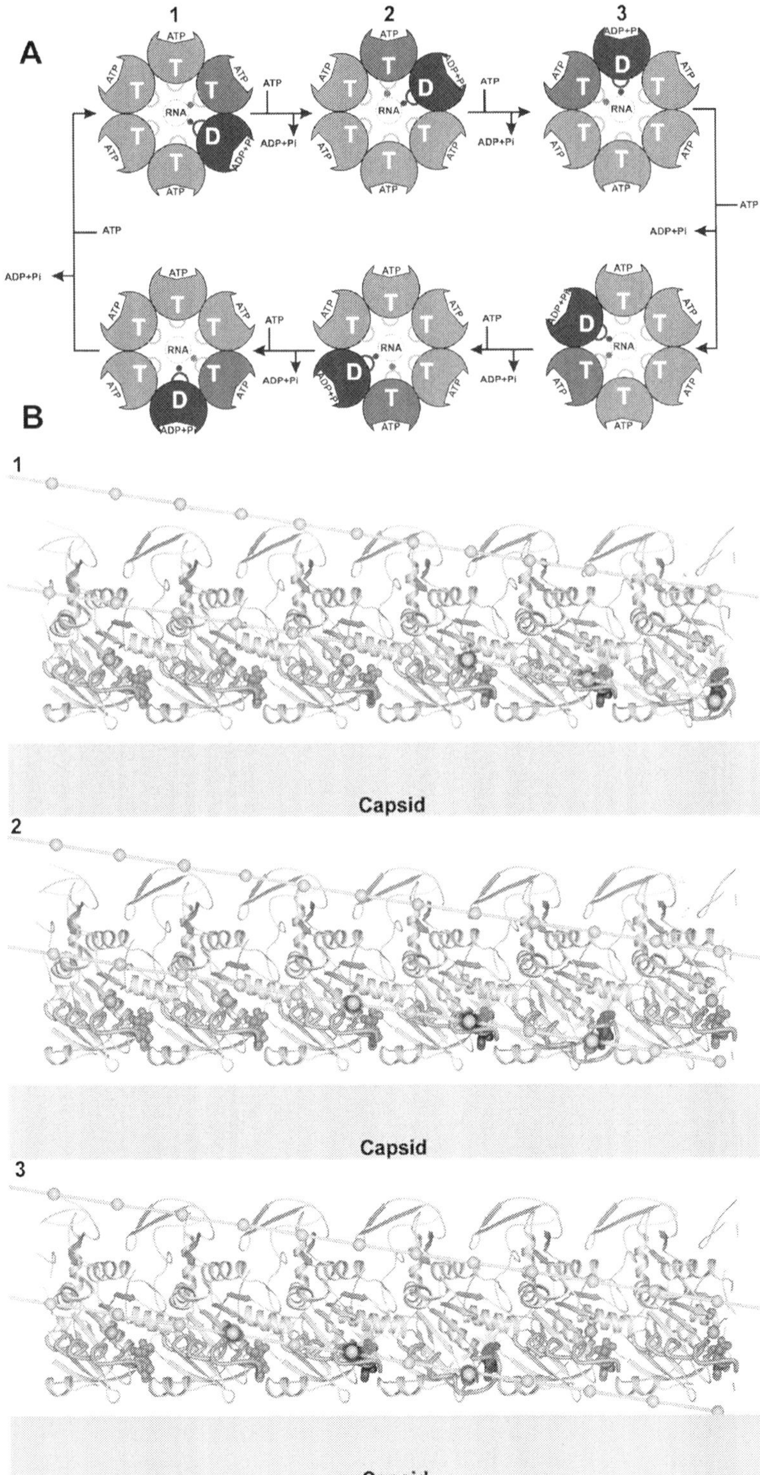
A
1
2
3
ATP
T
D
RNA
ADP+Pi
B
1
Capsid
2
Capsid
3
Capsid

replicative helicase RepA. Similarity is also found between P4 and both α and β subunits of F1-ATPase complex, which is at the heart of chemiosmotic energy conversion in mitochondria, chloroplasts and bacteria. Finally, P4 displays a nucleotide binding domain with a similar fold to several members of the AAA protein family, the closest being NSF (Table 15.1). This is an essential family of specialized chaperone-like enzymes that participate in membrane-fusion and trafficking, organelle biogenesis, proteolysis and protein folding. Conservation of the structural motifs in the nucleotide binding domain suggests a similar transmission system for driving translocation and mechanical force generation.

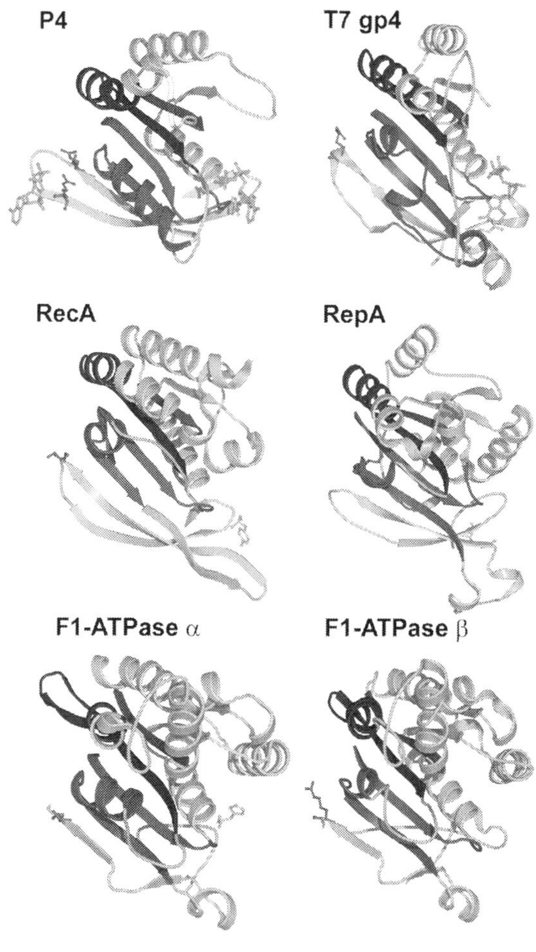

Figure 15.5 Structural superimposition of the nucleotide binding domains of P4 and structurally related proteins: T7 gp4 helicase (T7 gp4), *E. coli* RecA (RecA), regulatory protein RepA (RepA), Thermophilic Bacillus Ps3 F1-ATPase subunit α (F1-ATPase α) and β (F1-ATPase β). The structural superimposition was obtained with SHP (Stuart *et al.*, 1979). The colour scheme for conserved motifs (H1-H4) is the same used in the sequence alignment in Figure 2A. Alignment statistics are reported in Table 15.1 PDB deposition numbers are listed in the legend to Table 15.1. (See also Plate 15.5.)

Table 15.1 RMS deviations, Z scores and structural alignment statistics for P4 and structurally related proteins as obtained from a DALI search (Holm and Sander 1995)

	Z score	RMSD (Å)	n. of residues aligned	Sequence identity (%)
T7 helicase	12.2	3.7	175	15
RecA	10.2	4.4	169	12
RepA	9.3	3.7	168	17
F1-ATPase (α)	8.5	3.8	172	14
F1-ATPase (β)	8.1	4.2	176	12
TrwB	7.3	3.9	155	18
NSF	7.1	2.9	119	14

PDB deposition numbers: T7 gp4 helicase (1CR1), *E. coli* RecA (2REB), Thermophilic Bacillus Ps3 F1-ATPase subunit α (1SKYB) and β (1SKYE), regulatory protein RepA (1G8Y), Bacterial Conjugation Protein TRwB (1E9R), n-ethylmaleimide-sensitive fusion NSF (1D2N).

Relation to other viral packaging ATPases

The role of P4 in the packaging of genomic plus-strand RNA in the *Cystoviridae* resembles that of the portal complex encountered in the dsDNA bacteriophages. While P4 acts as both the NTPase and the portal ring, the dsDNA portal complex relies upon a transiently associated terminase complex for providing the required NTPase activity (Catalano *et al.*, 1995; Kanamaru *et al.*, 2004). It has been proposed that a conserved common ATPase catalytic centre in the N-terminal half of the large terminase subunits from dsDNA phages and herpesvirus constitutes the ATPase motor (U_L15 in HSV-1, gp17 in bacteriophage T4, gp16 in Φ29, gpP in P2, gpA in λ) (Bain *et al.*, 2001; Catalano 2000; Goetzinger and Rao 2003; Guo *et al.*, 1986; Linderoth *et al.*, 1991; Przech *et al.*, 2003; White *et al.*, 2003). The functional signatures and the structural framework are remarkably well conserved among the terminases, despite the lack of significant sequence similarity. These terminases were predicted to contain RecA-like ATPase domains consisting of 6 β strands (Goetzinger and Rao 2003). The hypothesis has been successfully verified by combinatorial mutagenesis of the putative ATPase motifs in T4 gp17 (Kanamaru *et al.*, 2004) and by recent determination of bacteriophage T4 terminase structure (Sun *et al.*, 2007).

The P4 ring and a typical portal complex share the circular topology with central channel. However, while the P4 is a hexamer, dsDNA portal complexes are usually dodecameric. Another major difference is that while packaging ATPases in *Cystoviridae* are structural components of both the procapsid and the mature capsid, terminases are only transiently associated with procapsids and are not present in mature virions. Further, DNA viruses generally package only a single DNA molecule, while dsRNA viruses are faced with a segmented genome and the need to package exactly one copy of each genomic segment.

Relation to other dsRNA packaging systems

It is now clear that most segmented dsRNA viruses share a common core structure (Bamford *et al.*, 2005). However, biochemical findings indicate that significant differences may exist between the packaging processes of *Cystoviridae* and other dsRNA viruses (Jayaram *et al.*, 2002; Taraporewala and Patton 2004). The ϕ6 system shows that assembly can proceed by formation of procapsids followed by packaging, like in many DNA bacteriophages, rather than the more concerted mechanism postulated for L-A, in which the structure is thought to form around the RNA (Fujimura *et al.*, 1990; Fujimura *et al.*, 1986). The non-structural rotavirus protein NSP2 has been implicated in RNA packaging (Taraporewala and Patton 2004). Although NSP2 possesses ATPase activity, it has been shown that its helicase activity is ATPase independent and probably relies on the formation of a stoichiometric complex with ssRNA (Taraporewala *et al.*, 1999). The protein is an octamer with a predominantly α-helical fold belonging to the HIT protein family (Jayaram *et al.*, 2002). This protein is not structurally related to the *Cystoviridae* packaging motor P4 and the mechanism of RNA packaging in rotavirus remains elusive.

Conclusions

The mechanism of DNA and RNA packaging in viruses is a fascinating biological problem. During packaging a complex, metabolically active oligonucleotide is translocated into the confines of an empty capsid and condensed into a highly ordered structure of near crystalline density. Recent evidence suggests that the phage ϕ29 DNA packaging motor is one of the strongest force-generating molecular devices reported to date (Smith *et al.*, 2001). P4, the packaging ATPase of ϕ12 responsible for generating the force used for RNA translocation, is the first of such motors to be visualized in atomic details. Given that the purified (recombinant) P4 hexamers from dsRNA bacteriophages are capable of RNA translocation in the absence of other viral components they constitute the simplest viral packaging motors. These motors are amenable to a broad range of structural, biochemical and biophysical methods, which have provided the basis for the proposed mechanism.

Acknowledgements

E.J.M. was supported by the UK Royal Society fellowship. R.T. held a research fellow post funded by the Academy of Finland.

References

Ahnert, P., Picha, K.M., and Patel, S.S. (2000). A ring-opening mechanism for DNA binding in the central channel of the t7 helicase-primase protein. EMBO J. *19*, 3418–27.

Bain, D.L., Berton, N., Ortega, M., Baran, J., Yang, Q., and Catalano, C.E. (2001). Biophysical characterization of the DNA binding domain of gpnu1, a viral DNA packaging protein. J. Biol. Chem. *276*, 20175–81.

Bamford, D.H., Grimes, J.M., and Stuart, D.I. (2005). What does structure tell us about virus evolution? Curr. Opin. Struct. Biol. *15*, 655–63.

Benevides, J.M., Juuti, J.T., Tuma, R., Bamford, D.H., and Thomas, G.J., Jr. (2002). Characterization of subunit-specific interactions in a double-stranded RNA virus, Raman difference spectroscopy of the phi6 procapsid. Biochemistry *41*, 11946–53.

Catalano, C.E. (2000). The terminase enzyme from bacteriophage lambda, A DNA-packaging machine. Cell. Mol. Life Sci. *57*, 128–48.

Catalano, C.E., Cue, D., and Feiss, M. (1995). Virus DNA packaging. The strategy used by phage lambda. Mol. Microbiol. *16*, 1075–86.

Chemla, Y.R., Aathavan, K., Michaelis, J., Grimes, S., Jardine, P.J., Anderson, D.L., and Bustamante, C. (2005). Mechanism of force generation of a viral DNA packaging motor. Cell *122*, 683–92.

de Haas, F., Paatero, A.O., Mindich, L., Bamford, D.H., and Fuller, S.D. (1999). A symmetry mismatch at the site of RNA packaging in the polymerase complex of dsRNA bacteriophage phi6. J. Mol. Biol. *294*, 357–72.

Fernández-Tornero, C., López, R., García, E., Giménez-Gallego, G., and Romero, A. (2001). A novel solenoid fold in the cell wall anchoring domain of the pneumococcal virulence factor lyta. Nat. Struct. Biol. *8*, 1020–4.

Fujimura, T., Esteban, R., Esteban, L.M., and Wickner, R.B. (1990). Portable encapsidation signal of the l-a double-stranded RNA virus of *S. cerevisiae*. Cell *62*, 819–28.

Fujimura, T., Esteban, R., and Wickner, R.B. (1986). *In vitro* l-a double-stranded RNA synthesis in virus-like particles from saccharomyces cerevisiae. Proc. Natl Acad. Sci. USA *83*, 4433–7.

Goetzinger, K.R., and Rao, V.B. (2003). Defining the ATPase center of bacteriophage T4 DNA packaging machine, Requirement for a catalytic glutamate residue in the large terminase protein gp17. J. Mol. Biol. *331*, 139–54.

Gottlieb, P., Qiao, X., Strassman, J., Frilander, M., and Mindich, L. (1994). Identification of the packaging regions within the genomic RNA segments of bacteriophage phi 6. Virology *200*, 42–7.

Gottlieb, P., Strassman, J., Frucht, A., Qiao, X.Y., and Mindich, L. (1991). *In vitro* packaging of the bacteriophage phi 6 ssRNA genomic precursors. Virology *181*, 589–94.

Guo, P., Grimes, S., and Anderson, D. (1986). A defined system for *in vitro* packaging of DNA-gp3 of the bacillus subtilis bacteriophage phi 29. Proc. Natl Acad. Sci. USA *83*, 3505–9.

Holm, L., and Sander, C. (1995). Dali, A network tool for protein structure comparison. Trends Biochem. Sci. *20*, 478–80.

Huiskonen, J.T., Jäälinoja, H.T., Briggs, J.A.G., Fuller, S.D., and Butcher, S.J. (2007). Structure of a hexameric RNA packaging motor in a viral polymerase complex. J. Struct. Biol. *158*, 156–64.

Jayaram, H., Taraporewala, Z., Patton, J.T., and Prasad, B.V. (2002). Rotavirus protein involved in genome replication and packaging exhibits a hit-like fold. Nature *417*, 311–5.

Jenkins, R.H., Tuma, R., Juuti, J.T., Bamford, D.H., and Thomas, G.J., Jr. (1999). A novel raman spectrophotometric method for quantitative measurement of nucleoside triphosphate hydrolysis. Biospectroscopy *5*, 3–8.

Juuti, J.T., and Bamford, D.H. (1995). RNA binding, packaging and polymerase activities of the different incomplete polymerase complex particles of dsRNA bacteriophage phi 6. J. Mol. Biol. *249*, 545–54.

Juuti, J.T., Bamford, D.H., Tuma, R., and Thomas, G.J., Jr. (1998). Structure and NTPase activity of the RNA-translocating protein (p4) of bacteriophage phi 6. J. Mol. Biol. *279*, 347–59.

Kainov, D.E., Lisal, J., Bamford, D.H., and Tuma, R. (2004). Packaging motor from double-stranded RNA bacteriophage phi12 acts as an obligatory passive conduit during transcription. Nucleic Acids Res. *32*, 3515–21.

Kainov, D.E., Pirttimaa, M., Tuma, R., Butcher, S.J., Thomas, G.J., Jr., Bamford, D.H., and Makeyev, E.V. (2003). RNA packaging device of double-stranded RNA bacteriophages, possibly as simple as hexamer of p4 protein. J. Biol. Chem. *278*, 48084–91.

Kanamaru, S., Kondabagil, K., Rossmann, M.G., and Rao, V.B. (2004). The functional domains of bacteriophage T4 terminase. J. Biol. Chem. *279*, 40795–801.

Liao, J.C., Jeong, Y.J., Kim, D.E., Patel, S.S., and Oster, G. (2005). Mechanochemistry of T7 DNA helicase. J. Mol. Biol. *350*, 452–75.

Linderoth, N.A., Ziermann, R., Haggard-Ljungquist, E., Christie, G.E., and Calendar, R. (1991). Nucleotide sequence of the DNA packaging and capsid synthesis genes of bacteriophage P2. Nucleic Acids Res. *19*, 7207–14.

Lisal, J., Kainov, D.E., Bamford, D.H., Thomas, G.J., Jr., and Tuma, R. (2004). Enzymatic mechanism of RNA translocation in double-stranded RNA bacteriophages. J. Biol. Chem. *279*, 1343–50.

Lisal, J., Kainov, D.E., Lam, T.T., Emmett, M.R., Wei, H., Gottlieb, P., Marshall, A.G., and Tuma, R. (2006). Interaction of packaging motor with the polymerase complex of dsRNA bacteriophage. Virology *351*, 73–9.

Lisal, J., Lam, T.T., Kainov, D.E., Emmett, M.R., Marshall, A.G., and Tuma, R. (2005). Functional visualization of viral molecular motor by hydrogen-deuterium exchange reveals transient states. Nat. Struct. Mol. Biol. *12*, 460–6.

Lisal, J., and Tuma, R. (2005). Cooperative mechanism of RNA packaging motor. J. Biol. Chem. *280*, 23157–64.

Mancini, E.J., Kainov, D.E., Grimes, J.M., Tuma, R., Bamford, D.H., and Stuart, D.I. 2004a. Atomic snapshots of an RNA packaging motor reveal conformational changes linking ATP hydrolysis to RNA translocation. Cell *118*, 743–55.

Mancini, E.J., Kainov, D.E., Wei, H., Gottlieb, P., Tuma, R., Bamford, D.H., Stuart, D.I., and Grimes, J.M. 2004b. Production, crystallization and preliminary x-ray crystallographic studies of the bacteriophage phi 12 packaging motor. Acta Crystallogr. D *60*, 588–90.

Menz, R.I., Walker, J.E., and Leslie, A.G. (2001). Structure of bovine mitochondrial F1-ATPase with nucleotide bound to all three catalytic sites, Implications for the mechanism of rotary catalysis. Cell *106*, 331–41.

Mindich, L. (1999). Precise packaging of the three genomic segments of the double-stranded RNA bacteriophage phi6. Microbiol. Mol. Biol. Rev. *63*, 149–60.

Mindich, L. (2004). Packaging, replication and recombination of the segmented genome of bacteriophage phi6 and its relatives. Virus Res. *101*, 83–92.

Ojala, P.M., Juuti, J.T., and Bamford, D.H. (1993). Protein p4 of double-stranded RNA bacteriophage phi 6 is accessible on the nucleocapsid surface, Epitope mapping and orientation of the protein. J. Virol. *67*, 2879–86.

Paatero, A.O., Mindich, L., and Bamford, D.H. (1998). Mutational analysis of the role of nucleoside triphosphatase p4 in the assembly of the RNA polymerase complex of bacteriophage phi6. J. Virol. *72*, 10058–65.

Paatero, A.O., Syvaoja, J.E., and Bamford, D.H. (1995). Double-stranded RNA bacteriophage phi 6 protein p4 is an unspecific nucleoside triphosphatase activated by calcium ions. J. Virol. *69*, 6729–34.

Pirttimaa, M., Paatero, A.O., Frilander, M., and Bamford, D.H. (2002). Nonspecific ntpase p4 of dsRNA bacteriophage phi6 polymerase complex is required for ssRNA packaging and transcription. J. Virol. *76*, 10122–10127.

Pirttimaa, M.J., and Bamford, D.H. (2000). RNA secondary structures of the bacteriophage phi6 packaging regions. RNA *6*, 880–9.

Poranen, M.M., and Tuma, R. (2004). Self-assembly of double-stranded RNA bacteriophages. Virus Res. *101*, 93–100.

Przech, A.J., Yu, D., and Weller, S.K. (2003). Point mutations in exon i of the herpes simplex virus putative terminase subunit, ul15, indicate that the most conserved residues are essential for cleavage and packaging. J. Virol. *77*, 9613–21.

Qiao, X., Casini, G., Qiao, J., and Mindich, L. (1995). *In vitro* packaging of individual genomic segments of bacteriophage phi 6 RNA, Serial dependence relationships. J. Virol. *69*, 2926–31.

Qiao, X., Qiao, J., and Mindich, L. (1997). Stoichiometric packaging of the three genomic segments of double-stranded RNA bacteriophage phi6. Proc. Natl Acad. Sci. USA *94*, 4074–9.

Smith, D.E., Tans, S.J., Smith, S.B., Grimes, S., Anderson, D.L., and Bustamante, C. (2001). The bacteriophage phi29 portal motor can package DNA against a large internal force. Nature *413*, 748–52.

Story, R.M., and Steitz, T.A. (1992). Structure of the RecA protein-ADP complex. Nature *355*, 374–6.

Story, R.M., Weber, I.T., and Steitz, T.A. (1992). The structure of the *E. coli* RecA protein monomer and polymer. Nature *355*, 318–25.

Stuart, D.I., Levine, M., Muirhead, H., and Stammers, D.K. (1979). Crystal structure of cat muscle pyruvate kinase at a resolution of 2.6 Å. J. Mol. Biol. *134*, 109–42.

Sun, Y., Qiao, X., and Mindich, L. (2004). Construction of carrier state viruses with partial genomes of the segmented dsRNA bacteriophages. Virology *319*, 274–9.

Sun, S., Kondabagil, K., Gentz, P.M., Rossman, M.G., and Rao, V.B. (2007). The structure of the ATPase that powers DNA packaging into bacteriophage T4 procapsids. Mol. Cell. , 943–9.

Taraporewala, Z., Chen, D., and Patton, J.T. (1999). Multimers formed by the rotavirus nonstructural protein nsp2 bind to RNA and have nucleoside triphosphatase activity. J. Virol. *73*, 9934–43.

Taraporewala, Z.F., and Patton, J.T. (2004). Nonstructural proteins involved in genome packaging and replication of rotaviruses and other members of the *Reoviridae*. Virus. Res. *101*, 57–66.

White, C.A., Stow, N.D., Patel, A.H., Hughes, M., and Preston, V.G. (2003). Herpes simplex virus type 1 portal protein ul6 interacts with the putative terminase subunits ul15 and ul28. J. Virol. *77*, 6351–8.

Structure and Function of the Rotavirus NSP2 Octamer, an Essential Component of the Viroplasm

16

Zenobia F. Taraporewala, Mukesh Kumar, B.V. Venkataram Prasad, and John T. Patton

Abstract

Viruses in the family *Reoviridae* produce cytoplasmic inclusion bodies (viroplasms) that serve as sites of genome packaging, replication and early stages of virion assembly. To support the complex events that occur in the viroplasm, the resident viral proteins must be dynamic and capable of multiple protein–protein and protein–RNA interactions. In rotavirus-infected cells, nucleation of viroplasms involves two non-structural proteins, NSP2 and NSP5. NSP2 is a single-stranded (ss) RNA-binding protein with helix-destablizing, NTPase and RTPase activities. The protein self-assembles into doughnut-shaped octamers made up of two stacked tetramers. Electropositive grooves that span the surface of the octamer competitively bind two ligands, ssRNA and NSP5, providing a mechanism for modulating the activity of NSP2 between two functions. The NSP2 monomer contains a histidine-triad (HIT)-like fold that is shared by a large ubiquitous class of cellular nucleotide-binding proteins, one not previously seen in a viral protein. The HIT-like motif of the monomer is contained within a deep cleft that functions as the catalytic site for the hydrolytic activities of the protein. In this chapter, we review the current information on the structure of the NSP2 octamer, its catalytic activities, and the nature of its interactions with NTPs, ssRNA and other viral proteins. Based on these properties, we discuss potential roles for NSP2 in virus replication. Finally, the features of NSP2 are contrasted with the inclusion-forming proteins of other members of the *Reoviridae*.

Introduction

Rotavirus is a major pathogen in the large and diverse group of segmented double-stransded (ds)RNA viruses that belong to the *Reoviridae* family. The rotavirus virion is made up of three concentric layers of protein that surround a genome of eleven segments of dsRNA (Estes, 2001). The innermost protein layer is made up of 60 dimers of the core lattice protein, VP2, arranged with T = 1 geometry (Lawton *et al.*, 1997b). A copy of the RNA-dependent RNA polymerase, VP1 (Valenzuela *et al.*, 1991), and the mRNA capping enzyme, VP3 (Lawton *et al.*, 1997b; Liu *et al.*, 1992; Patton and Chen, 1999), is anchored to the interior side of the VP2 pentamers that make up the T = 1 lattice. The dsRNA segments are distributed within the viral core such that each is associated with one of the VP1–VP3 enzyme complexes (Gouet *et al.*, 1999; Pesavento *et al.*, 2001).

The intermediate protein layer of the capsid is made up of trimers of VP6, arranged with T = 13 geometry. Cores surrounded by the VP6 layer are referred to as double-layered particles (DLPs). The antigenicity of the VP6 component is used to classify rotaviruses into different groups (A to G) (Kapikian, 2001). It is generally true that rotaviruses belonging to the same group can undergo reassortment to produce progeny viruses with mixed genome constellations, while rotaviruses belonging to different groups cannot. The outermost protein layer of the virion consists of trimers of the glycoprotein VP7, arranged with T = 13 symmetry, and the multimeric spike protein VP4 (Prasad *et al.*, 1988).

During entry of rotavirus into the host cell, the outer protein layer of the virion is removed, leaving a DLP (Estes, 2001). The VP1–VP3 components of the DLP then direct transcription of the genome segments, producing eleven capped messenger RNAs [(+)RNAs] (McCrae and McCorquodale, 1983). The (+)RNAs are extruded from the DLP through channels at the fivefold axes that transverse both the VP2 and VP6 protein layers (Lawton *et al.*, 1997a). The (+)RNA products of transcription serve as templates for protein synthesis, producing six structural and six non-structural proteins (Estes, 2001). Two of the non-structural proteins, NSP2 and NSP5, interact with each other to form viroplasms, structures that serve as viral factories supporting genome packaging and replication, and the assembly of progeny cores and DLPs (Brookes *et al.*, 1993; Miller *et al.*, 2003; Patton *et al.*, 2006; Petrie *et al.*, 1982; Petrie *et al.*, 1984; Rhim *et al.*, 1962; Silvestri *et al.*, 2004). These functions require the accumulation of numerous viral components in the viroplasm, including VP1, VP2, VP3 and VP6 as well as viral (+)RNAs. These functions also require the viroplasm to have the necessary fluidity to allow the interactions needed for the assembly of progeny particles and for their subsequent migration out of the inclusion.

The (+)RNAs that serve as templates for dsRNA synthesis are made by DLPs located within the viroplasm (Silvestri *et al.*, 2004). In contrast to (+)RNAs that are in the cytosol, those that are in the viroplasm are not susceptible to degradation by the RNA-induced silencing complexes (RISC) of the cellular RNA interference system. Thus, viroplasms have been referred to as safe houses, sites in which the viral genetic material can be replicated without attack by host antiviral mechanisms. During the replication cycle, the many small punctate-like viroplasms that initially appear give way to fewer large inclusions late in infection (Eichwald *et al.*, 2004b). Numerous endoplasmic-reticulum (ER) vesicles usually surround mature viroplasms. Progeny DLPs that leave the viroplasm move to such vesicles where during budding they acquire an outer VP4-VP7 protein shell (Cuadras *et al.*, 2006).

The synthesis of the dsRNA genome in viroplasms is mediated by replication intermediates (RIs) with a core-like component (Gallegos and Patton, 1989). Characterization of such RIs indicates that as viral (+)RNAs are packaged into core-like structures, the (+)RNAs are simultaneously replicated into dsRNA molecules. This linkage between packaging and replication results from the VP2-dependent nature of the catalytic activity of VP1 (Patton *et al.*, 1997; Tortorici *et al.*, 2003). In contrast to the 5′-ends of the (+)RNAs, the 5′-ends of the minus-strand RNAs [(−) RNA] contained in the dsRNA products of replication lack a cap structure. Instead, the (−)RNAs end with β and α phosphate groups (ppRNA) (McCrae and McCorquodale, 1983). The unexpected absence of a γ-phosphate at the 5′-end of the (−) RNA suggests that this group is targeted by an

RNA triphosphatase (RTPase) activity at some point following minus-strand initiation. Whether the RTPase activity that removes the γ-phosphate of the (−)RNA is the same activity that removes the 5′-terminal phosphates from nascent (+)RNA transcripts prior to capping (guanylylation) is unknown. The RTPase activity that is used in the capping of rotavirus transcripts must be a structural protein, since DLPs can accurately cap (+)RNAs *in vitro* in the absence of non-structural proteins.

Viroplasm-building blocks

Several studies have provided evidence indicating that NSP2 and NSP5 are the essential building blocks of the rotavirus viroplasm. Perhaps the most straightforward are results showing that when transiently expressed together, but not individually, NSP2 and NSP5 will form viroplasm-like structures in uninfected cells (Fabbretti *et al.*, 1999). Also, multiple studies have shown that the knockdown of either NSP2 or NSP5 expression in rotavirus-infected cells using short-interfering RNAs (siRNAs) leads to a loss of viroplasm formation (Campagna *et al.*, 2005; Lopez *et al.*, 2005; Silvestri *et al.*, 2004). The structure and activities of the NSP2-building block is the subject of this article, and are reviewed below.

The structure and activities of the other viroplasm building block, NSP5, are poorly understood. NSP5 is a dimeric acidic protein that undergoes two post-translation modifications in the infected cell: O-linked glycosylation and phosphorylation (Gonzalez and Burrone, 1991; Poncet *et al.*, 1997; Welch *et al.*, 1989). Variation in the extent of phosphorylation yields different NSP5 isomers *in vivo*, ranging from a 28-kDa hypophosphorylated form to 32–34-kDa hyperphosphorylated forms (Afrikanova *et al.*, 1998). Hyperphosphorylation of the protein is mediated by host cellular kinases (CKI and CKII), and occurs on serine and threonine residues (Eichwald *et al.*, 2004a; Eichwald *et al.*, 2002). Interestingly, NSP5 hyperphosphorylation only takes place in the presence of NSP2 (Carpio *et al.*, 2004). Presumably, NSP2 interacts with NSP5, triggering structural changes in the latter that expose residues critical to hyperphosphorylation (Eichwald *et al.*, 2004a). Since NSP2 localizes with NSP5 to viroplasms, it follows that accumulation of NSP5 in these inclusions would be correlated with the hyperphosphorylation of the protein (Poncet *et al.*, 1997). Despite the extensive phosphorylation of NSP5 that occurs in the infected cells, initial knockdown studies with siRNAs targeting the cellular transcripts that encode cellular kinases, suggest that this post translation modification is not essential for productive viral infection (Desselberger, 2007).

NSP2 function *in vivo*

NSP2 is a 35 kDa non-structural protein encoded by genome segment 8 (gene 8) of the simian SA11 strain of group A rotavirus. Several lines of evidence have suggested a role for NSP2 in rotavirus genome packaging and replication: (i) NSP2 localizes to sites of viral replication and assembly in the infected cell, vis-à-vis, the viroplasm (Petrie *et al.*, 1982; Petrie *et al.*, 1984), (ii) NSP2 is a component of RIs recovered from infected cells that exhibit replicase activity *in vitro* (Gallegos and Patton, 1989; Helmberger-Jones and Patton, 1986; Patton and Gallegos, 1988), and (iii) NSP2 is physically associated with partially replicated viral RNA (Aponte *et al.*, 1996) and with the viral polymerase VP1 (Kattoura *et al.*, 1994).

Until recently, the only *in vivo* studies that assessed the significance of NSP2 in rotavirus infection relied on the use of *tsE*, an SA11 mutant with a temperature-sensitive lesion in the NSP2 gene (Chen *et al.*, 1990; Ramig and Petrie, 1984). Cells infected with *tsE* and maintained at the non-permissive temperature contain few viroplasms, lack RIs with replicase activity and produce virus particles that are mostly empty, i.e. lack dsRNA. These characteristics led to the suggestion that NSP2 plays an essential role in RNA packaging and minus-strand synthesis. With the more contemporary loss-of-function approach made possible with siRNAs, Silvestri *et al.* (2004) recently demonstrated that NSP2 expression was essential for several aspects of rotavirus replication *in vivo*. Specifically, SA11-infected cells containing an siRNA targeting NSP2 (+)RNAs not only failed to support viroplasm formation, but also failed to synthesize viral dsRNA or assemble progeny virus particles. Subsequent studies with an NSP2-dependent complementation system showed that the effect of the siRNA could be reversed by providing to the cell, plasmid-derived NSP2 transcripts that were resistant to degradation by the siRNA (Taraporewala *et al.*, 2006). Taken together, the *in vivo* studies indicate that NSP2 plays a critical role in supporting the formation of viroplams, and several viroplasm-associated processes.

Activities of purified NSP2

NSP2 is a multimeric highly basic protein (pI 9.02) that binds single-stranded (ss)RNA non-specifically and with high affinity (Kattoura *et al.*, 1992; Taraporewala *et al.*, 1999). The protein displays very little, if any, affinity for dsRNA. Early studies analyzing the stoichiometry of RNA-binding activity by NSP2 using electrophoretic mobility shift assay (EMSA) showed that one pmol of RNA probe could form a higher order complex with up to 10 pmol of protein (Taraporewala *et al.*, 1999). This was one of the first indications that NSP2 formed a multimer capable of interacting with RNA. EMSA also showed that these NSP2 multimers bind to ssRNA cooperatively, with the binding of one multimer promoting the binding of another. The cooperativity in the binding of NSP2 multimers to ssRNA is likely critical to the helix-destabilizing activity that has been identified for the protein. Such an activity may be particularly important for removing secondary structures in viral (+)RNAs that could impede their use as templates for genome packaging and replication. Unlike the ATP-dependent activity of typical helicases, the helix-destabilizing activity of NSP2 is Mg^{2+}- and ATP-independent, and non-directional (Kadare and Haenni, 1997; Taraporewala and Patton, 2001). NSP2 also lacks the conserved motifs, such as the classic Walker motifs, that are hallmarks of many viral helicases. The helix-destabilizing activity of the NSP2 multimer, in fact, is more like that reported for the single-stranded DNA-binding proteins (SSBs) of some dsDNA viruses (Boehmer and Lehman, 1993; Monaghan *et al.*, 1994; Rochester and Traktman, 1998; Soengas *et al.*, 1995; Tsurumi *et al.*, 1998).

Structure of the NSP2 octamer

Initial insights into the size and shape of the NSP2 multimer were achieved by analyzing bacterial-expressed recombinant NSP2 using a variety of techniques including electron microscopy (EM), analytical ultracentrifugation, and dynamic light scattering. The findings of these analyses revealed (i) that NSP2 has an intrinsic capacity to self-assemble into highly stable homomultimers (~12S) with a molecular mass consistent with that of an octamer (Schuck *et al.*, 2001) and (ii) that the NSP2 multimer has an overall doughnut-shaped appearance.

X-ray crystallography has since been used to determine the structure of the NSP2 multimer to a resolution of 2.6 Å (Jayaram *et al.*, 2002) and more recently, to 2.2 Å (Taraporewala *et al.*, 2006). These studies have shown that the NSP2 multimer represents an octameric unit, with 4-2-2 symmetry, formed by the tail-to-tail stacking of two tetramers (Figure 16.1) (Jayaram *et al.*, 2002). The doughnut-shaped appearance results from a large 35 Å central hole in the octamer. A substantial amount of the surface area of each monomer is buried (985 $Å^2$ at the twofold axis, and 710 $Å^2$ at the fourfold axis) in the octamer assembly. Many of these solvent-inaccessible surface residues of the monomers are involved in intersubunit interactions, a feature probably contributing to the remarkable stability of the octamer. Recent cryo-EM analysis confirms that the solid and solution structures of the NSP2 octamer are the same (Jiang *et al.*, 2006). Overall, the quaternary organization of the octamer is unique and one not previously reported for a viral protein.

Structure of the NSP2 monomer

The NSP2 monomer contains two distinct structural domains (Figure 16.2A), the N-terminal domain (residues 1 to 141) and the C-terminal domain (residues 151–313). These domains are connected by a loop that is 10 residues long (Jayaram *et al.*, 2002). The arrangement of the N-terminal domain relative to the C-terminal domain creates an intervening cleft that is ~25 Å deep. The N-terminal domain is rich in α-helices and contains a putative RNA-binding site that is formed by a highly basic 24-residue loop. Although not all the residues in the loop are strictly conserved, the charges in this loop are conserved among group A, B, and C rotaviruses (Taraporewala and Patton, 2004). Three prolines at the beginning of this loop are conserved among diverse rotaviruses suggesting that these residues play a particularly important role in imparting conformation to this element. Upon assembly of NSP2 monomers to form an octamer, the loops come together in a manner that position the charged residues along four prominent grooves, 30 Å wide and 25 Å deep (Figure 16.2C), that span the surface of the octamer. The dimension and the electropositive nature of the grooves make them suitable for binding ssRNA. The concentration of the positively charged residues along the periphery of the octamer give it a ring-like appearance (Figure 16.1C), leading to the proposal that ssRNA wraps around the octamer, reminiscent of the binding of ssDNA to SSBs (Raghunathan *et al.*, 2000). The prediction that these grooves serve as RNA-binding sites has been confirmed by cryo-EM analysis (Jiang *et al.*, 2006) and by site-directed mutagenesis (Vasquez-Del Carpio *et al.*, 2006).

Interestingly, the electropositive residues that line the grooves form not only part of the RNA-binding site of the octamer, but also constitute part of the site at which NSP5 binds to the octamer (Jiang *et al.*, 2006). The overlapping functions of the grooves in binding NSP5 and RNA raises the possibility that one ligand regulates the interaction of the second ligand with the octamer. Because viroplasm nucleation occurs early in the infection cycle and is a prerequisite for packaging and dsRNA synthesis, it may be presumed that initial interactions with NSP2 are likely to involve NSP5 and only later to involve ssRNA. If this is the case, then ssRNA is the regulator of NSP2-NSP5 interactions.

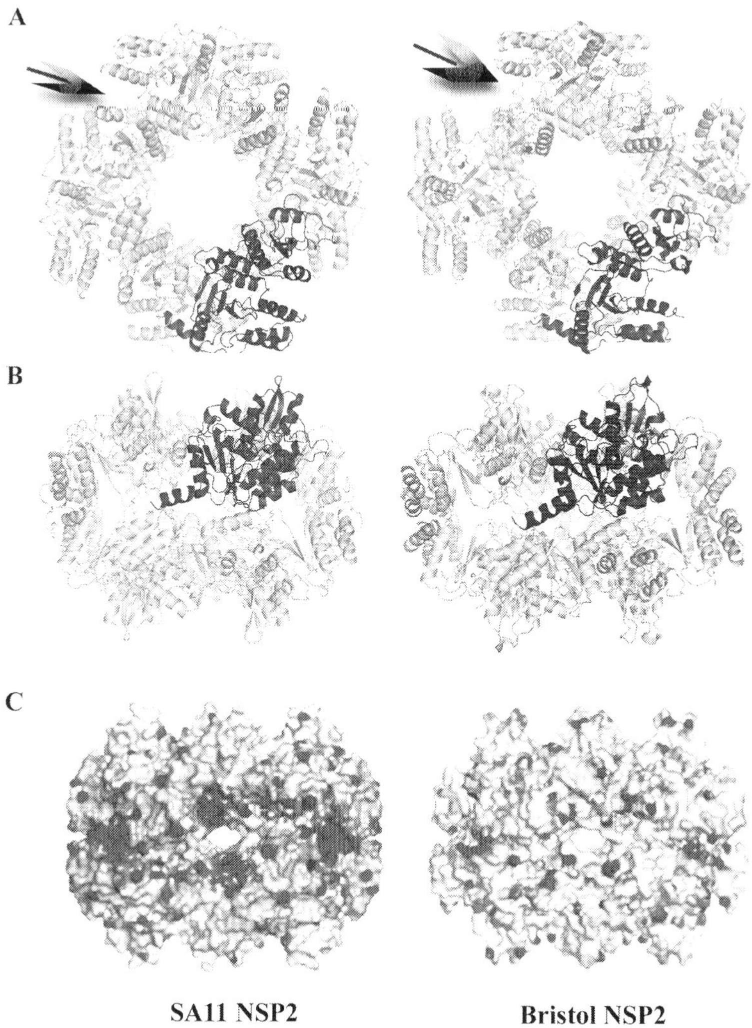

Figure 16.1 Structure of the NSP2 octamer. (A, B) Ribbon representation of the NSP2 octamer of group A SA11 (2.2 Å) and group C Bristol (2.8 Å) as viewed from the fourfold axis (A) and one of the twofold axes (B) of symmetry. The octamer represents two tetramers, stacked tail-to-tail. One monomer has been coloured in black. The thin and thick arrows in (A) reflect the narrowness of the mouth of the cleft in the SA11 octamer compared to Bristol. (C) Distribution of surface charges in the NSP2 octamer. A surface representation of the NSP2 octamer, viewed from a twofold axis of symmetry [the other twofold axis of symmetry is shown in (B)] with basic residues highlighted in dark. Unlike Bristol NSP2, clustering of basic residues is evident along the periphery of the SA11 octamer.

HIT-like fold of NSP2

NSP2 exhibits an NTPase activity that allows the protein to hydrolyze the terminal phosphoanhydride bond in any of the four NTPs, resulting in the release of NDP and free P_i (Taraporewala *et al.*, 1999). Structural studies have shown that NSP2 binds nucleotides

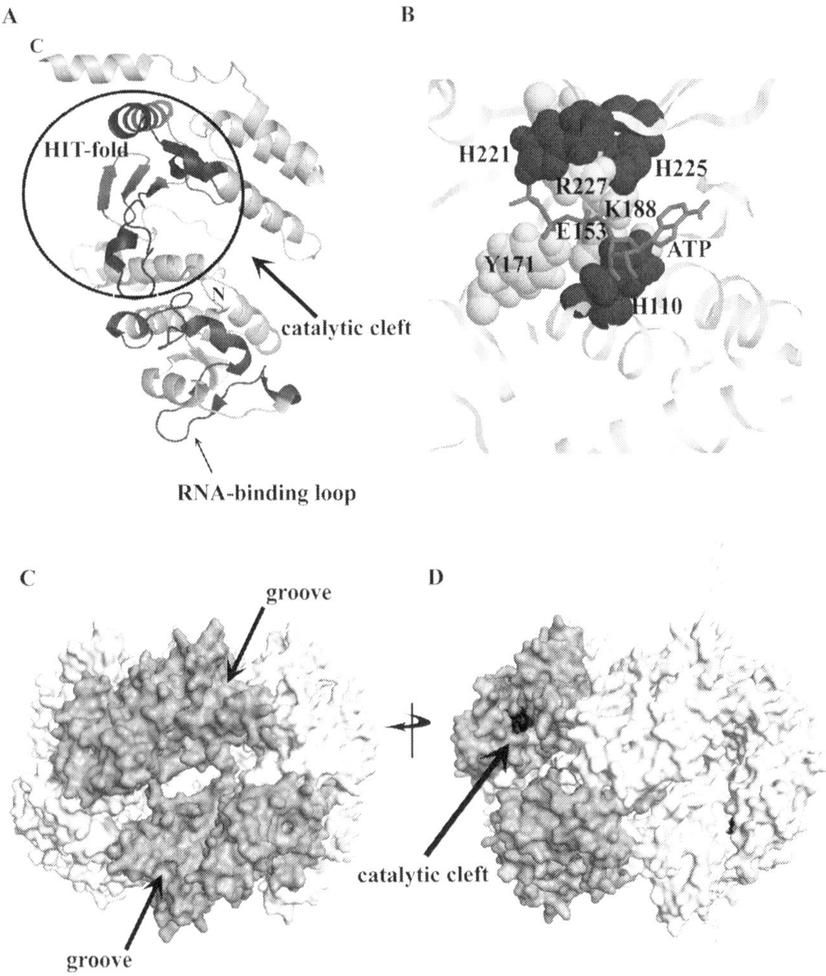

Figure 16.2 X-ray structure of the NSP2 monomer and ligand-binding sites. (A) Ribbon representation of the SA11 NSP2 monomer showing its 2-domain organization. The N-terminal domain includes a highly basic RNA-binding region and the C-terminal domain contains a HIT-fold. Arrows indicate locations of the inter-domain 25 Å catalytic cleft and the highly basic RNA-binding region. (B) The figure depicts a close-up view of the catalytic cleft of the NSP2 monomer with a docked NTP (model generated by Autodock). The Cα backbone of the NSP2 monomer with residues in the cleft that are involved in NTP-binding and hydrolysis [H110, H225 and H227, the histine triad (in dark grey); K188, R227] and Mg^{2+} coordination (Y171, E153) identified by space-filling. (C, D) Side-view of the octamer in surface representation with two interacting monomers (each in opposing tetramer units) highlighted in grey. One of the four grooves running diagonally across the octamer at a twofold axis of symmetry is indicated. These grooves represent binding sites for ssRNA and NSP5. (D) Adjacent to either side of the groove but hidden deep [and not visible from the groove face of the octamer, as in (C)] in the structure of the octamer are the catalytic clefts, the sites of NTP and pppRNA hydrolysis. The deep catalytic cleft can be seen (residues H221, H225 and R227 highlighted in a dark shade of grey colour) when the octamer is rotated ~45°C along the y-axis relative to the position from where the groove can be visualized (C).

within the deep cleft that separates the N- and C-terminal domains of the monomer (Figure 16.2A and B) (Kumar *et al.*, in press). The primary residues that participate in nucleotide binding are positioned in the C-terminal domain of NSP2 (Jayaram *et al.*, 2002). This domain is made up of twisted anti-parallel β-sheets surrounded by α-helices and shows a HIT (histidine triad)-like fold (Jayaram *et al.*, 2002) that is common among the HIT family of nucleotide-interacting proteins (Brenner, 2002; Lima *et al.*, 1997). NSP2 lacks a signature HIT motif (HØHØHØØ, where Ø is a hydrophobic residue) yet the fold of the C-terminal domain exhibits significant structural similarity with a prototypic member of a HIT subfamily, specifically, protein kinase C-interacting protein (PKCI). Based upon structural superposition of NSP2 and PKCI, H221 was suggested as the first histidine of a HIT-like catalytic triad, with H225 representing the second, and possibly the one that carries out the nucleophilic attack (Figure 16.2B). Based on the strength of ordered density peaks noted in the solved structure of NSP2 in the vicinity of H225, the conserved residues E153 and Y171 were proposed to coordinate the binding of Mg^{2+}, an essential cofactor for the NTPase activity of NSP2 (Jayaram *et al.*, 2002).

Co-crystallization of NSP2 with nucleotide analogues has shown that H225 is indeed the catalytic residue that carries out nucleophilic attack on the β-γ phosphoanhydride bond of the NTP and the residue that undergoes transient phosphorylation during NTP hydrolysis (Kumar *et al.*, in press). In the vicinity of H225, a cluster of conserved charged residues, K188, H221, K223 and R227, are also involved in nucleotide binding, specifically, in forming hydrogen bonds with the oxygen residues of the three phosphates groups of the substrate. The recognition of the triphosphate moiety by these residues and the lack of any base recognition in the cleft allows the protein to non-specifically hydrolyze nucleotides.

X-ray crystallography indicates that nucleotide binding results in small changes to the Cα backbone of NSP2. Specifically, the C-terminal domain of the monomer moves towards the N-terminal domain (Kumar *et al.*, in press). Similarly, analytic ultracentrifugation indicates that the NSP2 octamer undergoes nucleotide-induced shifts in conformation, possibly moving from an open to a more condensed form (Schuck *et al.*, 2001). Such behaviour suggests that the NSP2 octamer may cycle between different conformations in the cell. Although the significance of such oscillations is unknown, the different protein conformations may dictate which ligands (NTP, RNA) and binding partners (NSP5, VP1) can interact with the octamer, thereby potentially defining the function of the octamer.

The contribution of the HIT-like motif in the cleft to NTP binding and hydrolysis was confirmed by targeted mutagenesis (Carpio *et al.*, 2004). Specifically, mutation of residues in the cleft that constitute part of the HIT-like motif (H221, H225, K188 and R227) and that are predicted to coordinate Mg^{2+} binding (E153 and Y171) have detrimental effects on the NTPase activity of the octamer. Computer-based ATP-docking simulations have revealed that the electropositive nature of the cleft supports NTP binding, an interaction guided by the phosphate backbone and not the sugar-base moiety of the substrate (Figure 16.2B) (Carpio *et al.*, 2004). The lack of affinity for either a base or sugar suggests that the triphosphate linkage at the 5′ end of an RNA molecule (pppRNA) could also serve as a substrate for the hydrolytic activity of NSP2. Indeed, NSP2 was shown to hydrolyze the γ-phosphate from a pppRNA by a Mg^{2+}-dependent mechanism, utilizing the same active site as used in NTP hydrolysis (Vasquez-Del Carpio *et al.*, 2006). Thus, the HIT-like motif of NSP2, H^{221}-G-(K/H)-ϕ–H^{225}- ϕ–R-V, can accommodate and hydrolyze two different substrates that have a common triphosphate linkage, giving the octamer both an NTPase and RTPase activity.

For NSP2 to function as an RTPase in the viroplasm where the concentration of NTP would greatly exceed that of pppRNA, the affinity of NSP2 for its RNA substrate would need to be higher than for its NTP substrate. In fact, an analysis of the kinetics of the two activities have shown that although they have similar V_{max} values, the K_m of the RNA substrate is lower than that of the NTP substrate (Vasquez-Del Carpio *et al.*, 2006). The significant preference of NSP2 for pppRNA over NTP is probably a reflection of the high affinity of the protein for ssRNA and the close proximity of the RNA-binding site (grooves) and the catalytic clefts in the octamer (Figure 16.2C and D).

Binding of an NTP or a pppRNA to the cleft is followed by hydrolysis of the β-γ phosphoanhydride bond of the substrate and the autophosphorylation of the protein, via covalent linkage of the liberated γ-phosphate group (Taraporewala *et al.*, 1999) (Del Carpio *et al.*, 2004). Examination of mutant species of NSP2 with defects in NTPase activity have shown that the accumulation of phosphorylated NSP2 during NTP hydrolysis is disproportionately low relative to the amount of product generated (NDP + P_i). Furthermore, these analyses have shown that if the NTPase activity of NSP2 is abruptly stopped during incubation with NTP, the protein undergoes rapid dephosphorylation (Carpio *et al.*, 2004). Thus, phosphorylated NSP2 represents a short-lived intermediate formed during the NTPase reaction. The covalently bound phosphate is released from the protein via a Mg^{2+}-independent mechanism that requires a second nucleophilic attack. Kumar *et al.* (in press) have reported that this second attack, in the presence of an acceptor NDP, results in the transfer of the released phosphate to the NDP molecule to form an NTP. Thus, in addition to its NTPase and RTPase activities, NSP2 possesses an NDP-kinase activity. In a mechanism that parallels other cellular NDP kinases, the residue in NSP2 that undergoes phosphorylation is the same residue (H225) that mediates the catalytic attack on the NTP. Hence, the NDP-kinase activity of NSP2 is an extension of its NTPase activity. The function of the NDP-kinase activity is unknown, but perhaps is needed for regenerating and balancing NTPs pools in the viroplasm.

The intertwined nature of the NTPase, RTPase and NDP-kinase activities of NSP2 are unique, unlike that seen for any other viral or cellular protein. Although a HIT-like motif is critical for each of these activities, both the oligomeric status and mechanism of the hydrolysis of the NSP2 multimer is quite different from those of other HIT proteins. Notably, unlike NSP2 octamers, which can hydrolyze either an NTP or a pppRNA via a Mg^{2+}-dependent mechanism to liberate γ-phosphate, most HIT proteins are dimers that cleave dinucleotide polyphosphates (e.g. ApppA) via a cation-independent mechanism to form a nucleotidylated intermediate. Subsequently, a second nucleophilic attack typically liberates the bound monophosphate nucleotide from the HIT protein (Brenner, 2002; Brenner *et al.*, 1999; Lima *et al.*, 1997). Because NSP2 efficiently binds photoreactable nucleotide analogues even in the absence of Mg^{2+} (Vasquez-Del Carpio *et al.*, unpublished results), docking of an NTP into the cleft of NSP2 appears to be a cation-independent event. However, since Mg^{2+} can effect the conformation of an NTP, the presence of the cation may reposition or bend the phosphate groups of the substrate within the cleft, such that hydrolysis is possible (Lima *et al.*, 1997; McIntosh *et al.*, 2003).

Conservation of octamer architecture

Given the importance of the hydrolytic activity of NSP2 in virus replication, it is not surprising that residues making up the HIT-like motif are conserved among rotavirus strains (Figure 16.3). The only other notable region of sequence conservation includes those electropositive residues in the N-terminal domain that contribute to the RNA-binding activity of the protein. Because the octamer is the functional form of NSP2 (Taraporewala *et al.*, 2001), one would predict that its structure and that of the monomer would also be conserved. Indeed, this prediction was recently confirmed when the NSP2 structure of a group C strain of human rotavirus (Bristol) was solved by X-ray crystallography (Taraporewala *et al.*, 2006). Despite sharing only 57% sequence similarity and 35% sequence identity with SA11 (group A) NSP2, Bristol NSP2 was determined to assemble into doughnut-shaped octamers like those formed by the SA11 protein (Figure 16.1). Moreover, a comparison of the Cα backbones of the SA11 and Bristol NSP2 monomers showed that they are quite similar, with a root mean square deviation of only 1.8 Å. A close match was also noted in the organization of HIT-like motif within the clefts of the two proteins (His 225 and His 221 of SA11 NSP2 and His 222 and His 218 of Bristol NSP2), implying that Bristol NSP2 has the same catalytic activities as SA11 NSP2. Indeed, biochemical analyses have established that Bristol NSP2 is a ssRNA-binding protein with both NTPase and RTPase activities.

Although the overall structures of Bristol and SA11 NSP2 are nearly identical, there are two minor differences (Taraporewala *et al.*, 2006). One exists in the loop formed by residues corresponding to 170 to 185 of the SA11 protein. This loop projects off the C-terminal domain of the monomer, and forms portions of the floor of the grooves that span the octamer surface. In comparison to the SA11 loop, the Bristol loop is three residues shorter, lacking the equivalent of residues 183 to 185. This shortening of the Bristol loop is correlated with a slight shift in the overall positioning of the C-terminal domain, such that the mouth of the catalytic cleft of the Bristol monomer is of greater dimension than that of the SA11 cleft (Figure 16.1). The implications of this size change on the NTPase and RTPase activities of the protein are unknown.

The second notable difference between Bristol and SA11 NSP2 is in the distribution of electrostatic charges on the octamer surface (Taraporewala *et al.*, 2006). In contrast to the SA11 octamer, there is no significant clustering of positively charged residues along the grooves of the Bristol octamer (Figure 16.1C). These grooves are known sites of interaction between the octamer and NSP5 and RNA, ligands that are highly acidic in nature (Jiang *et al.*, 2006). The difference in the charge distribution along the grooves suggests that the interaction of the Bristol octamer with its NSP5 ligand is different than the interaction of the SA11 octamer with its NSP5 ligand. Thus, the interactions between homologous forms of NSP2 and NSP5 are probably specific, driven by unique charge complementarity. This idea has been supported by studies showing that co-expression of Bristol NSP2 with SA11 NSP5 fails to drive assembly of viroplasm-like structures. In addition, NSP2-dependent complementation assays have shown that Bristol NSP2 cannot replace SA11 NSP2 in supporting genome replication or virus assembly (Taraporewala *et al.*, 2006). Thus, although the octameric structure and HIT-dependent activities of rotavirus NSP2 are conserved among rotavirus groups, other features of the protein also dictate their functionality within the infected cell.

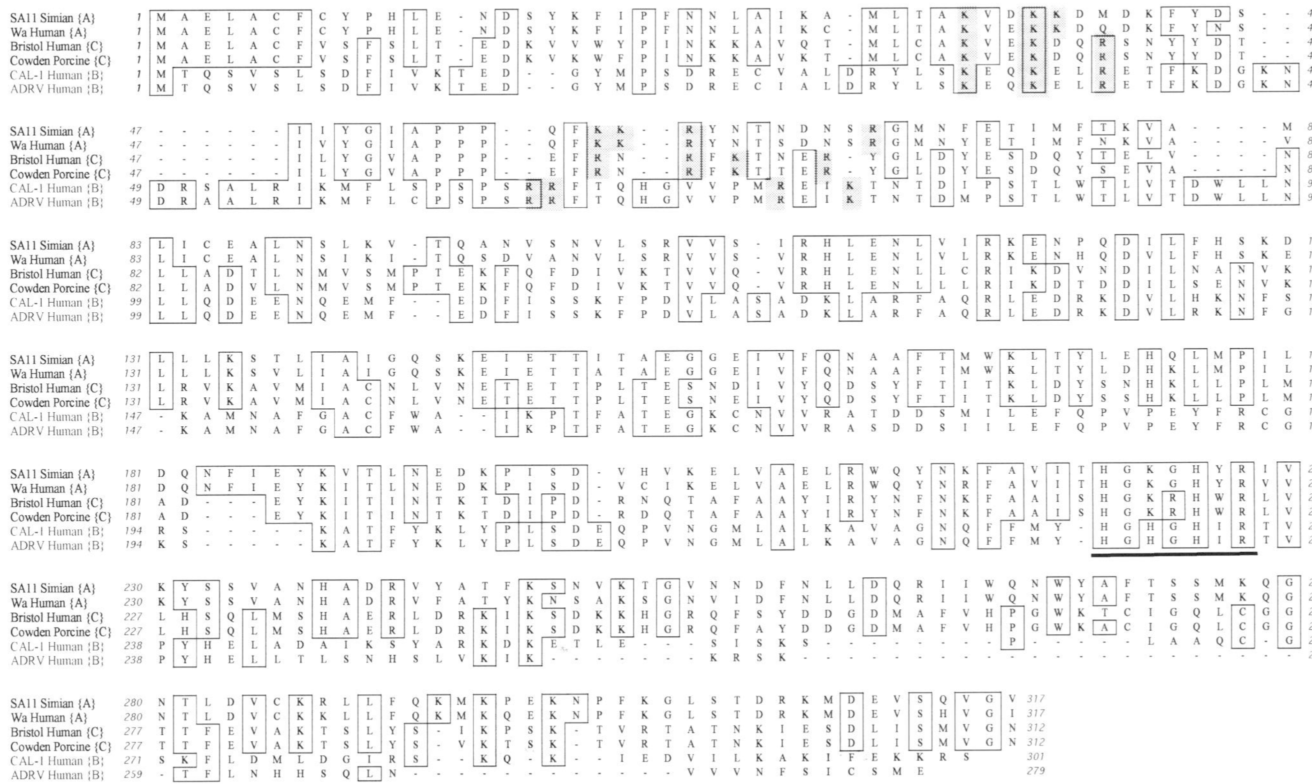

Figure 16.3 NSP2 sequence analysis. Alignment of NSP2 sequences of representative group A, B, and C viruses using ClustalW (Thompson *et al.*, 1994). Residues that make up the HIT-like motif (^{221}H-G-K/H-G/R-H-X-R-X-^{229}V, where X is a variable residue) are underlined. Conserved basic residues of the RNA-binding region (residues 38–77) are highlighted in grey. Genbank Accession numbers of the strains are as follows: SA11 simian (group A), AAA47298; Wa human (A), AAA47301; Bristol human (C), AJ132205; Cowden porcine (C), CAA46742; CAL-1 human (B), AAF72868; ADRV human (B), AAA47328.

Reovirus and BTV inclusion-forming proteins

Most of what is known about the replication cycles of the *Reoviridae* comes from the characterization of rotavirus, reovirus and bluetongue virus (BTV). Analysis of these viruses suggests that all members of this family produce electron-dense inclusion bodies, like that of the rotavirus viroplasm. In contrast to rotavirus, which requires two non-structural proteins (NSP2 and NSP5) to form a viroplasm, BTV requires but one, NS2 (Thomas *et al.*, 1990). One the other hand, the formation of viroplasms by the mammalian orthoreovirus involves two non-structural protein, μNS, σNS, and the core structural protein μ2 (Broering *et al.*, 2002), (Becker *et al.*, 2001; Becker *et al.*, 2003; Mbisa *et al.*, 2000) and (Mbisa *et al.*, 2000; Parker *et al.*, 2002). Although the inclusion-forming proteins of rotavirus, BTV, and reovirus are functional equivalents, all promoting viroplasm formation, they share no sequence homology and none outside rotavirus NSP2 contains the HIT-like motif. However, comparison of the inclusion-forming proteins of rotavirus (NSP2 and NSP5), BTV (NS2), and reovirus (μNS, σNS, μ2) do reveal three common characteristics.

(1) The inclusion-forming proteins of these viruses all include at least one with ssRNA-binding activity. For rotavirus, NSP2 has strong affinity for ssRNA, while NSP5 only weakly binds this ligand (Kattoura *et al.*, 1992; Taraporewala *et al.*, 1999; Vende *et al.*, 2002). For reovirus and BTV, the ssRNA-binding proteins are σNS and NS2, respectively (Gillian and Nibert, 1998) (Huismans *et al.*, 1987). All these ssRNA-binding proteins have been identified as components of RNA-protein complexes involved in genome replication in infected cells (Antczak and Joklik, 1992; Helmberger-Jones and Patton, 1986; Huismans *et al.*, 1987). Although NSP2, NS2 and σNS have been described as non-specific ssRNA-binding proteins, BTV NS2 has also been reported to specifically recognize structural elements contained in its folded viral (+)RNA (Lymperopoulos *et al.*, 2006; Lymperopoulos *et al.*, 2003).

(2) The inclusion-forming proteins all include a homomultimeric component: NSP2 for rotavirus, NS2 for BTV, and σNS for reovirus (Jayaram *et al.*, 2002; Kattoura *et al.*, 1994; Schuck *et al.*, 2001; Taraporewala *et al.*, 1999). Based on velocity sedimentation, the NS2 multimer is predicted to contain 6–8 subunits and the σNS multimer has 4–8 subunits (Gillian and Nibert, 1998) (Taraporewala *et al.*, 2001; Uitenweerde *et al.*, 1995). Given that the NSP2, NS2, and σNS multimers are relatively large structures (4–8 subunits) formed by ssRNA-binding proteins, it can be anticipated that they all will be multivalent in their RNA-binding activity, allowing them to interact simultaneously with multiple RNA molecules. The fact that these multimers also display cooperativity in their RNA-binding activities suggests that the structures may promote duplex unwinding and stabilize single-stranded regions of (+)RNAs, features characteristics of SSBs (Gillian *et al.*, 2000; Taraporewala and Patton, 2001). Such activities may help the multimers concentrate and retain (+)RNAs in viroplasms and may promote packaging by relaxing RNA secondary structures that could impede the translocation of (+)RNAs into cores.

Additional insight into the structure of BTV NS2 has recently come from the X-ray crystallography of an N-terminal fragment of the protein (residues 1–177) (Butan *et al.*, 2004). This region of the protein has been shown through biochemical assays to include non-specific ssRNA-binding activity. The atomic structure indicates that the N-terminal domain of NS2 supports multimerization and contains a unique β-sandwich topology unlike that seen in the NSP2 octamer. However, like in the NSP2 octamer, the RNA-binding sites of the NS2 multimer are constituted by electropositive amino acids present at monomer-monomer interfaces. The difference in the structures of the N-terminal domain of NSP2 and NS2, combined with the absence of a HIT-like motif in NS2, suggests that in their entirety, the architecture and composition of the NSP2 and NS2 multimers will differ dramatically.

(3) The inclusion-forming proteins of rotavirus, reovirus and BTV all include a component that is capable of hydrolyzing phosphoanhydride bonds. Like rotavirus NSP2, reovirus μ2 exhibits Mg^{2+}-dependent NTPase and RTPase activities (Kim *et al.*, 2004; Taraporewala *et al.*, 1999). In contrast, BTV NS2 is a phosphohydrolase, capable of cleaving all three phosphate groups from an NTP (Horscroft and Roy, 2000; Taraporewala *et al.*, 2001). Although the function of the catalytic activities of NSP2, NS2, and μ2 are not known, the fact that the inclusion-forming proteins of rotavirus, BTV, and reovirus all include one that can hydrolyze phosphate residues suggests that these activities have a common critical role in viroplasm function. Since mutant species of rotavirus NSP2 and reovirus μ2 remain capable of supporting viroplasm formation, even though defective in NTPase activity, the purpose of the catalytic activity is probably not connected to the protein–protein interactions that drive inclusion assembly (Carpio *et al.*, 2004; Kim *et al.*, 2004). Instead, the common function of the NTP-hydrolyzing proteins may be to provide the energy needed in some step of RNA packaging (e.g. translocation) or replication. Certainly, the observation that the inclusion forming proteins have ssRNA-binding activity, are components of RIs, and can interact with viral RNA polymerases is consistent with this possibility (Kattoura *et al.*, 1994; Aponte *et al.*, 1996; Kim *et al.*, 2004). Alternatively, NTP hydrolysis may drive conformation changes in NSP2, NS2, and σNS multimers, as part of a mechanism for controlling their interaction with ligands, and in turn their functions within the viroplasm. Whether or not RTPase and NDP-kinase activities are common properties of the inclusion-forming proteins is not known.

Summary

Structure and function analyses of NSP2 have provided exceptional insights into the multifaceted nature of the protein. NSP2 is involved at several steps in the replication cycle: interaction with NSP5 to establish viroplasms, interaction with viral (+)RNA in support of RNA packaging and replication, and HIT-dependent hydrolysis of NTP and/or pppRNA to create energy, rebalance NTP pools, or drive motor functions. Its dynamic and multivalent structure allows the NSP2 octamer to multitask and simultaneously interact with many ligands: NTP, RNA, NSP5 and VP1. Competition between NSP5 and (+)RNA for the grooves of the octamer likely modulate its function between viroplasm building and RNA replication. The fact that NSP2 has retained this unique octameric architecture even among distantly related rotavirus strains further hints towards the importance of this multimeric organization. The symmetry of the NSP2 octamer and the opposing directionality of the tetramers in each octamer unit, are features reminiscent of proteins delivering a work function (motor/packaging proteins). Yet the unusual collection of biochemical properties determined for NSP2 raises questions as to the potential target and the mechanism of such work. A better understanding of the role of NSP2 will come from studies pursuing additional insight into the structural and biochemical impact of NSP2 interactions with its ligands. Particularly important is determining the physical location of the NSP2 octamer in core-like RIs, structures in the viroplasm that concurrently support genome packaging and replication.

Acknowledgements

We acknowledge support from the Intramural Research Program of National Institute of Allergy and Infectious Diseases, NIH (to ZFT and JTP), and from NIH grant AI36040 (to BVVP).

References

Afrikanova, I., Fabbretti, E., Miozzo, M.C., and Burrone, O.R. (1998) Rotavirus NSP5 phosphorylation is up-regulated by interaction with NSP2. J. Gen. Virol. *79*(Pt 11), 2679–86.

Antczak, J.B., and Joklik, W.K. (1992) Reovirus genome segment assortment into progeny genomes studied by the use of monoclonal antibodies directed against reovirus proteins. Virology *187*(2), 760–76.

Aponte, C., Poncet, D., and Cohen, J. (1996) Recovery and characterization of a replicase complex in rotavirus-infected cells by using a monoclonal antibody against NSP2. J. Virol. *70*(2), 985–91.

Becker, M.M., Goral, M.I., Hazelton, P.R., Baer, G.S., Rodgers, S.E., Brown, E.G. Coombs, K.M., and Dermody, T.S. (2001) Reovirus sigmaNS protein is required for nucleation of viral assembly complexes and formation of viral inclusions. J. Virol. *75*(3), 1459–75.

Becker, M.M., Peters, T.R., and Dermody, T.S. (2003) Reovirus sigma NS and mu NS proteins form cytoplasmic inclusion structures in the absence of viral infection. J. Virol. *77*(10), 5948–63.

Boehmer, P.E., and Lehman, I.R. (1993) Herpes simplex virus type 1 ICP8: helix-destabilizing properties. J. Virol. *67*(2), 711–5.

Brenner, C. (2002) Hint, Fhit, and GalT: function, structure, evolution, and mechanism of three branches of the histidine triad superfamily of nucleotide hydrolases and transferases. Biochemistry *41*(29), 9003–14.

Brenner, C., Bieganowski, P., Pace, H.C., and Huebner, K. (1999) The histidine triad superfamily of nucleotide-binding proteins. J Cell Physiol *181*(2), 179–87.

Broering, T.J., Parker, J.S., Joyce, P.L., Kim, J., and Nibert, M.L. (2002) Mammalian reovirus nonstructural protein microNS forms large inclusions and colocalizes with reovirus microtubule-associated protein micro2 in transfected cells. J. Virol. *76*(16), 8285–97.

Brookes, S.M., Hyatt, A.D., and Eaton, B.T. (1993) Characterization of virus inclusion bodies in bluetongue virus-infected cells. J. Gen. Virol. *74*(Pt 3), 525–30.

Butan, C., Van Der Zandt, H., and Tucker, P.A. (2004) Structure and assembly of the RNA binding domain of bluetongue virus non-structural protein 2. J. Biol. Chem. *279*(36), 37613–21.

Campagna, M., Eichwald, C., Vascotto, F., and Burrone, O.R. (2005) RNA interference of rotavirus segment 11 mRNA reveals the essential role of NSP5 in the virus replicative cycle. J. Gen. Virol. *86*(Pt 5), 1481–7.

Carpio, R.V., Gonzalez-Nilo, F.D., Jayaram, H., Spencer, E., Prasad, B.V., Patton, J.T., and Taraporewala, Z.F. (2004) Role of the histidine triad-like motif in nucleotide hydrolysis by the rotavirus RNA-packaging protein NSP2. J. Biol. Chem. *279*(11), 10624–33.

Chen, D., Gombold, J.L., and Ramig, R.F. (1990) Intracellular RNA synthesis directed by temperature-sensitive mutants of simian rotavirus SA11. Virology *178*(1), 143–51.

Cuadras, M.A., Bordier, B.B., Zambrano, J.L., Ludert, J.E., and Greenberg, H.B. (2006) Dissecting rotavirus particle-raft interaction with small interfering RNAs: insights into rotavirus transit through the secretory pathway. J. Virol. *80*(8), 3935–46.

Desselberger, U. (2007) 9th dsRNA virus sumposium. Future Vir *2*(2), 145–152.

Eichwald, C., Jacob, G., Muszynski, B., Allende, J.E., and Burrone, O.R. (2004a) Uncoupling substrate and activation functions of rotavirus NSP5: phosphorylation of Ser-67 by casein kinase 1 is essential for hyperphosphorylation. Proc. Natl Acad. Sci. USA *101*(46), 16304–9.

Eichwald, C., Rodriguez, J.F., and Burrone, O.R. (2004b) Characterization of rotavirus NSP2/NSP5 interactions and the dynamics of viroplasm formation. J. Gen. Virol. *85*(Pt 3), 625–34.
Eichwald, C., Vascotto, F., Fabbretti, E., and Burrone, O.R. (2002) Rotavirus NSP5: mapping phosphorylation sites and kinase activation and viroplasm localization domains. J. Virol. *76*(7), 3461–70.
Estes, M.K. (2001) Rotaviruses and their replication, 1747–1785 pp. 4th ed. Fields Viorlogy, edited by H.M. Knipe D., *et al.* Lippincott Williams & Wilkins, Philadelphia.
Fabbretti, E., Afrikanova, I., Vascotto, F., and Burrone, O.R. (1999) Two non-structural rotavirus proteins, NSP2 and NSP5, form viroplasm-like structures *in vivo*. J. Gen. Virol. 80 (Pt 2), 333–9.
Gallegos, C.O., and Patton, J.T. (1989) Characterization of rotavirus replication intermediates: a model for the assembly of single-shelled particles. Virology *172*(2), 616–27.
Gillian, A.L., and Nibert, M.L. (1998) Amino terminus of reovirus nonstructural protein sigma NS is important for ssRNA binding and nucleoprotein complex formation. Virology *240*(1), 1–11.
Gillian, A.L., Schmechel, S.C., Livny, J., Schiff, L.A., and Nibert, M.L. (2000) Reovirus protein sigmaNS binds in multiple copies to single-stranded RNA and shares properties with single-stranded DNA binding proteins. J. Virol. *74*(13), 5939–48.
Gonzalez, S.A., and Burrone, O.R. (1991) Rotavirus NS26 is modified by addition of single O-linked residues of N-acetylglucosamine. Virology *182*(1), 8–16.
Gouet, P., Diprose, J.M., Grimes, J.M., Malby, R., Burroughs, J.N., Zientara, S., Stuart, D.I., and Mertens, P.P. (1999) The highly ordered double-stranded RNA genome of bluetongue virus revealed by crystallography. Cell 97(4), 481–90.
Helmberger-Jones, M., and Patton, J.T. (1986) Characterization of subviral particles in cells infected with simian rotavirus SA11. Virology *155*(2), 655–65.
Horscroft, N.J., and Roy, P. (2000) NTP binding and phosphohydrolase activity associated with purified bluetongue virus non-structural protein NS2. J. Gen. Virol. *81*(Pt 8), 1961–5.
Huismans, H., van Dijk, A.A., and Bauskin, A.R. (1987) *In vitro* phosphorylation and purification of a nonstructural protein of bluetongue virus with affinity for single-stranded RNA. J. Virol. *61*(11), 3589–95.
Jayaram, H., Taraporewala, Z., Patton, J.T., and Prasad, B.V. (2002) Rotavirus protein involved in genome replication and packaging exhibits a HIT-like fold. Nature *417*(6886), 311–5.
Jiang, X., Jayaram, H., Kumar, M., Ludtke, S.J., Estes, M.K., and Prasad, B.V. (2006) Cryoelectron microscopy structures of rotavirus NSP2-NSP5 and NSP2-RNA complexes: implications for genome replication. J. Virol. *80*(21), 10829–35.
Kadare, G., and Haenni, A.L. (1997) Virus-encoded RNA helicases. J. Virol. *71*(4), 2583–90.
Kapikian, A.Z., Hoshino, Y., and R. M. Chanock. (2001) Rotaviruses, 1787–1834 pp. 4 ed. Field's Virology edited by D.M. Knipe, and Howley, Peter M. 2. 2 vols. Lippincott Williams & Wilkins, Philadelphia.
Kattoura, M.D., Chen, X., and Patton, J.T. (1994) The rotavirus RNA-binding protein NS35 (NSP2) forms 10S multimers and interacts with the viral RNA polymerase. Virology *202*(2), 803–13.
Kattoura, M.D., Clapp, L.L., and Patton, J.T. (1992) The rotavirus nonstructural protein, NS35, possesses RNA-binding activity *in vitro* and *in vivo*. Virology *191*(2), 698–708.
Kim, J., Parker, J.S., Murray, K.E., and Nibert, M.L. (2004) Nucleoside and RNA triphosphatase activities of orthoreovirus transcriptase cofactor mu2. J. Biol. Chem. *279*(6), 4394–403.
Kumar, M., Jayaram, H., Vasquez-Del Carpio, R, Jiang, X., Taraporewala, Z.F., Jacobson, R.H., Patton, J.T., and Prasad, B.V.V. (2007). Crystallographic and biochemical analysis of rotavirus NSP2 with nucleotides reveals NDP kinase like activity. J. Virol. (*in press*).
Lawton, J.A., Estes, M.K., and Prasad, B.V. (1997a) Three-dimensional visualization of mRNA release from actively transcribing rotavirus particles. Nat. Struct. Biol. *4*(2), 118–21.
Lawton, J.A., Zeng, C.Q., Mukherjee, S.K., Cohen, J., Estes, M.K., and Prasad, B.V. (1997b) Three-dimensional structural analysis of recombinant rotavirus-like particles with intact and amino-terminal-deleted VP2: implications for the architecture of the VP2 capsid layer. J. Virol. *71*(10), 7353–60.
Lima, C.D., Klein, M.G., and Hendrickson, W.A. (1997) Structure-based analysis of catalysis and substrate definition in the HIT protein family. Science *278*(5336), 286–90.
Liu, M., Mattion, N.M., and Estes, M.K. (1992) Rotavirus VP3 expressed in insect cells possesses guanylyltransferase activity. Virology *188*(1), 77–84.
Lopez, T., Rojas, M., Ayala-Breton, C., Lopez, S., and Arias, C.F. (2005) Reduced expression of the rotavirus NSP5 gene has a pleiotropic effect on virus replication. J. Gen. Virol. *86*(Pt 6), 1609–17.

Lymperopoulos, K., Noad, R., Tosi, S., Nethisinghe, S., Brierley, I., and Roy, P. (2006) Specific binding of Bluetongue virus NS2 to different viral plus-strand RNAs. Virology *353*(1), 17–26.

Lymperopoulos, K., Wirblich, C., Brierley, I., and Roy, P. (2003) Sequence specificity in the interaction of Bluetongue virus non-structural protein 2 (NS2) with viral RNA. J. Biol. Chem. *278*(34), 31722–30.

Mbisa, J.L., Becker, M.M., Zou, S., Dermody, T.S., and Brown, E.G. (2000) Reovirus mu2 protein determines strain-specific differences in the rate of viral inclusion formation in L929 cells. Virology *272*(1), 16–26.

McCrae, M.A., and McCorquodale, J.G. (1983) Molecular biology of rotaviruses. V. Terminal structure of viral RNA species. Virology *126*(1), 204–12.

McIntosh, D.B., Clausen, J.D., Woolley, D.G., MacLennan, D.H., Vilsen, B., and Andersen, J.P. (2003) ATP binding residues of sarcoplasmic reticulum Ca2+-ATPase. Ann N Y Acad Sci 986, 101–5.

Miller, C.L., Broering, T.J., Parker, J.S., Arnold, M.M., and Nibert, M.L. (2003) Reovirus sigma NS protein localizes to inclusions through an association requiring the mu NS amino terminus. J. Virol. *77*(8), 4566–76.

Monaghan, A., Webster, A., and Hay, R.T. (1994) Adenovirus DNA binding protein: helix destabilising properties. Nucleic Acids Res *22*(5), 742–8.

Parker, J.S., Broering, T.J., Kim, J., Higgins, D.E., and Nibert, M.L. (2002) Reovirus core protein mu2 determines the filamentous morphology of viral inclusion bodies by interacting with and stabilizing microtubules. J. Virol. *76*(9), 4483–96.

Patton, J.T., and Chen, D. (1999) RNA-binding and capping activities of proteins in rotavirus open cores. J. Virol. *73*(2), 1382–91.

Patton, J.T., and Gallegos, C.O. (1988) Structure and protein composition of the rotavirus replicase particle. Virology *166*(2), 358–65.

Patton, J.T., Jones, M.T., Kalbach, A.N., He, Y.W., and Xiaobo, J. (1997). Rotavirus RNA polymerase requires the core shell protein to synthesize the double-stranded RNA genome. *J. Virol.* **71**(12), 9618–26.

Patton, J.T., Silvestri, L.S., Tortorici, M.A., Vasquez-Del Carpio, R., and Taraporewala, Z.F. (2006) Rotavirus genome replication and morphogenesis: role of the viroplasm. Curr. Top. Microbiol. Immunol. 309, 169–87.

Pesavento, J.B., Lawton, J.A., Estes, M.E., and Venkataram Prasad, B.V. (2001) The reversible condensation and expansion of the rotavirus genome. Proc. Natl Acad. Sci. USA *98*(4), 1381–6.

Petrie, B.L., Graham, D.Y., Hanssen, H., and Estes, M.K. (1982) Localization of rotavirus antigens in infected cells by ultrastructural immunocytochemistry. J. Gen. Virol. *63*(2), 457–67.

Petrie, B.L., Greenberg, H.B., Graham, D.Y., and Estes, M.K. (1984) Ultrastructural localization of rotavirus antigens using colloidal gold. Virus. Res. *1*(2), 133–52.

Poncet, D., Lindenbaum, P., L'Haridon, R., and Cohen, J. (1997) In vivo and *in vitro* phosphorylation of rotavirus NSP5 correlates with its localization in viroplasms. J. Virol. *71*(1), 34–41.

Prasad, B.V., Wang, G.J., Clerx, J.P., and Chiu, W. (1988) Three-dimensional structure of rotavirus. J. Mol. Biol. *199*(2), 269–75.

Raghunathan, S., Kozlov, A.G., Lohman, T.M., and Waksman, G. (2000) Structure of the DNA binding domain of E. coli SSB bound to ssDNA. Nat. Struct. Biol. *7*(8), 648–52.

Ramig, R.F., and Petrie, B.L. (1984) Characterization of temperature-sensitive mutants of simian rotavirus SA11: protein synthesis and morphogenesis. J. Virol. *49*(3), 665–73.

Rhim, J.S., Jordan, L.E., and Mayor, H.D. (1962) Cytochemical, fluorescent-antibody and electron microscopic studies on the growth of reovirus (ECHO 10) in tissue culture. Virology 17, 342–55.

Rochester, S.C., and Traktman, P. (1998) Characterization of the single-stranded DNA binding protein encoded by the vaccinia virus I3 gene. J. Virol. 72(4), 2917–26.

Schuck, P., Taraporewala, Z., McPhie, P., and Patton, J.T. (2001) Rotavirus nonstructural protein NSP2 self-assembles into octamers that undergo ligand-induced conformational changes. J. Biol. Chem. *276*(13), 9679–87.

Silvestri, L.S., Taraporewala, Z.F., and Patton, J.T. (2004) Rotavirus replication: plus-sense templates for double-stranded RNA synthesis are made in viroplasms. J. Virol. *78*(14), 7763–74.

Soengas, M.S., Gutierrez, C., and Salas, M. (1995) Helix-destabilizing activity of phi 29 single-stranded DNA binding protein: effect on the elongation rate during strand displacement DNA replication. J. Mol. Biol. *253*(4), 517–29.

Taraporewala, Z., Chen, D., and Patton, J.T. (1999) Multimers formed by the rotavirus nonstructural protein NSP2 bind to RNA and have nucleoside triphosphatase activity. J. Virol. 73(12), 9934–43.

Taraporewala, Z.F., Chen, D., and Patton, J.T. (2001) Multimers of the bluetongue virus nonstructural protein, NS2, possess nucleotidyl phosphatase activity: similarities between NS2 and rotavirus NSP2. Virology *280*(2), 221–31.

Taraporewala, Z.F., Jiang, X., Vasquez-Del Carpio, R., Jayaram, H., Prasad, B.V., and Patton, J.T. (2006) Structure-function analysis of rotavirus NSP2 octamer by using a novel complementation system. J. Virol. *80*(16), 7984–94.

Taraporewala, Z.F., and Patton, J.T. (2001) Identification and characterization of the helix-destabilizing activity of rotavirus nonstructural protein NSP2. J. Virol. *75*(10), 4519–27.

Taraporewala, Z.F., and Patton, J.T. (2004) Nonstructural proteins involved in genome packaging and replication of rotaviruses and other members of the *Reoviridae*. Virus. Res. *101*(1), 57–66.

Thomas, C.P., Booth, T.F., and Roy, P. (1990) Synthesis of bluetongue virus-encoded phosphoprotein and formation of inclusion bodies by recombinant baculovirus in insect cells: it binds the single-stranded RNA species. J. Gen. Virol. 71 (Pt 9), 2073–83.

Thompson, J.D., Higgins, D.G., and Gibson, T.J. (1994) CLUSTAL W: improving the sensitivity of progressive multiple sequence alignment through sequence weighting, position-specific gap penalties and weight matrix choice. Nucleic Acids Res *22*(22), 4673–80.

Tortorici, M.A., Broering, T.J., Nibert, M.L., and Patton, J.T. (2003). Template recognition and formation of initiation complexes by the replicase of a segmented double-stranded RNA virus. *J. Biol. Chem. 278*(35), 32673–82.

Tsurumi, T., Kishore, J., Yokoyama, N., Fujita, M., Daikoku, T., Yamada, H., Yamashita, Y., and Nishiyama, Y. (1998) Overexpression, purification and helix-destabilizing properties of Epstein-Barr virus ssDNA-binding protein. J. Gen. Virol. 79 (Pt 5), 1257–64.

Uitenweerde, J.M., Theron, J., Stoltz, M.A., and Huismans, H. (1995) The multimeric nonstructural NS2 proteins of bluetongue virus, African horsesickness virus, and epizootic hemorrhagic disease virus differ in their single-stranded RNA-binding ability. Virology *209*(2), 624–32.

Valenzuela, S., Pizarro, J., Sandino, A.M., Vasquez, M., Fernandez, J., Hernandez, O., Patton, J., and Spencer, E. (1991) Photoaffinity labeling of rotavirus VP1 with 8-azido-ATP: identification of the viral RNA polymerase. J. Virol. *65*(7), 3964–7.

Vasquez-Del Carpio, R., Gonzalez-Nilo, F.D., Riadi, G., Taraporewala, Z.F., and Patton, J.T. (2006) Histidine triad-like motif of the rotavirus NSP2 octamer mediates both RTPase and NTPase activities. J. Mol. Biol. *362*(3), 539–54.

Vende, P., Taraporewala, Z.F., and Patton, J.T. (2002) RNA-binding activity of the rotavirus phosphoprotein NSP5 includes affinity for double-stranded RNA. J. Virol. *76*(10), 5291–9.

Welch, S.K., Crawford, S.E., and Estes, M.K. (1989) Rotavirus SA11 genome segment 11 protein is a nonstructural phosphoprotein. J. Virol. *63*(9), 3974–82.

Function and Structure of Rotavirus NSP3

17

Michelle M. Becker, Stefan T. Arold, Stephen. K. Burley, Rahul C. Deo, Caroline M. Groft, Damien Vitour, and Didier Poncet

Abstract

Rotaviruses encode a protein, NSP3, known to intercept the cellular function of translation, such that capped, but not polyadenylated, viral mRNAs are preferentially translated in the presence of cellular mRNAs optimized for translation by the cellular machinery. The generalities of these interactions have been determined through molecular biology, but the specifics necessary for NSP3 to carry out the misdirection of the components of the translation apparatus are revealed only through detailed structural analysis. In this chapter, the interactions of NSP3 with eukaryotic initiation factor 4G and with the 3′ end of viral RNA, as well as the contacts required to form NSP3 homodimers are discussed. In addition, a novel NSP3-interacting protein is presented and this protein, dubbed RoXaN (Rotavirus: X protein associated with NSP3), has no known function, however the potential contact between NSP3 and RoXaN bears intriguing similarities to interactions between other cellular proteins involved in cytoskeletal rearrangements through paxillin. All of this indicates that a single viral protein may have multiple functions during the viral life cycle through interactions with various cellular proteins and pathways, and these are revealed in intricate detail through structural studies.

Introduction

Most viruses contain a limited genome, yet are capable of entering a cell and diverting the cell's genetic program to that of the virus, with the goal of producing an abundance of progeny virions. In the study of viruses and the proteins they produce, insights into cellular activities are often unearthed. Most cellular events are quite complex and can involve tens to hundreds of proteins that are expressed or modified at different times during development or in reaction to stimuli. Viral strategies frequently involve the utilization of one or two proteins to stop or divert cellular activities. By studying the interaction of the viral and cellular proteins, the complexity of eukaryotic processes can be simplified and examined with greater facility or, alternatively, previously unknown cellular processes may be discovered.

These statements are true for many viruses, including rotavirus, and for one rotaviral protein in particular. Rotaviruses encode non-structural protein NSP3, which is a dimer and binds a group-specific consensus tetranucleotide located at the 3′ end of rotavirus single-stranded (ss) RNAs. In addition, NSP3 interacts with eukaryotic initiation factor (eIF) 4G. This discovery puts this viral protein squarely in the middle of the cellular process of protein translation and has led to a variety of clues, in addition to strengthening what was already known, about how cells translate their own mRNAs and how rotaviruses interfere and manipulate cells to translate viral mRNA. Recently, NSP3 was also found to interact with another cellular protein, called RoXaN (Rotavirus: X protein associated with NSP3). At this time, the only known function of RoXaN is its association with NSP3, thus providing a handle with which to commence examination of the function of an unknown, and perhaps, important cellular protein.

The identification of the interactions between viral and cellular proteins and RNA, and the ensuing biological implications, are important and are greatly enhanced by structural data, furthering comprehension of both the function of viral and cellular proteins and processes in which they are involved. In this chapter, we will focus on the previously determined structures of NSP3 in complex with its various binding partners, and on the predicted structures of the newly identified NSP3-interacting protein, RoXaN.

Rotavirus NSP3 protein

Group A, B, and C rotaviruses all encode NSP3 (Both *et al.*, 1984; Qian *et al.*, 1991; Eiden, 1993). There is some variability in the size of NSP3 from different groups, but even within group A, NSP3 ranges in size from 306 to 315 amino acids, depending on the rotavirus strain from which it originates. Group A NSP3 from rotavirus Simian Agent 11 (SA11) is a 315 amino acid, 36kDa protein (Both *et al.*, 1984) with three structurally and biochemically distinct domains. The amino-terminal portion of NSP3 binds to a group-specific tetranucleotide consensus sequence found at the 3′ end of rotaviral mRNAs (Poncet *et al.*, 1994; Deo *et al.*, 2002). This activity has been mapped to the region including amino acids 1–150. RNA-binding activity has been delimited to a minimal region within this portion of the protein via cross-linking and gel retardation assay (Piron *et al.*, 1999), and is required for translation of viral mRNAs both *in vitro* and *in vivo* (Vende *et al.*, 2000). The stoichiometry of binding was predicted to be a single RNA ligand per NSP3 dimer (Piron *et al.*, 1999). Only four nucleotides are necessary for high-affinity binding, although these nucleotides must be located at the 3′ end of the RNA (Poncet *et al.*, 1994; Vende *et al.*, 2000). In the case of group A rotavirus, the consensus sequence consists of UGACC. There are several confirmed exceptions to this rule, four in NSP1-encoding gene 5, one in VP2-encoding gene 2 (SA11), and one in NSP3-encoding gene 7 (SA11) (Patton *et al.*, 2001; Kearney *et al.*, 2004).

The middle portion of the protein, amino acids 150–250, consists of a region predicted to form a coiled-coil structure and is required for the dimerization of NSP3, which is necessary to enable NSP3 to interact with either viral mRNA or its cellular protein partners. The remaining carboxy-terminal third of the protein, amino acids 206–313, has been found, initially by yeast two-hybrid analysis, to interact with a cellular protein, eIF4G (Piron *et al.*, 1998), a critical component of protein translation initiation and most likely, re-initiation.

Therefore, this viral protein, NSP3, is capable of three different, concurrent, intermolecular interactions, first with viral mRNA, secondly with itself in forming a functional homodimer, and thirdly with a cellular protein critical in protein translation. Further examination of each of these interactions, especially with the detail offered through structural information, will generate a clearer understanding of the role played by this unique protein in viral translation and replication, and cellular translation.

Rotavirus NSP3 and cellular translation

Most cellular mRNAs are appended with a 5′ cap structure and a 3′ poly(A) tail, which are added cotranscriptionally (Figure 17.1, left). Following processing, mRNAs are exported from the nucleus through the nuclear pore. In the cytoplasm, eIF4E binds to the 5′ cap structure and poly(A) binding protein (PABP) binds to the 3′ poly(A) sequence. In the cytoplasm, the scaffold protein eIF4G binds to both eIF4E and PABP, circularizing the mRNA (Figure 17.1, left). The formation of this complex initiates a cascade of interactions, starting with the association of eIF4G and the multiprotein complex eIF3, followed by the interaction of the small ribosome subunit with eIF3, and ending with a complete ribosome

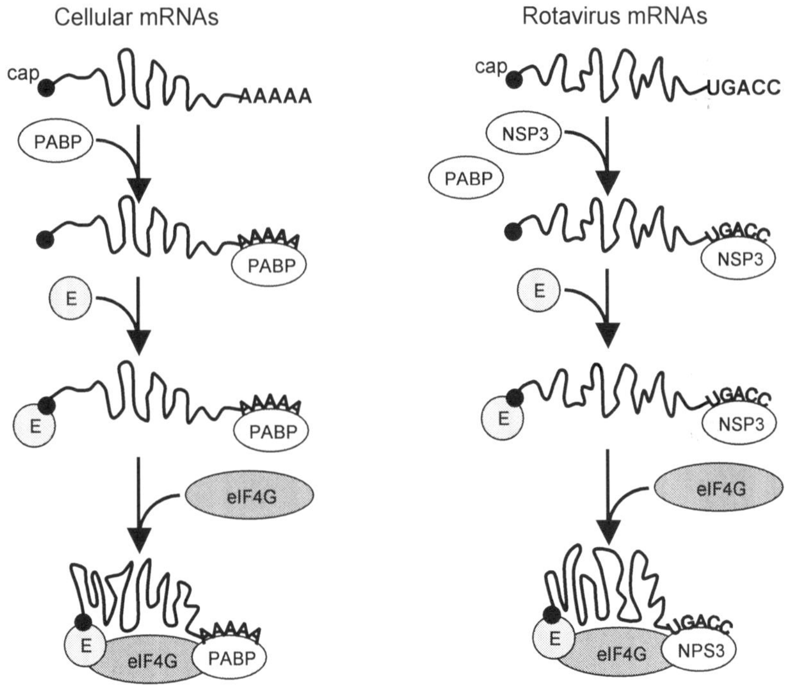

Figure 17.1 Comparison of translation of cellular and rotaviral mRNAs. (Left) By the time they reach the cytoplasm, most cellular mRNAs are capped and have a poly(A) tail. The cap structure is bound by eIF4E and the poly(A) tail by PABP. These two proteins both interact with a bridging protein, eIF4G, which leads to the circularization of the mRNA. Upon formation of this complex, a cascade of interactions is initiated, leading to the commencement of translation. (Right) In contrast, rotavirus mRNAs are capped, but not polyadenylated. The viral mRNAs contain a group-specific tetranucleotide consensus sequence at the 3′ end, which is bound by rotavirus non-structural protein NSP3. The viral mRNA cap is recognized by eIF4E. NSP3 is capable of interacting with eIF4G, thus both sequestering this essential protein away from complexes containing cellular mRNAs and inserting viral mRNAs into the protein interaction cascade leading to translation by the cellular machinery.

bound to an initiation codon and loaded with an initiator met-tRNA (Preiss and Hentze, 2003). Circularization of the mRNA seems particularly important for efficient and accurate translation initiation when competition exists between mRNAs (Preiss and Hentze, 1998), or when ribosomes or initiation factors are in limited supply (Proweller and Butler, 1997; Michel *et al.*, 2001). Interestingly, the cap structure, the methyl-7-guanosine linked to the 5′ end of the mRNA, was first discovered using reovirus ssRNAs, (Furuichi *et al.*, 1975) but reovirus mRNAs lack a 3′ poly(A) sequence. Taking into account the importance of the 3′ poly(A) tail in this process, the question arises of how a non-polyadenylated mRNA synthesized directly in the cytoplasm can be successfully inserted into this cascade of events and compete efficiently with cellular polyadenylated mRNAs for translation during viral infection. Reoviruses and rotaviruses are classified in the same family and have the same basic mRNA structure, containing a 5′ cap structure, but no poly(A) sequence at the 3′ end, so the question applies to both viruses (Estes, 2001; Nibert and Schiff, 2001). For rotaviruses, the answer is at least partly explained by the interaction of NSP3 and eIF4G. Rotavirus NSP3 enhances viral and inhibits cellular translation by simultaneously binding to the 3′ conserved sequence of rotavirus mRNAs (Poncet *et al.*, 1993) and to eIF4G (Piron *et al.*, 1998). NSP3 acts as a functional homologue of the cellular protein PABP, inserting non-polyadenylated rotavirus message into the poly(A)-dependent cellular translational machinery (Figure 17.1, right). Because NSP3 binds to eIF4G in the same region as PABP, but with higher affinity (Imataka *et al.*, 1998; Piron *et al.*, 1998), NSP3 is able to inhibit cellular translation by sequestering eIF4G, a critical component of the translational apparatus. The viral mRNA-NSP3-eIF4G complex that is formed effectively cloaks the viral mRNA and permits it to be recognized by the other components of the cellular translational machinery, thus allowing viral proteins to be efficiently produced at the expense of cellular proteins (Figure 17.1, right).

Recognition of the rotavirus mRNA 3′ consensus by NSP3

The X-ray crystal structure of the amino-terminus of NSP3 of group A SA11 (NSP3-N) in complex with the group A rotavirus mRNA 3′ consensus sequence has been determined to 2.45 Å (Deo *et al.*, 2002). NSP3-N forms a heart-shaped, asymmetric homodimer, with a central RNA binding cleft. The amino-terminal segment of each monomer, consisting of approximately 75 residues, is formed by three α-helices, while the carboxy-terminal segment of each monomer, approximately 50 residues, includes 4 α-helices and 3 β-strands (Figure 17.2A). The NSP3-N homodimer is asymmetric with the two monomers differing partially in secondary structure element composition and more dramatically in the spatial arrangement of these elements. In fact, the two monomers cannot be superimposed with one another. It is likely that a combination of extensive intermolecular packing and RNA binding contribute to this asymmetry.

The NSP3-N dimerization interface is very extensive, with more than one-third of the residues in each monomer participating in intermolecular contacts. In fact, nearly all secondary structural elements in the two monomers contribute to NSP3-N homodimerization. By examining the composition of the group A dimerization interface, some predictions can be made about dimerization of the RNA-binding domain in the other rotavirus groups. Virtually all of the 79 residues comprising the dimer interface are con-

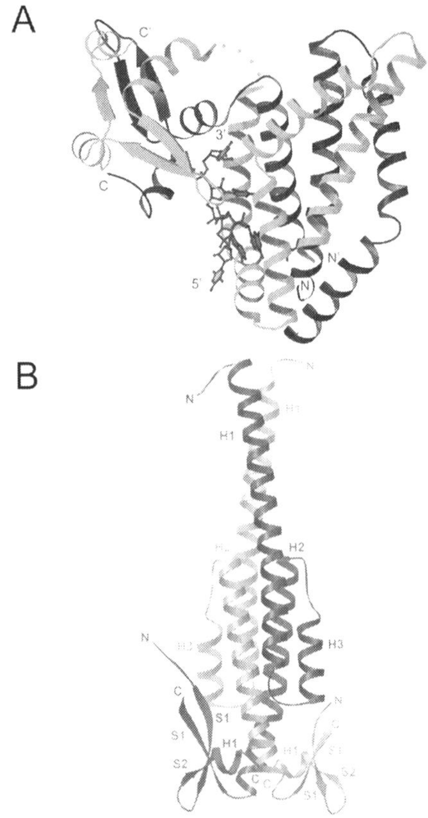

Figure 17.2 A. Structure of SA11 NSP3-N in complex with the rotavirus mRNA 3′ consensus sequence at a resolution of 2.45 Å. RIBBONS drawing showing two monomers of NSP-N. The amino- and carobxy-termini for each monomer are labelled, as are the 5′ and 3′ ends of the bound ssRNA sequence. NSP3-N forms a heart-shaped, asymmetric homodimer, with a central RNA binding cleft. The amino-terminal segment of each monomer, consisting of approximately 75 residues, is formed by three α-helices, while the carboxy-terminal segment of each monomer, approximately 50 residues, includes 4 α-helices and 3 β-strands.

served among different group A rotaviruses (Deo *et al.*, 2002). There is somewhat lower conservation between distinct rotavirus groups. However, even when differences exist, a non-conservative change observed at one amino acid position is often offset by a reciprocal change in the residue with which it interacts. For example, a group A Ser/Val interaction observed in the NSP3-N homodimer is Val/Ser in group C. At another position, an Asn/Arg interaction has been replaced in group B rotavirus by an Arg/Asp interaction. Other substitutions leading to an increase in the residue size at one position are offset by a change to a smaller residue in its interaction partner. Additionally, non-polar residue pairs have replaced pairs of interacting polar residues at certain positions. The preservation of these intermolecular interactions implies considerable evolutionary pressure to conserve the structure of the NSP3-N homodimer. These reciprocal and compensatory substitutions may have functional significance, as they may prevent the formation of non-functional NSP3-N heterodimers upon co-infection of one cell by different groups of rotavirus.

To achieve viral mRNA circularization, NSP3 must recognize the mRNA 3′ consensus sequence amidst a formidable number of cytoplasmic RNAs. Asymmetric homodimerization of NSP3-N creates a single, highly basic RNA-binding tunnel composed of residues from both monomers, which confirms previous work suggesting that full-length NSP3 is dimeric and that the protein-to-RNA stoichiometry is 2:1 (Piron *et al.*, 1999). The RNA-binding tunnel is markedly asymmetric, with different secondary structure elements from each monomer contributing to RNA recognition. Our structure reveals that specificity for rotaviral mRNAs is partially achieved by formation of an RNA-binding close-ended tunnel, which excludes internal mRNA sequences. In addition, NSP3-N employs a combination of hydrogen bonds, van der Waals contacts, salt bridges, and stacking interactions that appear ideally suited for binding the consensus tetranucleotide.

In the X-ray structure, experimental electron density was visible for five nucleotides: 5′-Ura-Gua-Ade-Cyt-Cyt-3′ (referred to as Ura-1, Gua-2, Ade-3, Cyt-4, and Cyt-5), which correspond to the mRNA 3′ end consensus sequence of group A rotavirus (Deo *et al.*, 2002). The RNA-binding tunnel, which lies deep within the homodimer, creates a 'dead end' for the terminal nucleotide. The two cytidines at the 3′ end of the consensus sequence, Cyt-4 and Cyt-5, bind predominantly to one monomer, while the second makes the majority of the contacts with the two purines (Figure 17.2A). No protein-RNA interactions are seen for Ura-1, although there are salt bridges and hydrogen bonds with the phosphate group linking Ura-1 and Gua-2. The 3′ cytidine, Cyt-5, is completely hidden inside the RNA-binding tunnel.

Although each group of rotaviruses utilizes a different 3′ end mRNA consensus sequence (group A: 5′-Gua-Ade-Cyt-Cyt-3′; group B: 5′-Cyt/Ade-Cyt/Ade-Cyt-Cyt-3′; group C: 5′-Gua-Gua-Cyt-Ura-3′), NSP3s from all three rotavirus groups probably employ a common mode of RNA-binding. The majority of residues involved in polar interactions with the phosphate groups or 2′-hydroxyl groups of the RNA are conserved.

Among group A rotaviruses, base specifying residues are conserved. There is also significant conservation of these same residues between groups A and C, which share a similar mRNA 3′ end consensus sequence. Since the group A and C NSP3-N domains share a high degree of sequence identity and are, therefore, likely to have highly similar protein backbone positions, differences in a small number of RNA-binding residues may explain the preference of group C NSP3 for uracil vs. cytidine (Cyt-5) at the terminal position (Deo *et al.*, 2002).

NSP3s from group B rotaviruses show substantial divergence at some of the base-specifying residues. Given the low sequence identity between the group A and B rotavirus NSP3 RNA-binding domains, attempts to explain the differences in RNA recognition specificity are made difficult by the likely differences in protein backbone positions between the proteins. However, there are some indications that all NSP3 proteins employ a similar mode of binding since four key RNA-binding residues are invariant across all rotavirus strains examined (Deo *et al.*, 2002).

To the best of our knowledge, the RNA recognition strategy of NSP3 is unlike that of any other RNA-binding protein. Comparison with the PABP-RNA co-crystal structure

(Deo *et al.*, 1999) reveals that NSP3 fulfils many of the same functional roles as PABP using an entirely different strategy of mRNA 3′ end recognition. PABP uses multiple RNA-recognition motifs (RRMs) to bind the elongated poly(A) tail segment, with multiple copies of PABP binding a single mRNA poly(A) tail. On the contrary, NSP3 uses only a single RNA-binding domain for the purpose of enclosing the rotavirus 3′ consensus sequence. This essential difference may reflect how regulated mRNA stability, which is of significant importance for eukaryotic translational control, likely has a more limited role in the control of rotaviral mRNA translation. In the degradation mechanism of many eukaryotic mRNAs, the first step of mRNA decay involves a gradual 3′-5′ exonucleolytic digestion of the poly(A) tail (reviewed in Wilusz *et al.*, 2001). The presence of multiple bound copies of PABP would permit an enhancement of translation initiation dependent on the length of the poly(A) tail of a specific mRNA (Preiss *et al.*, 1998). Moreover, the multidomain nature of PABP would allow transient dissociation of one or more domains to allow exonuclease access to the poly(A) tail. NSP3 probably serves a less subtle role during the rotavirus life cycle by shielding rotaviral mRNAs from host cell exonucleases and stimulating translation for the duration of the infection.

Replication of the double-stranded rotaviral RNA genome involves negative-strand synthesis from mRNA templates using the 3′ end consensus sequence to initiate replication (reviewed in Patton and Spencer, 2000). This process does not require NSP3 (Bican *et al.*, 1982), and the structure and biophysical characterization of the NSP3-N/RNA complex (Deo *et al.*, 2002) suggest that NSP3 could even interfere with rotaviral genome replication. High-affinity RNA binding (K_d = 79 nM) and the slow dissociation rate ($t_{1/2}$ = 8 hours) raise the question of how a free 3′ end could be generated for priming of negative-strand RNA synthesis. The co-crystal structure of NSP3 suggests active separation of NSP3 from its consensus sequence may be required for viral RNA replication. This could be achieved through proteolysis and/or phosphorylation of one or more critical RNA-binding segments of the protein, which may cause a significant decrease in affinity of the protein for RNA, permitting initiation of negative-strand synthesis by the viral RNA polymerase.

NSP3 is a PABP functional homologue

The carboxy-terminal domain of group A NSP3 (NSP3-C) competes with PABP for the same segment of eIF4G (Imataka *et al.*, 1998; Piron *et al.*, 1998), and this activity is identical for full length NSP3 and for NSP3-C (Piron *et al.*, 1999). Addition of NSP3-C to cell extracts decreases translation of luciferase reporter mRNAs containing the rotaviral 3′ consensus sequence, a poly(A) tail, or an internal consensus sequence shown to be inaccessible for binding by NSP3 (Vende *et al.*, 2000), presumably because NSP3-C sequesters eIF4G so that PABP cannot bind and continue translation initiation. Quantitation via isothermal titration calorimetry (ITC) of the binding between fragments of eIF4GI [eIF4G1(132–160)] and NSP3-C(206–315) (estimated 10–50 nM) and between the same fragment of eIF4GI and the first two RRMs of hPABP(1–190) (27+/−1 μM) supported this theory (Groft and Burley, 2002). During group A rotavirus infection, NSP3 accumulates in higher order complexes with eIF4G, while PABP disappears from the translation machinery (Piron *et al.*, 1998). Eventually, NSP3 is thought to shut down host cell protein production (Michel *et al.*, 2000) by displacing PABP and utilizing the available pool of eIF4G for translation of rotaviral mRNAs bearing the 3′ tetranucleotide consensus sequence.

NSP3-C homodimer structure and symmetry

NSP3-C is a rod-shaped symmetrical homodimer that contains two eIF4G-binding sites that, in the crystal structure, are both bound to fragments of eIF4G (Groft and Burley, 2002). Each monomer consists of three α-helices and a terminal β-strand, which forms a three-stranded anti-parallel β-sheet with the eIF4G fragment (Figure 17.2B). The NSP3-C monomers are largely symmetric and contain a dimeric, parallel coiled-coil, which follows the heptad repeat pattern found in other coiled-coil interactions. However, NSP3-C uses all of its secondary structural elements to promote dimerization, not just the coiled-coil α-helix. In this way, NSP3-C is unlike many well-characterized dimeric coiled-coil proteins, which contain separate, globular domains at the coiled-coil amino- or carboxy-termini. In all, over 50% of the residues in each NSP3-C monomer directly contribute to dimer formation, which involves every secondary structure element, including the carboxy-terminal β-strand. The total solvent-accessible surface area buried upon dimerization (1800 $Å^2$/monomer) is about twice that for an average homodimer of the same molecular weight (Jones and Thornton, 1996). Therefore, extensive homodimerization is seen both with the amino-terminus of NSP3 (Figure 17.2A) (Deo *et al.*, 2002) and the carboxy-terminus of NSP3 (Figure 17.2B) (Groft and Burley, 2002). It is probable that functional NSP3 exists as a homodimer *in vivo* and that labelling the intervening sequence a 'dimerization domain' is a misnomer.

The two eIF4G-binding pockets of the NSP3-C homodimer are virtually identical, and are hydrophobic with a high proportion of leucines and aromatic residues, consistent with a typical protein-binding surface. Each binding pocket consists of residues from both NSP3-C monomers. In the cell, NSP3-C probably only binds to one copy of eIF4G, as there is presumably only one copy of eIF4G per eIF4F complex. However, two identical ligand-binding sites would increase the likelihood that NSP3-C will form a productive complex with the eIF4G subunit of eIF4F. Intermolecular contacts extend to the very carboxy-terminus of NSP3-C (Groft and Burley, 2002), which explains why a carboxy-terminal truncation at residue 292 abolishes eIF4G binding and NSP3-C mediated inhibition of cellular mRNA translation (Vende *et al.*, 2000).

Key interactions between NSP3-C and eIF4G include five eIF4G hydrophobic residues. eIF4GI residues Ile140 and Ile142 insert into a hydrophobic niche created by aromatic residues in NSP3-C (Phe296, Phe299, Tyr308 and Tyr310). Ile152, Ile156, and Met157 from the eIF4GI α-helix contribute additional van der Waals interactions, primarily with NSP3-C residues Tyr308, Met298 and Leu302 (Groft and Burley, 2002).

Mapping group A NSP3 conserved sequences onto the molecular surface of NSP3-C revealed a conserved patch on the eIF4G binding surface, suggesting that other group A NSP3 proteins recognize eIF4G in the same manner. An additional conserved region extends above the peptide binding pocket and includes an absolutely conserved, solvent exposed tryptophan side chain (Trp247) (Groft and Burley, 2002). This hydrophobic/acidic patch could make additional contacts with a portion of eIF4G that was not included in the crystal structure. Alternatively, this conserved surface feature could serve as a binding site for other cellular and/or viral factors.

Very little homology exists between the carboxy-terminal domains of group A rotavirus NSP3 and those of NSP3 proteins from other groups. This is somewhat surprising, because the group A NSP3-C/eIF4G interaction mediates host cell translation takeover. The carboxy-terminal region of group C rotavirus NSP3 is shorter and shares limited homology with group A NSP3. Informative alignments cannot be constructed due to large gaps, although there is modest similarity in the region corresponding to the group A NSP3-C third α-helix and first β-strand. It is possible that the eIF4G binding site is constructed similarly in group A and group C NSP3, because both proteins, as well as PABP, bind the same eIF4G fragment (Piron *et al.*, 1998). NSP3 from group B shares no sequence homology outside the amino-terminal domain with either group A or group C NSP3, and there is no biochemical evidence that group B NSP3 binds to eIF4G. It appears likely that group B rotaviruses use a distinct mechanism to manipulate host-cell translation.

NSP3-C, PABP and eIF4G binding

Like NSP3-N, NSP3-C shares no secondary structural homology with PABP, even though the two bind the same target sequence of eIF4G (Deo *et al.*, 1999; Deo *et al.*, 2002; Groft and Burley, 2002). PABP interactions with both human and yeast eIF4G proteins have been mapped to PABP-RRM2 (Imataka *et al.*, 1998; Kessler and Sachs, 1998). The dorsal surface of PABP-RRM2 (that is, the surface opposite the RNA-binding surface), exhibits several features that resemble the NSP3-C peptide-binding pocket, namely a hydrophobic groove between two α-helices, a patch of aromatic residues, a serial arrangement of hydrophobic residues, and a solvent-exposed lysine (Deo *et al.*, 1999; Groft and Burley, 2002). Alanine point mutations of the corresponding residues in a human PABP construct containing the first two RRMs (residues 1–190) were used in ITC experiments to assess changes in protein–protein affinity with an eIF4G peptide fragment [eIF4G1(132–160)]. Several of the proposed analogous residues did have a deleterious effect on hPABP-eIF4G binding, specifically, residues K157, M161 and L162 of hPABP. Further, mutation of an acidic residue in close proximity to the proposed binding surface to an alanine showed only minimal disruption of binding (Groft and Burley, 2002). Additional support for analogous recognition strategies of NSP3 and PABP for eIF4G came from alanine point mutations of four hydrophobic residues of eIF4GI that make multiple contacts with NSP3-C in the crystal structure (Ile140, Ile142, Ile152 and Ile155). When assessed for PABP RRM1/2 (1–190) binding using ITC experiments, none of the mutant peptides exhibited detectable binding, indicating that eIF4G residues integral for NSP3 binding are also important for binding PABP (Groft and Burley, 2002).

Using ITC assays, it has been demonstrated that NSP3-C and PABP have homologous residues that bind eIF4G, arranged in approximately the same manner on the molecular surface. However, PABP and NSP3 bind the eIF4G fragment with considerably different binding affinities, specifically 27 μM for PABP/eIF4G and 10–50 nM for NSP3-C/eIF4G. This difference allows NSP3 to sequester eIF4G from cellular mRNAs and compete PABP away from potentially viable mRNA circularization events, thereby downregulating host cell translation. It is imperative that NSP3 bind eIF4G tightly, because there is only one NSP3 dimer bound per viral mRNA and there are multiple copies of PABP bound to a given poly(A) tail, each of which could bind eIF4G. Weak binding of eIF4G by PABP would allow productive interactions with eIF4G only in the presence of high local concentrations of PABP, such as when PABP is bound to the 3′ end of an mRNA. NSP3 is able to take advantage of this subtle regulatory interaction, presumably because subtlety is not needed for the viral life cycle. Rotavirus infection is lytic, and is completed in a number of hours. Although there is an abundance of PABP in the cell, the high affinity of NSP3 for both its own mRNA and for eIF4G presumably allows direction of eukaryotic translation machinery toward viral mRNAs and inhibition of formation of the normal eukaryotic mRNA closed loop structure, even relatively early during the course of infection (Piron *et al.*, 1998).

Comparison of the NSP3-C structure with that of RRM2 of PABP suggests two possible explanations for the differing affinities despite similar recognition strategies. First, NSP3-C presents a deep hydrophobic pocket, supporting a large number of protein–protein contacts and efficient solvent exclusion. It is possible that solvent cannot effectively compete for binding given such a buried interface. In contrast, the PABP binding surface is relatively flat. Although there is a hydrophobic groove, this groove is not as deep as that of NSP3-C. Secondly, the last 10–12 residues of NSP3-C hydrogen bond extensively with the eIF4G fragment, creating a three-stranded, anti-parallel β-sheet. This hydrogen bond network may serve to decrease the off-rate of the NSP3-C/eIF4G interaction, which would be necessary for competition with the highly abundant cellular factor PABP. The differences in affinity for eIF4G could reflect a contrast between the importance of regulated mRNA metabolism and translational control in eukaryotic cells and the limited importance of such control during the rotaviral life cycle. It should be noted, however, that without a PABP/eIF4G structure, assumptions about the details of PABP/eIF4G binding remain speculative.

Because much work on mRNA circularization and translation has been done in the yeast system, it is worth noting that yeast employ a distinct PABP/eIF4G recognition strategy. Although the RRM2 of yeast PABP interacts with eIF4G, the RRM2 of human PABP cannot substitute for its yeast counterpart *in vivo*, and human PABP does not bind yeast eIF4G (Otero *et al.*, 1999). Moreover, the region of yeast eIF4G responsible for interactions with yeast PABP differs from the fragment of human eIF4G responsible for PABP and NSP3 interaction (Tarun *et al.*, 1997; Imataka *et al.*, 1998; Piron *et al.*, 1998). Disruption of the interaction between human PABP and human eIF4G by changing Lys157 to Ala, the corresponding residue in yeast PABP, suggests that the residue at position 157 may distinguish the surface of yeast PABP from that of higher eukaryotes. Further *in vivo* analyses will be necessary to confirm this hypothesis (Otero *et al.*, 1999).

A novel cellular partner for NSP3

Recently, we have identified a new cellular partner for rotavirus NSP3. This protein, called RoXaN, is predicted to contain 977 amino acids and includes several recognizable motifs and domains (Vitour *et al.*, 2004). In the amino-terminal portion of the protein two or three tetratricopeptide (TPR) domains are found and five zinc finger motifs are situated in the carboxy-terminal region (Vitour *et al.*, 2004). Between these two domains, a LD motif is found, which corresponds to the consensus sequence LDxLLxxL, first identified in paxillin and shown to mediate the interaction of paxillin with elements of the cytoskeleton (Brown *et al.*, 1996; Tumbarello *et al.*, 2002). RoXaN was first identified by yeast two-hybrid screening for proteins capable of interacting with NSP3. Using deletion mutants, *in vitro* binding assays, two-hybrid and sandwich-hybrid analyses, it has been shown that RoXaN interacts with the coiled-coil intermediate region of NSP3, and that this interaction is compatible with the association between NSP3 and eIF4G. The region of RoXaN that interacts with NSP3 has been defined to be a short stretch of 10 amino acids, well conserved within a region of low homology between RoXaN and another similar protein found on a different chromosome, identified as RoXaN II (Vitour *et al.*, 2004). This 10 amino acid recognition sequence conforms to the consensus for the LD motif (Turner, 2000; Tumbarello *et al.*, 2002). Comparison of the interaction of the LD domain in RoXaN with the parallel coiled coil in NSP3 to the binding of the LD domain in paxillin to the FAT (focal adhesion targeting) domain of focal adhesion kinase (FAK), which consists of a four antiparallel alpha helix bundle (Hoellerer *et al.*, 2003), may shed light on the structural convergence between NSP3 and the FAT domain in FAK.

Function and molecular architecture of LD motif interactions

LD motifs were first identified in the paxillin superfamily (Brown *et al.*, 1996; Brown *et al.*, 1998), but later also in other, unrelated proteins such as the papilloma virus E6 (Tong and Howley, 1997; Vande Pol *et al.*, 1998; reviewed in Tumbarello *et al.*, 2002). Sequence analysis defined the minimal core sequence of LD motifs to be LDxLL. Through binding assays, LD motifs revealed themselves to be recognition sequences mediating protein–protein interactions, similar to, for example, proline-rich motifs or motifs containing phosphorylated tyrosine or serine residues (Brown *et al.*, 1998). Ligands that recognize LD motifs include the carboxy-terminal region of FAK (Hildebrand *et al.*, 1995), vinculin (Turner *et al.*, 1990), GIT-1 (Premont *et al.*, 2000), and Integrin-linked Kinase (ILK) (Nikolopoulos and Turner, 2001). These proteins are all implicated in sub-membrane signalling cascades via integrins at sites where the cell attaches to the extracellular matrix. The regions of these proteins that interact with the LD motif share functional and structural similarities: all recruit the proteins to which they belong to sites of integrin signalling, and all contain (FAK, Vinculin) (Bakolitsa *et al.*, 1999; Hoellerer *et al.*, 2003), or are predicted to contain (GIT-1, ILK), an architecture centred on a four-helix bundle. Recent structural analysis reveals with atomic detail how LD motifs 2 and 4 of paxillin bind to the carboxy-terminal domain of FAK, the FAT domain. The four-helix FAT domain displays two binding sites, one formed by helices 1 and 4, and one formed by helices 2 and 3. Both binding sites display the same architecture, typified by a hydrophobic patch between the two helices, surrounded by basic residues (Hoellerer *et al.*, 2003; Gao *et al.*, 2004).

Both LD peptides interact with the two sites on FAT in a similar fashion: The peptides form an amphipathic helix, with the hydrophobic residues (ELDxLL) stacking onto the greasy patch of the binding site within FAT, whereas the negative charges of the LD motifs (ELDxLL) interact with the positive charges of the FAT binding site. *In vitro* binding assays demonstrated an affinity of 3–10 mM for both LD peptides and both binding sites. A paxillin fragment comprising LD motifs 2, 3 and 4, however, displayed a tenfold stronger interaction (Thomas *et al.*, 1999). Therefore, both structural and binding data imply that the specificity of the LD-ligand interactions is not imposed on the level of the LD motif sequence, but requires other factors, such as interactions between full-length proteins, subcellular co-localization or other, unknown cellular components. Although the molecular architecture of the FAT-paxillin LD interaction is likely to be prototypical for the above mentioned LD ligands, counterexamples exist: Actopaxin interacts with LD motifs via its CH domains (composed of a helix-loop motif) (Nikolopoulos and Turner, 2001), whereas the LD-binding bovine papilloma virus E6 protein contains sequence motifs predicted to form two zinc fingers (Tong *et al.*, 1997). Thus, it is very likely that actopaxin and E6 have evolved different LD binding strategies.

LD interaction between RoXaN and NSP3

The LD motif of RoXaN has been shown to bind to the central coiled-coil region of NSP3. On NSP3, this binding surface is situated roughly equidistant from the amino- and carboxy-terminal sites where NSP3 interacts with viral mRNA and eIF4G, respectively (Vitour *et al.*, 2004). A coiled-coil architecture, even though constructed by parallel helices, is in principle capable of offering LD binding sites very similar to those formed between the anti-parallel helices of the FAT domain. The predicted twofold symmetry of the NSP3 coiled-coil region, being formed by two identical monomers, makes it likely that one NSP3 dimer offers two LD-motif binding sites. Considering that RoXaN probably dimerizes through its own carboxy-terminal coiled-coil region, the interaction between RoXaN and NSP3 could lead to either a heterotetrameric complex composed of one RoXaN dimer and one NSP3 dimer, or to a multimeric aggregate where one RoXaN dimer binds to two different NSP3 dimers. These two models differ in the functional implications of this interaction.

Similarities between paxillin and RoXaN

The similarities between the LD-motif-containing proteins paxillin and RoXaN are intriguing: Both proteins display their LD motifs (one in RoXaN, five in paxillin) within a flexible and unstructured region, as predicted by sequence analysis, of their amino-terminal half. The amino-termini of both proteins also contain other protein–protein recognition sequences: proline-rich motifs for both, TPR and phospho-tyrosine motifs for RoXaN and paxillin, respectively (Schaller, 2001; Vitour *et al.*, 2004). Consequently, both proteins are likely to function as scaffolds for the assembly of multiprotein complexes. Finally, the carboxy-termini of both paxillin and RoXaN contain zinc finger repeats. In paxillin these arrange in pairs to form LIM motifs, which target paxillin to focal adhesions (FAs) by binding to a yet undefined protein target (Schaller, 2001). The function of the multiple zinc-finger motifs of RoXaN remains elusive. From their sequence, it seems possible that they bind RNA, rather than DNA, but protein ligands can actually not be excluded (Vitour *et al.*, 2004). Together, a large number of similarities exist, but the protein complexes they stimulate, as well as their subcellular localization could be different.

In uninfected cells, dimeric PABP links mRNA to eIF4G for translation, and interacts with paxillin (Woods *et al.*, 2002). In infected cells, NSP3 displaces PABP to connect viral mRNA with eIF4G (Piron *et al.*, 1998). Additionally, in rotavirus-infected cells, NSP3 interacts with RoXaN, perhaps in a manner parallel to the interaction of PABP with paxillin. In this way, it is conceivable that NSP3 could substitute RoXaN for paxillin.

This possibility raises two questions: (i) Given the presence of highly similar LD motifs in paxillin and RoXaN, what are the factors that allow NSP3 to substitute RoXaN for paxillin and (ii) what are the functional reasons for NSP3 to prefer RoXaN rather than paxillin? The answers to these questions, once unravelled by future research, are likely to illuminate, not only aspects of NSP3 function, but also features of the specificity and selectivity of LD-motif interactions, and, finally, the role of RoXaN in normal cellular physiology and during rotavirus infection.

Conclusion

The elucidation of the structure of rotavirus NSP3 alone and in conjunction with its binding partners leads to a clearer understanding of how these interactions occur, and in turn, instructs the continuing examination of the roles of these proteins. Discerning structural aspects of previously determined associations of NSP3, such as dimerization and binding to viral mRNA and cellular eIF4G, may lead to design and testing of specific molecules potentially capable of interrupting these interactions. Examining structures of similar molecules and interactions where the role of one of the proteins is not well understood, such as that between NSP3 and RoXaN, can lead to the formation of testable hypotheses in the process of determining the functions of these proteins and their binding partners. Fine details revealed by structural evidence often lead to a better understanding of important biological interactions, unveiling hidden corners in both viral and cellular biology. Information relating to the interface of cellular and viral biology is vital, since the manner in which viruses disrupt, interfere, and manipulate cellular systems often gives invaluable insight into normal cellular processes that may be otherwise difficult to dissect. In addition, these data broaden the knowledge base and new structural information leads to other avenues of investigation in both cellular and viral biology, assisting in the goal of continually learning so as to improve human health.

References

Bakolitsa, C., de Pereda, J.M., Bagshaw, C.R., Critchley, D.R., and Liddington, R.C. (1999). Crystal structure of the vinculin tail suggests a pathway for activation. Cell. *99*, 603–613.

Bican, P., Cohen, J., Charpilienne, A., and Scherrer, R. (1982). Purification and characterization of bovine rotavirus cores. J. Virol. *43*, 1113–1117.

Both, G.W., Bellamy, A.R., and Siegman, L.J. (1984). Nucleotide sequence of the dsRNA genomic segment 7 of Simian 11 rotavirus. Nucleic Acids Res. *12*, 1621–1626.

Brown, M.C., Curtis, M.S., and Turner, C.E. (1998). Paxillin LD motifs may define a new family of protein recognition domains. Nat. Struct Biol. *5*, 677–678.

Brown, M.C., Perrotta, J.A., and Turner, C.E. (1996). Identification of LIM3 as the principal determinant of paxillin focal adhesion localization and characterization of a novel motif on paxillin directing vinculin and focal adhesion kinase binding. J. Cell Biol. *135*, 1109–1123.

Deo, R.C., Bonanno, J.B., Sonenberg, N., and Burley, S.K. (1999). Recognition of polyadenylate RNA by the poly(A)-binding protein. Cell. *98*, 835–845.

Deo, R.C., Groft, C.M., Rajashankar, K.R., and Burley, S.K. (2002). Recognition of the rotavirus mRNA 3′ consensus by an asymmetric NSP3 homodimer. Cell. *108*, 71–81.

Eiden, J.J. (1993). Gene 5 of the IDIR agent (group B rotavirus) encodes a protein equivalent to NS34 of group A rotavirus. Virology. *196*, 298–302.

Estes, M.K. (2001). Rotaviruses and their Replication. Virology. D.M. Knipe and P.M. Howley, Eds. Lippincott, Williams, and Wilkins. Philadelphia, PA. p. 1679–1728.

Furuichi, Y., Morgan, M., Muthukrishnan, S., and Shatkin, A.J. (1975). Reovirus messenger RNA contains a methylated, blocked 5′-terminal structure, m-7G(5′)ppp(5′)G-MpCp. Proc. Natl Acad. Sci. USA *72*, 362–366.

Gao, G., Prutzman, K.C., King, M.L., Scheswohl, D.M., DeRose, E.F., London, R.E., Schaller, M.D., and Campbell, S.L. (2004). NMR solution structure of the focal adhesion targeting domain of focal adhesion kinase in complex with a paxillin LD peptide, evidence for a two-site binding model. J. Biol. Chem. *279*, 8441–8451.

Groft, C.M., and Burley, S.K. (2002). Recognition of eIF4G by rotavirus NSP3 reveals a basis for mRNA circularization. Mol. Cell. *9*, 1273–1283.

Hildebrand, J.D., Schaller, M.D., and Parsons, J.T. (1995). Paxillin, a tyrosine phosphorylated focal adhesion-associated protein binds to the carboxyl terminal domain of focal adhesion kinase. Mol. Biol. Cell. *6*, 637–647.

Hoellerer, M.K., Noble, M.E., Labesse, G., Campbell, I.D., Werner, J.M., and Arold, S.T. (2003). Molecular recognition of paxillin LD motifs by the focal adhesion targeting domain. Structure (Camb). *11*, 1207–1217.

Imataka, H., Gradi, A., and Sonenberg, N. (1998). A newly identified N-terminal amino acid sequence of human eIF4G binds poly(A)-binding protein and functions in poly(A)-dependent translation. EMBO J. *17*, 7480–7489.

Jones, S., and Thornton, J.M. (1996). Principles of protein–protein interactions. Proc. Natl Acad. Sci. USA *93*, 13–20.

Kearney, K., Chen, D., Taraporewala, Z.F., Vende, P., Hoshino, Y., Tortorici, M.A., Barro, M., and Patton, J.T. (2004). Cell-line-induced mutation of the rotavirus genome alters expression of an IRF3-interacting protein. EMBO J. *23*, 4072–4081.

Kessler, S.H., and Sachs, A.B. (1998). RNA recognition motif 2 of yeast Pab1p is required for its functional interaction with eukaryotic translation initiation factor 4G. Mol. Cell Biol. *18*, 51–57.

Michel, Y.M., Borman, A.M., Paulous, S., and Kean, K.M. (2001). Eukaryotic initiation factor 4G-poly(A) binding protein interaction is required for poly(A) tail-mediated stimulation of picornavirus internal ribosome entry segment-driven translation but not for X-mediated stimulation of hepatitis C virus translation. Mol. Cell. Biol. *21*, 4097–4109.

Michel, Y.M., Poncet, D., Piron, M., Kean, K.M., and Borman, A.M. (2000). Cap-Poly(A) synergy in mammalian cell-free extracts. Investigation of the requirements for poly(A)-mediated stimulation of translation initiation. J. Biol. Chem. *275*, 32268–32276.

Nibert, M.L., and Schiff, L.A. (2001). Reoviruses and their Replication. Virology. D.M. Knipe and P.M. Howley, Eds. Lippincott, Williams, and Wilkins. Philadelphia, PA. p. 1679–1728.

Nikolopoulos, S.N., and Turner, C.E. (2001). Integrin-linked kinase (ILK) binding to paxillin LD1 motif regulates ILK localization to focal adhesions. J. Biol. Chem. *276*, 23499–23505.

Otero, L.J., Ashe, M.P., and Sachs, A.B. (1999). The yeast poly(A)-binding protein Pab1p stimulates *in vitro* poly(A)-dependent and cap-dependent translation by distinct mechanisms. EMBO J. *18*, 3153–3163.

Patton, J.T., and Spencer, E. (2000). Genome replication and packaging of segmented double-stranded RNA viruses. Virology. *277*, 217–225.

Patton, J.T., Taraporewala, Z., Chen, D., Chizhikov, V., Jones, M., Elhelu, A., Collins, M., Kearney, K., Wagner, M., Hoshino, Y., and Gouvea, V. (2001). Effect of intragenic rearrangement and changes in the 3′ consensus sequence on NSP1 expression and rotavirus replication. J. Virol. *75*, 2076–2086.

Piron, M., Delaunay, T., Grosclaude, J., and Poncet, D. (1999). Identification of the RNA-binding, dimerization, and eIF4GI-binding domains of rotavirus nonstructural protein NSP3. J. Virol. *73*, 5411–5421.

Piron, M., Vende, P., Cohen, J., and Poncet, D. (1998). Rotavirus RNA-binding protein NSP3 interacts with eIF4GI and evicts the poly(A) binding protein from eIF4F. EMBO J. *17*, 5811–5821.

Poncet, D., Aponte, C., and Cohen, J. (1993). Rotavirus protein NSP3 (NS34) is bound to the 3′ end consensus sequence of viral mRNAs in infected cells. J. Virol. *67*, 3159–3165.

Poncet, D., Laurent, S., and Cohen, J. (1994). Four nucleotides are the minimal requirement for RNA recognition by rotavirus non-structural protein NSP3. EMBO J. *13*, 4165–4173.

Preiss, T., and Hentze, M.W. (1998). Dual function of the messenger RNA cap structure in poly(A)-tail-promoted translation in yeast. Nature. *392*, 516–520.

Preiss, T., and Hentze, M.W. (2003). Starting the protein synthesis machine, eukaryotic translation initiation. Bioessays. *25*, 1201–1211.

Preiss, T., Muckenthaler, M., and Hentze, M.W. (1998). Poly(A)-tail-promoted translation in yeast, implications for translational control. Rna. *4*, 1321–1331.

Premont, R.T., Claing, A., Vitale, N., Perry, S.J., and Lefkowitz, R.J. (2000). The GIT family of ADP-ribosylation factor GTPase-activating proteins. Functional diversity of GIT2 through alternative splicing. J. Biol. Chem. *275*, 22373–22380.

Proweller, A., and Butler, J.S. (1997). Ribosome concentration contributes to discrimination against poly(A)-mRNA during translation initiation in Saccharomyces cerevisiae. J. Biol. Chem. *272*, 6004–6010.

Qian, Y.A., Jiang, B.M., Saif, L.J., Kang, S.Y., Ojeh, C.K., and Green, K.Y. (1991). Molecular analysis of the gene 6 from a porcine group C rotavirus that encodes the NS34 equivalent of group A rotaviruses. Virology. *184*, 752–757.

Schaller, M.D. (2001). Paxillin, a focal adhesion-associated adaptor protein. Oncogene. *20*, 6459–6472.

Tarun, S.Z., Jr., Wells, S.E., Deardorff, J.A., and Sachs, A.B. (1997). Translation initiation factor eIF4G mediates *in vitro* poly(A) tail-dependent translation. Proc. Natl Acad. Sci. USA *94*, 9046–9051.

Thomas, J.W., Cooley, M.A., Broome, J.M., Salgia, R., Griffin, J.D., Lombardo, C.R., and Schaller, M.D. (1999). The role of focal adhesion kinase binding in the regulation of tyrosine phosphorylation of paxillin. J. Biol. Chem. *274*, 36684–36692.

Tong, X., and Howley, P.M. (1997). The bovine papillomavirus E6 oncoprotein interacts with paxillin and disrupts the actin cytoskeleton. Proc. Natl Acad. Sci. USA *94*, 4412–4417.

Tong, X., Salgia, R., Li, J.L., Griffin, J.D., and Howley, P.M. (1997). The bovine papillomavirus E6 protein binds to the LD motif repeats of paxillin and blocks its interaction with vinculin and the focal adhesion kinase. J. Biol. Chem. *272*, 33373–33376.

Tumbarello, D.A., Brown, M.C., and Turner, C.E. (2002). The paxillin LD motifs. FEBS Lett. *513*, 114–118.

Turner, C.E. (2000). Paxillin and focal adhesion signalling. Nat. Cell Biol. *2*, E231–236.

Turner, C.E., Glenney, J.R., Jr., and Burridge, K. (1990). Paxillin, a new vinculin-binding protein present in focal adhesions. J. Cell Biol. *111*, 1059–1068.

Vande Pol, S.B., Brown, M.C., and Turner, C.E. (1998). Association of Bovine Papillomavirus Type 1 E6 oncoprotein with the focal adhesion protein paxillin through a conserved protein interaction motif. Oncogene. *16*, 43–52.

Vende, P., Piron, M., Castagne, N., and Poncet, D. (2000). Efficient translation of rotavirus mRNA requires simultaneous interaction of NSP3 with the eukaryotic translation initiation factor eIF4G and the mRNA 3′ end. J. Virol. *74*, 7064–7071.

Vitour, D., Lindenbaum, P., Vende, P., Becker, M.M., and Poncet, D. (2004). RoXaN, a novel cellular protein containing TPR, LD, and zinc finger motifs, forms a ternary complex with eukaryotic initiation factor 4G and rotavirus NSP3. J. Virol. *78*, 3851–3862.

Wilusz, C.J., Wang, W., and Peltz, S.W. (2001). Curbing the nonsense, the activation and regulation of mRNA surveillance. Genes Dev. *15*, 2781–2785.

Woods, A.J., Roberts, M.S., Choudhary, J., Barry, S.T., Mazaki, Y., Sabe, H., Morley, S.J., Critchley, D.R., and Norman, J.C. (2002). Paxillin associates with poly(A)-binding protein 1 at the dense endoplasmic reticulum and the leading edge of migrating cells. J. Biol. Chem. *277*, 6428–6437.

Analyses of Rotavirus NSP4 Genetic Groups, Structure, and Function

18

Judith M. Ball, Rebecca D. Parr, and Clarence E. Schutt

Abstract

NSP4 is a non-structural, multifunctional glycoprotein with roles in rotavirus (RV) morphogenesis, pathogenesis, and intracellular signalling events. Although traditionally classified as an integral ER glycoprotein, new data show NSP4 localizes to multiple intracellular sites outside of the ER. Novel NSP4–protein and –lipid interactions have been unveiled, which contribute to RV pathogenesis, and NSP4 structure and intracellular transport. Complimentary to these reports, several new functions have been disclosed, primarily through the use of silencing mRNA techniques. NSP4 contributes to viroplasm formation, distribution of viral proteins in infected cells, and regulation of viral transcription. As one of several virulence factors, NSP4 induces diarrhoea by promoting a signalling event at the cell surface resulting in chloride secretion. NSP4 sequences are divided into distinct genetic groups that vary both in size and intracellular interactions. The crystal structure of NSP4 residues 95–137 discloses an extended, helical domain that folds as a coiled-coil and encompasses several binding sites. With the C-terminal structure in place, dissection of the many links between NSP4 structure–function and the complex interplay between NSP4 and host cell proteins are ongoing. This chapter summarizes NSP4 genetic variations and groupings, roles in RV replication and pathogenesis, and the potential impact of NSP4 structure on function.

Introduction

Rotavirus (RV) non-structural protein 4 (NSP4) is a product of gene segment 10 and traditionally has been classified as an endoplasmic reticulum (ER) transmembrane glycoprotein. NSP4 is synthesized as a primary translation product of 175 amino acids (aa) with an apparent molecular weight (*Mr*) of 20 kD, is cotranslationally glycosylated to 29 kD, and undergoes subsequent oligosaccharide processing to the mature 28 kD protein (Ericson *et al.*, 1983). Despite the presence of three, N-terminal hydrophobic domains, termed H1, H2, H3, NSP4 appears to span the ER membrane once (Chan *et al.*, 1988; Bergmann *et al.*, 1989). An early study of NSP4 topology using cell-free translation of gene 10 transcripts in the presence of microsomal membranes followed by protease digestion indicated H3 is the transmembrane (TM) domain (Chan *et al.*, 1988). Subsequent analyses by Bergman, *et al.* (1989) using site-directed mutagenesis and *in vitro* translation conclude H2 is the TM domain. H2 or a region overlapping H2 is now generally accepted to function as the NSP4 membrane anchor such that an N-terminal 22 aa segment remains in the lumen of the ER and residues 45–175 form an extended, cytoplasmic domain (Figure 18.1). Two *N*-linked, high mannose glycosylation sites are located within the short ER luminal domain at aa 8 and 18 (Both *et al.*, 1983). These carbohydrate moieties remain sensitive to endoglycosidase H digestion (oligosaccharide processing stops at Man-$_8$GlcNAc) accounting for the ER glycoprotein designation in these early studies (Ericson *et al.*, 1983;Both *et al.*, 1983;Kabcenell and Atkinson, 1985;Bergmann *et al.*, 1989;Taylor *et al.*, 1992). The isolation of a NSP4, C-terminal fragment containing residues 112–175 from culture medium of baculovirus-gene 10 recombinant infected *Spodoptera frugiperda* (Sf9) insect cells and RV-infected mammalian cells (Zhang *et al.*, 2000) indicates the C-terminal region of NSP4 is released from the ER, while the N-terminal domain containing both glycosylation sites, as well as the transmembrane region, is retained at the ER. Hence, an intracellular proteolytic cleavage of NSP4 at residue 112 would permit the extended C-terminal cytoplasmic domain to undergo intracellular transport and release. The precise transport pathway and method of exocytosis remains unknown, although the trafficking of the cleaved product is blocked by drugs that disrupt the cytoskeleton, and the polypeptide fails to utilize the classical ER–Golgi secretory pathway (Zhang *et al.*, 2000). Recent data show full-length NSP4 also traffics from the ER to the cell periphery by a Golgi-bypassing mechanism (Parr *et al.*, 2006; Berkova *et al.*, 2006).

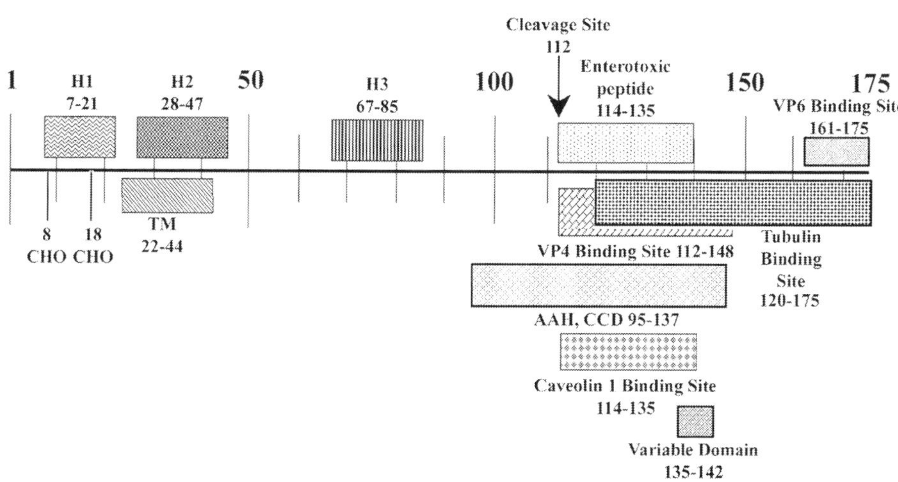

Figure 18.1 Linear schematic of structural and functional domains of NSP4. The 175 residues of NSP4 has three N-terminal hydrophobic domains, H1 (aa 7–21), H2 (aa 28–47), and H3 (aa 67–85) (Chan *et al.*, 1988). Glycosylation sites (CHO) are shown at aa 8 and 18. There is a single transmembrane domain (TM) resulting in a short luminal region and an extended cytoplasmic domain. The precise location of the TM domain slightly varies with the RV strain. AAH, CCD indicates the amphipathic α-helical region (aa 95–137) that folds as a coiled-coil domain and functions in oligomerization (Taylor *et al.*, 1996; Bowman *et al.*, 2000). The C-terminal residues contain the immature viral particle binding site (VP6 binding site, aa 161–175), the tubulin-binding domain (residues 120–175) and the VP4-binding site (aa 112–148) (Maass and Atkinson., 1990; Taylor *et al.*, 1992, 1993; Au *et al.*, 1993; O'Brien *et al.*, 2000; Xu *et al.*, 2000). The enterotoxic peptide (114–135) is indicated, as well as the cleavage site (↓) at aa 112 (Ball *et al.*, 1996; Halailel *et al.*, 2000; Zhang *et al.*, 2000). A variable domain localizes to residues 135–142 based on sequence comparisons (Horie *et al.*, 1997; Lin *et al.*, 2003).

The interest in RV NSP4 significantly expanded in 1996 with the report of NSP4 as the first viral enterotoxin (Ball *et al.*, 1996). Purified SA11 NSP4 and a synthetic peptide corresponding to residues 114–135 ($NSP4_{114-135}$) of the 175 aa protein induce diarrhoea in rodents in the absence of histological alterations and potentiates the release of Cl^- ions across the intestinal mucosa of mice as measured in Ussing chambers, and thus, fulfils the definition of an enterotoxin (Sears *et al.*, 1995). Diarrhoea induction by NSP4 and its enterotoxic peptide likely accounts for the early fluid loss (diarrhoea) post RV infection, prior to ultrastructure damage to the intestinal mucosa (Osborne *et al.*, 1988). During later stages of RV infection, villus blunting seen in some species contributes to malabsorption and diarrhoea, independent of the presence of NSP4 (Barnes, 1996). Diarrhoea induction by multiple strains of RV NSP4 or NSP4 fragments has been confirmed in neonatal mice (Zhang *et al.*, 1998; Horie *et al.*, 1999; Zhang *et al.*, 2000; Sasaki *et al.*, 2001; Mori *et al.*, 2002a). NSP4 from the homologous murine RV strain EW (genotype D, Table 18.1) also induces diarrhoea when aa 86–175 are expressed as a fusion protein and administered intraperitoneally (IP) to neonatal mice (Horie *et al.*, 1999). The effect of the fusion partner is unclear; however, the authors note that conformation of NSP4 may have been altered and consequently, diarrhoea may not have been as severe. Although group C RV NSP4 is truncated by 25 residues when compared to group A NSP4, aa 55–150 of the human RV strain Ehime 9301 (group C, genotype C) similarly induces diarrhoea in suckling mice (Sasaki *et al.*, 2001). Lastly, Mori *et al* (2002a) show NSP4 from an avian RV, PO-13 (genotype E) induces diarrhoea in neonatal mice. In the avian RV study, a difference was noted in enterotoxic activity when NSP4 was bound to a fusion partner, and that the enterotoxic activity was localized to residues 109–135, similar to that seen with mammalian group A NSP4 (Ball *et al.*, 1996; Sasaki *et al.*, 2001). Taken together, these studies show that the C-terminal region of NSP4 from different genotypes (Table 18.1) and from RV group C induces diarrhoea in suckling mice. The kinetics of diarrhoea induction is very similar in all of the studies and in agreement with the first report by Ball *et al.* (1996).

NSP4: one of many RV virulence factors

In addition to NSP4, VP3, VP4, VP7, NSP1 and NSP2 have been implicated in RV virulence. It is clear that more than one gene product is responsible for RV pathogenesis and that no single viral component accounts for pathogenesis in all situations (Burke and Desselberger, 1996). Alterations in NSP4 residues 130–141 from a pair of virulent and attenuated porcine RV (OSU) have been purified and introduced IP into suckling mice (Zhang *et al.*, 1998). In this study, only NSP4 from the virulent strain induces diarrhoea illustrating NSP4 may be associated with virulence of the porcine RV (Zhang *et al.*, 1998). Using a gnotobiotic piglet model and single-gene reassortants between a pathogenic porcine strain (SB-1) and a human RV strain (DS-1) that does not cause disease in piglets, Hoshino *et al.*, (1995) show genes encoding for VP3, VP4, VP7, and NSP4 confer a pathogenic phenotype. NSP1 and NSP2 also have been associated with virulence in the mouse model (Broome *et al.*, 1993).

Full-length NSP4 and the enterotoxic peptide (aa 114–135) function as a non-competitive inhibitor of the Na^+-D-glucose symporter (SGLT1) (Halaihel *et al.*, 2000). By specifically inhibiting Na^+-coupled solute transport across young rabbit jejunal brush border membranes, these data indicate water reabsorption would be inhibited by the

Table 18.1 NSP4 genotypes/genetic groups+

Genetic group I[a]	Genotype A[b]	KUN allele[c]	KUN group[d]	Genotype A*[e]	Genetic group II[a]	Genotype B[b]	Wa allele[c]	Wa group[d]	Genotype B*[e]	Genetic group III[a]	Genotype C[b]	AU-1 allele[c]	AU-1 group[d]	Genotype C*[e]	Genotype D*[e]	Genotype E[f]
		KUN/H	KUN/H	KUN/H	Wa/H/Sy	Wa/H/Sy	Wa/H/Sy	Wa/H/Sy	Wa/H/Sy			AU-1/H	AU-1/H	AU-1/H	**EW/M**	**PO-13/Av**
SA11/S	**SA11/S**	**SA11/S**	**SA11/S**	**SA11/S**	Ug71/H/A		**OSU/P**	**OSU/P**	**OSU/P**	RRV/S	RRV/S	RRV/S	RRV/S	RRV/S	EHP/M	TY-1/Av
S2/H/Sy	S2/H/Sy	S2/H/Sy	S2/H/Sy	S2/H/Sy	116E/H/A	ST3/H/A	ST3/H/A	ST3/H/A	ST3/H/A			FRV64/F	02-92/H	02-92/H	EC/M	TY-3/Av
UK/B/cc	UK/B/cc	UK/B/cc	CHW17/H	UK/B	GIMR7/H		AU32/H	AU32/H	AU32/H			RS15/C		RS15/C		CH-1/Av
NCDV/B		NCDV/B	NCDV/B	NCDV/B	YM/P	YM/P	YM/P	Ai-75/H	YM/P			FRV-1/F		FRV-1/F		AvRV1/Av[g]
RV5/H/Sy	RV5/H/Sy	RV5/H/Sy		RV5/H/Sy	RV4/H/Sy	RV4/H/Sy	RV4/H/Sy	Ai-39/H	RV4/H/Sy					Cu-1/C		
H120/H	1076/H/A	1076/H/A		1076/H/A	M37/H/A	M37/H/A	M37/H/A	ITO/H	M37/H/A					FRV64/F		
Th44/H	E210/H/Sy	E210/H/Sy		E210/H/Sy	U1274/H	RV3/H/A		CH927A/H	RV3/H/A					Bristol/H		
A28/H		A28/H		A28/H	CDC125/H	VA70/H/Sy		TK15/H	VA70/H/Sy					208/H[g]		
Ph49/H	E201/H/Sy			E201/H/Sy	RIHI4/H			CH-55/H	H-1/E					**Ehime9301/H[g]**		
Ph52/H				B223/B	Ug70/H			MO/H	A411/P					Cowden/P[g]		
				BRV033/B	Br1067/H			CH927B/H	A131/P					FRV303/F[g]		
				H-2/E	H150/H			J-12/H	A253/P					FRV313/F[g]		
				F1-23/E	Th36/H			HW2/H	Gottfried/P					FRV348/F[g]		
				R-2/L	Th41/H			K08/H	1095/H[g]					FRV381/F[g]		
				Fl-14/E	Ug72/H			TK28/H	116E/H[g]					FRV70/F[g]		
				BAP-2/L				CH-32/H	AI-36/H[g]					GRV/O[g]		
				ALA/L				Ai-36/H	261/H[g]							
				C-11/L				Ai-53/H	262/H[g]							

+Virus strain/species (H = human, S = simian, B = bovine, E = equine, P = porcine, F = feline, C = canine, L = lapine, M = murine, O = ovine, Av = avian)/ symptomatic (Sy) or asymptomatic (A) infection in natural host. * indicates the same nomenclature was used to distinguish the genotypes that was utilized in Kirkwood and Palombo, 1997 (b above). Shaded cells indicate strains that were sequenced in the individual studies. The thick vertical lines divide the different NSP4 genotypes. **Bold and underlined** strains: C-terminal regions induces diarrhoea in suckling mice following intraperitoneal administration.

(**a**) From Cunliffe *et al*, 1997; data derived from Protdist and Kitsch modules of the Phylip 3.5 program; compared 26 strains of human and animal viruses of diverse G and P genotypes; Wa was used as the reference strain for aa alignments. (**b**) From Kirkwood and Palombo, 1997; data derived from Clustal W analysis, version 8.1; phylogenetic tree constructed by neighbour-joining method; Northern hybridization analyses combined with NSP4 gene and aa sequence comparisons. (**c**) From Horie *et al*, 1997; compared sequences of 22 strains from 6 species. (**d**) From Cao *et al*, 1999; all human strains in the Wa allele are serotype G3; 20 new NSP4 sequences compared with 10 known sequences of human and animal strains. (**e**) From Ciarlet *et al*, 2000; data derived from the PAUP program, version 9, rooted with the Bristol rotavirus C NSP4 sequence; 42 human and animal NSP4 sequences were compared. (**f**) From Mori *et al*, 2002 (vs res); phylogenetic tree was determined by the neighbour-joining method using the Clustal X program; TY-1 and –3 isolated from turkeys, CH-1 from chickens, PO-13 from a pigeon. (**g**) From Lin and Tian (2003); examined 176 NSP4 sequences; note that the table is not inclusive – strains followed by ‘g’ indicate designation according to Lin and Tian (2003).

NOTE: sources or newly determined NSP4 sequences are included in the referenced manuscripts a–g.

enterotoxic peptide in the absence of histological lesions. Taken together, it appears that in addition to promoting fluid loss, NSP4 may prevent fluid reabsorption, both functions contributing to diarrhoea.

NSP4 alters intracellular calcium homeostasis

Several reviews summarize the roles of NSP4 in pathogenesis, including the mobilization of intracellular calcium ($[Ca^{2+}]$i) critical to viral assembly and diarrhoea induction (Michelangeli *et al.*, 1991; Michelangeli *et al.*, 1995; Burke and Desselberger, 1996; Tian *et al.*, 1996a; Estes and Morris, 1999; Ruiz *et al.*, 2000; Lundgren and Svensson, 2001; Morris and Estes, 2004; Ball *et al.*, 2005). The physical integrity of RV particles and specific protein conformations are Ca^{2+}-dependent (Cohen *et al.*, 1979; Shahrabadi and Lee, 1986) and increased $[Ca^{2+}]$i is seen following viral protein synthesis (Michelangeli *et al.*, 1991). Moreover, the exogenous addition of purified NSP4 or $NSP4_{114-135}$ mobilizes $[Ca^{2+}]$i in human intestinal HT-29 cells or Sf9 insect cells by phospholipase C (PLC) activation and inositol trisphosphate (IP3) production (Tian *et al.*, 1995; Dong *et al.*, 1997). The activity of exogenously added NSP4 appears to be protein-mediated as trypsin treatment of mammalian cells prior to the addition of NSP4 ablates $[Ca^{2+}]$i mobilization (Dong *et al.*, 1997). It remains unclear if NSP4 functions by binding a specific G-protein-coupled receptor (GPCR) resulting in transmission of a signal across the membrane, or if NSP4 binds a surface receptor and is subsequently internalized allowing the enterotoxic domain to interact with signalling molecules localized to the cytoplasmic face of the plasma membrane (PM) or functions by multiple mechanisms. Enhanced mobilization of $[Ca^{2+}]$i by 'cross-talk' of several GPCR molecules has been shown in other systems (Werry *et al.*, 2003). It is clear that NSP4 activity is more complex than first realized.

The endogenous expression of NSP4 in Sf9 insect cells increases cytosolic Ca^{2+}, but the mechanism appears to be distinct from the pathway activated by exogenously added NSP4. When NSP4 is expressed in insect cells using a baculovirus expression system, addition of a PLC inhibitor (U7221) fails to influence $[Ca^{2+}]$i mobilization. Furthermore, ionomycin or thapsigargin treatment of Sf9 cells expressing NSP4 results in a significantly more rapid decrease in Ca^{2+} release when compared to control cells, indicating NSP4 expression alters ER permeability to Ca^{2+} (Tian *et al.*, 1994; Tian *et al.*, 1995). Changes in Ca^{2+} permeability also have been shown in the PM of virus-infected cells (Michelangeli *et al.*, 1991; Newton *et al.*, 1997). Likewise, the endogenous expression of NSP4-enhanced green fluorescent protein (eGFP) fusion protein in mammalian cells mobilizes $[Ca^{2+}]$i by a PLC-independent pathway (Berkova *et al.*, 2003). Perhaps NSP4 mobilizes Ca^{2+} at two points in the RV replication cycle as proposed by Berkova *et al.* (2003). The first alteration of Ca^{2+} levels could occur during viral replication at which time intracellular NSP4 chronically elevates $[Ca^{2+}]$i by a PLC-independent pathway. The second Ca^{2+} event then would occur when the active NSP4 segment (aa 112–175) is released from the cells and subsequently binds a specific surface receptor to induce a transient, but acute Ca^{2+} response (Zhang *et al.*, 2000). Additional studies are required to verify this model.

Electrophysiological analysis of mouse, unstripped intestinal mucosa shows NSP4 mimics the cholinergic agonist, carbochol, by potentiating cAMP-mediated secretory Cl^- currents (Ball *et al.*, 1996). Further, NSP4 induces diarrhoea in cystic fibrosis transmembrane conductance regulator (CFTR)-depleted neonatal mice indicating NSP4 diarrhoeagenic activity is CFTR independent (Morris *et al.*, 1999). In this study, large and small intestinal crypt cells from normal and CFTR null mice are isolated and compared to correlate the effects of NSP4 [Ca^{2+}]i mobilization to fluid transport. The results show Ca^{2+} signalling does not regulate the age-dependent Cl^- secretion in normal intestinal crypts, and that NSP4-induced [Ca^{2+}]i mobilization is independent of cell differentiation and/or crypt age. Conversely, NSP4 elicits an age- and Ca^{2+}-dependent anion PM permeability as measured by I^- ion influx (Morris *et al.*, 1999). Thus, fluid loss by a Ca^{2+}-sensitive, Cl^--conductive mechanism in small intestinal crypts would occur independent of CFTR function, which could explain the NSP4 induction of diarrhoea in CFTR-depleted mouse pups (Angel *et al.*, 1998).

NSP4 binding domains and upregulation of cellular proteins

It is well established that NSP4 functions as an intracellular receptor to subviral double-layered particles (DLPs). Upon binding of VP6 (major structural protein of the DLPs) to the cytoplasmic domain of NSP4, the DLPs bud into the ER and acquire the outer coat protein, VP7, perhaps the spike protein, VP4, and a transient lipid envelope. The DLP-binding domain has been mapped to the NSP4 C-terminal residues 161–175 (Taylor *et al.*, 1992; Au *et al.*, 1993; Taylor *et al.*, 1996; O'Brien *et al.*, 2000) (Figure 18.1). Binding of the DLPs is dependent on the conformation integrity of the NSP4 C-terminal binding domain and is enhanced in the presence of divalent cations (Au *et al.*, 1993; O'Brien *et al.*, 2000). However, membrane association does not appear to be required for binding as soluble fragments of the NSP4 cytoplasmic domain efficiently bind VP6 (Taylor *et al.*, 1996). Additional data show that the binding is cooperative and that the NSP4 binding domain is presented as a protease-resistant coiled-coil tetramer that interacts with the VP6-binding sites on the DLP (O'Brien *et al.*, 2000). The transiently acquired membrane is subsequently lost while VP7 and possibly VP4 (see below) are retained by an unknown mechanism(s) as the viral particles mature (Meyer *et al.*, 1989; O'Brien *et al.*, 2000). Glycosylation of NSP4 and cellular Ca^{2+} are both essential for removal of the ER envelope (Ericson *et al.*, 1983; Petrie *et al.*, 1983) in that there is an accumulation of enveloped particles when glycosylation is blocked by and/or Ca^{2+} is depleted using specific drugs (Sabara *et al.*, 1982; Shahrabadi and Lee, 1986; Poruchynsky *et al.*, 1991). Further, it has been suggested that NSP4 plays a role in the removal of the ER envelope (Tian *et al.*, 1996b).

Because NSP4 localizes to the ER and functions as an intracellular receptor, molecular chaperones involved in protein processing (folding, transition, retention) in the ER have been examined for NPS4 association and binding to VP7, the outer capsid protein added to the subviral particles in the lumen of the ER during viral morphogenesis. VP7 requires Ca^{2+} and N-linked glycosylation for correct folding and interacts with protein disulfide isomerase, a chaperone that catalyzes disulfide bonds (Dormitzer and Greenberg, 1992; Mirazimi *et al.*, 1998). Further, the two N-linked oligosaccharides of NSP4 seem to be critical for RV assembly (Petrie *et al.*, 1983; Estes and Cohen, 1989). Hence, calnexin, an ER-transmembrane chaperone, was evaluated for NSP4 and VP7 association. NSP4 transiently associates with calnexin when the N-linked oligosaccharides are partially trimmed (di- or mono-glycosylated), but the association is lost upon complete enzymatic removal of the NSP4 glycan moieties (Mirazimi *et al.*, 1998). Inhibition of glycosylation with tunicamycin confirms the glycosylation dependence of the interaction and is consistent with the time-dependence of the NSP4-calnexin association (Mirazimi *et al.*, 1998). No link between calnexin and VP7 is seen despite the interaction of VP7 with other chaperones.

Other glucose-regulated chaperone proteins (GRP) found in the ER lumen, such as BiP (GRP78) and endoplasmin (GRP94), are upregulated by RV infection or expression of NSP4 in MA104 cells (Xu *et al.*, 1998). NSP4-specific immunoblots of 2-dimensional gels reveal the presence of multiple forms of NSP4, similar to that seen with NSP5 (Blackhall *et al.*, 1997; Xu *et al.*, 1998) and VP6 (Emslie *et al.*, 2000). The multiple spots corresponding to NSP4 likely represent the different oligomeric and/or glycosylated states of NSP4 or possible truncations. Additional analyses reveal the lack of a direct interaction between NSP4 and BiP or endoplasmin, but NSP4 expression results in a three- to four-fold increase in the GRPs mRNA levels and transcriptional activity. The authors conclude that the notable increase of BiP and endoplasmin elicited by NSP4 expression may be due to the possible disruption of ER integrity, in particular with regards to Ca^{2+} homeostasis (Xu *et al.*, 1998). High ER Ca^{2+} levels are required for RV maturation (Mirazimi and Svensson, 1998) and BiP plays a role in Ca^{2+} storage in the ER (Lievremont *et al.*, 1997). Lastly, BiP and endoplasmin interact with VP4 and VP7 (based on immunoprecipitation experiments) (Xu *et al.*, 1998), suggesting that both chaperones are likely involved in the folding of these structural proteins and in viral morphogenesis in the ER.

In addition to the VP6-binding domain, the C-terminal region of NSP4 contains a VP4-binding site (aa 112–148, Figure 18.1) (Maass and Atkinson, 1990). NSP4 has been shown to form large heteroligomers with both VP4 and VP7 (Maass and Atkinson, 1990). NSP4 folds primarily as a tetramer although higher *Mr* complexes are frequently seen; SDS-resistant dimers are the likely intermediates of the tetrameric and other multimeric forms (Maass and Atkinson, 1990). The VP7, VP4, NSP4 heteroligomers likely are important to RV morphogenesis. Notably, VP7 utilizes intramolecular disulfide bonds to oligomerize (Gorziglia *et al.*, 1985), contains carbohydrate of the high-mannose type akin to NSP4 (Kabcenell and Atkinson, 1985), and is an ER luminal glycoprotein.

In recent years, the intracellular location and transport of VP4 have become less clear. In contrast to VP7, VP4 is a cytoplasmic protein and has been localized between the periphery of the viroplasm (site of DLP assembly in the cytoplasm) and the ER (Petrie *et al.*, 1984). However, VP4 is closely associated with VP7 as demonstrated with cross-reactive monoclonal antibodies (Chen *et al.*, 1992). VP4 co-localizes with β-tubulin and localizes to the PM in association with VP7 early post infection (Nejmeddine *et al.*, 2000). These data suggest a second maturation event secondary to that which occurs in the ER. The combined evidence of (i) NSP4 aa 112–175 being released from the ER (Zhang *et al.*, 2000), (ii) a specific interaction of the NSP4 enterotoxic domain with model membranes resembling caveolae (Huang *et al.*, 2001, 2004), and (iii) the presence of a caveolin-1 binding domain in the NSP4 C-terminal cytoplasmic region (Parr *et al*, 2006) strongly suggest this secondary maturation event is not initiated by NSP4 receptor activity or dependent on NSP4 ER localization. Rather, this second maturation event likely occurs at or near the PM and may involve NSP4 intracellular transport from the ER to the PM in association with VP4, VP7 and/or caveolin-1, all of which is facilitated by the cytoskeleton. Perhaps the final morphogenesis of RV virions requires NSP4 and VP4 to exit the ER in association with membrane rafts and then traffic to cell membranes for viral maturation and release. The presence of NSP4-specific antibodies in RV-infected children supports a model of intact NSP4 being released from the cell for processing by immune cells (Ishida *et al.*, 1997; Johansen *et al.*, 1999; Ray *et al.*, 2003).

NSP4 sequence variation and genetic groups/genotypes

NSP4 sequences from human and animal RV strains have been carefully examined and organized into distinct genetic groups or genotypes (Cunliffe *et al.*, 1997; Kirkwood and Palombo, 1997; Horie *et al.*, 1997a; Cao *et al.*, 1999; Ciarlet *et al.*, 2000) (Table 18.1). To evaluate the extent of NSP4 sequence variation between human RV field isolates, reference strains, and vaccine candidates, Cunliffe *et al.* (1997) sequenced fragments of 16 NSP4 genes of varying G and P types from six different countries. These data combined with eleven known NSP4 sequences reveal 3 distinct genetic groups, termed I, II and III, based on phylogenetic analyses. Strains within a group appear highly conserved with < 10% divergence at the nucleotide and aa level, but distantly related to strains in the other groups. Interestingly, Cunliffe *et al.* (1997), and Kirkwood and Palombo (1997) found genetic group III consists of only the rhesus RV vaccine strain, RRV. In the Cunliffe *et al.* (1997) study, NSP4 from the human RV strains examined fall within genetic groups I (10 strains) and II (16 strains).

NSP4 from 7 newly sequenced standard human RV strains and two cell-culture adapted RV isolates subsequently were examined by Northern blot hybridization analyses against another 11 animal and human strains (Kirkwood and Palombo, 1997). Upon computer evaluation of these sequences, two major NSP4 genotypes designated as genotype A and B emerged amongst the human and animal viruses. The third genotype (C) consisting of only RRV shows 81–85% nucleotide homology with the other NSP4s (Table 18.1). Comparisons of the deduced aa sequences disclose a clustering of divergence in 3 regions, the N-terminus found in the ER lumen, the VP4-binding region (residues 112–148) (Au *et al.*, 1993), and the subviral particle binding domain (aa 161–175) (O'Brien *et al.*, 2000). The human sequences examined correspond to isolates from both symptomatic and asymptomatic infections, however no correlation between the genotype and the virulence of the strain appears to exist. Kirkwood and Palombo (1997) note an apparent correlation between the VP6 (capsid protein composing the outer layer of double layered particles) subgroup specificity and the NSP4 genotype. This genetic linkage between the genes encoding NSP4 and VP6 was confirmed by Iturriza-Gomara, *et al.* (2003) using reassortants and common human genotypes. Linkage was absolute between the NSP4 genotype and the VP6 subgroups, which suggests co-evolution of the two genes (Iturriza-Gòmara *et al.*, 2003). Alternatively, the linkage may be more functional since NSP4 acts as an intracellular receptor (Meyer *et al.*, 1989; Au *et al.*, 1993; Taylor *et al.*, 1993; Taylor *et al.*, 1996), binding VP6 of DLPs and facilitating their budding into the ER.

Independent comparisons of the phylogenic trees generated by these first studies show NSP4 (gene 10) from group A RVs S2, RV5, UK, and SA11 are in genetic group I or genotype A, whereas Wa, RV4, M37, and YM are classified in genetic group II or genotype B (Table 18.1). The grouping of the animal and human RV strains into just two groups is surprising and indicates an evolutionary relationship between the animal and human gene 10. This is particularly evident with the simian strain SA11, which is distinct from the human genogroups with the exception of gene 10 (Nakagomi and Nakagomi, 1991). These data suggest gene 10 diverged independent of the other RV gene segments. In this study, NSP4 genetic groups I, II and III correspond to the KUN, Wa and AU-1 alleles identified by Horie *et al.* (1997a). Likewise, Cao *et al.* (1999) note three NSP4 genetic groups formed by 20 newly sequenced human G3 strains that were isolated in China and Japan. These genetic groups have been designated as the KUN, Wa, and AU-1 groups in keeping with the Horie *et al.* (1997a, 1997b) designation (Cao *et al.*, 1999). The NSP4 and VP4 genes lack the typical geographic divergence in RV isolates from Japan and China (Cao *et al.*, 1999), in contrast to the previously analyzed VP7 gene sequences (Wen *et al.*, 1997).

To determine the interspecies relatedness of NSP4 from animal strains, an additional 15 NSP4 genes from animal RVs were sequenced and compared to the sequences of known isolates (Ciarlet *et al.*, 2000). This study discloses several important findings: (i) NSP4 genes from 15 animal viruses consist of 750 or 751 base pairs and encode 175 aa, similar to other RV strains. (ii) A new and distinct fourth genotype is composed exclusively of murine sequences (genotype D). (iii) There is an apparently defined evolution pattern within individual species, as the NSP4 gene clustering correlates to the species of origin. Specifically, lapine, bovine and equine strains cluster to genotype A, whereas the porcine stains belong to genotype B, and the canine, feline and simian strains cluster to genotype C. And (iv) deduced aa sequence variations concentrate near variable regions noted earlier, i.e. at aa 135–142, the putative VP4-binding domain (residues 112–148), and at the DLP-binding region (residues 161–175) with two aa (169 and 174) consistently differing between genotypes A and B (Ciarlet *et al.*, 2000) (Figure 18.1).

Group A RVs are known to be the primary agents of severe diarrhoea in infants and young children, while group C RVs are diagnosed less often but seen more frequently during family and community outbreaks (Matsumoto *et al.*, 1989; Caul *et al.*, 1990;Oishi *et al.*, 1993; Kapikian and Chanock, 1996). Group C RV strains Bristol and Ehime 9301 are truncated at 150 aa (25 residues shorter than group A NSP4s), 96% identical at the nucleotide and aa level to each other, and appear to be structurally similar to group A RV NSP4s despite the length difference (Horie *et al.*, 1997b). The first hydrophobic segment, H1, appears to be missing in the shorter group C NSP4s. However, NSP4 from the group C RV is structurally similar to those of group A with N-terminal hydrophobic domains and a cytoplasmic coiled-coil region (Horie *et al.*, 1997b; Mori *et al.*, 2001; Mori *et al.*, 2002b). Presumably, group C viruses bud into the ER during viral morphogenesis akin to group A viruses and this process is mediated by NSP4.

A fifth NSP4 genetic group (E) consisting of the newly isolated and sequenced avian RVs has been introduced (Mori *et al.*, 2001; Mori *et al.*, 2002b). The avian NSP4s are 6–7 aa shorter than those of the mammalian proteins, have two predicted N-terminal hydrophobic domains (residues 27–47 and 67–85), and three putative glycosylation sites at aa 3, 8, and 17 (Mori *et al.*, 2001). NSP4 aa identities between the turkey and pigeon strains (TY-1, TY-3, and PO-13) are 90–97% with low identity to mammalian NSP4s (32–36%) (Ito *et al.*, 2001). It is interesting that the enterotoxin domain (aa 114–135, Ball *et al.*, 1996) is conserved among the avian RV NSP4s and that the homology to the mammalian strains is higher in this region (50–59%) when compared to the intact protein (Mori *et al.*, 2002b). Also maintained in the avian NSP4s are the VP4- and VP6-binding domains with homologies of 75–97% and 83–100%, respectively, when compared to the mammalian NSP4s (Mori *et al.*, 2002b). The heptad repeats in the NSP4 cytoplasmic domain are conserved in the avian RV strains, but may not form a coiled-coil structure (Bowman *et al.*, 2000; Mori *et al.*, 2002b; Lin and Tian, 2003). Lastly, the chicken RV, CH-1, only displays 78–79% sequence identity with the other avian strains and may be classified as a sixth genotype (Mori *et al.*, 2002b).

Mohan and Atreya (2000) show that gene 10 of the licensed Chinese lamb RV vaccine strain has significant identity with HI23 (93.71%), a group A equine RV strain (Mohan and Atreya, 2000). Based on phylogenetic analyses with 60 NSP4 sequences and reactivity with a group A monoclonal antibody, the Chinese vaccine lamb strain groups with genotype A. The authors also note that the enterotoxic region of NSP4 varies to the same degree as other RVs supporting the idea that NSP4 alone is not solely responsible for the attenuated phenotype, and pose the possibility that other RV proteins contribute to attenuation by blocking the NSP4 toxic domain (Mohan and Atreya, 2000). There is consistently a lack of correlation between NSP4 sequence variation and RV attenuation, yet NSP4 from all 5 genotypes induce diarrhoea in suckling mice (Ball *et al.*, 1996; Zhang *et al.*, 1998; Horie *et al.*, 1999; Mori *et al.*, 2001; Mori *et al.*, 2002a).

In 2003, computer analysis of NSP4 sequences from group A, B and C RV strains confirmed the 5 genotypes of NSP4 and showed the presence of 14 absolutely conserved residues in NSP4 from group A RVs, independent of genotype (Lin and Tian, 2003). These conserved residues are localized at the two glycosylation sites, the middle of the TM domain, in the VP4-binding site (7 conserved aa), and across the enterotoxic region (5 aa) (Lin and Tian, 2003). Examination of the NSP4 sequences from genotypes A and B also reveals common variable regions at residues 70–78, 91–98, 135–142, and at the C-terminus. The variable region encompassed by aa 135–142 within the VP4-binding domain (Figure 18.1) occurs in all genotypes (Lin and Tian, 2003). Variation within the enterotoxic region of NSP4 may explain the different clinical outcomes of natural RV infections (Cunliffe *et al.*, 1997).

Overall, NSP4 appears less divergent than other RV genes with both species-specific and interspecies relatedness. Other than the NSP4 sequence variations of different genotypes, NSP4 appears to maintain biological properties varying only in the host range. Kirkwood *et al.* (1997) hypothesize NSP4 genes of human serotype G2 RVs are derived from animal RVs. Species and interspecies relatedness also have been shown for both VP4 and VP7 (Nishikawa *et al.*, 1989; Gorziglia *et al.*, 1990). Although derived from multiple hosts, genotypes A, B, and C have approximately equal genetic distances, whereas the genetic distances of genotypes D and F of single hosts are twofold and fourfold greater, respectively (Lin and Tian, 2003). These data indicate an early evolutionary split of the avian RVs from the mammalian viruses and a later split of RVs infecting mice and mammals (Lin and Tian, 2003). The early split of the avian host is consistent with reports comparing other avian RV genes to the corresponding mammalian genes (Ito *et al.*, 1995; Ito *et al.*, 1997; Ito *et al.*, 2001).

Silencing RV gene 10 expression

Recent studies show silencing the expression of NSP4 following RV infection results in a 75% reduction in viral progeny and the redistribution of VP6 throughout the cell (Lopez *et al.*, 2005). These data clearly demonstrate a critical role of NSP4 in viral morphogenesis. In a separate study, when NSP4 is silenced there is a decrease in mature viroplasms and a reduced number of packaged virions with a concomitant increase in the assembly of empty viral particles, in particular DLPs (Silvestri *et al.*, 2005). Further, a large increase in plus-strand RNA synthesis (viral transcription) is seen late in infection. The authors reason that the reduction in viroplasm maturation results in the disorganization and lack of coordination of core assembly and genome packaging. Similar to the observations of Lopez *et al.* (2005), a redistribution of viral proteins in the cells is noted with silencing NSP4 (Silvestri *et al.*, 2005). Likewise, a third study reports a reduction in RV particles upon silencing NSP4 (Cuadras *et al.*, 2006). Taken together, there is no doubt NSP4 functions at several stages of the viral life cycle varying from viroplasm assembly to viral transcription.

NSP4–membrane and membrane microdomain interactions

NSP4 has a diverse and apparently assay-dependent effect on biological membranes when expressed in mammalian or insect cells, or bacteria. When NSP4 is expressed by a dual recombinant vaccinia virus system in kidney cells, there is a decrease of PM integrity as reflected by release of Cr^{51} and lactate dehydrogenase into the medium (Newton *et al.*, 1997). The disruption of the PM is also reflected by an increase in cell death. Analysis of NSP4 deletion mutants expressed by the vaccinia virus recombinant expression system localizes a putative cytotoxic region to residues 55–72 at the membrane proximal region in the cytoplasmic tail that may alter PM integrity (Newton *et al.*, 1997). The authors note that the putative cytotoxic region is predicted to fold as a cationic amphipathic helix similar to that of other membrane disrupting peptides (Epand *et al.*, 1995; Polozov *et al.*, 1998; Browne *et al.*, 2000). Tian *et al.* (1995) show expression of NSP4 in Sf9 insect cells fails to alter PM permeability to divalent cations. The use of insect rather than mammalian cells may explain this discrepancy. Another report notes that the inducible expression of NSP4 in *E. coli* perturbs the bacterial membrane as measured by a decrease in optical density post induction and release of [^{3}H]-uridine (Browne *et al.*, 2000). This study further localizes the bacterial membrane- altering domain of NSP4 to aa 48–91, which overlaps the membrane destabilizing region described by Newton *et al.* (1997) in mammalian cells, and notes that the bacterial inner membrane is similar to the ER membrane since both have a low cholesterol content (Browne *et al.*, 2000). However, no consideration was given to the phospholipid content of the membranes, which could influence both binding and membrane destabilization (Finean, 1973). The primary phospholipid found in the rough ER is 1-palmitoyl-2-oleoyl-sn-glycero-3-phosphocholine, similar to that of bacterial membranes (McMurray, 1973).

Apical addition of NSP4 to polarized kidney cells results in an increase in paracellular leakage of molecules with a *Mr* of 20kD, alteration of F-actin, and a significant decrease in transepithelial resistance (TER) (Tafazoli *et al.*, 2001). The effect(s) of NSP4 on the paracellular leakage varies as to whether the protein is added apically or basolaterally and if the cells are polarized as monitored by high TER. In undifferentiated cells, exogenous addition of NSP4 blocks targeting of a key tight junction protein, Zonula Occludens-1 (ZO-1), and delays development of the polarized phenotype (Tafazoli *et al.*, 2001). This block in transport is likely due to the disorganization of the apical F-actin. By disrupting the cytoskeleton, normal cellular transport pathways also would be inhibited or altered. However, inhibition of transport was not directly tested in this study. It is reasonable to speculate that alteration of ZO-1 and/or other tight junction proteins would also alter membrane permeability properties given the tight association of the proteins with the membrane. A report of a similar cell differentiation-dependent effect on the cytoskeleton of intestinal epithelial cells post RV infection notes disruption of the vimentin network is only seen in undifferentiated cells, whereas the differentiated epithelial cells display a Ca^{2+}-dependent microtubule disassembly (Brunet *et al.*, 2000b). When NSP4 is removed from the medium, TER is restored suggesting that NSP4 is not killing the cells or disrupting the PM, but does not preclude alteration in permeability. The kinetics of NSP4 activity is compatible with NSP4 binding to an apical receptor and inducing a signal transduction pathway (Tafazoli *et al.*, 2001). Dong *et al.* (1997) show that exogenous addition of NSP4 to human intestinal cells activates the PLC pathway with subsequent production of IP3 and diacylglycerol (DAG). Elevated IP3 increases $[Ca^{2+}]i$ whereas elevated DAG increases phophokinase C that has been shown to decrease TER in kidney cells. RV infection has been reported to perturb microvillus F-actin organization, which would influence cell differentiation (Obert *et al.*, 2000; Brunet *et al.*, 2000a). Tafazoli *et al.* (2001) suggest that the cytoskeletal perturbation seen in RV-infected cells is due to NSP4. This is reasonable given a report by Xu *et al.* (2000) in which NSP4 blocks ER to Golgi trafficking by interacting with microtubules (Xu *et al.*, 2000). A specific interaction between NSP4 and α-and β-tubulin has been delineated to the C-terminal 54 aa of NSP4 (Xu *et al.*, 2000). It would be interesting to determine if the same or an overlapping domain inhibits ZO-1 lateral transport.

Circular dichroism (CD) and cross-linking analyses have shown that NSP4 oligomerizes into tetramers (Maass and Atkinson, 1990; Taylor *et al.*, 1996). The CD spectrum of the NSP4 cytoplasmic tail (aa 86–175) in physiological buffer displays an α-helical content of 22% that fold as a coiled-coil (Taylor *et al.*, 1996). The coiled-coil structure of residues 95–137 has been confirmed by crystallographic studies (Bowman *et al.*, 2000). Other amphipathic helical regions have been shown in NSP4 outside of the extended, C-terminal amphipathic helix by CD studies (Jagannath *et al.*, 2006). CD experiments also have been utilized to evaluate the interaction (s) of intact NSP4 and NSP4-specific synthetic peptides with model membranes of different composition and curvature (Huang *et al.*, 2001; Huang *et al.*, 2004). CD analysis of purified, full-length NSP4 in physiological buffer shows NSP4 contains 33% helical structure with 31% β-strand and 12% β-turn, consistent with the 22% helical content of residues 86–175 and the enterotoxic peptide, $NSP4_{114-135}$, likewise adopts a secondary structure with a high helical content in buffer calculated at 37% (Huang *et al.*, 2001). When placed in a hydrophobic solvent (50% trifluoroethanol), the α-helical content of $NSP4_{114-135}$ increases to 91% indicating the peptide may favour the more hydrophobic environment of membranes to promote structural changes. The isodichroic point at 205 nm suggests that the structural transformation is from random coil to α-helix (Huang *et al.*, 2001). These structural alterations were utilized to determine if NSP4 or $NSP4_{114-135}$ bind lipid vesicles of different curvature and composition. CD analyses of NSP4 and $NSP4_{114-135}$ with large unilamellar vesicles (LUV) of 100–120 nm and small unilamellar vesicles (SUV) of 20–30 nm of different compositions indicate a specific interaction of defined membranes (Huang *et al.*, 2001;

Huang *et al.*, 2004). Full-length NSP4 preferentially partitions with anionic SUVs (high membrane curvature) containing 35 mol % cholesterol as shown by significant conformational changes (43% helical) elicited upon binding. Similarly, $NSP4_{114-135}$ becomes more helical upon interacting with anionic SUVs with increasing cholesterol content. When compared to the peptide in aqueous buffer, the helical content is twofold greater in the presence of anionic SUVs with 50 mol % cholesterol, but cholesterol in the absence of negative phospholipids is insufficient for the peptide interaction (Huang *et al.*, 2001). An independent filtration binding assay confirm these results demonstrating NSP4 and $NSP4_{114-135}$ interact exclusively with lipids and preferentially interact with vesicles that are highly curved, rich in anionic phospholipids and cholesterol, and thus resemble caveolae microdomains (Anderson *et al.*, 1992; Brown and Rose, 1992; Rothberg *et al.*, 1992; Anderson, 1998). The anionic phospholipid preference of NSP4 suggests the membrane-NSP4 interaction is ionic. The structural basis of NSP4 and $NSP4_{114-135}$ interactions with highly curved membranes likely includes the more mobile packing and lower surface pressure of the phospholipids in the outer leaflet and the packing constraint in the inner leaflet (Machida and Ohnishi, 1980) resulting in an exposed hydrophobic core that may facilitate hydrophobic interactions. It is interesting to note that the enterotoxic peptide is amphipathic and contains a hydrophobic and charged helical face. Highly curved membrane regions are prevalent in the intestinal microvillus (Lipka *et al.*, 1995; Boffelli *et al.*, 1997), the site of NSP4 enterotoxic activity (Chan *et al.*, 1988). Additional NSP4 peptides, including $NSP4_{120-147}$ and $NSP4_{131K}$, a mutant 114–135 peptide in which tyrosine (aa 131) is replaced with a positively charged lysine, similarly interact with anionic and cholesterol-rich SUVs when analyzed by CD and filtration binding assays (Huang *et al.*, 2004). These data correlate with the finding that NSP4 is associated with rafts (Triton-X resistant membrane fractions) in RV-infected cells (Sapin *et al.*, 2002). Taken together it is clear that NSP4-membrane association is more complex than first realized and that additional studies are needed to dissect the precise interactions and effects.

NSP4 is known to bind microtubules, causing a redistribution of COP1 and ERGIC-53, two markers of trafficking vesicles that cycle between the ER and Golgi (Xu *et al.*, 2000). NSP4 co-localizes with these markers and redistributes them throughout the cell, which can be visualized as radial projections extending from the ER toward the PM and away from the Golgi (Xu *et al.*, 2000). Co-localization of the NSP4-GFP fusion protein with the autophagosome marker LC3 has been reported (Wang and Klionsky, 2003; Berkova *et al.*, 2006). Autophagosomes typically traffic via a cytoplasm to vacuole or lysosomes pathway and bypasses the Golgi (Abeliovich *et al.*, 2000). Taken together, these studies suggest that NSP4 utilizes an unconventional vesicular trafficking pathway upon exiting the ER and likely utilizes multiple mechanisms. One potential mechanism for such transit is the cell caveolae membrane system.

In summary, several reports have implicated the association of NSP4 with lipid rafts. First, a preferential interaction between NSP4 and $NSP4_{114-135}$ with anionic, cholesterol-rich, caveolae-like model membranes has been shown by CD and a filter-binding assay (Huan *et al.*, 2001, 2004). Second, RV particles, VP4 (the RV spike protein), and NSP4 are associated with detergent resistant membranes in RV-infected cells (Sapin *et al.*, 2002). Third, infectious RV particles and NSP4 are detected in Triton X-100-insoluble fractions and are sensitive to drugs that compromise lipid raft integrity (Cuadras and Greenberg,

2003). Fourth, the interaction of NSP4 with the extracellular matrix proteins, laminin β-3 and fibronectin, indicate NSP4 traffics to enteric basolateral domains (Boshuizen *et al.*, 2004). Fifth, we show co-localization of NSP4 with caveolin-1 at both intracellular sites and at the cell surface of RV-infected MDCK and Caco-2 cells; we also report a direct interaction between NSP4 and caveolin-1 by yeast two-hybrid, *in vitro* binding, and co-immunoprecipitation assays (Parr *et al.*, 2006). Other studies confirm the presence of NSP4 at multiple locations in RV-infected cells (Xu *et al.*, 2000; Cuadras and Greenberg, 2003; Berkova *et al.*, 2006; Parr *et al.*, 2006;). Taken together these data strongly support a role of caveolar membranes and caveolin-1, the structural protein of caveolae, in the unconventional transport of NSP4 and possibly RV particles from the ER to the PM. Recently, Storey *et al.* (2007) demonstrated the presence of full-length, endoglycosidase-sensitive NSP4 in isolated PM caveolae. The isolated caveolae fraction is free of ER and other organelles markers, but contains caveolin-1, flotillin, and GM1, marker proteins of caveolae. Together with the direct binding of NSP4 and caveolin-1 (Parr *et al.*, 2006) the relevance of NSP4 association with caveolin and caveolae domains are significantly strengthened.

Placement of NSP4 at PM caveolae could explain the signalling events induced by the endogenous expression or exogenous addition of the enterotoxin. Caveolin-1 is found in the intestine, interacts with cholesterol, and functions to transport *de novo* synthesized cholesterol to and from the ER and PM caveolae as well as organizing signalling molecules, including those involving Ca^{2+}, at the cell surface (Field *et al.*, 1998; Hailstones *et al.*, 1998; Cheng *et al.*, 2006). Caveolae are known to regulate Ca^{2+} homeostasis and caveolin-1 functions to organize signalling molecules at the PM (Anderson, 1998; Anderson and Jacobson, 2002; Mehta and Malik, 2006). We originally proposed that endogenously expressed NSP4 would directly interact with signaling molecules bound to the caveolin-1 scaffolding domain (Williams and Lisanti, 2004). However, our data reveal that NSP4 binds the N- and C-termini of caveolin-1 (Mir *et al.*, 2007). We map the NSP4 binding site to caveolin-1 aa 2-31 and 161-178, which is an unusual interaction for caveolin-1. The biological significance of NSP4 binding caveolin-1 at both termini is not known. Perhaps the cytoplasmic arms of caveolin-1 bind the enterotoxic region to promote transport to the scaffolding domain resulting in the release of relevant signal transducers held in an inactive state (Williams and Lisanti, 2004). Exogenous addition could function more at the lipid level such that phosphatidylinositol bisphosphate (PIP2) signalling lipids within PM caveolae translocate to the cytofacial leaflet and become available for the activated PLC.

Additional properties and functions of NSP4 are listed in Table 18.2.

NSP4 structure: symmetry and quasi-symmetry in the oligomerization domain of NSP4

The oligomerization domain of NSP4 (aa 95 to 137) contains novel elements of structure when compared with other members of viral receptors, such as influenza haemagglutinin (Skehel and Wiley, 1998). While it was expected on the basis of previous sequence analysis of heptad repeats that NSP4 would oligomerize via coiled-coil interactions (Taylor *et al.*, 1996), it came as a surprise that a divalent cation is buried in the inter-helical contact region. There was little precedent for this form of structure stabilization of coiled-coils, necessitating an independent determination of a strontium-bound form of the structure

Table 18.2 Selected properties of NSP4 and $NSP4_{114–135}$

Property	Reference
NSP4 from avian and animal RV induce similar diarrhoea responses.	Horie *et al.*, 1999; Morrris and Estes 2004; Sasaki *et al.*, 2001
Associated with virulence based on studies of viral reassortants, and virulent vs. avirulent clones.	Zhang *et al.*, 1998; Hoshino *et al.*, 1995
Associated with diarrhoea-induction in adults infected with group B human RV.	Ishino *et al.*, 2006
NSP4 induces a protective immune response in mice and elicits a humoral and cell-mediated immune response in infected and vaccinated children and animals.	Johansen *et al.*, 1999; Ray 2003; Rodriguez-Diaz *et al.*, 2004; Izzi *et al.*, 2005; Yuan 2004; Choi *et al.*, 2005
Transport is inhibited by nocodazole and cytochalasin D, but NOT brefeldin A, indicating a non-classical secretory pathway.	Zhang *et al.*, 2000
NSP4-EGFP is found in vesicular structures, co-localizes with the LC3 autophagosome marker, and is brefeldin A insensitive.	Berkova *et al.*, 2006
Upregulates the synthesis of a translationally controlled tumour protein.	Xu *et al.*, 1999
NSP4-EGFP expressing cells loaded with Fura 2 show [Ca^{2+}] mobilization that is PLC-independent.	Berkova *et al.*, 2003
Binds the extracellular matrix protein laminin-β3 and fibronectinin in the basement membrane of infected neonatal mice.	Boshuizen *et al.*, 2004
Induces nitric oxide.	Rodriguez-Diaz *et al.*, 2006
Functions in RV replication, viroplasm development and particle assembly.	Lopez *et al.*, 2005; Sapin *et al.*, 2002; Silvestri *et al.*, 2005
Induces a Th1 response when administered as a fusion protein with shiga toxin B or HIV tat.	Choi 2005, Kim *et al.*, 2004
Secreted from the apical surface in polarized cells.	Bugarcic *et al.*, 2006
Present in isolated plasma membrane caveolae.	Storey *et al.*, 2007
Remodels actin organization via cofilin.	Berkova *et al.*, 2007
Activity is distinct from bacterial enterotoxins.	Lorrot and Vasseur, 2007
Infection with a human strain expressing NSP4 and VP4 of porcine origin.	Mascarenhas *et al.*, 2007
Group A RV strain identified with independent NSP4 and VP6 segregation.	Ghosh *et al.*, 2007
Associate with detergent resistant membranes (DRM) in RV-infected cells and is sensitive to drugs that compromise lipid raft integrity.	Cuadras *et al.*, 2003; Storey *et al.*, 2007

to establish that the cation was indeed Ca^{2+} and not Mg^{2+} (Bowman *et al.*, 2000). Detailed analysis of the helical packing interactions N-terminal to the buried Ca^{2+} ion reveal an unusual quasi-symmetry in which a 4-helix bundle is formed from two dimeric units, each of which contains α-helices locally interacting as if they were members of a trimer (Figure 18.2). The dimer of 'trimer-like' dimers arrangement is not found on the C-terminal side of the cation layer, which displays true fourfold symmetry.

An analysis of the crystal structure of the oligomerization domain of NSP4 is presented in the original publication (Bowman *et al.*, 2000). High quality crystals of a synthetic peptide corresponding to aa 95 to 137 were obtained using standard crystallization protocols. Crystallographic analysis of small protein structures has been considerably improved and simplified by the use of seleno-derivatized specimens, isostructural to native methionines, because they diffract slightly differently at different X-ray wavelengths which enables phases to be extracted for the X-ray amplitudes used in calculating electron density maps (Hendrickson, 1991). Therefore, the peptide was synthesized with a selenomethionine at position 112 in the sequence. The initial electron density maps were sufficiently clear and continuous for the helical backbones to be traced, as well as most of the aa side-chains for the two α-helices in the asymmetric unit. The space group twofold axis of symmetry generates a 4-helix bundle from two asymmetric units. A high electron density site is located above the glutamate-120 layer, where it was clear that alternating glutamate-120 residues around the bundle axis point toward and away from the cation site. The outward pointing

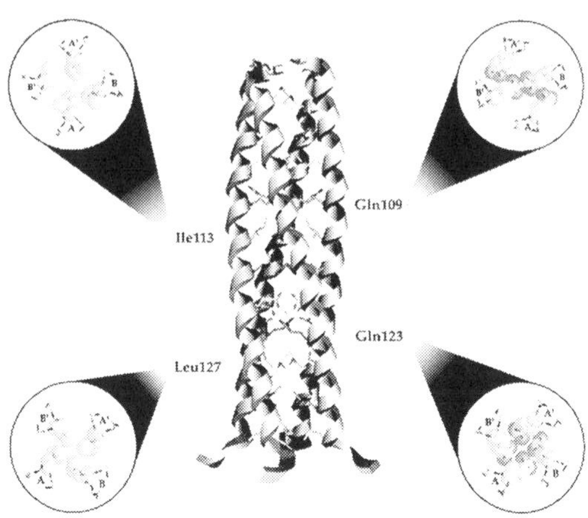

Figure 18.2 The oligomerization domain (residues 95–137) of NSP4 is shown in a ribbon representation, with identical polypeptide chains in the tetramer related by the crystallographic twofold axis being similarly shaded. The side chains of the a and d residues of the heptad repeat are depicted as sticks. The calcium ion is coordinated by the side chains of the Glu120 and Gln123 layers. The packing arrangement of selected layers from the homo-tetrameric structure is shown as CPK representations of cross-sectional slices, viewed from the C-terminus down the central coiled-coil axis. Note the fourfold symmetry of the layers C-terminal to the calcium-binding site (Gln123, Leu127), compared to the N-terminal layers, Gln109 and Ile113, which display a quasi-twofold symmetry axis and a helical packing characteristic of a trimeric coiled-coil (See Bowman *et al.*, 2000, for a thorough analysis and discussion). Gln, glutamine; Glu, glutamate; Ile, isoleucine; Leu, leucine.

glutamates are seen forming salt-bridges with arginine-119 on neighbouring helices in the bundle, whereas one oxygen from each of the two inward pointing carboxylates complete the ligation shell with two water molecules and the 4 overlaying glutamine-123 oxygen atoms.

In analyzing the detailed conformation of coiled-coil structures it is customary to begin with the identification of heptad repeats, that is the specific aa residues found in the first (denoted 'a') and the fourth (denoted 'd') positions in a block of seven helical residues (Crick, 1953). The structural basis for this formalism is that the pitch of an α-helix is such that every residue is rotated 100 degrees about the helical axis compared with its preceding neighbour along the polypeptide chain. When helical super-coiling is taken into account, the 'a' and 'd' positions lie on top of one another forming a continuous ridge on the inside of the helical bundle. The presence of hydrophobic residues along these ridges strongly predisposes the protein to helical bundling. This is driven by the burying of the ridges in the interior of the helical bundle, as is the case of close-packed hydrophobic cores in the interior of globular proteins, the well-known 'hydrophobic effect'. It has been established that the identity of non-polar amino acids in the 'a' and 'd' positions dictates the symmetry of the parallel coiled-coil bundle. When large hydrophobic residues are found along the ridge, especially leucine and isoleucine, tetramers are preferred (Kohn and Hodges, 1998).

The oligomerization domain of NSP4 has three interior polar layers at the 'a' and 'd' position. Core polar layers have been seen in all viral fusion proteins whose structures have been determined to date. Several of these structures contain bound Cl^- ions or small networks of water molecules (see Bowman *et al.*, 2000 for references). NSP4 possesses a layer of glutamine-109 residues, which in other structures is sufficient for Cl^- binding, although only water molecules were seen. The most striking feature of the structure of NSP4 is the presence of two polar layers, glutamate-120 and glutamine-123, which coordinate a Ca^{2+} ion. The possibility that the electron density represented a Mg^+ ion was ruled out by an examination of the lengths of the metal-oxygen bonds, and the observed isomorphism of an independently determined structure of strontium-bound NSP4, a structure expected to have similar cation binding geometry. A formalism has been developed for classifying local symmetry relations in neighbouring helices in a coiled-coil bundle (Harbury *et al.*, 1993). Application of these rules (Bowman *et al.*, 2000) reveal that the helices N-terminal to the Ca^{2+} binding site, pack locally as if they were members of a 3-helix bundle, whereas those C-terminal to the site has pure fourfold character.

These structural features are consistent with Ca^{2+} ions acting as triggers controlling the oligomerization state of NSP4. In the absence of Ca^{2+}, the polar layers in the interior would likely be energetically unfavourable to bundle formation. Once liganded to Ca^{2+}, not only would the structure be stabilized by strong electrostatic interactions in the core, but also by salt-bridges formed between the two twisted out glutamates-120 and arginines-119. These transitions are accommodated by a local distortion of the helices from a tetramer to a dimer of threefold-like dimers, a good example of quasi-equivalence in a functional context, reminiscent of transitions in the assembly of icosahedrally symmetric simple viruses (Harrison *et al.*, 1978). The surface of the coiled-coil between residues 119 and 123, unique to NSP4 amongst viral fusion proteins, is a sound candidate for modulation by other proteins in the complex maturation pathway of this pathogen.

The structure–function connection

The ER-localized receptor activity of NSP4 and the specific transport of residues 112–175 and full-length NSP4 out of the ER indicate that the toxin must be tightly regulated to accomplish its multiple and diverse functions. The C-terminus of NSP4 contains binding sites for VP4, VP6, calnexin, mictrotubules, and caveolin-1. No doubt additional interactions in the cell will be disclosed with further study. Mechanisms in which key receptor domains are exposed or hidden in concert within the NSP4 multimeric structure may be due to structural alterations induced by varying Ca^{2+} levels or to interactions with intracellular proteins such as caveolin-1 or chaperones. Alternatively, controlling the proportion of NSP4 molecules that are cleaved and released, or regulating the timing of full-length NSP4 release and transport, could regulate NSP4 function. Because of the many interactions of NSP4 with viral proteins, perhaps the stage of viral infection plays a role in regulation. However, these possibilities do not explain how NSP4 is transported to diverse intracellular sites at specific times post infection to accomplish specific purposes in the virus life cycle. The association of NSP4 with LC3-positive autophagosomes that fail to reach the PM, but contribute to viroplasm formation, suggests the presence of multiple, distinct pools of NSP4 in the cell. These unique pools of protein could explain how NSP4 radiates from the ERGIC along microtubules, associate with raft membranes and caveolin-1, and localize to autophagosomes and viroplasms in the course of rotaviral infections. It is reasonable to propose NSP4 is transported by multiple cellular pathways to different intracellular locations for distinct purposes. More than one pathway may direct NSP4 to the same intracellular site dependent on the infecting viral strain, time post infection, and the cell environment. The amphipathic character of NSP4 cytoplasmic domains clearly contribute to the interactions with cellular proteins and membranes, providing a format for both ionic and hydrophobic binding events, and may contribute to the enterotoxic activity. The N-terminal region of NSP4 is hydrophobic and could interact with interior membrane leaflet lipids. Thus, the overall structure of NSP4 could be unique when associated with membranes versus when present as a soluble protein or complex. The role of diverse multimers of NSP4 is unclear. It would be interesting to disclose if a tetramer functions differently from a dimer or monomer of NSP4 or if different aa side chains are exposed when NSP4 folds as a tetramer versus other multimeric complexes. Although NSP4 only is glycosylated at 2 N-terminal sites, Western blot analyses of RV-infected cells reveal unglycosylated, mono-glycosylated, and di-glycosylated forms of NSP4 are present during the course of infection. How does glycosylation influence structure and/or function? How do lipid binding, glycosylation states, and multiple oligomeric forms orchestrate NSP4 structure and function? It will be exciting to watch the unfolding of new and novel insights on NSP4 regulation, function and structure at different sites in the cell.

Reference

Abeliovich, H.D., Dunn Jr., W.A., Kim, J., and Klionsky, D.J. (2000). Dissection of autophagosome biogenesis into distinct nucleation and expansion steps. J. Cell Biology *151*, 1025–1034.

Anderson, R.G.W., Kamen, B.A., Rothberg, K.G., and Lacey, S.W. (1992). Potocytosis, sequestration and transport of small molecules by caveolae. Science 255, 410–411.

Anderson,R.G.W. (1998). The caveolae membrane system. Ann. Rev. Biochem. *67*, 199–225.

Anderson, R.G.W., and Jacobson, K. (2002). A role for lipid shells in targeting proteins to caveolae, rafts, and other lipid domains. Science. *296*, 1821–1825.

Angel, J., Tang, B.L., Feng, N., Greenberg, H.B., and Bass, D. (1998). Studies of the role for NSP4 in the pathogenesis of homologous murine rotavirus diarrhea. J. Infect. Dis. *177*, 455–458.

Au, K.S., Mavoungou, E., and Estes, M.K. (1993). A subviral particle binding domain on the rotavirus nonstructural glycoprotein NSP28. Virology *194*, 165–673.

Ball, J.M., Tian, P., Zeng, C.Q.Y., Morris, A.P., and Estes, M.K. (1996). Age-dependent diarrhea induced by a rotaviral nonstructural glycoprotein. Science. *272*, 101–104.

Ball, J.M., Mitchell, D.M., Gibbons, T.F., and Parr, R.D. (2005). Rotavirus NSP4, A multifunctional viral enterotoxin. Viral Immunol. *18*, 27–40.

Barnes, G.L. (1996). Viral Infections. In, Pediatric Gastrointestinal Disease. A.W. Walker, P.R. Durie, J.R. Hamilton, J.A. Walker-Smith, and J.B. Watkins, ed. St. Louis, MO, Mosby, 654–663.

Bergmann, C.C., Maass, D., Poruchynsky, M.S., Atkinson, P.H., and Bellamy, A.R. (1989). Topology of the non-structural rotavirus receptor glycoprotein NS28 in the rough endoplasmic reticulum. EMBO J. *8*, 1695–1703.

Berkova, Z., Morris, A., and Estes, M.K. (2003). Cytoplasmic calcium measurement in rotavirus enterotoxin-enhanced green fluorescent protein (NSP4-EGFP) expressing cells loaded with Fura-2. Cell Calcium *34*, 55–68.

Berkova, Z., Crawford, S.E., Trugnan, G., Yoshimori, T., Morris, A.P., and Estes, M.K. (2006). Rotavirus NSP4 induces a novel vesicular compartment regulated by calcium and associated with viroplasms. J. Virol. *80*, 6061–6071.

Berkova, Z., Crawford, S.E., Blutt, S.E., Morris, S.E., Morris, A.P., and Estes, M.K. (2007). Expression of rotavirus NSP4 alters the actin network organization through the actin remodeling protein cofilin. J. Virol. *81*, 3545–3553.

Blackhall, J., Fuentes, A., Hansen, K., and Magnusson, G. (1997). Serine protein kinase activity associated with rotavirus phosphoprotein NSP5. J. Virol. *71*, 138–144.

Boffelli, D., Weber, F.E., Compassi, S., Werder, M., Schulthess, G., and Hauser, H. (1997). Reconstitution and further characterization of the cholesterol transport activity of the small-intestinal brush border membrane. Biochemistry *36*, 10784–10792.

Boshuizen, J.A., Rossen, J.W.A., Sitaram, C.K., Kinenai, F.F.P., Simons-Oosterhuis, Y., Lafferber, J.A., Büller, H.A., and Einerhand, A.W.C. (2004). Rotavirus enterotoxin NSP4 binds to the extracellular matrix proteins laminin-β3 and fibronectin. J. Virol. *78*, 10045–10053.

Both, G.W., Siegman, L.J., Bellamy, A.R., and Atkinson, P.H. (1983). Coding assignment and nucleotide sequence of simian rotavirus SA11 gene segment *10*, Location of glycosylation sites suggests the signal peptide is not cleaved. J. Virol. *48*, 335–339.

Bowman, G.D., Nodelman, I.M., Levy, O., Lin, S.L., Tian, P., Zamb, T.J., Udem, S.A., Venkataraghavan, B., and Schutt, C.E. (2000). Crystal structure of the oligomerization domain of NSP4 from rotavirus reveals a core metal-binding site. J. Mol. Biol. *304*, 861–871.

Broome, R.L., Vo, P.T., Ward, L.A., Clark, H.F., and Greenberg, H.B. (1993). Murine rotavirus genes encoding outer capsid proteins VP4 and VP7 are not major determinants of host restriction and virulence. J. Virol. *76*, 2448–2455.

Brown, D.A., and Rose, J.K. (1992). Sorting of GPI-anchored proteins to glycolipid-enriched membrane subdomains during transport to the apical cell surface. Cell. *68*, 533–544.

Browne, E.P., Bellamy, A.R., and Taylor, J.A. (2000). Membrane-destabilizing activity of rotavirus NSP4 is mediated by a membrane-proximal amphipathic domain. J. Gen. Virol. *81*, 1955–(1959).

Brunet, J.-P., Cotte-Laffiette, J., Linxe, C., Quero, A.-M., Geniteau-Legendre, M., and Servin, A. 2000a. Rotavirus infection induces an increase in intracellular calcium concentration in human intestinal epithelial cells, role in microvillar actin alteration. J. Virol. *74*, 2323–2332.

Brunet, J.-P., Jourdan, N., Cotte-Laffiette, J., Linxe, C., Géniteau-Legendre, M., Servin, A., and Quéro, A.-M. 2000b. Rotavirus infection induces cytoskeleton disorganization in human intestinal cells, Implication of an increase in intracellular calcium concentration. J. Virol. *74*, 10801–10806.

Bugarcic, A., and Taylor, J.A. (2006). Rotavirus nonstructural glycoprotein NSP4 is secreted from the apical surfaces of polarized epithelial cells. J. Virol. *80*, 12343–12349.

Burke, B., and Desselberger, U. (1996). Rotavirus pathogenicity. Virology *218*, 299–305.

Cao, X.R., Akihara, S., Fang, Z.Y., Nakagomi, O., and Ushijima, H. (1999). Genetic variation in the VP4 and NSP4 genes of human rotavirus serotype 3 (G3 type) isolated in China and Japan. Microbiol. Immunol. *43*, 171–175.

Caul, E.O., Ashley, C.R., Darville, J.M., and Bridger, J.C. (1990). Group C rotavirus associated with fatal enteritis in a family outbreak. J. Med. Virol. *30*, 201–205.

Cuadras, M.A., and Greenberg, H.B. (2003). Rotavirus infectious particles use lipid rafts during replication for transport to the cell surface *in vitro* and *in vivo*. Virology *313*, 308–321.

Cuadras, M.A., Bordier, B.B., Zambrano, J.L., Ludert, J.E., and Greenberg, H.B. (2006). Dissecting rotavirus particle-raft interaction with small interfering RNAs, Insights into rotavirus transit through the secretory pathway. J. Virol. *80*, 3935–3946.

Chan, W.K., Au,K.S., and Estes, M.K. (1988). Topography of the simian rotavirus nonstructural glycoprotein (NS28) in the endoplasmic reticulum membrane. Virology *164*, 435–442.

Chen, D., Estes, M.K., and Ramig, R.F. (1992). Specific interactions between rotavirus outer capsid proteins VP4 and VP7 determine expression of a cross-reactive, neutralizing VP4-specific epitope. J. Virol. *66*, 432–439.

Cheng, Z.J., Singh, R.D., Marks, D.L., and Pagano, R.E. (2006). Membrane microdomains, caveolae, and caveolar endocytosis of sphingolipids (review). Mol. Mem. Biol. *23*, 101–110.

Choi, N.-W., Estes, M.K., and Langridge, W.H. (2005). Oral immunization with a shigat toxin B subunit: rotavirus NSP4 (90) fusion protein protects mice against gastroenteritis. Vaccine *23*, 5168–5176.

Ciarlet, M., Liprandi, F., Conner, M.E., and Estes, M.K. (2000). Species specificity and interspecies relatedness of NSP4 genetic groups by comparative NSP4 sequences analyses of animal rotaviruses. Arch. Virol. *145*, 371–383.

Cohen, J., Laporte, J., Charpilienne, A., and Scherrer, R. (1979). Activation of rotavirus RNA polymerase by calcium chelation. Arch. Virol. *60*, 177–186.

Crick, F.H.C. (1953). The packing of alpha-helices, simple coiled-coils. Acta Crystallog. *6*, 689–697.

Cuadras, M.A., and Greenberg, H.B. (2003). Rotavirus infectious particles use lipid rafts during replication for transport to the cell surface *in vitro* and *in vivo*. Virology *313*, 308–321.

Cunliffe, C., Woods, P.A., Leite, J.P.G., Das, B.K., Ramachandran, M., Bhan, M.K., Hart, C.A., Glass, R.I., and Gentsch, J.R. (1997). Sequence analysis of NSP4 gene of human rotavirus allows classification into two main genetic groups. J. Med. Virol. 53, 41–50.

Dong, Y., Zeng, C.Q.Y., Ball, J.M., Estes, M.K., and Morris, A.P. (1997). The rotavirus enterotoxin NSP4 mobilizes intracellular calcium in human intestinal cells by stimulating phospholipase C-mediated inositol 1,4,5-phosphate production. Proc. Natl Acad. Sci. USA *94*, 3960–3965.

Dormitzer, P.R., and Greenberg, H.B. (1992). Calcium chelation induces a conformational change in recombinant herpes simplex virus-1-expressed rotavirus VP7. Virology *189*, 828–832.

Emslie, K.R., Molloy, M.P., Barardi, C.R., Jardine, D., Wilkins, M.R., Bellamy, A.R., and Williams, K.L. (2000). Serotype classification and characterisation of the rotavirus SA11 VP6 protein using mass spectrometry and two-dimensional gel electrophoresis. Funct. Int. Genom. *2*, 12–24.

Epand, R.M., Shai, Y., Segrest, J.P., and Anantharamaiah,G.M. (1995). Mechanisms for the modulation of membrane bilayer properties by amphipathic helical peptides. Biopolymers (Peptide Science). *37*, 319–338.

Ericson, B.L., Graham, D.Y., Mason, B.B., Hanssen, H.H., and Estes, M.K. (1983). Two types of glycoprotein precursors are produced by the simian rotavirus SA11. Virology *127*, 320–332.

Estes, M.K., and Cohen, J. (1989). Rotavirus gene structure and function. Microbiol. Rev. 53, 410–449.

Estes, M.K., and Morris, A.P. (1999). A Viral Enterotoxin, a New Mechanism of Virus-Induced Pathgenesis. In, Mechanism in the pathogenesis of enteric Diseases P.S. Paul, D.H.Francis, ed. New York, Kluwer Academics/Plenum Publishers, 73–82.

Field, F.J., Born, E., Murthy, S., and Mathur, S.N. (1998). Caveolin is present in intestinal cells, Role in cholesterol trafficking? J. Lipid Res. *39*, 1938–(1950).

Finean, J.B. (1973). Phospholipids in Biological Membranes in the Study of Phospholipid-Protein Interactions. In, Form and Function of Phospholipids. G.B. Ansell, J.N. Hawthorne, and R.M.C. Dawson, ed., New York, Elsevier Scientific Publishing Company, 171–203.

Ghosh, S., Varghese, V., Samajdar, S., Bhattacharya, S.K., Kobayashi, N., and Naik, T.N. (2007). Evidence for independent segregation of the VP6- and NSP4-encoding genes in procine group A rotavirus G6P[13] strains. Arch. Virol. *152*, 423–429.

Gorziglia, M., Larrea, C., Liprandi, F., and Esparza, J. (1985). Biochemical evidence for the oligomeric (possible trimeric) structure of the major inner capsid polypeptide (45kD) of rotaviruses. J. Gen. Virol. *66*, 1889–(1900).

Gorziglia, M., Nishikawa, K., Hoshino, Y., and Taniguchi, K. (1990). Similarity of the outer capsid protein VP4 of the Gottfried strain of porcine rotavirus to that of asymptomatic human rotavirus strains. J. Virol. *64*, 414–418.

Hailstones, D., Sleer, L.S., Parton, R.G., and Stanley, K.K. (1998). Regulation of caveolin and caveolae by cholesterol in MDCK cells. J. Lipid Res. *39*, 369–379.

Halaihel, N., Lievin, V., Ball, J.M., Estes, M.K., Alvarado, F., and Vasseur, M.Direct inhibitory effect of rotavirus NSP4(114–135) peptide on the Na^+-D-glucose symporter of rabbit intestinal brush border membrane. (2000). J. Virol. *74*, 9464–9470.

Harbury, P.B., Zhang, T., Kim, P.S., and Alber, T. (1993). A switch between two, three, and four-stranded coiled coils in GCN4 leucine zipper mutants. Science. *262*, 1401–1407.

Harrison, S.C., Olson, A.J., Schutt, C.E., and Winkler, F.K. (1978). Tomato Bushy Stunt Virus at 2.9 Å Resolution. Nature *276*, 368–373.

Hendrickson, W.A. (1991). Determination of macromolecular structures from anomalous diffraction of synchrotron radiation. Science. *254*, 51–58.

Horie, Y., Masamune, O., and Nakagomi, O. 1997a. Three major alleles of rotavirus NSP4 proteins identified by sequence analysis. J. Gen. Virol. *78*, 2341–2346.

Horie, Y., Nakagomi, T., Oseto, M., Masamune, O., and Nakagomi, O. 1997b. Conserved structural features of nonstructural glycoprotein NSP4 between group A and group C rotaviruses. Arch. Virol. *142*, 1865–1872.

Horie,Y., Nakagomi, O., Nakagomi, T., Suzuki, Y., Oka, T., Sasaki, S., Matsuda, Y., and Watanabe, S. (1999). Diarrhea induction by rotavirus NSP4 in the homologous mouse model system. Virology *262*, 398–407.

Hoshino, Y., Saif, L.J., Kang, S.Y., Sereno, M.M., Chen, W.K., and Kapikian, A.Z. (1995). Identification of group A rotavirus genes associated with virulence of a porcine rotavirus and host range restriction of a human rotavirus in the gnotobiotic piglet model. Virology *209*, 274–280.

Huang, H., Schroeder, F., Zeng, C.Q.Y., Estes, M.K., Schoer, J., and Ball, J.M. (2001). Membrane interactions of a novel viral enterotoxin, Rotavirus nonstructural glycoprotein NSP4. Biochemistry *38*, 13231–13243.

Huang, H., Schroeder, F., Estes, M.K., McPherson, T., and Ball, J.M. (2004). The interaction(s) of rotavirus NSP4 C-terminal peptides with model membranes. Biochem. J. *380*, 723–733.

Ishida, S.-I., Feng, N., Gilbert J.M., Tang, B.L., and Greenberg, H.B. (1997). Immune responses to individual rotavirus proteins following heterologous and homologous rotavirus infection in mice. J. Infect. Dis. *175*, 1317–1323.

Ishino, M., Mise, K., Takemura, H., Ahmed, M.U., Alam, M.M., Naik, T.N., and Kobayashi, N. (2006). Comparison of NSP4 protein between group A and B human rotaviruses: detection of novel diarrhea-causing sequences in group B NSP4. Arch. Virol. *151*, 173–182.

Ito, H., Minamoto, N., Sasaki, I., Goto, H., Sugiyama, M., Kinjo, T., and Sugita, S. (1995). Sequence analysis of cDNA for the VP6 protein of group A avian rotavirus, A comparison with group A mammalian rotaviruses. Arch. Virol. *140*, 605–612.

Ito, H., Minamoto, N., Hiraga, and S., Sugiyama, M. (1997). Sequence analysis of the VP6 gene of group A turkey and chicken rotaviruses. Virus Res. *47*, 79–83.

Ito, H., Sugiyama, M., Masubuchi, K., Mori, Y., and Minamoto, N. (2001). Complete nucleotide sequence of a group A avian rotavirus genome and a comparison with its counterparts with mammalian rotavirus. Virus Res. *75*, 123–138.

Itturriza-Gòmara, M., Anderton, E., Kang, G., Gallimore, C., Phillips, W., Desselberger, U., and Gray, J. (2003). Evidence for genetic linkage between the gene segments encoding NSP4 and VP6 proteins in common and reassortant human rotavirus strains. J. Clin. Microbiol. *41*, 3566–3573.

Jagannath, M.R., Kesavulu, M.M., Deepa, R., Narayan Sastri, P., Senthil Kumar, S., Suguna, K., and Durgo Rao, C. (2005). N- and C-terminal cooperation in rotavirus enterotoxin, Novel mechanism of modulation of the properties of a multifunctional protein by a structurally and functionally overlapping conformational domain. J. Virol. *80*, 412–425.

Johansen, K., Hinkula, J., Espinoza, F., Levi, M, Zeng, C.Q.Y., Ruden, U., Vesikari, T., Estes, M.K., and Svensson, L. (1999). Humoral and cell-mediated immune responses in humans to the NSP4 enterotoxin of rotavirus. J. Med. Virol. *59*, 369–377.

Kabcenell, A.K., and Atkinson, P.H. (1985). Processing of the rough endoplasmic reticulum membrane glycoproteins of rotavirus SA11. J. Cell Biol. *101*, 1270–1280.

Kapikian, A.Z., Chanock, R.M. (1996). Rotaviruses. In, Fields Virology B.N. Fields, and D.M. Knipe, ed., New York, Raven Press, 1657–1708.

Kim, T.G., Befus, N., and Langridge, W.H. (2004.) Co-immunization with an HIV-1 Tat transduction peptide-rotavirus enterotoxin fusion protein stimulates Th1 mucosal immune response in mice. Vaccine *22*, 431–438.

Kirkwood, C.D., and Palombo, E.A. (1997). Genetic characterization of the rotavirus nostructural protein, NSP4. Virology *236*, 258–265.

Kohn, W.D, and Hodges, R.S. (1998). De novo design of alpha-helical coiled-coils and bundles, models for the development of protein-design principles. Trends Biotechnol. *16*, 379–389.

Lievremont, J.-P., Rizzuto, R., Hendershot,L., and Meldolesi, J. (1997). BiP, a major chaperone of the endoplasmic reticulum lumen, plays a direct and important role in the storage of the rapidly exchanging pool of Ca^{2+}. J. Biol. Chem. *272*, 30873–30879.

Lin, S.L., and Tian, P. (2003). Detailed computational analysis of a comprehensive set of group A rotavirus NSP4 proteins. Virus Genes *26*, 271–282.

Lipka,G., Schulthess,G., Thurnhofer,H., Wacker,H., Wehrli,E., Zeman,K., Weber,F.E., Hauser,H. (1995). Characterization of lipid exchange proteins isolated from small intestinal brush border membrane. J. Biol. Chem. *270*,5917–5925.

Lopez,T., Camacho,M., Zayas,M., Najera,R., Sanchez,R., Arias,C.F., Lopez,S. (2005). Silencing the morphogenesis of rotavirus, J. Virol. *79*,184–192.

Lorrot, M., and Vasseur, M. (2007). How do the rotavirus NSP4 and bacterial enterotoxins lead differently to diarrhea? Virol. J. *4*, 31.

Lundgren,O., Svensson,L. (2001). Pathogenesis of rotavirus diarrhea. Microbes Infect. 3,1145–1156.

Maass, D., and Atkinson, P.H. (1990). Rotavirus proteins VP7, NS28, and VP4 form oligomeric structures. J. Virol. *64*, 2641.

Machida, K., and Ohnishi, S.I. (1980). Effect of bilayer membrane curvature on activity of phosphatidylcholine exchange protein. Bioch. Biophys. Acta *596*, 201–209.

Mascarenhas, J.D., Leite, G.J.P., Lima, J.C., Heinemann, M.B., Oliveira, D.S., Araujo, I.T., Soares, L.S., Gusmao, R.H., Gabbay, Y.B., and Linhares, A.C. (2007). Detection of a neonatal human rotavirus strain with VP4 and NSP4 genes of porcine origin. J. Med. Microbiol. *56*, 524–532.

Matsumoto, K., Hatano, M., Kobayashi, K., Hasegawa, A., Yamazaki, S., Nakata, S., Chiba, S., and Kimura, Y. (1989). An outbreak of gastroenteritis associated with acute rotaviral infection in school children. J. Infect. Dis. *160*, 611–615.

McMurray, W.C. (1973). Phospholipids in Subcellular Organelles and Membranes. In, Form and Function of Phospholipids. G.B. Ansell, J.N. Hawthorne, and R.M.C. Dawson, ed. New York, Elsevier Scientific Publishing Company, 205–252.

Mehta, D., and Malik, A.B. (2006). Signaling mechanisms regulating endothelial permeability. Physiol. Rev. *86*, 279–367.

Meyer, J.C., Bergmann, C.C., and Bellamy, A.R. (1989). Interaction of rotavirus cores with the nonstructural glycoprotein NS28. Virology *171*, 98–107.

Michelangeli, F., Ruiz, M.-C., Del Castillo, J.R., Ernesto Ludert, J., Liprandi, F. (1991). Effect of rotavirus infection on intracellular calcium homeostasis in cultured cells. Virology *181*, 520–527.

Michelangeli, F., Liprandi, F., Chemello, M.E., Ciarlet, M., and Ruiz, M.-C. (1995). Selective depletion of stored clcium by thapsigarrgin blocks rotavirus maturation but not the cytopathic effect. J. Virol. *69*, 3838–3847.

Mir, K., Parr, R.D., Schroeder, F., and Ball, J.M. (2007). Rotavirus NSP4 binds both the amino- and carboxyl- termini of caveolin-1. Viral Res. *126*, 106–115.

Mirazimi, A., Nilsson, M., and Svensson, L. (1998). The molecular chaperone calnexin interacts with the NSP4 enterotoxin of rotavirus *in vivo* and *in vitro*. J. Virol. *72*, 8705–8709.

Mirazimi, A., Svensson, L. (1998). Carbohydrates facilitate correct disulfide bond formation and folding of rotavirus VP7. J. Virol. *72*, 3887–3892.

Mohan, K.V.K., and Atreya, C.D. (2000). Comparative sequence analysis identified mutations outside the NSP4 cytotoxic domain of tissue-culture adapted ATCC-Wa strain of human rotavirus and a novel inter-species variable domain in its C-terminus. Arch. Virol. *145*, 1789–1799.

Mori, Y., Sugiyama,M., Takayama,M., Atoji, Y., Masegi, T., and Minamoto, N. (2001). Avian-to-mammal transmission of an avian rotavirus, Analysis of its pathogenicity in a heterologous mouse model. Virology *288*, 63–70.

Mori, Y., Borgan, M.A., Ito,N., Sugiyama, M., and Minamoto, N. 2002a. Diarrhea-inducing activity of avian rotavirus NSP4 glycoproteins, which differ greatly from mammalian rotavirus NSP4 glycoproteins in deduced amino acid sequence, in suckling mice. J. Virol. *76*, 5829–5834.

Mori, Y., Borgan, M.A., Ito, N., Sugiyama, M., and Minamoto, N. 2002b. Sequential analysis of nonstructural protein NSP4s derived from group A avian rotaviruses. Virus Res. *89*, 145–151.

Morris, A.P., Scott, J.K., Ball, J.M., Zeng, C.Q.Y., O'Neal, W.K., and Estes, M.K. (1999). NSP4 elicits age-dependent diarrhea and Ca^{2+}-mediated I^- influx into intestinal crypts of CF mice. Am. J. Physiol. *277*, G431-G444.

Morris, A.P., and Estes, M.K. (2004). Microbes and microbial toxins, Paradigms for microbial-mucosal interactions. VIII. Pathological consequences of rotavirus infection and its enterotoxin. Am. J. Physiol. *281*, G303-G310.

Nakagomi, O., and Nakagomi, T. (1991). Genetic diversity and similarity among mammalian rotaviruses in relation to interspecies transmission of rotavirus. Arch. Virol. *120*, 43–55.

Nejmeddine, M., Trugnan, G., Sapin, C., Kohli, E., Svensson, L., Lopez, S., and Cohen, J. (2000). Rotavirus spike protein VP4 is present at the plasma membrane and is associated with microtubules in infected cells. J. Virol. *74*,3313–3320.

Newton, K., Meyer, J.C., Bellamy, A.R., and Taylor, J.A. (1997). Rotavirus nonstructural glycoprotein NSP4 alters plasma membrane permeability in mammalian cells. J. Virol. *71*, 9458–9465.

Nishikawa, K., Hochino, Y., Taniguchi, K., Green, K.Y., Greenberg, H.B., Kapikian, A.Z., Chanock, R.M., and Gorziglia, M. (1989). Rotavirus VP7 neutralization epitopes of serotype 3 strains. Virology *171*, 503–515.

O'Brien, J.A., Taylor, J.A., and Bellamy, A.R. (2000). Probing the structure of rotavirus NSP4, a short sequence at the extreme C-terminus mediates binding to the inner capsid particle. J. Virol. *74*, 5388–5394.

Obert, G., Peiffer, I., Servin, A. (2000). Rotavirus-induced structural and functional alterations in tight junctions of polarized intestinal Caco-2 cell monolayers. J. Virol. *74*, 4645–4651.

Oishi, I., Yamazaki, K., and Minekawa, Y. (1993). An occurrence of diarrheal cases associated with group C rotaviruses in adults. Microbiol. Immunol. *37*, 505–509.

Osborne, M.P., Haddon, S.J., Spencer, A.J., Collins, J.E., Starkey, W.G., Wallis, T.S., Clarke, G.J., Worton, K.J., Candy, D.C., and Stephen, J. (1988). An electron microscopic investigation of time-related changes in the intestine of neonatal mice infected with murine rotavirus. J. Ped. Gastroenterol. Nutr. *7*, 236–248.

Parr, R.D., Storey, S.M., Mitchell, D.M., McIntosh, A.L., Zhou,M., Mir, K.D., and Ball, J.M. (2006). The rotavirus enterotoxin NSP4 directly interacts with the caveolar structural protein caveolin-1. J. Virol. *80*, 2842–2854.

Petrie, B.L., Estes, M.K., and Graham, D.Y. (1983). Effects of tunicamycin on rotavirus morphogenesis and infectivity. J. Virol. *46*, 270–274.

Petrie, B.L., Greenberg, H.B., Graham, D.Y., and Estes, M.K. (1984). Ultrastructural localization of rotavirus antigen using colloidal gold. Virus Res.*1*, 133–152.

Polozov, I.V., Polozov, A.I., Mishra, V.K., Anantharamaiah, G.M., Segrest, J.P., and Epand, R.M. (1998). Studies of kinetics and equilibrium membrane binding of class A and class L model amphipathic peptides. Bioch. Biophys. Acta *1368*, 343–354.

Poruchynsky, M.S., Mass, D.R., and Atkinson, P.H. (1991). Calcium depletion blocks the maturation of rotavirus by altering the oligomerization of virus-encoded proteins in the ER. European J. Cell Biol. *114*, 651–661.

Ray,P., Malik, J., Singh, R.K., Bhatnagar, S., Bohl, R., Kumar, R., and Bhan, M.K. (2003). Rotavirus nonstructural protein NSP4 induces heterotypic antibody responses during natural infection in children. J. Infect. Dis. *187*, 1786–1793.

Rothberg, K.G., Heuser, J.E., Donzell, W.C., Ying, Y.-S., Glenney, J.R., and Anderson, R.G.W. (1992). Caveolin, a protein component of caveolae membrane coats. Cell *68*, 673–682.

Rodriguez-Diaz, J., Monedero, V., Perez-Martinez, G., and Buesa, J. (2004). Single-chain variable fragment (scFv) antibodies against rotavirus NSP4 enterotoxiin generated by phage display. J. Virolog. Methods *121*, 231–238.

Ruiz, M.-C., Cohen, J., and Michelangeli, F. (2000). Role of Ca^{2+} in the replication and pathogenesis of rotavirus and other viral infections. Cell Calcium *28*, 137–149.

Sabara, M., Babiuk, L.A., Gilchrist, J., and Miisra, V. (1982). Effects of tunicamycin on rotavirus assembly and infectivity. J. Virol. *43*, 1082–1090.

Sapin, C., Colard, O., Delmas, O., Tessier, C., Breton, M., Enouf, V., Chwetzoff, S., Ouanich, J., Cohen, J., Wolf, C., and Trugnan, G. (2002). Rafts promote assembly and atypical targeting of a nonenveloped virus, rotavirus, in Caco-2 cells. J. Virol. *76*, 4591–4602.

Sasaki,S., Horie,Y., Nakagomi,T., Oseto,M., and Nakagomi,O. (2001). Group C rotavirus NSP4 induces diarrhea in neonatal mice. Arch. Virol. *146*, 801–806.
Sears, C.L., Guerrant, R.L., and Kaper, J.B. (1995). Enteric Bacterial Toxins. In, Infections of the Gastrointestinal Tract. M.J. Blaser, P.D. Smith, J.I. Ravdin, and H.B. Greenberg, R.L., ed. Guerrant New York, Raven Press, Ltd. 617–634.
Shahrabadi, M.S., and Lee, P.W. (1986). Bovine rotavirus maturation is a calcium-dependent process. Virology *152*,.298–307.
Skehel, J.J. & Wiley, D.C. (1998). Coiled-coil in both intracellular vesicle and viral membrane fusion. Cell. 95,.871–874.
Silvestri, L.S., Tortorici, M.A., Vasquez-Del Carpio, R., and Patton, J.T. (2005). Rotavirus glycoprotein NSP4 is a modulator of viral transcription in the infected cell. J. Virol. *79*, 15165–15174.
Storey, S.M., Schroeder, F., and Ball, J.M. (2007). Full-length, glycosylated NSP4 is localized to plasma membrane caveolae by a novel raft isolation technique. J. Virol. *8,1* 5472–5483.
Tafazoli, F., Zeng, C.Q.Y., Estes, M.K., Magnusson, K.-E., and Svensson, L.(2001). NSP4 enterotoxin of rotavirus induces paracellular leakage in polarized epithelial cells. J. Virol. *75*,1540–1546.
Taylor, J.A., Meyer, J.C., Legge, M.A., O'Brien, J.A., Street, J.E., Lord, V.J., Bergmann, C.C., and Bellamy, A. R. (1992). Transient expression and mutational analysis of the rotavirus intracellular receptor, the C-Terminal methionine residue is essential for ligand binding. J. Virol. *66*, 566–3572.
Taylor, J.A., O'Brien, J.A., Lord, V.J., Meyer, J.C., and Bellamy, A.R. (1993). The RER-localized rotavirus intracellular receptor, A truncated purified soluble form is multivalent and binds virus particles. Virology *194*, 07–814.
Taylor, J.A., O'Brien, J.A., and Yeager, M. (1996). The cytoplasmic tail of NSP4, the endoplasmic reticulum-localized nonstructural glycoprotein of rotavirus, contains a distinct virus binding and coiled-coil domains. EMBO J. 15; 4469–4476.
Tian, P., Ball, J.M., Zeng, C.Q.Y., and Estes, M.K. 1996a. Rotavirus protein expression is important for virus assembly and pathogenesis. Arch. Virol. *12*, 69–77.
Tian, P., Ball, J.M., Zeng, C.Q.Y., and Estes, M.K. 1996b. The rotavirus nonstructural glycoprotein NSP4 possesses membrane destabilization activity. J. Virol. *70*, 6937–6981.
Tian, P., Estes, M.K., Hu, Y., Ball, J.M., Zeng, C.Q.Y., and Schilling, W.P. (1995). The rotavirus nonstructural glycoprotein NSP4 mobilizes Ca^{2+} from the endoplasmic reticulum. J. Virol. *69*,5763–5772.
Tian, P., Hu, Y., Schilling, W.P., Lindsay, D.A.E.J., and Estes, M.K. (1994). The nonstructural glycoprotein of rotavirus affects intracellular calcium levels. J. Virol. *68*, 251–257.
Vizzi, E., Calvino, E., Gaonzalez, R., Perez-Schael, I., Ciarlet, M., Kang, G., Estes, M.K., Liprandi, F., and Ludert, J.E. (2005). Evaluation of serum antibody responses against the rotavirus nonstructural protein NSP4 in children after rhesus rotavirus tetravalent vaccination or natural infection. Clin. Diag. Lab. Immunol. *12*, 1157–1163.
Wang, C.-W., and Klionsky, D. J. (2003). The molecular mechanism of autophagy. Mol. Med. *9*,65–76.
Wen, L., Nakayama, M., Yamanishi, Y., Nishio, O., Fang, Z.Y., Nakagomi, O., Araki, K., Nishimura, S., Hasegawa, A., Müller, W.E.G., and Ushijima, H. (1997). Genetic variation in the VP7 gene of human rotavirus serotype 3 (G3 type) isolated in China and Japan. Arch. Virol. *142*, 1481–1489.
Werry, T.D., Wilkinson, G.F., and Willars, G.B. (2003). Mechanisms of cross-talk between G protein-coupled receptors resulting in enhanced release of intracellular Ca^{2+}. Biochem. J. *374*, 281–296.
Williams, T.M., and Lisanti, M.P. (2004). The Caveolin genes: from cell biology to medicine. Ann. Med. 36, 584–95.
Xu, A., Bellamy, A.R., and Taylor, J.A. (1998). Bip (GRP78) and endoplasmin (GRP94) are induced following rotavirus infection and bind transiently to an endoplasmic reticulum-localized virion component. J. Virol. *72*,9865–9872.
Xu, A., Bellamy, A.R., and Taylor, J.A. (1999). Expression of translationally controlled tumour protein is regulated by calcium at both the transcriptional and post-transcriptional levels. Biochem. J. *342*, 683–689.
Xu, A., Bellamy, A.R., and Taylor, J.A. (2000). Immobilization of the early secretory pathway by a virus glycoprotein that binds to microtubules. EMBO J. *19*, 6465–6474. (2000).
Yuan, L., Honma, S., Ishida, S.-I., Yan, X.-Y., Kapikian, A.Z., and Hoshino, Y. (2004). Species-specific but not genotype-specific primary and secondary isotype-specific NSP4 antibody responses in gnotobiotic calves and piglets infected with homologous host bovine (NSP4[A]) or porcine (NSP4[B]) rotavirus. Virology *330*, 92–104.

Zhang, M., Zeng, C.Q.Y., Dong, Y., Ball, J.M., Saif, L.J., Morris, A.P., and Estes, M.K. (1998). Mutations in rotavirus nonstructural glycoprotein NSP4 are associated with altered virus virulence. J. Virol. *72*, 3672.

Zhang, M., Zeng, C.Q.Y., Morris, A.P., and Estes, M.K. (2000). A functional NSP4 enterotoxin peptide secreted from rotavirus-infected cells. J. Virol. *74*, 11663–11670.

Part III

RNA Replication

The Infectious Reovirus RNA–Reverse Genetics System: The Assembly of the Reovirus Genome

19

Michael R. Roner and Wolfgang K. Joklik

Abstract

Viruses of the *Reoviridae* family, which encompasses nine genera and more than 200 members, possess genomes that comprise ten or twelve double-stranded (ds)RNA segments. Using reovirus as a model, we have developed a functional reverse genetics system for these viruses, by means of which alterations introduced artificially into their genomes are transferred to the genomes of viable infectious virus particles. The system consists of (a) the plus strands of nine wild type reovirus genome segments; (b) single-stranded (ss)RNA transcripts generated from a genetically modified cDNA that replace the wild type tenth genome segment; and (c) a cell line transformed to express the protein normally encoded by the tenth genome segment. With this system, we have generated ST3 reoviruses with an engineered S2 genome segment, an M1 genome segment, and both S2 and M1 genome segments. Both the S2 and M1 engineered genes contain a functional CAT gene. The viruses are stable, replicate in cells that have been transformed to express the S2 gene product protein σ2, the M1 gene product protein μ2 or both, and express high levels of CAT activity. This technology, which we are extending to the orbiviruses and rotavirus, provides a powerful system for basic studies of dsRNA virus replication and modification.

Introduction

The reovirus genome comprises 10 segments of double-stranded (ds)RNA that are transcribed non-conservatively into plus-sense single-stranded (ss)RNA within cores from which they are extruded via 12 icosahedrally distributed projections/spikes. These plus strands are translated into the reovirus-encoded complement of proteins and they are also transcribed into minus strands with which they remain associated, thereby generating the progeny dsRNA genome segments (Joklik, 1985; Joklik and Roner, 1996).

Several years ago, we demonstrated that reovirus RNA is infectious (Roner, Sutphin, and Joklik, 1990). The basic features of the infectious reovirus RNA system are as follows.

The infectious reovirus RNA system

Monolayers of L929 mouse embryo fibroblasts are lipofected with the 10 species of ss and/or dsRNA of reovirus serotype 3 (ST3 virus) strain Dearing together with a reticulocyte lysate in which ss or melted dsRNA has been translated for 60 min (that is, a primed rabbit reticulocyte lysate (PRRL)). Eight hours later, the cells are infected with reovirus

ST1 or ST2 as helper virus. Virus is harvested 24 to 48 hours later and plaqued either in the presence of antiserum against ST1 virus if it is used as a helper virus, or without antiserum if the helper virus is ST2 virus (since ST3 virus forms plaques after 5 days, whereas ST2 virus forms plaques only after 12 days). Under the conditions described (Roner, Sutphin, and Joklik, 1990), about 4% of cells yield ST3 virus. Each cell that yields virus yields about 5 PFU.

The following aspects of this system should be noted:

(a) It is essential to use high quality, intact, ss and dsRNA species. The quality of ssRNA is tested by its ability to yield all viral protein species in cell-free protein synthesizing systems. The quality of dsRNA is judged by the nature of its electrophoretic pattern. Lipofection should be carried out exactly as described, particularly with respect to the amounts/cell lipofected (Roner, Sutphin, and Joklik, 1990).

(b) Helper virus is essential; no infectious ST3 virus is formed if helper virus is omitted. In practice, ST2 virus is preferred because when it is used as helper virus, isolation of ST3 virus does not require the use of antiserum. UV-inactivated ST2 virus fails to support the formation of infectious ST3 virus. No reassortants containing ST2 genome segments have ever been observed among the yield, although many ST3-ST2 reassortants replicate as rapidly as ST3 virus, that is, form plaques within 5 days (Moody and Joklik, 1989). The nature of the function performed by the helper virus is not known. It may induce the functioning of essential cellular factors (though it is not clear why UV-inactivated virus should be unable to do this also), or it may provide transfunctional proteins during the initial stages of the series of reactions initiated by naked RNA that culminate in the formation of infectious virus. However, there is clearly no, or minimal, interaction between assortment complexes of ST3 and ST2 helper virus since there are no reassortants among the progeny.

(c) Both ss and dsRNA are infectious, that is, will yield infectious virus. However, the yields resulting from lipofection of dsRNA are 10 to 20 times higher than those following lipofection of ssRNA, and when ss and dsRNA are lipofected together, yields are ten times higher still. The function of the dsRNA is to increase the infectiousness of ssRNA, as demonstrated using ST3 virus and the monoreassortant containing 9 ST3 genome segments and the ST2 S2 genome segment (Moody and Joklik, 1989) each in the form of ssRNA and dsRNA.

    The fact that ss and dsRNA together are 10 to 20 times more infectious than dsRNA alone and 100 times more infectious than ssRNA alone is intriguing. Clearly the two sets of segments interact intimately. Not only can they enhance each other's infectiousness, but they can also diminish it; for if ST2 ss/ST3 dsRNA or the reverse combination is used, in the absence of PRRLs so as to avoid ambiguity, no infectious virus is formed, no matter what the identity of the helper.

(d) Lipofection of primed rabbit reticulocyte lysates (PRRLs) is not essential, but increases virus yields 10- to 100-fold. It is in PRRLs that the proteins necessary for RNA transcription, as well as, conceivably, a significant proportion of the structural proteins necessary for virus particle formation, are generated. However, the function of PRRLs is not limited to supplying viral proteins; for PRRLs in which ssRNA is destroyed following translation (as demonstrated by loss of ability to form more protein) are unable to boost infectious virus formation. Thus, the active principle in PRRLs is not merely newly synthesized viral protein; it may be RNA-protein intermediary-complexes of unknown composition.

(e) These properties of the basic infectious reovirus RNA system focus attention on the genome segment assortment process as the most critical stage during the reovirus multiplication cycle, that is, the formation of complexes that comprise one of each of the ten genome segments. It is known that it is not possible to introduce ST2 genome segments into the ST3 genome by injecting either the ss or the ds forms of single ST2 genome segments into cells infected with ST3 virus, by transfecting or lipofecting them, or by transcribing them intracellularly from appropriate plasmids (unpublished results). By contrast, the yields of virus in cells infected with mixtures of reovirus particles belonging to two different serotypes comprise roughly 15 reassortants that contain the genome segments of the two parents in approximately Poissonian distribution. Interestingly, when cores of the two serotype viruses are lipofected into cells (that is, under conditions when they possess the same specific infective as infectious virus particles) 10 to 15 of the progeny are again reassortants, but now they are all monoreassortants (Moody and Joklik, 1989). And when mixtures of ST3 and ST2 dsRNA are lipofected into cells together with appropriate PRRLs, 10–15% of the virus yields (which are only about one-tenth of that obtained with ST3 reovirus alone) again comprises reassortants, and all are monoreassortants that contain a single ST2 genome segment, two of which (S2 and Ml) are preferred (Roner, Sutphin, and Joklik, 1990).

Irrespective of the uncertainty of the precise nature of helper and PRRL fraction and of the enhancement of ssRNA infectivity by dsRNA, it is clear that the infectious reovirus RNA system permits the introduction of heterologous genetic information into the, reovirus genome and, presumably, into the genome of any member of the *Reoviridae* family. It will permit the identification of encapsidation and assortment signals on genome segments, functional and regulatory domains of viral proteins, the development of members of the *Reoviridae* family as expression vectors, and the identification of efficient non-pathogenic vaccine strains for members of the *Reoviridae* family.

The construction of a reovirus double temperature-sensitive mutant

In order to validate the system with respect to its ability to introduce functional heterologous genome segments into the reovirus genome, it was used to introduce genome segments with one or two mutations. The segments chosen were M2 and S2, the former derived from ST3 strain Dearing tsA 201, the latter from ST3 strain Dearing tsC 447 (Fields and Joklik, 1969). The tsA 201 m2 RNA differs from wt m2 RNA in two locations, namely G944 to U and C972 to U (unpublished results); the tsC 447 s2 RNA differs from wt s2 RNA in Al 166 to G (Cashdollar *et al.*, 1982; Wiener, McLaughlin, and Joklik, 1989).

The tsA/C double mutant was constructed from two populations of RNA (Roner *et al.*, 1997). The first was the ssRNA population transcribed by cores of tsA 201 from which the s2 RNA species had been removed. This was achieved by adding to 2 pmol of RNA, 10 pmol of the oligodeoxyribonucleotide complementary to residues 774–760 of reovirus

ST3 s2 RNA (Cashdollar *et al.*, 1982) together with 5 units of RNase H. After 20 min incubation at 50°C, the RNA preparation was extracted three times with phenol and precipitated with sodium acetate. The second population comprised the small (s) size class RNA species of the RNA transcribed by reovirus ST3 tsC 447 cores. These two RNA populations were lipofected into mouse L929 fibroblasts using the experimental conditions of the infectious reovirus RNA system. Progeny virus formed very small plaques after 12 days at 30°C; at that temperature ST2 virus formed plaques within 16 days. The significant feature of the double mutant construction reported here is that it eliminates the necessity for selecting the desired progeny because the genome segments that were to be replaced were removed completely or absent from the accepting genome segment population (Roner *et al.*, 1997).

The efficiency of plaquing at 30°/39°C for wt ST3 strain Dearing virus is 1.12; for tsA 201, 165; for tsC 447, 1450; and for tsA/C it is 450 000. The one-step growth curves of reovirus ST3 strain Dearing, tsA 201, tsC 447 and tsA/C are shown in Figure 19.1. As a result of its tremendous temperature sensitivity, tsA/C is essentially unable to replicate at body temperature. In spite of this, tsA/C elicits the formation of very significant amounts of antibody: the neutralization titre of antiserum elicited by it three weeks after infection is about one-half of that elicited by wt virus (Roner *et al.*, 1997). This property, together with the fact that it differs minimally from the wild-type (wt) strain, makes it a potentially extremely safe, very attractive vaccine strain.

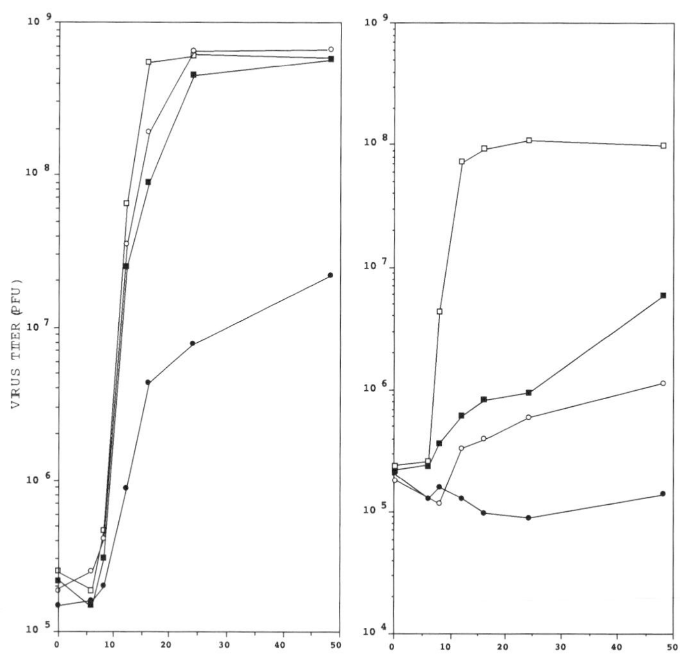

Figure 19.1 One-step growth curves of wt reovirus ST3, ts A 201, tsC 447, and tsA/C at 30 C(A) and 39 C(B) in mouse fibroblast monolayers. ■ (wt virus), □ (tsA 201), O (tsC 447), ● (tsA/C).

Identification of signals required for the insertion of heterologous genome segments into the reovirus genome

The genomes of mammalian reoviruses are assembled extremely efficiently: the ratio of the number of total to infectious reovirus particles is close to 1 (Spendlove, McClain, and Lennette, 1970). These viruses can be grouped into three serotypes that possess homologous genome segment sets. It is not only the sequences of the genome segments that encode the proteins that possess the type- specific epitopes that differ significantly, but the sequences of the other nine genome segments differ very significantly also. In cells simultaneously infected with viruses of any two of the three serotypes, up to 15% of the progeny are intertypic reassortants that contain all possible combinations of parental genome segments. It turns out, however, that not all genome segments in reassortants are wild type. The first indication that this might be so was obtained when it was found unexpectedly that the infectious reovirus RNA system was unable to incorporate a heterologous genome segment into a homologous genome segment set. As described above, if a mixture of two complete genome segment sets, say that of a ST2 virus and that of a ST3 virus, are lipofected into cells under standard system conditions, that is, with appropriate PRRLS, 10–15% of the yield are monoreassortants. However, when it was attempted to introduce the ST2 Sl genome segment into the ST3 genome segment set, which would be easy to detect immunologically, results were uniformly and completely negative.

The insertion of ST2 genome segments into the ST3 genome: the mutant ST3 S4 genome segment

The use of nine genome segment set acceptors was therefore explored (Roner *et al.*, 1995). Such sets can be obtained by selectively destroying single genome segments (as was done for the generation of the double ts mutant above), or by transcribing the nine genome segments present in cores of defective interfering (DI) virus particles. ST3 DI virus lacks an intact LI genome segment and instead contains a truncated version of it that is about one half its normal length of 3860 bp; and ST1 DI virus lacks an intact Ml genome segment and contains instead a truncated version of it that is ~700 bp long (instead of 2204 bp) (Zou and Brown, 1992). Preparations of DI virus contain helper virus, the proportion of which is a function of the multiplicity used for passaging; clearly cells in which DI particles are generated must contain a minimum of functional wt virus particles. In the experiments about to be described, ST3 DI virus preparations were used that contained about 1 wt virus particles. When the ability of ST3 DI ssRNA to generate infectious virus in conjunction with the l size class segments of wt ST3 ssRNA was tested, no infectious virus was formed. This suggested the possible presence of 'incompatible' mutations in one or more of the nine intact genome segments of DI virus. When this was tested directly, it was found that wt ST3 1 size class RNA was able to generate infectious virus in conjunction with DI virus m size class RNA and wt ST3 s size class RNA, demonstrating the absence of such mutations in m size class DI genome segments and in wt ST3 s size class ssRNA; but it was unable to generate infectious virus in conjunction with m size class wt ST3 RNA and s size class DI virus RNA (Roner *et al.*, 1995).

The four s size class genome segments of DI ST3 virus were therefore sequenced. No mutations were found in the sl, s2 and s3 genome segments, but two mutations were found in s4 ssRNA: G74 to A (silent) and G624 to A (aspartic acid to lysine). The presence of an S4 genome segment with these two mutations in the ST3 genome (a) permits the acceptance of truncated, that is, heterologous, genome segments into the ST3 genome and (b) renders the wt ST3 genome non-infectious (Roner *et al.*, 1995).

This situation is analogous to the introduction of ST1 or ST2 genome segments into the ST3 genome using ten genome segment set acceptors, that is, via infection of cells with infectious parents for the formation of intertypic reassortants. Here also such reassortment is efficient in cells infected with the two parents, but is not affected by the infectious reovirus RNA system in which mixtures such as ST2 1 + ST3 m +ST3 s size class RNAs, in conjunction with appropriate PRRLs and helper virus, fail to yield infectious virus. The s4 genome segments of intertypic reassortants, both monoreassortants and multireassortants (that is, each serotype represented by at least 2 genome segments) were therefore sequenced; and in all were found the same two mutations at G7410 A and G624 to A. Thus these two mutations in the ST3 s4 genome segment function to permit the introduction into the ST3 genome of RNA molecules that differ from the wt sequence both in size and in sequence (Roner *et al.*, 1995).

These results demonstrate that for heterologous genome segments to be accepted into the ST3 genome, a mutant S4 genome segment is required. The question then arises as to why the introduction of heterologous genome segments like intertypic reassortment proceeds efficiently in cells infected with the two parents, but is not affected by the infectious reovirus RNA system. The most likely reason for this is lack of adequate amounts of genome segments with the signals that permit the introduction of heterologous genome segments. Indeed, in the infectious reovirus RNA system only about 100 molecules of each genome segment are lipofected into cells, whereas in infected cells at least one hundred times more molecules of each genome segment are transcribed.

Another question that arises is why the genome segment set ST3 L size class + ST3 M size class + ST3 S size class containing the S4 genome segment with the G74 to A and G624 to A mutations is *not* infectious, whereas sets in which the ST3 L or ST3 M size classes *are* replaced by either ST1 or ST2 size class segments together with the ST3 S size class containing the S4 genome segment with the two mutations are infectious (they would be viable reassortants). The mutated S4 genome segment is obviously functional (in the reassortants) and present in the genome segment set of small proportions of ST3 virus populations, otherwise reassortants could not arise. The answer to this question is not yet apparent.

The insertion of ST3 genome segments into the ST1 genome: the existence of mutations in the inserted genome segments

The overwhelming importance of sequence recognition in genome segment reassortment, that is, in the introduction of heterologous genome segments into genome segment sets, is illustrated by examination of the sequences of the heterologous genome segments that are actually incorporated. The reassortants that were examined in this analysis were the result of ST1 × ST3 crosses (Roner *et al.*, 1995). Not surprisingly, the nature of the signals required for the insertion of ST3 genome segments into the ST1 genome is different from those required for the insertion of ST 2 genome segments into the ST3 genome, for which it is presence of the G74 to A and G624 to A mutations in the ST3 S4 genome segment. For the insertion of ST3 genome segments into the ST1 genome no mutations in the ST1

S4 genome segment are required, nor are there any in the ST1 DI S4 genome segment. On the other hand, the incoming ST3 genome segments are all variants of their wt forms (Table 19.1), the Sl genome possessing four mutations, the Ml genome segment possessing three mutations, and the S2, S3 and S4 genome segments possessing one mutation each. In each case, the effect of the mutation is to change a residue where the ST3 genome segments differ from that of the ST1 genome segment — that is, the effect is to render the incoming ST3 genome segment more similar to the ST1 genome segment that it replaces. Interestingly, six of these mutations are silent; four cause amino acid changes, all of them conservative. One consequence of the fact that the introduction of ST3 genome segments into the ST1 genome segment set requires no mutation(s) in any ST1 genome segment is that crosses of STl/ST3 monoreassortants (one ST3 segment) yield 50 double reassortants and 50 wt ST1 virus. This is obviously impossible for crosses between ST3/ST2 monoreassortants (one ST2 segment) in which the virus yield comprises only double reassortants and no wt ST3 virus, because all monoreassortants possess the mutated S4 genome segment (Roner *et al.*, 1995).

The conclusion from all these investigations is that the introduction of genome segments into the reovirus genome is controlled primarily by sequence.

The generation of mutant ST3 S4 genome segments

It has been pointed out above that populations of ST3 reovirus must contain virus particles that possess the S4 genome segments with the G74 to A and G624 to A mutations. How rapidly, and to what extent, do these mutations accumulate in plaque-purified virus populations?

It turns out that the G624 to A mutation disrupts an Mnl restriction endonuclease cleavage site (Roner *et al.*, 1995). It is therefore possible to test for it by determining whether the appropriate PCR fragment is cleaved by Mnl and to adapt this test to provide a semi-

Table 19.1 Mutations in ST3 genome segments in monoreassortants containing nine ST1 genome segments and one ST3 genome segment

Monoreassortant	Mutations
ST1-ST3 M1	G462 to A (R to Q)
	U985 to G (silent)
	G1129 to U (M to I)
ST1-ST3 S1	C77 to U (A to V)
	G108 to U (V to F)
	G378 to C (silent)
	G504 to A (silent)
ST1-ST3 S2	U702 to C (silent)
ST1-ST3 S3	G699 to A (silent)
ST1-ST3 S4	C383 to U (silent)

Amino acids are in single-letter code.

quantitative estimate of the proportion of this mutant in virus populations. Examination of a series of virus stocks led to the following conclusions. The G624 to A mutation is not detectable immediately following plaque purification. The mutation accumulates both under conditions of high-multiplicity passage such as are employed for the generation of DI virus stocks (multiplicities of infection of 50 to 100 PFU per cell), as well as under conditions employed for routine passaging of virus stocks (multiplicities of infection of about 10 PFU per cell). The actual proportions of mutants in virus populations varies depending on the number of passages following plaque purification and on the multiplicity used for each passage. Virus stocks containing 25 to 50 mutants are still perfectly functional because the sole effect of their presence is that the specific infectivity of the stocks would be slightly lower, which would not be noticed under normal conditions. Thus, normal stocks of reovirus ST3, such as would be used routinely, are likely to contain about 25 non-infectious particles, the genomes of which contain S4 genome segments with the G624 to A mutation and probably also with the G74 to A mutation, the presence of which has not yet been established because this can be done only by sequencing.

It would be predicted from this analysis that the lower the proportion of G624 to A mutations in a given virus stock, the lower would be the proportion of intertypic reassortants in ST3 × ST2 crosses. This prediction holds true. With 'normal' ST3 virus stocks, the proportion of reassortants is usually around 15; with 'young' virus stocks (less than 4 passages following plaque purification) the percentage is about 4; and virus immediately following plaque purification yields less than 1 reassortants when crossed with ST2 virus under exactly the same experimental conditions (Roner *et al.*, 1995).

Are the sequence alterations with essential roles in assortment required in genome segments *or in the proteins that they encode?*

The question arises as to whether the sequence alterations with essential roles in reassortment are required in genome segments for recognition by other genome segments, or in the proteins that they encode.

In the case of ST3 × ST1 reassortants, the changes are clearly essential in the genome segments because at least one half of them are silent. In ST3 × ST2 crosses, however, the situation is different: here there is no question of recognition of incoming molecules because the one mutation that is not silent, G624 to A, facilitates admission into the ST3 genome of ST2 genome segments as well as of the truncated homologous LI genome segment. Does this mutation encode a protein with a significantly different structure and function? In order to test this, mouse L929 fibroblasts were infected with wt ST1 or ST3 virus or with monoreassortants, two of the 9 ST3–1 ST2 series and two of the 9 ST1–1 ST3 series (Roner *et al.*, 1995). Samples of these cells were then also infected with recombinant vaccinia virus into the TK gene of which the ST3 S4 genome segment had been cloned so as to be under control of the phage T7 and T7 polymerase gene promoters. Reovirus yields were measured after 48 hours (Table 19.2). The results indicate that over expression of the wt ST3 σ3 protein has no effect on the yield of ST3 virus, but markedly inhibits the replication of ST1 and of the monoreassortants of both series – that is, it interferes with the function of heterologous σ3 regardless of whether it is from a different serotype or contains the G624 to A (aspartic acid to lysine) mutation. This suggests a functional role for σ3 in genome segment reassortment, which is not unexpected in view of the fact

Table 19.2 Demonstration of interference between wt ST3 σ3 and ST1 σ3 and ST3 σ3 possessing the Asp-198 to Lys mutation

Vaccinia virus		Virus titres 48 hours after infection					
VV-s3	VTF7-3	ST3 with ST2 S1	ST3 with ST2 S2	ST1 with ST3 M1	ST1 with ST3 S1	ST3	ST1
+	–	3.6×10^3	1.2×10^5	2.3×10^6	3.6×10^7	3.3×10^7	2.6×10^8
–	+	3.3×10^3	1.5×10^5	2.2×10^6	3.4×10^7	3.2×10^7	2.4×10^8
+	+	2.0×10^1	1.6×10^4	1.1×10^4	2.4×10^6	2.9×10^7	2.4×10^6

VV-σ3 is a recombinant vaccinia virus expressing the reovirus σ3 protein with transcription under control of the T7 promoter. VTF7–3 is a recombinant vaccinia virus expressing T7 RNA polymerase.

that σ3 associates with reovirus mRNA molecules as soon as they are transcribed and is a component of the ssRNA-containing complexes that are precursors of the complexes in which progeny dsRNA genome segments are generated and assembled (Antczak and Joklik, 1992).

The surprising conclusions from these results are that (i) the mode of acceptance of ST2 genome segments into the ST3 genome differs fundamentally from that for the acceptance of ST3 genome segments into the ST1 genome (bearing in mind that ST3 and ST1 diverged long after their precursor diverged from ST2); and (ii) that the introduction of 'foreign' genetic information into the reovirus genome involves not only wt genome segments, but also mutant or variant genome segments that are minor population components.

Reovirus as a vector: incorporation of the CAT gene into the reovirus genome

The infectious reovirus RNA system can be used to generate a variety of novel viruses through selection of the nature of the RNA species that are lipofected. Variants of a wt virus, such as the ST3 Dearing strain, can be constructed, like the double ts mutant described above, and immunologic variants; reassortants can be constructed containing any desired combination of genome segments of parents of any reovirus serotype; and foreign genes can be inserted into any genome segment and introduced into the reovirus genome, thereby transforming reovirus into a gene vector, as described in the following section.

Reovirus vectors can be used in either of two ways. The first employs the vector virus under conditions where it cannot replicate. This involves infecting cells with the vector virus at high multiplicities which requires large amounts of the virus. The second way involves using cells in which the vector virus can multiply, and that means cells in which the protein encoded by the genome segment into which the foreign gene is cloned is provided. The missing protein is then supplied either by co-infecting the cells with wt virus, or using cells in which the missing protein is expressed from a plasmid.

In the work to be described (Roner and Joklik, 2001), the CAT gene was cloned into the ST3 strain Dearing s2 genome segment which encodes protein σ2, retaining 198 bp of its 5′-end and 284 bp of its 3′- end. The T7 polymerase promoter was placed at the 5′- end and the T7 terminator at the 3′-end; and the HDV ribozyme sequence was inserted at the 3′-terminus of the S2-CAT RNA to ensure the presence of the universal reovirus RNA 3′-terminal sequence, -UCAUC-. This RNA, s2-CAT RNA, was 1234 nucleotides long; 97 nucleotides shorter than wt s2 RNA. It encodes a σ2-CAT fusion protein that possesses 60 σ2 amino acids at its N-terminus and does not exercise protein σ2 function, but possesses CAT catalytic activity. To generate CAT-expressing reovirus, ST3.S2.CAT, the s2-depleted set of nine ST3 virus ss RNA species was supplemented with s2.CAT RNA transcribed from the pS2CAT construct and inserted into the reovirus genome using the infectious reovirus RNA system. A PRRL in which the mixture of 9 ST3 virus genome segments together with s2-CAT was translated was used, and ST2 virus was the helper virus. It permitted the vector virus to replicate and form plaques. Interestingly, the S2.CAT genome segment was accepted into the ST3 genome segment about 5 times less efficiently than the wt S2 genome segment despite possessing the latter's 198 bp at its 5′-end and 284 bp at its 3′-end.

It should be pointed out that for the purpose of demonstrating the feasibility of inserting the CAT gene into the S2 genome segment and then introducing the resulting construct into the ST3 genome, the length of the S2.CAT construct was designed to be approximately the same as that of the intact S2 genome segment; in fact, it was only 97 nucleotides shorter. Subsequently the minimum lengths of the S2 5′- and 3′-sequences necessary for acceptance into the ST3 virus genome were determined (Roner, Bassett, and Roehr, 2004; Roner and Roehr, 2006). These lengths turned out to be 96 bp at the 5′-end and 98 nucleotides at the 3′-end (Table 19.3). Interestingly, it appears that an -A-U-U-sequence at positions 79–81 is required for introduction into the ST3 genome; mutating it in any way leads to loss of activity (Roner, Bassett, and Roehr, 2004).

The ST3.S2.CAT virus was grown in L-ST3.S2 cells which express the σ2 protein. These cells were prepared by transforming L929 fibroblasts with ppAPr-1-neo that contains the S2 cDNA coding sequence under the control of the β-actin promoter. They express sufficient levels of protein σ2 to support the replication of the ST3 virus temperature-sensitive mutant tsC447 at 39°C.

The ST3.S2.CAT virus was characterized according to several criteria. Figure 19.2 presents an electropherogram of its genome segments; and Table 19.4 presents the kinetics of CAT formation in ST3.S2.CAT virus – infected L929 and L-ST3.S2 cells. ST3.S2.CAT virus clearly generates abundant amounts of the catalytically active CAT fusion protein.

Recently we have extended this work and have used *in vitro* engineered and transcribed reovirus m1 and s2 ssRNAs to demonstrate that the nucleotides used to identify these ssRNAs are localized to the 5′ and not the 3′ termini (Roner and Steele, 2007). The M1 dsRNA of our M1-CAT virus retains 124 nucleotides from the wildtype M1 gene preceding the CAT gene and 172 nucleotides from the wt gene following the CAT gene. The engineered M1-CAT ssRNA is 1048 nucleotides in length, much shorter than the wild type M1 at 2304 nucleotides. We have used a set of chimeric s2.m1 ssRNAs to localize the packaging signals within these RNAs. By packaging signals we mean that the presence of these signals in engineered ssRNAs results in these ssRNAs being replicated to dsRNA and packaged into progeny virus. An engineered ssRNA with a 5′ sequence identical with the wt s2 ssRNA, supported by a 3′ sequence from either the m1 or s2 ssRNA, is incorpo-

Table 19.3 Summary of the biological activities of recombinant ssRNAs used to drive the reovirus infectious RNA system

cDNA Clone	Length of 5′ S2 sequence	Length of 3′ S2 sequence	CAT activity [pg/ml]	ssRNA detected	dsRNA detected	Engineered RNA incorporated into infectious virus
pS2CAT198	198	284	64	+	+	+
pS2CAT96	96	284	64	+	+	+
pS2CAT3b	96	200	32	+	+	+
pS2CAT3c	96	150	32	+	+	+
pS2CAT3d	96	100	64	+	+	+
pS2CAT3e	96	50	64	+	−	−
pS2CAT3l	96	97	32	+	−	−
pS2CAT3m	96	98	64	+	+	+
pS2CAT3n	96	99	64	+	+	+
pS2CAT198l	198	97	64	+	−	−
pS2CAT198k	198	96	64	+	−	−

Recombinant ssRNAs were generated from each cDNA with the indicated length of reovirus S2 5′ and 3′ sequences flanking the CAT gene. These ssRNAs were then used to supplement a set of 9 ssRNAs and the now completed set of 10 ssRNAs used in the infectious RNA system to generate recombinant reoviruses containing this S2 CAT genome segment. The minimal 5′ ssRNA sequence capable of being replicated to dsRNA and incorporated in a recombinant reovirus is 96 nucleotides, the minimal 3′ sequence is 98 nucleotides (Roner, Bassett, and Roehr, 2004; Roner and Roehr, 2006).

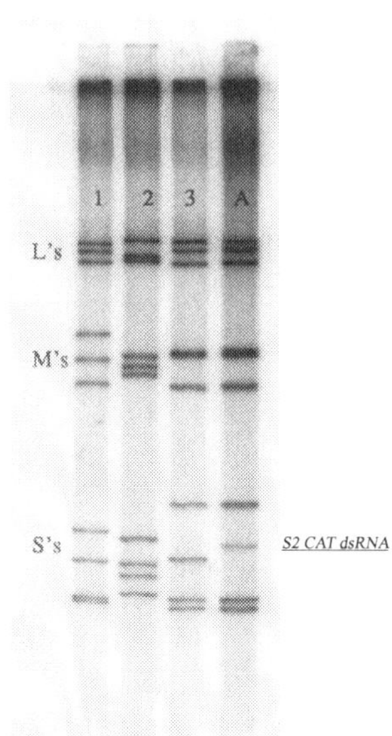

Figure 19.2 Electropherograms of the genome segments of ST1, ST2, ST3 virus, and the ST3. S2.CAT virus. Electrophoresis was carried out in an SDS/PAGE. Lane 1, ST1 virus; lane 2, ST2 virus; lane 3, ST3 virus; lane 4, ST3.S2.CAT virus.

Table 19.4 Kinetics of CAT formation in ST3.S2.CAT virus-infected L929 and L-ST3.S2 cells

Time (h)	A	B	C	D
0	< 4	< 4	< 4	< 4
4	< 4	16	< 4	16
8	< 4	64	< 4	500
12	< 4	64	< 4	500
16	< 4	8	< 4	1200
24	< 4	4	< 4	1800
36	< 4	4	< 4	2000
48	< 4	4	< 4	2000

A, ST3 virus in L929 cells; B, ST3.S2.CAT virus in L929 cells; C, ST3 virus in ST3.S2 cells; D, ST3.S2.CAT virus in ST3.S2 cells.

Numbers refer to CAT enzyme levels in terms of pg/ml.

rated into a virus as an S2 dsRNA. Likewise, a ssRNA with an m1 5′ end is incorporated as an M1 dsRNA (Roner and Steele, 2007) (Tables 19.5 and 19.6).

We have demonstrated that the reovirus system can be used to create a two-CAT gene engineered virus from two *in vitro* transcribed ssRNAs (Roner and Steele, 2007). The 5′ nucleotide sequence flanking the CAT gene is used to identify the ssRNAs as 's2' or 'm1' and leads to the replication of the ssRNAs to dsRNA and incorporation into an infectious reovirus.

Neither the $^{5'}$S2-CAT-M1$^{3'}$ (ssRNA with the 5′ s2 minimum 96 nucleotides fused to the CAT gene followed by the 3′ m1 minimum 172 nucleotides) nor the $^{5'}$M1-CAT-S2$^{3'}$ (ssRNA with the 5′ m1 minimum 124 nucleotides fused to the CAT gene followed by the 3′ s2 minimum 98 nucleotides) ssRNA is able to take the place of both the s2 and m1 ssRNAs and generate a virus with two copies of either of these genes, even when the proteins encoded by both genes are present. This result supports our hypothesis that the reovirus

Table 19.5 Structure of engineered CAT ssRNAs

	ssRNA number of nucleotides					
	5′		3′			Total length ssRNA
Template	s2	m1	s2	m1	CAT	(nucleotides)
$^{5'}$S2-CAT-S2$^{3'}$	96		98		752	946
$^{5'}$M1-CAT-M1$^{3'}$		124		172	752	1048
$^{5'}$S2-CAT-M1$^{3'}$	96			172	752	1020
$^{5'}$M1-CAT-S2$^{3'}$		124	98		752	974

Table summarizing structure of CAT ssRNAs.

Table 19.6 Use of chimeric ssRNAs to generate virus

Engineered RNA	Core-generated RNA set	Transformed cell line	Screening cell line	Progeny virus produced	CAT ssRNA detected in cells	CAT dsRNA detected in cells
$^{5'}$S2-CAT-S2$^{3'}$	9 ssRNAs (-s2)	σ2	σ2	$^{5'}$S2-CAT-S2$^{3'}$	Yes	Yes
$^{5'}$M1-CAT-M1$^{3'}$	9 ssRNAs (-m1)	μ2	μ2	$^{5'}$M1-CAT-M1$^{3'}$	Yes	Yes
$^{5'}$S2-CAT-M1$^{3'}$	9 ssRNAs (-m1)	μ2	μ2	No Virus	Yes	No
	9 ssRNAs (-s2)	σ2	σ2	$^{5'}$S2-CAT-M1$^{3'}$	Yes	Yes
	8 ssRNAs (-M1 and -s2)	σ2 and μ2	σ2 and μ2	No Virus	Yes	No
$^{5'}$M1-CAT-S2$^{3'}$	9 ssRNAs (-m1)	μ2	μ2	$^{5'}$M1-CAT-S2$^{3'}$	Yes	Yes
	9 ssRNAs (-s2)	σ2	σ2	No Virus	Yes	No
	8 ssRNAs (-M1 and -s2)	σ2 and μ2	σ2 and μ2	No Virus	Yes	No
$^{5'}$S2-CAT-S2$^{3'}$ and $^{5'}$M1-CAT-M1$^{3'}$	8 ssRNAs (-M1 and -s2)	σ2 and μ2	σ2 and μ2	$^{5'}$S2-CAT-S2$^{3'}$ $^{5'}$M1-CAT-M1$^{3'}$	Yes	Yes
$^{5'}$S2-CAT-M1$^{3'}$ and $^{5'}$M1-CAT-S2$^{3'}$	8 ssRNAs (-M1 and -s2)	σ2 and μ2	σ2 and μ2	$^{5'}$S2-CAT-M1$^{3'}$ $^{5'}$M1-CAT-S2$^{3'}$	Yes	Yes

A table summarizing the activity of the chimeric ssRNAs we constructed and tested for their ability to generate recombinant viruses. The engineered RNAs are shown, the acceptor sets of 8 or 9 RNAs with which the engineered RNAs were mixed with to generate virus, together with the initial cell line in which virus generation occurred and the cell line used to screen for progeny virus that was generated. The genotype of the inserted dsRNA in progeny virus produced as well as detection of the lipofected CAT ssRNA and any CAT dsRNA produced are indicated.

assortment process requires not only ten ssRNAs, but the signals from 10 ssRNAs. If this were incorrect, we would have expected a virus to be generated that had eight genes with the $^{5'}$S2-CAT-M1$^{3'}$ or the $^{5'}$M1-CAT-S2$^{3'}$ as the ninth gene. Alternatively, we could have generated a virus with eight genes and two copies of either the $^{5'}$S2-CAT-M1$^{3'}$ or the $^{5'}$M1-CAT-S2$^{3'}$ as the ninth and tenth genes.

When both the $^{5'}$S2-CAT-M1$^{3'}$ and the $^{5'}$M1-CAT-S2$^{3'}$ are added to a set of eight ssRNA depleted of the wt s2 and m1 ssRNAs, at least four outcomes were possible. Neither ssRNA could have been incorporated and no virus generated, this did not occur. Two copies of either CAT ssRNA could have been incorporated to generate a virus with 10 genome segments, but two copies of one, this did not occur. One copy of either CAT ssRNA could have been incorporated into a nine-genome segment stable reovirus, this did not occur. Additionally, one copy of each CAT ssRNA could have been incorporated into a 10-genome segmented virus, this is what we isolated. One copy of each ssRNA, the $^{5'}$S2-CAT-M1$^{3'}$ and $^{5'}$M1-CAT-S2$^{3'}$ were incorporated into a virus with eight wt genes. Based on our single gene replacement experiments in which the 5′ sequence determines the identity of the ssRNA and subsequent dsRNA, the ssRNA $^{5'}$S2-CAT-M1$^{3'}$ was incorporated in place of the wt S2 dsRNA and the ssRNA $^{5'}$M1-CAT-S2$^{3'}$ as the wt M1 dsRNA.

In summary, we have devised a powerful reverse genetics system for members of the *Reoviridae* family. Reverse genetics for members of the genus orbivirus and rotavirus should be valuable for the construction, identification and evaluation of vaccine virus strains for use in both humans and animals. Recently a reverse genetics system was described for rotavirus (Komoto, Sasaki, and Taniguchi, 2006). The genetic modifications that this system permits are mutations, deletions and insertions, and it possesses the advantage that the novel variants represent a very high (50% or more) proportion of the infectious reovirus RNA system's virus yields. The technology also permits reovirus to be used as an abundantly replicating non-pathogenic expression vector that could find wide applicability in gene therapy and tumour oncolysis. Finally, the technology provides a powerful system for basic studies on all aspects of reovirus replication, molecular biology and molecular genetics.

References

Antczak, J.B., and Joklik, W.K. (1992). Reovirus genome segment assortment into progeny genomes studied by the use of monoclonal antibodies directed against reovirus proteins. Virology *187*(2), 760–76.

Cashdollar, L.W., Esparza, J., Hudson, G.R., Chmelo, R., Lee, P.W., and Joklik, W.K. (1982). Cloning the double-stranded RNA genes of reovirus: sequence of the cloned S2 gene. Proc. Natl Acad. Sci. USA *79*(24), 7644–8.

Fields, B.N., and Joklik, W.K. (1969). Isolation and preliminary genetic and biochemical characterization of temperature-sensitive mutants of reovirus. Virology 37(3), 335–42.

Joklik, W.K. (1985). Recent progress in reovirus research. Annu. Rev. Genet. *19*, 537–75.

Joklik, W.K., and Roner, M.R. (1996). Molecular recognition in the assembly of the segmented reovirus genome. Prog. Nucleic Acid Res. Mol. Biol. 53, 249–81.

Komoto, S., Sasaki, J., and Taniguchi, K. (2006). Reverse genetics system for introduction of site-specific mutations into the double-stranded RNA genome of infectious rotavirus. Proc. Natl Acad. Sci. USA *103*(12), 4646–51.

Moody, M.D., and Joklik, W.K. (1989). The function of reovirus proteins during the reovirus multiplication cycle: analysis using monoreassortants. Virology *173*(2), 437–46.

Roner, M.R., Bassett, K., and Roehr, J. (2004). Identification of the 5′ sequences required for incorporation of an engineered ssRNA into the Reovirus genome. Virology *329*(2), 348–60.
Roner, M.R., and Joklik, W.K. (2001). Reovirus reverse genetics: Incorporation of the CAT gene into the reovirus genome. Proc. Natl Acad. Sci. USA *98*(14), 8036–41.
Roner, M.R., Lin, P.N., Nepluev, I., Kong, L.J., and Joklik, W.K. (1995). Identification of signals required for the insertion of heterologous genome segments into the reovirus genome. Proc. Natl Acad. Sci. USA *92*(26), 12362–6.
Roner, M.R., Nepliouev, I., Sherry, B., and Joklik, W.K. (1997). Construction and characterization of a reovirus double temperature-sensitive mutant. Proc. Natl Acad. Sci. USA *94*(13), 6826–30.
Roner, M.R., and Roehr, J. (2006). The 3′ sequences required for incorporation of an engineered ssRNA into the Reovirus genome. Virol. J. *3*, 1.
Roner, M.R., and Steele, B.G. (2007). Localizing the reovirus packaging signals using an engineered m1 and s2 ssRNA. Virology *358*, 89–97.
Roner, M.R., Sutphin, L.A., and Joklik, W.K. (1990). Reovirus RNA is infectious. Virology *179*(2), 845–52.
Spendlove, R.S., McClain, M.E., and Lennette, E.H. (1970). Enhancement of reovirus infectivity by extracellular removal or alteration of the virus capsid by proteolytic enzymes. J. Gen. Virol. *8*(2), 83–94.
Wiener, J.R., McLaughlin, T., and Joklik, W.K. (1989). The sequences of the S2 genome segments of reovirus serotype 3 and of the dsRNA-negative mutant ts447. Virology *170*(1), 340–1.
Zou, S., and Brown, E.G. (1992). Identification of sequence elements containing signals for replication and encapsidation of the reovirus M1 genome segment. Virology *186*(2), 377–88.

Genomic RNA Packaging and Replication in the *Cystoviridae* 20

Leonard Mindich

Abstract

The *Cystoviridae*, bacteriophages with genomes of three double-stranded (ds)RNA segments, produce empty procapsid structures that are capable of packaging transcripts of their genomic segments, synthesizing minus strands to form dsRNA and finally transcribing the genome and secreting the plus strands from the complex. All of these reactions can be carried out *in vitro* using particles assembled from proteins whose synthesis is directed by cDNA copies of the viral genomes and RNA transcribed from cDNA copies of the genomic segments. Packaging is dependent upon the specific binding of *pac* sequences near the 5′ ends of the transcripts to sites on the outside of the procapsids. Packaging is serially dependent in that the transcript of S is packaged first and then that of M and finally the transcript of the L segment. The *in vitro* packaging and replication system has facilitated studies of the rules of packaging, replication and recombination in this virus family. It has also led to the development of powerful reverse genetics technology resulting in the effective manipulation of the structure and content of the viral genomes.

Introduction

The *Cystoviridae* are a family of bacteriophages with genomes of three double-stranded (ds)RNA segments packaged inside a polyhedral core. The core is covered by a lipid-containing membrane, and in most cases by an interior shell composed of a single protein species (Mindich, Sinclair, and Cohen, 1976; Vidaver, Koski, and Van Etten, 1973). The polyhedral core has an arrangement of its major structural protein that is remarkably similar to that found in the *Reoviridae* and the *Totiviridae*, namely 120 molecules of a single protein species arranged as 12 pentamers and 20 trimers (de Haas *et al.*, 1999; Huiskonen *et al.*, 2006; Reinisch, Nibert, and Harrison, 2000). Twelve molecules of an RNA dependent RNA polymerase (RdRp) are located on the inner surface of the fivefold vertices. The *Reoviridae* contain 10, 11 or 12 genomic segments while the *Cystoviridae* contain three. The packaging of the genomic segments appears to be precise in the case of the *Reoviridae*, with one copy of each segment in each particle (Patton, 1990). The packaging in Φ6, the first representative of the *Cystoviridae*, has been shown to be very precise, with the efficiency of plating of purified virus being one plaque forming unit per virus particle (Day and Mindich, 1980).

The genomic packaging of most DNA viruses involves the entry of the single or double-stranded genome into a preformed particle. In the cases of dsDNA viruses there is usually a complex at one fivefold vertex that is able to recognize and transport the DNA into the particle, usually called the prohead. This complex, called terminase, is associated with a neck protein complex that is a symmetry mismatch with the portal vertex (Catalano, 2000; Lander *et al.*, 2006). The terminase components do not remain with the prohead after packaging. The genomic packaging of single-stranded (ss)RNA viruses involves the association of the genomic RNA with a capsid protein and the subsequent nucleation of capsid assembly around the RNA (Choi, Grantham, and Rao, 2000; Horn *et al.*, 2004). In the situations where the virus contains two different ssRNA molecules, as in the case of flock house virus, the assembly still involves the association of the RNA molecules with capsid protein and the subsequent nucleation of capsid assembly around the RNA. However the details of the assembly mechanism are not fully understood (Krishna, Marshall, and Schneemann, 2003).

The packaging of the genomes of the *Reoviridae* is presently a mystery. There is no *in vitro* packaging system and reverse genetics has only recently been reproducibly demonstrated (Komoto, Sasaki, and Taniguchi, 2006; Roner and Joklik, 2001). Some elements of the packaging process are understood in the *Totiviridae*, namely that packaging specificity is determined by the relationship between amino acid sequences on the polymerase and internal secondary structures in the RNA (Ribas and Wickner, 1996). It seems likely that packaging in this case is similar to that found for ssRNA viruses.

The *Cystoviridae*

There are presently 10 known members of the *Cystoviridae* and about half of them are almost identical to Φ6, the first isolate of the family (Mindich *et al.*, 1999). The three genomic segments contain from 13 to 18 genes arranged in a similar fashion in the different viruses (Figure 20.1). In all cases, the L segment carries the four genes coding for the components of the inner core or procapsid (Mindich *et al.*, 1988). The protein products

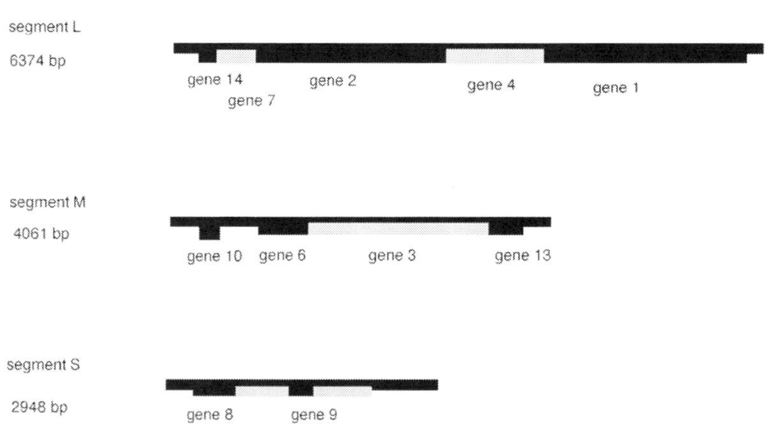

Figure 20.1 The genomic segments of Φ6. Proteins P7, P2, P4 and P1 comprise the inner core particle. Proteins P6 and P3 are the host attachment apparatus while P10 and P5 are involved in host cell lysis. P5 is also involved in host penetration. P9 is the major membrane protein, which is inserted into membrane through the action of P12. P8 forms a shell around the inner core particle. Each segment has a unique *pac* region near the 5′ end of the plus strand.

are; P1, the major structural protein of the procapsid; P2, the RNA dependent RNA polymerase (Butcher *et al.*, 2001); P4, the hexameric NTPase (Mancini *et al.*, 2004) and P7, an auxiliary component of the core. Segment M carries the genes for host attachment as well as a holin involved in cell lysis (Gottlieb *et al.*, 1988). Segment S carries genes for a protein, P8 that forms a shell around the inner core, a morphogenetic protein P12 that is necessary for membrane formation, a major membrane protein and a muramidase that is necessary for cell entry and for cell lysis (McGraw, Mindich, and Frangione, 1986). There are a small number of additional genes that are found on the various segments in the different viruses. They are not directly involved in genomic packaging or replication. Most of the known members of the *Cystoviridae* have base sequences very similar to those of Φ6, while three of the members are quite divergent. The proteins of Φ13 that are paralogues of those of Φ6 have about 30% identity and 50% similarity, while those of Φ12 and Φ8 have almost none except in the proteins P2 and P4 of the polymerase complex where the identity is of the order of 20%. Φ8, Φ12 and Φ13 attach to rough LPS of the host pseudomonads while the rest of the family attaches to a type IV pilus (Mindich *et al.*, 1999).

The proteins of the procapsid form a dodecahedral structure with the P4 hexamer rings located at the fivefold vertices (de Haas *et al.*, 1999; Huiskonen *et al.*, 2006) (Figure 20.2). These structures are produced in cells of *E. coli* or pseudomonads that carry cDNA copies of genomic segment L in plasmids. The empty particles show a compressed appearance in cryoelectron microscope images. This is in contrast to the images seen for cores that contain the three genomic segments (Huiskonen *et al.*, 2006) (Figure 20.2). The filled cores have a spherical shape with a greater interior volume. Purified empty procapsids produced in *E. coli* can package plus strand transcripts of the genomic segments when incubated in simple buffers with ATP (Gottlieb *et al.*, 1991). If the incubation medium contains the four NTPs the packaged RNA will serve as templates for the synthesis of minus strands to form dsRNA (Gottlieb *et al.*, 1990). The dsRNA under the same conditions will serve as template for the formation of plus strands that are extruded from the particles in a semi conservative reaction (Emori, Iba, and Okada, 1980; Usala, Brownstein, and Haselkorn, 1980).

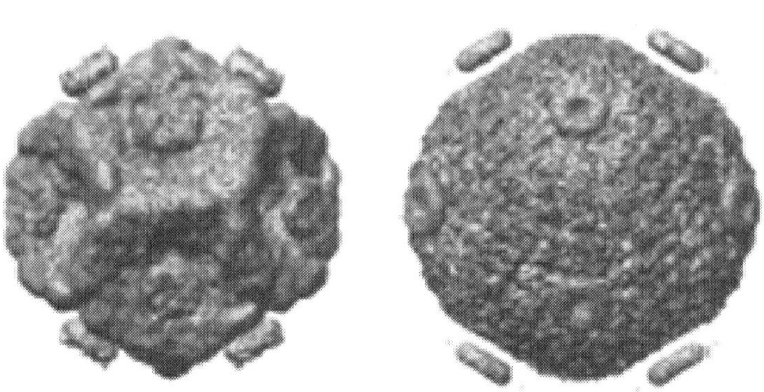

Figure 20.2 Cryoelectron microscopy reconstructions of empty procapsids (de Haas *et al.*, 1999) (left) and filled procapsids (Huiskonen *et al.*, 2006) (right). The structures on the five fold faces are the multimers of the NTPase, P4.

Genomic packaging in the *Cystoviridae*

Packaging of RNA was found to be serially dependent, in that the transcript of S could be packaged alone, but that of M could only be packaged after that of S (Figure 20.3). The transcript of L could only be packaged after that of M (Qiao *et al.*, 1995). The rules could be abrogated to some extent under various conditions. The transcript of L could be packaged to a small extent when just the S transcript was packaged. Minus strand synthesis occurred after the packaging of the L transcript; however in the presence of manganese ions minus strand synthesis did not require complete packaging (Frilander *et al.*, 1992; Gottlieb *et al.*, 1992).

Each plus strand has a region of about 200 nucleotides that is unique to that segment and about 50 nucleotides away from the 5′ end (Gottlieb *et al.*, 1994). This is called the *pac* region and it is necessary and sufficient for genomic packaging when linked to the consensus sequence at the 5′ end. The distance between the *pac* sequence and the consensus is critical for packaging (Qiao, Qiao, and Mindich, 1997b). The consensus region contains 18 nucleotides in Φ6 but only 7 in Φ8 (Figure 20.4). In all members of the family with the exception of Φ8, the second nucleotide is different in L from that in S and M. This difference results in reduced activity for transcription of the L segment relative to the others (Van Dijk, Frilander, and Bamford, 1995). This difference functions in the temporal regulation of message abundance. It is consistent with the demonstration that the ϕ6 polymerase starts synthesis at the second base of the template and the insertion of G is preferred (Butcher *et al.*, 2001). The structure of the ϕ6 polymerase has been determined in crystal forms with NTP binding and with DNA and RNA templates (Butcher *et al.*, 2001; Salgado *et al.*, 2004). Φ8 has its own mechanism for temporal regulation of message abundance (Mindich, unpublished results). The *pac* region has extensive secondary structure and the maintenance of the stem–loops appears to be important enough for the phages in the groups closest to Φ6, so that sequence changes in the stems of the *pac* region

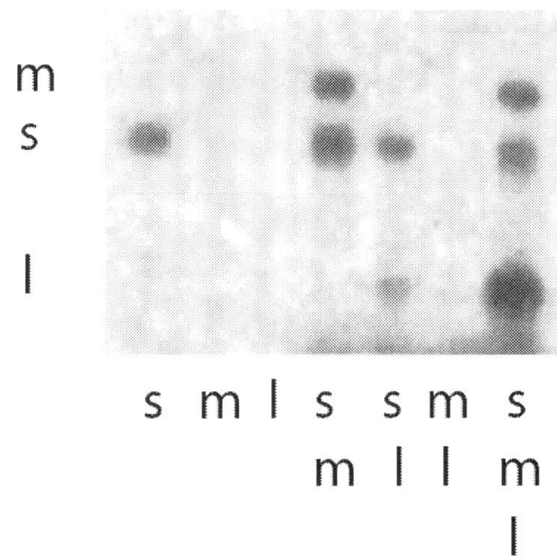

Figure 20.3 Serial dependence of genomic packaging. Packaging of radioactive plus strands of exact copies of genomic segments S and M and of a truncated segment L of Φ6. Radioactive transcripts of plasmids were incubated with procapsids, treated with RNase I and applied to a 2% agarose gel.

Φ6	l	GUAAAAAAACUUUAUAUA
	m,s	GGAAAAAAACUUUAUAUA
Φ13	l	GAUAAAAACUUUAUAUU
	m	GGAAAAAACUUAUAUUU
	s	GGAAAAAACUUUUAUAA
Φ12	l	ACAAUAAAAUAAUACAA
	m	GAAUUAAUUAAAUUUUA
	s	GAAUAAACUAAAUAAAG
Φ8	l	GAAAUUUAUUUAAAGCU
	m	GAAAUUUUCAAAGUCUU
	s	GAAAUUUUCAAAUCUUU

Figure 20.4 5′ sequences of plus strand transcripts of genomic segments.

are often compensated by complementary changes that maintain base pairing in the stem structures (Figure 20.5). The *pac* regions do not overlap with open reading frames (orfs). They usually terminate about 50 nucleotides before the first orf.

The 3′ end of the plus strands also shows considerable secondary structure (Mindich, 1999) (Figure 20.6). This seems to be important for the stability of the messages, since removing the structures results in the appearance of recombinants that have exchanged the 3′ ends with those of other genomic segments (Mindich *et al.*, 1994). The sequence at the very 3′ end of the transcripts is important in template function for minus strand synthesis. Although pure polymerase is able to use templates of various 3′ sequences as templates

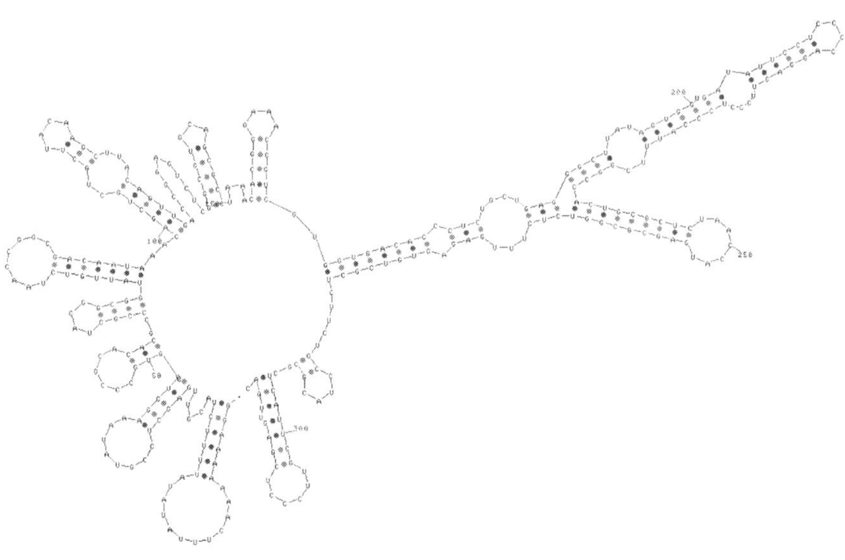

Figure 20.5 5′ sequence of the plus strand transcript of genomic segment M of Φ6. Secondary structure is predicted by the mfold program (Mathews *et al.*, 1999).

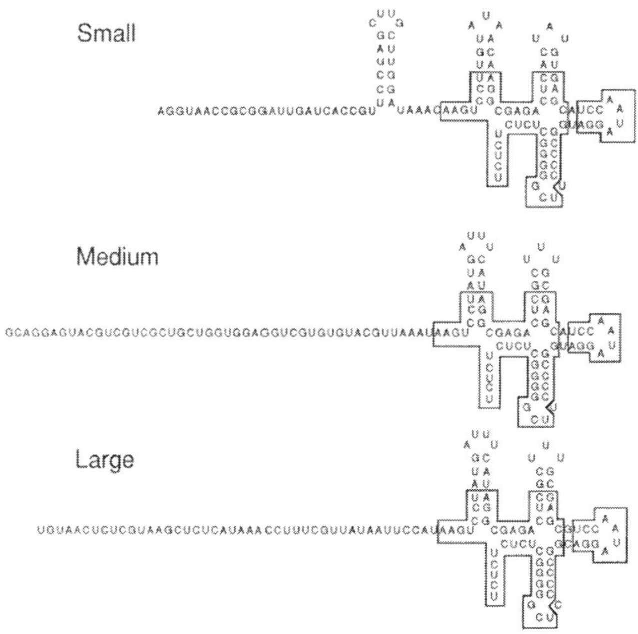

Figure 20.6 3′ sequences of plus strand transcripts of Φ6 genomic segments. The boxed regions are identical on all three segments.

for minus strand synthesis, the polymerase in the core particle is much more stringent (Makeyev and Bamford, 2000; Mindich *et al.*, 1994). In the case of Φ6, the 3′ sequence is CUCUCUCUCU. A sequence of GGUACCC works well but GAGCAGC, UCGACC, CUUAA or AAGCU does not (Mindich *et al.*, 1994). A transcript that is missing its normal 3′ end can be packaged and can promote the packaging of the next transcript in the packaging program and the turning on of minus strand synthesis, but it will usually not serve as template for its own minus strand synthesis (Frilander *et al.*, 1992). However, the strand can serve as a template for minus strand synthesis if template switching occurs so that a minus strand begun on another transcript switches so as to form a non-homologous crossover (Mindich, 1996).

Specific packaging in Φ6 takes place in the absence of P2, the RNA polymerase, but minus strand synthesis does not occur. Specific packaging also takes place in the absence of P7, an auxiliary protein; however the extent of packaging is diminished and minus strand synthesis occurs to a lesser extent. Packaging does not occur in the absence of P4, the hexameric NTPase. Packaging in ϕ6 takes place in the presence of ATP or a number of other NTPs that have been shown to be substrates for the NTPase activity of P4 (Gottlieb *et al.*, 1991). The stoichiometry of P4 is not completely clear. In Φ6 it appears that there are approximately 12 hexamers per virion; however the number is probably less in empty core particles. Mutants have been isolated that lose P4 so as to result in particles with only one hexamer. These particles are able to package RNA normally. It has also been shown that P4 can be removed from cores such that only one hexamer remains. These particles are also able to package the three RNA transcripts (Pirttimaa *et al.*, 2002). There is a suggestion that one of the P4 hexamers is integrated into the core in a manner that is different from the others. These results suggest that all three transcripts can enter the core through one portal. However, experiments with transcripts having large strong hairpin structures at the 3′ end indicate that packaging can occur at more than one portal. Molecules with strong 3′ hairpins are packaged from the 5′ end and the 3′ end remains outside and susceptible to RNase I cleavage (Qiao, Qiao, and Mindich, 1995). P4 hexamers of ϕ12 have been crystallized and their structures determined in various states of nucleotide binding (Mancini *et al.*, 2004). A model for the translocation of RNA has been proposed wherein ATP is bound at sites between subunits and successive hydrolysis results in movement of the RNA through the central pore of the hexamer (Kainov, Tuma, and Mancini, 2006). Cores that have packaged hairpin transcripts of S are still capable of packaging normal M transcripts. Core structures cannot be assembled in the absence of P1, the major structural protein of the procapsid. Many studies have indicated that P1 is the determinant of packaging specificity (Qiao, Qiao, and Mindich, 2003; Sun, Qiao, and Mindich, 2004b).

RNA binds to Φ6 procapsids (Juuti and Bamford, 1995). The binding is specific in the presence of ATP (Qiao, Qiao, and Mindich, 2003). Transcripts that have proper sized deletions between the *pac* region and the 5′ consensus can compete for packaging but are not themselves packaged. A molecule competing for the packaging of S will prevent the packaging of the M and L transcripts, while a molecule competing with M will prevent the packaging of L. These molecules exhibit binding behaviour that is consistent with the packaging properties of their normal counterparts. Empty procapsids bind transcripts of S, but not those of M. In the absence of ATP, the M transcripts bind almost as well as those of S. Non-packaging transcripts can be cross-linked to procapsids by UV treatment. The major region of cross-linking of the S transcript is between amino acids 98 and 155 in protein P1, which has a total of 769 amino acids (Qiao, Qiao, and Mindich, 2003).

As mentioned earlier, the *pac* region and the 5′ consensus sequence and the intervening sequence are necessary and sufficient for packaging. If a proper 3′ end is present the molecules will also serve as templates for minus strand synthesis. When packaging is studied with molecules that are smaller than normal, it is found that more than one molecule of that class is packaged in each particle. If the molecules are half size, two molecules are packaged; if they are one quarter size then approximately four molecules will be packaged. The increase in the number of molecules packaged is always in the class that is smaller than normal (Mindich, Qiao, and Qiao, 1995). Conversely, if the size of the molecule is increased, steps can be bypassed in the packaging program (Mindich, 1999). If the S transcript is constructed to be equal in size to the sum of the S and M transcripts, its packaging will allow the L transcript to be packaged without any M transcript being packaged. If the M transcript is lengthened to the size of the sum of M and L, then minus strand synthesis will be turned on despite the lack of any packaging of the L transcript. A transcript that is the size of the entire genome can be prepared with the *pac* region of S. This molecule can be packaged and will serve as a template for minus and plus strand synthesis.

The model

We have proposed a model for the genomic packaging in the *Cystoviridae* (Figure 20.7) (Mindich, 1999). We propose that the empty procapsid has binding sites on the outside

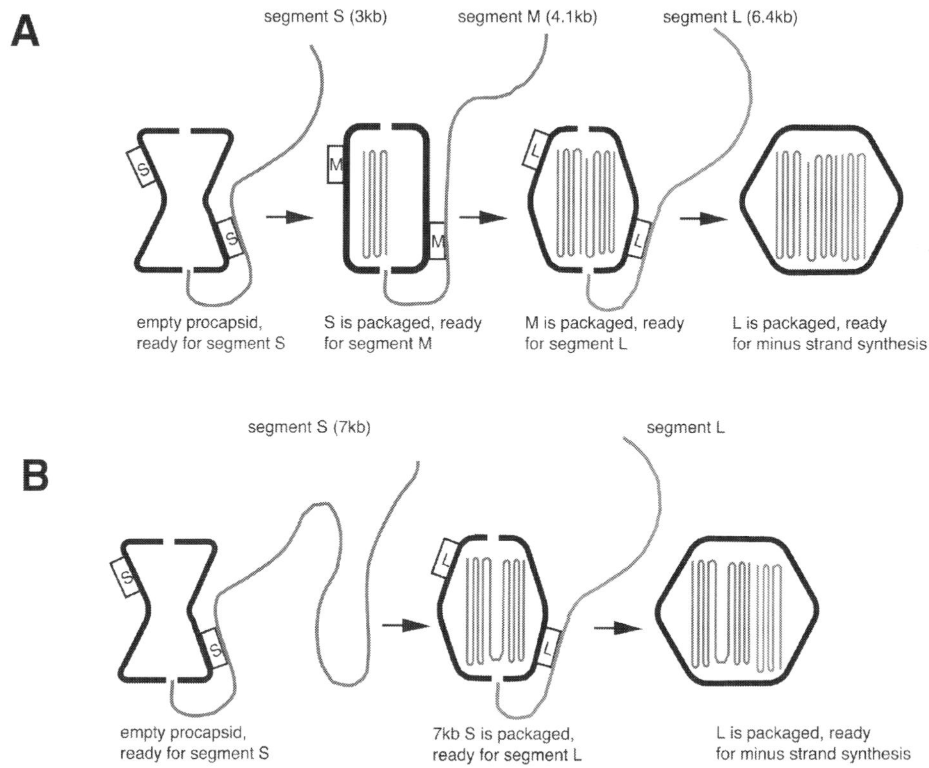

Figure 20.7 (A) The packaging model. The empty procapsid shows only binding sites for s. After a full size s is packaged, the s sites disappear and m sites appear. After a full size m is packaged, the m sites disappear and l sites appear. After a full size l is packaged, minus strand synthesis commences. After minus strand synthesis is completed, plus strand synthesis commences. (B) If segment s is of the size equal to the sum of both s and m, the s sites will disappear and the l sites will appear and segment l will be packaged without segment m.

for only the S transcript. The transcript binds to protein P1 in a manner such that the 5′ end is placed in the proximity of one of the P4 hexamers and through the energy generated by the hydrolysis of ATP the molecule is transported into the procapsid. The procapsid expands during this process and the conformation of the surface changes so that the binding sites for S are lost and those for M appear. If the molecule is smaller than the normal size, the expansion is not sufficient to change the surface conformation and the S binding sites remain so that another molecule can be packaged. The same considerations apply for the packaging of the M transcript and subsequently for the L transcript. In the case of an S transcript that is the size of S and M together, the procapsid expands to the state where the L binding sites appear (Figure 20.7B). The model predicts five different states of the core; three different exposures of the binding sites plus a signal to turn on minus strand synthesis and finally to turn on plus strand synthesis.

Reverse genetics

Reverse genetics has made it possible to investigate some of the aspects of packaging in *in vivo* systems. The first reverse genetics system utilized *in vitro* packaging and minus strand synthesis. Transcripts of cDNA copies of genomic segments were packaged in procapsids and minus strand synthesis was promoted. The particles were then coated with purified protein P8 and used to infect spheroplasts of the host bacteria (Olkkonen *et al.*, 1990). The Bamford laboratory had previously shown that nucleocapsids (cores covered by a P8 shell) could infect spheroplasts of the host pseudomonad. They then found that the P8 shell could be removed with EGTA, resulting in the loss of infectivity. Replacing the P8 shell restored infectivity of spheroplasts (Olkkonen, Ojala, and Bamford, 1991). With this process and with *in vitro* packaging of plasmid transcripts it was possible to construct phages with deletions or with the addition of reporter groups such as lacα or kanamycin resistance (*kan*) (Onodera *et al.*, 1992). The same system can be utilized with cores of Φ8 but in this case there is no need for the shell of P8 as Φ8 cores can penetrate membranes unaided (Sun *et al.*, 2003) (Figure 20.8). Deletion mutants could be maintained on strains carrying plasmids with genes that complemented them. Using deletion strains of phage it was observed that virus could acquire plasmid transcripts as substitutes for the segments bearing deletions, or in addition to those segments (Onodera *et al.*, 1995; Onodera *et al.*, 1998a). Ultimately, it was found that phage could be produced directly from plasmid transcripts by electroporating three non-replicating plasmids that contained cDNA copies of the three genomic segments into host strains carrying plasmids that expressed either T7 or SP6 RNA polymerase (Sun, Qiao, and Mindich, 2004a) (Figure 20.9). It is even possible to electroporate the three cDNA plasmids along with a non-replicating polymerase plasmid and recover large numbers of phage plaques.

The introduction of reporter groups into the phage genome opened another aspect of *Cystoviridae* biology to study. Φ6 is able to form a carrier state in a small fraction of infected cells (Cuppels, Vidaver, and Van Etten, 1979). This is usually difficult to study, but when the cells carry an antibiotic resistance gene in the phage genome it is easy to maintain the carrier state strains (Onodera *et al.*, 1992). Carrier state strains can be created by infecting

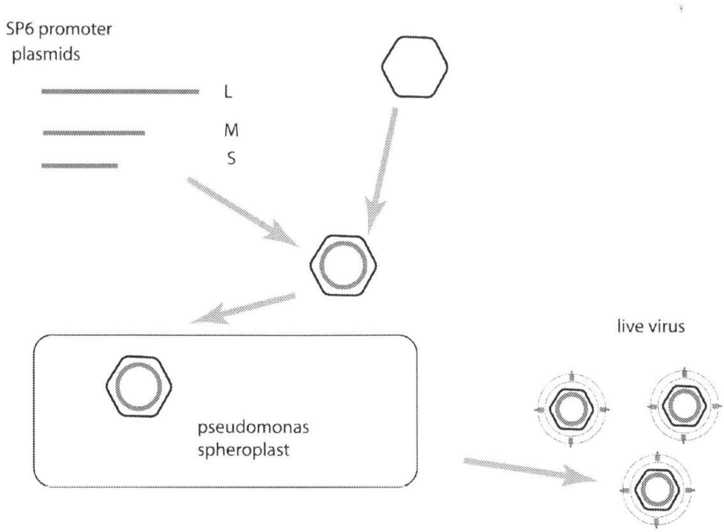

Figure 20.8 Formation of plaque forming virions by infecting spheroplasts with procapsids that have been packaged *in vitro* with transcripts of cDNA copies of the genomic segments.

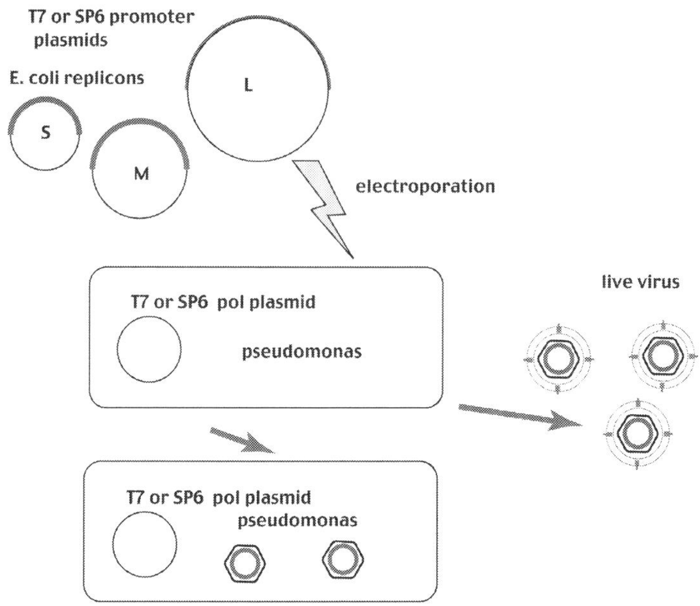

Figure 20.9 Formation of plaque forming virions or carrier states by electroporation of cDNA copies of genomic segments into cells expressing T7 or SP6 RNA polymerase. The ColEI replicons do not replicate in pseudomonads.

cells with phages carrying *kan* genes in their genomes or by electroporating non replicating cDNA plasmids into cells that are expressing T7 or SP6 RNA polymerase and selecting for kanamycin resistance (Sun, Qiao, and Mindich, 2004a) (Figure 20.9).

Carrier states

The carrier state system has presented several findings that seemed to contradict the model that we have proposed for the genomic packaging (Figure 20.10) (Sun, Qiao, and Mindich, 2004a). It is possible to establish carrier states with only segments S and L in both Φ6 and Φ8. Segment M is not necessary; however M is not lost easily from three segment carrier state cultures. It has been observed that segment L can be packaged in particles that have only packaged S, but this packaging is at a lower level than obtained when M is present (Figures 20.3 and 20.11A). It is also possible to establish carrier states with segments M and L. This would seem to be a clear conflict with the idea that the S transcript must be packaged first and that empty particles have binding activity only for the S transcript. The anomaly is resolved by the finding that the M:L carrier state formation is dependent upon a mutation taking place in the gene coding for P1. Procapsids prepared with mutant P1 show a different *in vitro* packaging behaviour than normal (Figure 20.11). These particles can package the S transcript alone; however they can package M and L without S and the M transcript actually competes with that of S for packaging and S is excluded when all three transcripts are present (Onodera *et al.*, 1998b). It appears that the mutant procapsids are behaving as if they have already packaged the S transcript. These experiments lend further support to the idea that P1 is the determinant of packaging specificity.

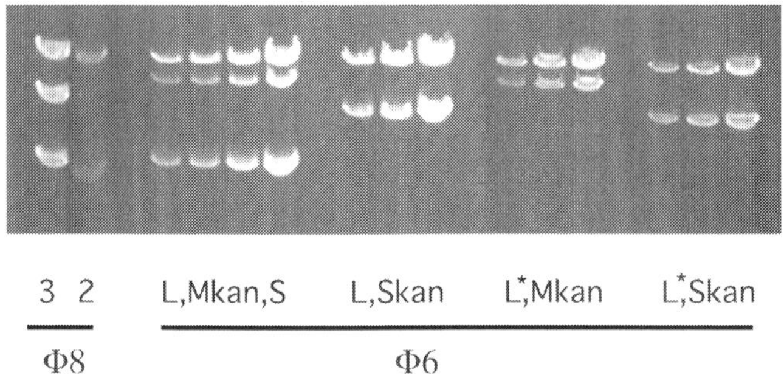

Figure 20.10 Agarose gel electrophoresis of dsRNA isolated from carrier state cells. Plasmids were electroporated into strain LM2691 and selection for kanamycin resistance was used to isolate carrier state colonies. Increasing amounts of RNA were applied for each Φ6 sample. L* is derived from mutant pLM3191.

Another test of the packaging rules was done with the *in vivo* plasmid transcript acquisition test. We stated earlier that deletion mutants could acquire plasmid transcripts as substitutes or additions to their deletion segments. A test was developed wherein the plasmid transcripts contained *pac* sequences of other classes. It was found that these transcripts were picked up by the phage but that they had undergone recombination to replace the heterologous *pac* sequence with the homologous one. These findings lend support to the idea that there is a specific quantity of material for each segment class and strong selection exists to maintain this condition (Onodera *et al.*, 1995).

A further use of the reverse genetics was the construction of live phages with combinations that were consistent with the packaging model (Qiao, Qiao, and Mindich, 1997b) (Figure 20.12). Examples are deletion mutants in which it is clear that a segment that is half size has packaged two molecules; phages that contain a chimeric segment SM with the *pac* sequence of S along with segment L, but phages that contain a chimeric segment MS with *pac* sequence of M along with segment L also carry a segment S, albeit that

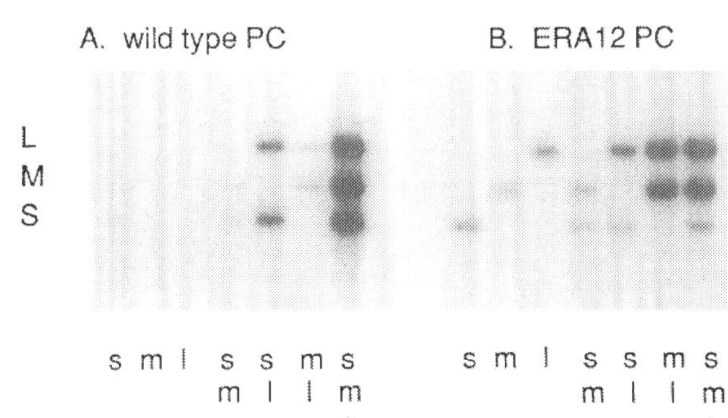

Figure 20.11 *In vitro* minus strand synthesis with procapsids containing wild type (A) and mutant (B) protein P1. Wild type procapsids require all three plus strands for minus strand synthesis while ERA12 procapsids need only L and M. ERA12 excludes the plus strand of S.

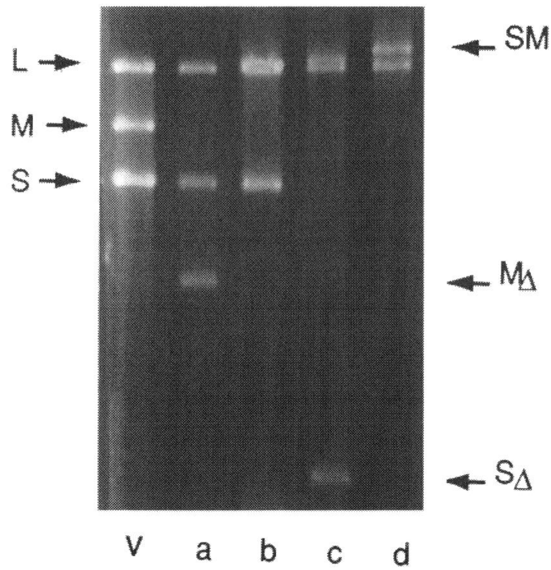

Figure 20.12 Agarose gel electrophoresis of dsRNA isolated from virions. Lane v shows the distribution of normal segments L, M and S of Φ6. Lane a shows dsRNA from bacteriophage Φ2007, which has a deletion in segment M. Lane b shows RNA from Φ2064 which has normal L, an MS chimera picked up from pLM1114 and a normal segment S. Lane c shows RNA from Φ2323, which contains normal L, the MS chimera shown in b and a deleted segment S that contains no genes and is only 798 bp. Lane d shows RNA from Φ2361, which contains normal L, a chimera of S and M, but no normal segment M or S.

segment can be very small and contain no genes. We have also created genomic segments that contain all the genes of the virus in one segment with the *pac* sequence of S (Figure 20.13). These phages plaque normally although the plaques are smaller than normal.

We have not been able to produce carrier state cultures with just segment L with L *pac* sequences; however we have produced carrier state cultures with segment L with S *pac* sequences. In some of the latter cases, the segments are normal size but in others the segments are the size of the entire phage genome in one piece, caused by the acquisition of a plasmid transcript that contains an RNA copy of the entire plasmid including the ampicillin resistance gene (Mindich, unpublished results).

Suppressor mutations

The *pac* regions of Φ6 are about 200 nucleotides in length. The consequences of changes in the sequence or deletions of portions of the region were investigated. It was found that many changes reduced the frequency of acquisition of transcript into phages carrying deletions in the target segments (Qiao *et al.*, 2003). In many of these cases, the altered RNA was packaged and replicated effectively once taken up by the phage without changes in the RNA or mutations affecting the proteins of the core particle. In some cases the altered RNA was acquired at very low frequencies and only concomitant with mutations in gene 1 coding for the major structural protein of the core. An example is the change of G142A in Φ6M (Figure 20.5). This change led to a reduction in acquisition of several orders of mag-

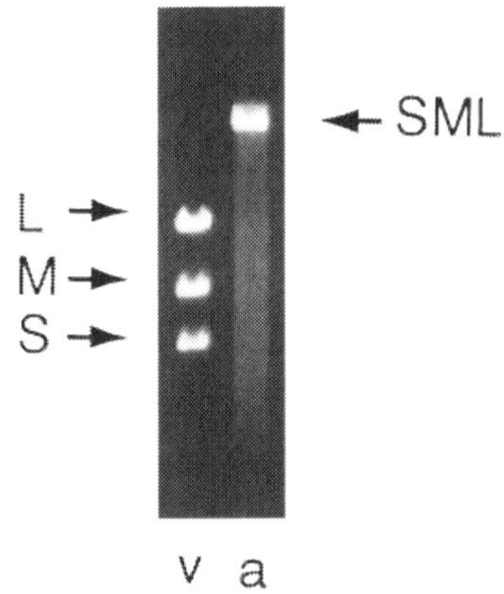

Figure 20.13 Agarose gel electrophoresis of dsRNA isolated from virions. Lane v shows the distribution of normal segments L, M and S. Lane a shows dsRNA from Φ2515, which contains the entire genome of Φ6 in one segment. The migration of the RNA indicates a size of about 14kbp.

nitude and the resulting phage had amino acid changes in protein P1 at positions E133G, Q384K, Q384R or E101G. The mutant phage still preferred wild type *pac* sequences but was able to package the altered RNA 100 times better than wild type phage. A change of GC99AU in Φ6S also led to a reduction in acquisition that was suppressed by a mutation in gene 1. This change at E390A resulted in a phage that packaged the altered RNA almost as well as wild type RNA.

Phages Φ7, Φ9, Φ10 and Φ11 are very similar to Φ6 and their *pac* sequences are very similar with only a few base changes (Mindich *et al.*, 1999). They can exchange segments with Φ6 without difficulty. Φ8 and Φ12 are more distant from Φ6 and they do not exchange segments with Φ6. Φ13 is between these two groups. It has the same type of attachment apparatus as Φ12 and Φ8, but it is closer to Φ6 in the sequence of other proteins. The *pac* sequences of Φ6 and Φ13 are very different; however Φ13 is able to acquire segment M of Φ6, thereby changing its host range and attachment mechanism (Mindich *et al.*, 1999; Qiao *et al.*, 2000). Φ6 is not able to acquire the M segment of Φ13 unless the *pac* sequence is changed to that of Φ6.

Recombination in the *Cystoviridae*

Recombination in the *Cystoviridae* is rare and homologous recombination is virtually non-existent because there is no extensive sequence similarity between the genomic segments and normally, there is only one genomic segment of each class in each particle. However, conditions can be established that promote heterologous and homologous recombination. If a plus strand that is missing its normal 3′ end is packaged, it will usually not serve as a template for minus strand synthesis. The other two molecules in the particle can serve as templates. In about one percent of the cases, there will be a rescue of the truncated segment by a template switching in which a minus strand begun on another template will land on the truncated plus strand and elongate to form a recombinant minus strand. The point of launching is not particularly localized, nor is the point of landing on the truncated strand (Qiao, Qiao, and Mindich, 1997a). The consequence is that the recombinant molecules are of varied sizes and most of them will be either too long to serve as proper genomic segments or will be too short, meaning that genes will be missing or cut. This recombination is heterologous in that there is little sequence identity between the donor and recipient strands. The average number of identical bases at the crossover point is 3.6, but the identity ranges from zero to 12. This recombination acts as a relatively inefficient repair mechanism. The mechanism of template switching is not clear. The Φ6 polymerase appears to be a closed structure (Butcher *et al.*, 2001) therefore it seems that template switching might be the result of a nascent minus strand annealing to a plus strand that has threaded through a polymerase without initiating minus strand synthesis (Figure 20.14). This could be the consequence of the plus strand lacking a proper 3′ sequence that would cause proper initiation. However, there is evidence that the polymerase can open to work on templates that do not have free ends (Ranjith-Kumar and Kao, 2006).

Homologous recombination can be observed by propagating phages that contain deletions on strains that produce plasmid transcripts that do not contain *pac* sequences but have the missing genes with overlap regions greater than 600 nucleotides. Under these conditions the phages can acquire the transcripts at very low frequencies and can only integrate them by recombination. If the region of overlap is long enough, about 600 nucleotides, the recombination will be homologous (Onodera, Sun, and Mindich, 2001). It seems that the acquisition of RNA that does not have *pac* sequences is possible, but

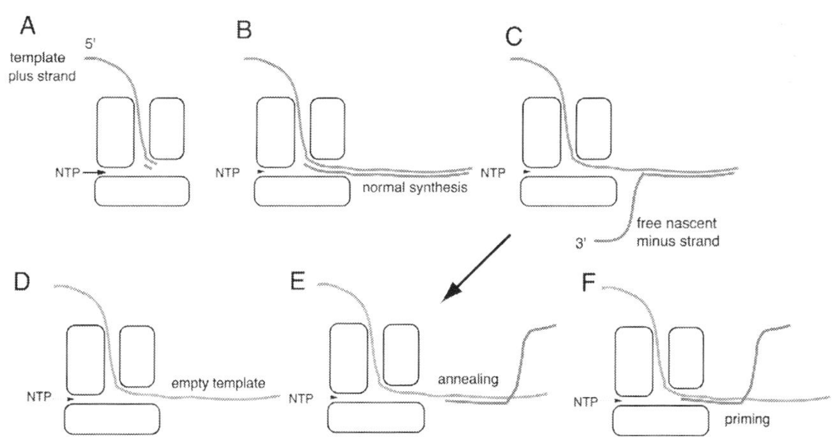

Figure 20.14 Proposed model for the process of template switching recombination in Φ8. Normally, the 3′ end of the template plus strand enters the polymerase and is arrested near the site of nascent minus strand chain formation (A); as the minus strand is synthesized the dsRNA leaves the polymerase through the exit pore (B). The 3′ end of the nascent chain can displace from the template (C). It can reanneal to the template or it can anneal to an empty template (E). A plus strand RNA that does not have a proper 3′ end can enter the polymerase, but the template is not arrested at the catalytic site (D). Instead, it passes through the polymerase and can scan backwards and forwards. When a nascent minus strand anneals to the empty template and the template moves back into the polymerase, the nascent chain can act as a primer to start minus strand synthesis on the empty template (F). The result is a recombinant minus strand.

extremely rare. However, this may be an important mechanism for the addition of new genes or the exchange of genetic material between distantly related phages.

The family of *Cystoviridae* is composed of many viruses that have little common sequence identity. There are, however, suggestions that particular genes or regions have been exchanged in the past. The most dramatic is the finding that the 3′ sequence of segment M of Φ13 is identical to that of Φ6. Although the 3′ sequences of Φ6 plus strands are very similar to each other, the sequence of the Φ13M 3′ is closer to Φ6M than either Φ6 S or L. The 3′ sequences of Φ13 S and L are somewhat similar to each other but very different from M (Figure 20.15) (Qiao *et al.*, 2000). Φ13 has significant similarity to Φ6 in the amino acid sequences of its core proteins and in protein P8. It has a very different attachment apparatus from Φ6 and therefore a recombination is probably responsible for this difference. Φ12 has a similar attachment strategy to that of Φ13 and there is similarity in the attachment proteins; however Φ12 and Φ13 do not share similarity in any of the other proteins. It appears that their attachment proteins have a common origin but the rest have their origins in common with Φ6, but with the Φ12 sequences diverging more than those of Φ13 from the Φ6 pattern.

Comparisons with the *Reoviridae*

Because there are striking similarities in the structures of the *Reoviridae* and the *Cystoviridae* it is tempting to explore the possibility that the two families employ similar mechanisms for genomic packaging. The similarities are the T = 1 asymmetric dimer arrangement of the major core proteins and the multisegmented dsRNA genome packaged precisely with one of each segment in each particle (Reinisch, Nibert, and Harrison, 2000). The arrangement of the polymerase molecules under the 12 fivefold vertices is another common feature. It also appears that genome assembly in all these cases involves single-stranded RNA that

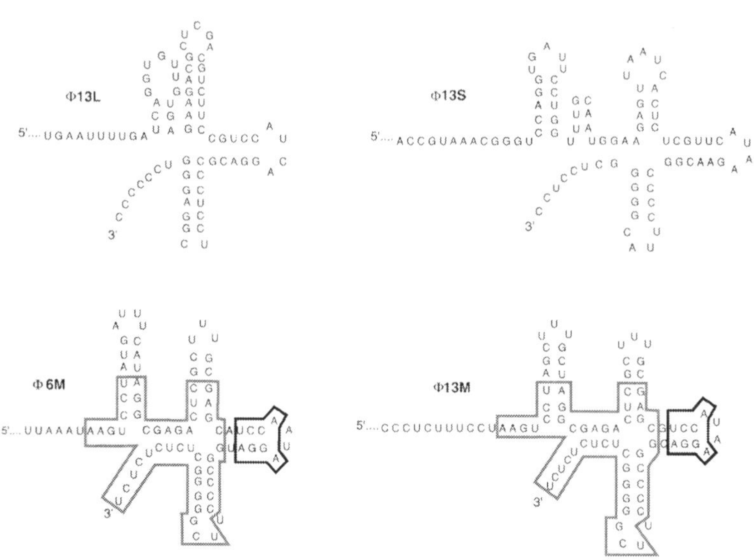

Figure 20.15 The secondary structure at the 3′ end of the Φ13M plus strand is strikingly similar to that of Φ6M. It is proposed that it was acquired by recombination.

becomes double-stranded in the assembly complex. A key element in the packaging mechanism of the *Cystoviridae* is the P4 NTPase hexamer that is the motor for the translocation of ssRNA into the core particle. There is the possibility that NSP2 of rotavirus might play a similar role to that of P4 (Jayaram *et al.*, 2002). NSP2 seems to form octamers rather than hexamers, but it has NTPase activity. There are non-structural proteins in the other members of the *Reoviridae* that might play the same role as NSP2. However, recent studies have shown that NSP2 is a NDP kinase and this role seems likely to be more important for RNA synthesis rather than packaging. Packaging in the *Cystoviridae* leads to resistance to RNase, while the evidence for assembly in the *Reoviridae* suggests that the RNA remains susceptible to RNase degradation after forming its first complex (Patton and Gallegos, 1990). This implies that the structure is open, but one could envision a capsid structure that is capable of assembling the genome and being open enough to allow RNase sensitivity. Another key element in the *Cystoviridae* packaging is the presence of *pac* sequences near the 5′ ends of the plus strands. So far, it has not been possible to set up an *in vitro* packaging system or a high yielding reverse genetics system (Komoto, Sasaki, and Taniguchi, 2006; Roner and Joklik, 2001) in the *Reoviridae* and therefore impossible to search for *pac* sequences. The roles of the 5′ and 3′ sequences in polymerase recognition has been studied in rotavirus but this might be unrelated to genomic packaging. In the case of the totiviruses it has been found that there are internal sequences that are necessary for packaging and that these sequences react with a part of the polymerase domain of gagpol; however, it is likely that genomic packaging in the *Totiviridae* does not involve packaging into preformed particles but instead involves the nucleation of capsid structure around the RNA. However, the packaging sequences in the *Totiviridae* are located within the *pol* gene reading frame (Ribas and Wickner, 1996). This might also be true for the *Reoviridae* in the sense that *pac* sequences might be embedded in the orfs of the genomic segments. It is of interest that sequences within the orfs of the influenza genomic segments are important for packaging (Muramoto *et al.*, 2006).

In conclusion, the mechanisms found in the *Cystoviridae* can help in posing questions about replication and packaging in the *Reoviridae* but it remains to develop *in vitro* packaging and/or reverse genetics in order to probe the mechanisms in the eukaryotic viruses.

Acknowledgement

This work was supported by grants GM34352 and GM31709 from the National Institutes of Health.

References

Butcher, S.J., Grimes, J.M., Makeyev, E.V., Bamford, D.H., and Stuart, D.I. (2001). A mechanism for initiating RNA-dependent RNA polymerization. Nature *410*, 235–240.

Catalano, C.E. (2000). The terminase enzyme from bacteriophage lambda: a DNA-packaging machine. Cell Mol. Life Sci. *57*(1), 128–48.

Choi, Y.G., Grantham, G.L., and Rao, A.L. (2000). Molecular studies on bromovirus capsid protein. Virology *270*(2), 377–85.

Cuppels, D.A., Vidaver, A.K., and Van Etten, J.L. (1979). Resistance to bacteriophage Φ6 by Pseudomonas phaseolicola. J. Gen Virol. *44*, 493–504.

Day, L.A., and Mindich, L. (1980). The molecular weight of bacteriophage Φ6 and its nucleocapsid. Virology *103*, 376–385.

de Haas, F., Paatero, A.O., Mindich, L., Bamford, D.H., and Fuller, S.D. (1999). A symmetry mismatch at the site of RNA packaging in the polymerase complex of dsRNA bacteriophage ϕ6. J. Mol. Biol. *294*, 357–372.

Emori, Y., Iba, H., and Okada, Y. (1980). Semi-conservative transcription of double-stranded RNA catalyzed by bacteriophage Φ6 RNA polymerase. J. Biochem. *88*, 1569–1575.

Frilander, M., Gottlieb, P., Strassman, J., Bamford, D.H., and Mindich, L. (1992). Dependence of minus strand synthesis upon complete genomic packaging in the dsRNA bacteriophage Φ6. J. Virol. *66*, 5013–5017.

Gottlieb, P., Metzger, S., Romantschuk, M., Carton, J., Strassman, J., Bamford, D.H., Kalkkinen, N., and Mindich, L. (1988). Nucleotide sequence of the middle dsRNA segment of bacteriophage Φ6: Placement of the genes of membrane-associated proteins. Virology *163*, 183–190.

Gottlieb, P., Qiao, X., Strassman, J., Frilander, M., and Mindich, L. (1994). Identification of the packaging regions within the genomic RNA segments of bacteriophage Φ6. Virology *200*, 42–47.

Gottlieb, P., Strassman, J., Frucht, A., Qiao, X., and Mindich, L. (1991). *In vitro* packaging of the bacteriophage Φ6 ssRNA genomic precursors. Virology *181*, 589–594.

Gottlieb, P., Strassman, J., Qiao, X., Frilander, M., Frucht, A., and Mindich, L. (1992). *In vitro* packaging and replication of individual genomic segments of bacteriophage Φ6 RNA. J. Virol. *66*, 2611–2616.

Gottlieb, P., Strassman, J., Qiao, X., Frucht, A., and Mindich, L. (1990). *In vitro* replication, packaging and transcription of the segmented dsRNA genome of bacteriophage Φ6: studies with procapsids assembled from plasmid encoded proteins. J. Bacteriol. *172*, 5774–5782.

Horn, W.T., Convery, M.A., Stonehouse, N.J., Adams, C.J., Liljas, L., Phillips, S.E., and Stockley, P.G. (2004). The crystal structure of a high affinity RNA stem–loop complexed with the bacteriophage MS2 capsid: further challenges in the modeling of ligand-RNA interactions. RNA *10*(11), 1776–82.

Huiskonen, J.T., de Haas, F., Bubeck, D., Bamford, D.H., Fuller, S.D., and Butcher, S.J. (2006). Structure of the bacteriophage phi6 nucleocapsid suggests a mechanism for sequential RNA packaging. Structure *14*(6), 1039–48.

Jayaram, H., Taraporewala, Z., Patton, J.T., and Prasad, B.V. (2002). Rotavirus protein involved in genome replication and packaging exhibits a HIT-like fold. Nature *417*(6886), 311–5.

Juuti, J.J., and Bamford, D.H. (1995). RNA binding, packaging and polymerase activities of the different incomplete polymerase complex particles of dsRNA bacteriophage Φ6. J. Mol. Biol. *249*, 545–554.

Kainov, D.E., Tuma, R., and Mancini, E.J. (2006). Hexameric molecular motors: P4 packaging ATPase unravels the mechanism. Cell Mol. Life Sci. *63*(10), 1095–105.

Komoto, S., Sasaki, J., and Taniguchi, K. (2006). Reverse genetics system for introduction of site-specific mutations into the double-stranded RNA genome of infectious rotavirus. Proc. Natl Acad. Sci. USA *103*(12), 4646–51.

Krishna, N.K., Marshall, D., and Schneemann, A. (2003). Analysis of RNA packaging in wild-type and mosaic protein capsids of flock house virus using recombinant baculovirus vectors. Virology *305*, 10–24.

Lander, G.C., Tang, L., Casjens, S.R., Gilcrease, E.B., Prevelige, P., Poliakov, A., Potter, C.S., Carragher, B., and Johnson, J.E. (2006). The structure of an infectious P22 virion shows the signal for headful DNA packaging. Science *312*(5781), 1791–5.

Makeyev, E.V., and Bamford, D.H. (2000). The polymerase subunit of a dsRNA virus plays a central role in the regulation of viral RNA metabolism. EMBO J. *19*(22), 6275–84.

Mancini, E.J., Kainov, D.E., Grimes, J.M., Tuma, R., Bamford, D.H., and Stuart, D.I. (2004). Atomic Snapshots of an RNA Packaging Motor Reveal Conformational Changes Linking ATP Hydrolysis to RNA Translocation. Cell *118*, 743–755.

Mathews, D.H., Sabina, J., Zuker, M., and Turner, D.H. (1999). Expanded Sequence Dependence of Thermodynamic Parameters Improves Prediction of RNA Secondary Structure. J. Mol. Biol. *288*, 911–940.

McGraw, T., Mindich, L., and Frangione, B. (1986). Nucleotide sequence of the small double-stranded RNA segment of bacteriophage Φ6: Novel mechanism of natural translational control. J. Virol. *58*, 142–151.

Mindich, L. (1996). Heterologous recombination in the segmented dsRNA genome of bacteriophage Φ6. Sem. Virol. *7*, 389–397.

Mindich, L. (1999). Precise packaging of the three genomic segments of the double-stranded-RNA bacteriophage Φ6. Microbiol. Mol. Biol. Rev. *63*, 149–160.

Mindich, L., Nemhauser, I., Gottlieb, P., Romantschuk, M., Carton, J., Frucht, S., Strassman, J., Bamford, D.H., and Kalkkinen, N. (1988). Nucleotide sequence of the Large dsRNA segment of bacteriophage Φ6: The genes specifying the viral replicase and transcriptase. J. Virol. *62*, 1180–1185.
Mindich, L., Qiao, X., Onodera, S., Gottlieb, P., and Frilander, M. (1994). RNA structural requirements for stability and minus strand synthesis in the dsRNA bacteriophage Φ6. Virology *202*, 258–263.
Mindich, L., Qiao, X., and Qiao, J. (1995). Packaging of multiple copies of small-sized genomic segments by bacteriophage Φ6. Virology *212*, 213–217.
Mindich, L., Qiao, X., Qiao, J., Onodera, S., Romantschuk, M., and Hoogstraten, D. (1999). Isolation of additional bacteriophages with genomes of segmented double-stranded RNA. J. Bacteriol. *181*, 4505–4508.
Mindich, L., Sinclair, J.F., and Cohen, J. (1976). The morphogenesis of bacteriophage Φ6: Particles formed by nonsense mutants. Virology *75*, 224–231.
Muramoto, Y., Takada, A., Fujii, K., Noda, T., Iwatsuki-Horimoto, K., Watanabe, S., Horimoto, T., Kida, H., and Kawaoka, Y. (2006). Hierarchy among viral RNA (vRNA) segments in their role in vRNA incorporation into influenza A virions. J. Virol. *80*(5), 2318–25.
Olkkonen, V.M., Gottlieb, P., Strassman, J., Qiao, X., Bamford, D.H., and Mindich, L. (1990). *In vitro* assembly of infectious nucleocapsids of bacteriophage Φ6: Formation of a recombinant double-stranded RNA virus. Proc. Natl Acad. Sci. USA *87*, 9173–9177.
Olkkonen, V.M., Ojala, P., and Bamford, D.H. (1991). Generation of infectious nucleocapsids by *in vitro* assembly of the shell protein onto the polymerase complex of the dsRNA bacteriophage Φ6. J. Mol. Biol. *218*, 569–581.
Onodera, S., Olkkonen, V.M., Gottlieb, P., Strassman, J., Qiao, X., Bamford, D.H., and Mindich, L. (1992). Construction of a transducing virus from dsRNA bacteriophage Φ6: establishment of carrier states in host cells. J. Virol. *66*, 190–196.
Onodera, S., Qiao, X., Qiao, J., and Mindich, L. (1995). Acquisition of a fourth genomic segment in bacteriophage Φ6: a bacteriophage with a genome of three segments of dsRNA. Virology *212*, 204–212.
Onodera, S., Qiao, X., Qiao, J., and Mindich, L. (1998a). Directed changes in the number of dsRNA genomic segments in bacteriophage Φ6. Proc. Natl Acad. Sci. USA *95*, 3920–3924.
Onodera, S., Qiao, X., Qiao, J., and Mindich, L. (1998b). Isolation of a mutant that changes genomic packaging specificity in Φ6. Virology *252*, 438–442.
Onodera, S., Sun, Y., and Mindich, L. (2001). Reverse genetics and recombination in Φ8, a dsRNA bacteriophage. Virology *286*, 113–118.
Patton, J.T. (1990). Evidence for equimolar synthesis of double-strand RNA and minus-strand RNA in rotavirus-infected cells. Virus Res. *17*, 199–208.
Patton, J.T., and Gallegos, C.O. (1990). Rotavirus RNA replication: single-stranded RNA extends from the replicase particle. J. Gen. Virol. *71*(Pt 5), 1087–94.
Pirttimaa, M.J., Paatero, A.O., Frilander, M., and Bamford, D.H. (2002). Nonspecific nucleoside triphosphatase P4 of double-stranded RNA bacteriophage Φ6 is required for single-stranded RNA packaging and transcription. J. Virol. *76*, 10122–10127.
Qiao, J., Qiao, X., Sun, Y., and Mindich, L. (2003). Isolation and analysis of mutants with altered packaging specificity in the dsRNA bacteriophage F6. J. Bacteriol. *185*, 4572–4577.
Qiao, X., Casini, G., Qiao, J., and Mindich, L. (1995). *In vitro* packaging of individual genomic segments of bacteriophage F6 RNA: serial dependence relationships. J. Virol. *69*, 2926–2931.
Qiao, X., Qiao, J., and Mindich, L. (1995). Interference of bacteriophage Φ6 genomic RNA packaging by hairpin structures. J. Virol. *69*, 5502–5505.
Qiao, X., Qiao, J., and Mindich, L. (1997a). An *in vitro* system for the investigation of heterologous RNA recombination. Virology *227*, 103–110.
Qiao, X., Qiao, J., and Mindich, L. (1997b). Stoichiometric packaging of the three genomic segments of dsRNA bacteriophage Φ6. Proc. Natl Acad. Sci. USA *94*, 4074–4079.
Qiao, X., Qiao, J., and Mindich, L. (2003). Analysis of the specific binding involved in genomic packaging of the dsRNA bacteriophage Φ6. J. Bacteriol. *185*(21), 6409–6414.
Qiao, X., Qiao, J., Onodera, S., and Mindich, L. (2000). Characterization of Φ13, a bacteriophage related to Φ6 and containing three dsRNA genomic segments. Virology *275*, 218–224.
Ranjith-Kumar, C.T., and Kao, C.C. (2006). Recombinant viral RdRps can initiate RNA synthesis from circular templates. Rna *12*(2), 303–12.
Reinisch, K.M., Nibert, M.L., and Harrison, S.C. (2000). Structure of the reovirus core at 3.6 A resolution. Nature *404*, 960–967.

Ribas, J.C., and Wickner, R.B. (1996). Saccharomyces cerevisiae L-BC double-stranded RNA virus replicase recognizes the L-A positive-strand RNA 3′ end. J. Virol. *70*(1), 292–7.
Roner, M.R., and Joklik, W.K. (2001). Reovirus reverse genetics: Incorporation of the CAT gene into the reovirus genome. Proc. Natl Acad. Sci. USA *98*, 8036–8041.
Salgado, P.S., Makeyev, E.V., Butcher, S.J., Bamford, D.H., Stuart, D.I., and Grimes, J.M. (2004). The structural basis for RNA specificity and Ca2+ inhibition of an RNA-dependent RNA polymerase. Structure *12*(2), 307–16.
Sun, Y., Qiao, X., and Mindich, L. (2004a). Construction of carrier state viruses with partial genomes of the segmented dsRNA bacteriophages. Virology *319*, 274–279.
Sun, Y., Qiao, X., and Mindich, L. (2004b). Construction of carrier state viruses with partial genomes of the segmented dsRNA bacteriophages. Virology *319*(2), 274–9.
Sun, Y., Qiao, X., Qiao, J., Onodera, S., and Mindich, L. (2003). Unique properties of the inner core of bacteriophage phi8, a virus with a segmented dsRNA genome. Virology *308*(2), 354–61.
Usala, S.J., Brownstein, B.H., and Haselkorn, R. (1980). Displacement of parental RNA strands during *in vitro* transcription by bacteriophage Φ6 nucleocapsids. Cell *19*, 855–862.
Van Dijk, A.A., Frilander, M., and Bamford, D.H. (1995). Differentiation between minus- and plus-strand synthesis: polymerase activity of dsRNA bacteriophage Φ6 in an *in vitro* packaging and replication system. Virology *211*, 320–323.
Vidaver, A.K., Koski, R.K., and Van Etten, J.L. (1973). Bacteriophage Φ6: a lipid-containing virus of Pseudomonas phaseolicola. J. Virol. *11*, 799–805.

Index

T

W

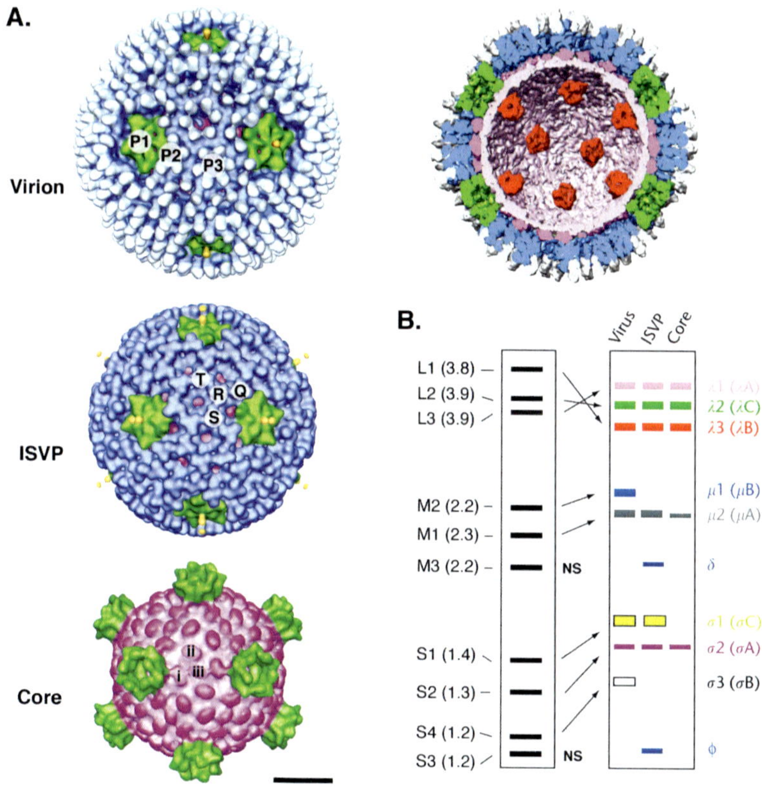

Plate 1.2 A) Surface-shaded, three-dimensional reconstructions of the mammalian reovirus (MRV) virion, intermediate subviral particle (ISVP), and core (left side from top to bottom, respectively), and a cross-section of the virion (top right). Features are colour-coded by protein as identified in B. Labels are shown for the P1, P2 and P3 channels on the virion, the Q, R, S and T positions of the μ1 trimers on the ISVP and the (i), (ii) and (iii) positions of the σ2 nodules on the core. Bar = 200 Å. B) Schematic representation of the electrophoretic patterns for the dsRNA segments with kilobase values (left), and the eight structural proteins (right). Homologous ARV proteins are indicated in parentheses (far right). (See also Figure 1.2.)

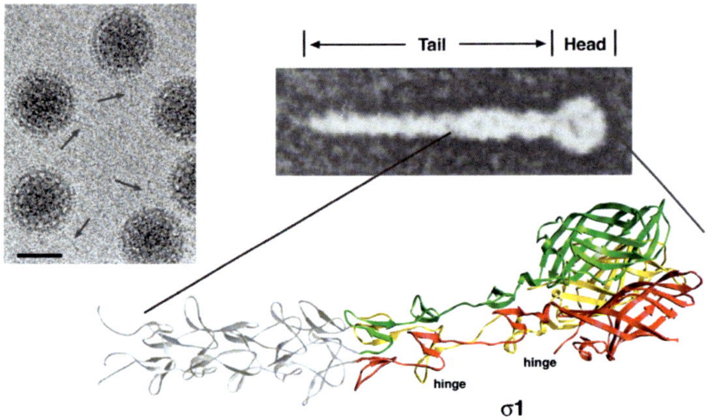

Plate 1.3 The cell attachment protein, σ1, mediates binding to receptors such as JAM and sialic acid. Cryo-electron micrograph of ISVPs (left) shows extended σ1 fibres (arrows). A two-dimensional average of negatively stained σ1 displays displays head and tail regions (upper right) (from Fraser et al., 1990, with permission). The head domain and tail fragment that was determined by X-ray crystallography is coloured by monomer (pdb 1FN9), and the extra predicted β-spirals in grey (bottom). Hinges would confer conformational flexibility. The remainder of the protein is not modelled. Bar = 750 Å for the micrograph. (See also Figure 1.3.)

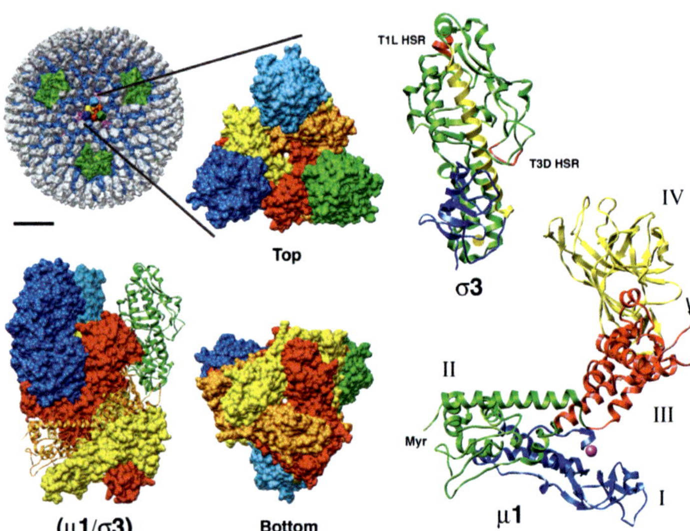

Plate 1.4 The major outer capsid proteins σ3 and μ1 confer stability to the virions. The left-hand panels are surface renderings of the μ1/σ3 heterohexamer, with an overview of the placement of one heterohexamer in the virion (upper left). The complex is viewed in different orientations with the μ1 subunits shown in red, orange and yellow and the σ3 subunits shown in dark blue, cyan, and green (pdb 1JMU). Note how each μ1 wraps completely around the complex. A closer view of the σ3 monomer (upper right) is coloured blue for the N-terminal small lobe, green for the C-terminal large lobe and loop to the small lobe, and yellow for the central helix. The proteolytically hypersensitive regions (HSR) are red. In μ1 (lower right), domains I to IV are coloured, blue, green, red and yellow, respectively. The μ1C/μ1N cleavage is identified with a magenta sphere at Asn42, and the arrow indicates the region for δ/ϕ cleavage. Bar = 200 Å for the virion; 25 Å for the crystal structures. (See also Figure 1.4.)

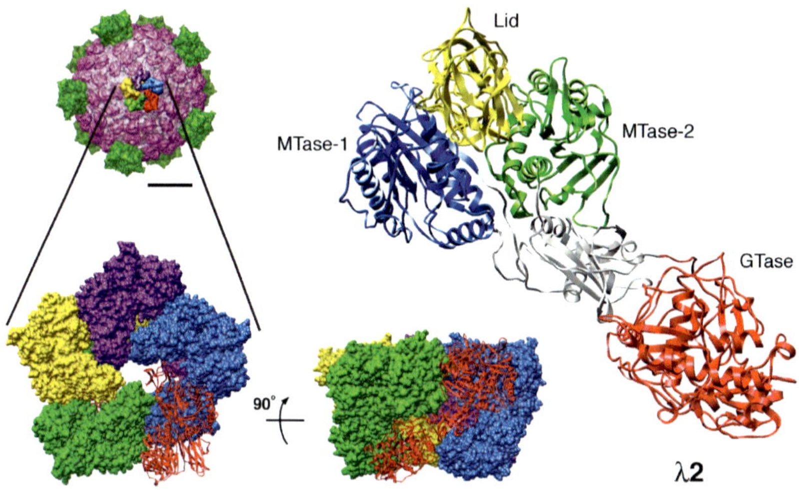

Plate 1.5 The λ2 turret protein is an RNA capping enzyme. The five subunits in the pentamer (bottom left) are coloured differently (pdb 1EJ6). A ribbon representation of the yellow monomer in the pentamer (right) shows the inner side of the protein. The seven domains are identified as red for the guanylyltransferase (GTase), blue and green for the methyltransferases (MTase), grey for the connecting domain, and yellow for the three Ig-like lid domains. Bar = 200 Å for the core particle; 40 Å for the crystal structures. (See also Figure 1.5.)

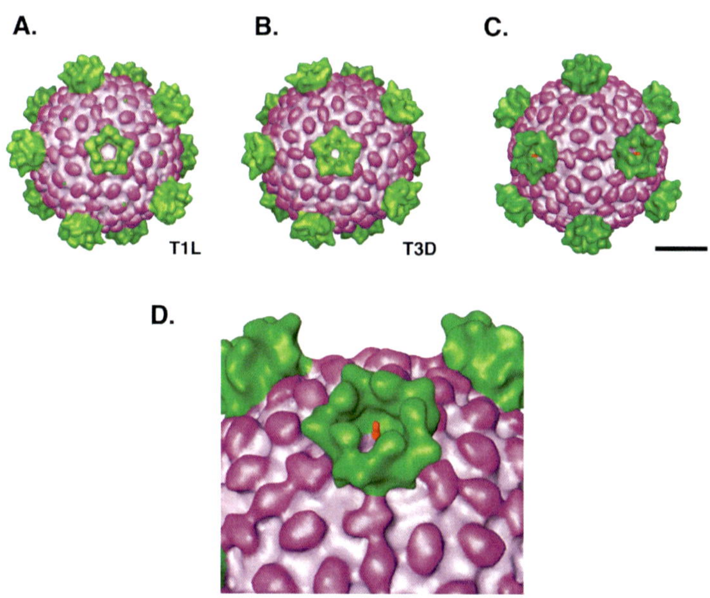

Plate 1.6 Surface-shaded, three-dimensional reconstructions of strain-specific conformational changes and of transcriptionally active reovirus core particles. Two strains of MRV display a different amount of opening of the λ2 turrets. T1L (A) opens the P1 channel much wider than T3D (B) by swinging the λ2 lid domain almost perpendicular to the turret. A reconstruction of active core particles viewed down the 2-fold symmetry axis (C) shows that the particles remain intact during transcription. The oblique, close-up view of the fivefold vertex position (D) displays a thread of density (red) exiting the λ2 pentamers, ascribed to newly synthesized single-stranded RNA. Bar = 200 or 75 Å for A-C and D, respectively. (See also Figure 1.6.)

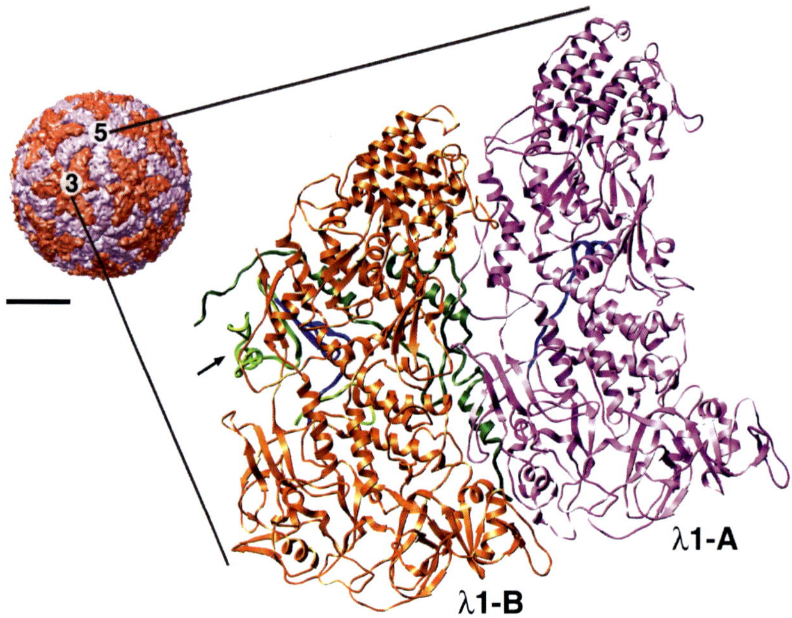

Plate 1.7 The λ1 core protein forms a fairly smooth shell for attachment of σ2. A surface-shaded view (left) showing the unique λ1-A (pink) and λ1-B (orange) subunits. Although the A and B subunits are very similar in structure (right), there is a subtle shift in the lower domains relative to each other (pdb 1EJ6). The additional amino acids observed in the crystal structure of λ1-B are in green, with the residues predicted to come from a neighbouring λ1-B subunit, 40-to-267, in dark green. Residues 130–159 (blue), which are essential for particle assembly, are located under the σ2 i and ii binding positions. The Zn-finger motif is at residues 181–208 (arrow). Bar = 200 Å. (See also Figure 1.7.)

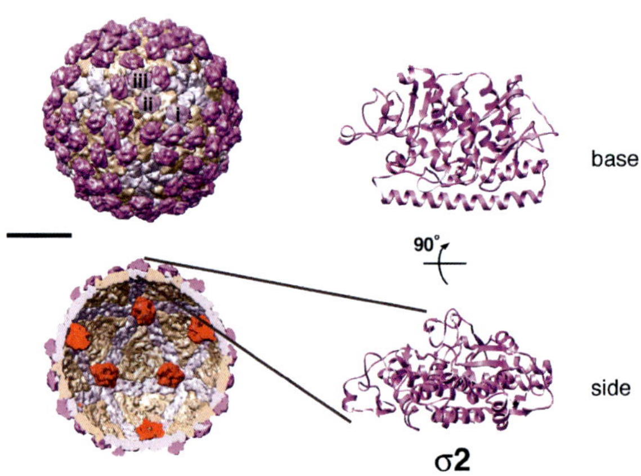

Plate 1.8 The σ2 nodules are required for core particle assembly. Surface-shaded view of the core capsid with the nodules viewed along the threefold axis (top left) and in cross-section (bottom left). Note that not all λ3 subunits display equally in the cross-section due to their off-axis positions. Close-up views of the protein from the base (top right) and side (bottom right) (pdb 1EJ6). Different residues on the base interact with the λ1 shell in three unique positions: (i) 60 around the fivefold axes, (ii) 60 around the threefold axes, and (iii) 30 on the 2-fold axis. Bar = 200 Å. (See also Figure 1.8.)

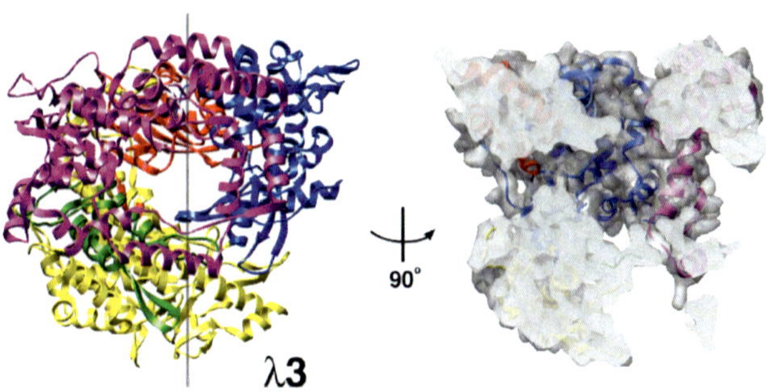

Plate 1.9 The λ3 RdRp mediates RNA synthesis. In the left panel, the three polymerase subdomains (red, blue and green) are bracketed by the N-terminal domain (yellow) and the C-terminal bracelet (pink) (pdb 1MUK). In the middle panel the monomer is rotated 90° and slabbed to display the channels in the same orientation as in the cross-section of the pseudoatomic model (right) (pdb 2CSE). The alignment of λ3 is based on fitting the crystal structure into a 7.6 Å resolution cryoEM density map (Zhang *et al.*, 2003) and positions λ3 slightly off centre from the fivefold axis. (See also Figure 1.9.)

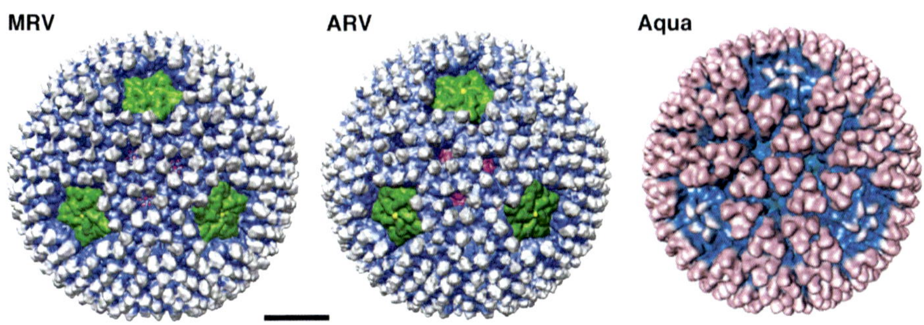

Plate 1.10 Comparison of reoviruses MRV with avian reovirus (ARV) and aquareovirus (AqRV) (Zhang *et al.*, 2005b). The structure of ARV (B) is very similar to MRV, except for the absence of density at low radii in the P1 and P2 channels (see Figure 1.2A). The closely related genus AqRV also has structural similarities (C; from Nason *et al.*, 2000, with permission), but the turrets appear narrower and have an open conformation. AqRV is coloured blue to radius 375 Å and pink for the outer radii. They are also missing the equivalent σ2 core nodules at the 2-fold symmetry axes (not shown). (See also Figure 1.10.)

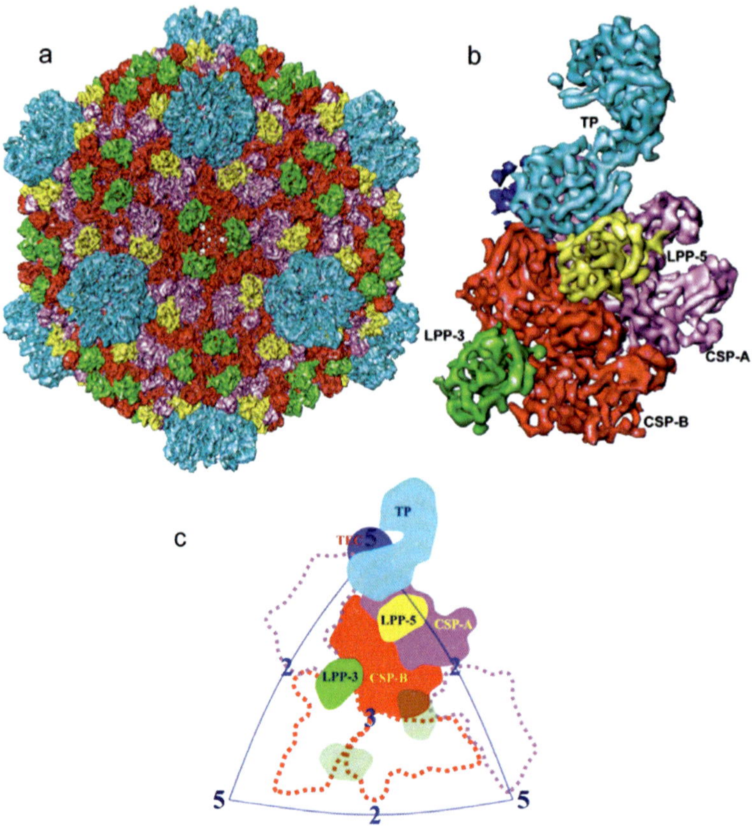

Plate 2.3 Structure organization and surface complementarity of CPV. (a) Shaded surface view of a CPV reconstruction at about 7-Å resolution as viewed along a threefold axis. Different molecules are coloured differently. (b) Close-up view of an asymmetric unit, which is made up of one TP (VP4), two copies of the CSP (VP1), and two copies of the LPP (VP7) (fivefold-proximal LPP-5 is in yellow; threefold-proximal LPP-3 is in green). (c) Schematic diagram illustrating the positions and interactions among the structural components with respect to the icosahedral five-, three-, and twofold axes. The blue triangle represents one triangular face of the icosahedron. (b) and (c) are adapted from (Zhou *et al.*, 2003) with permission from the publisher. (See also Figure 2.3.)

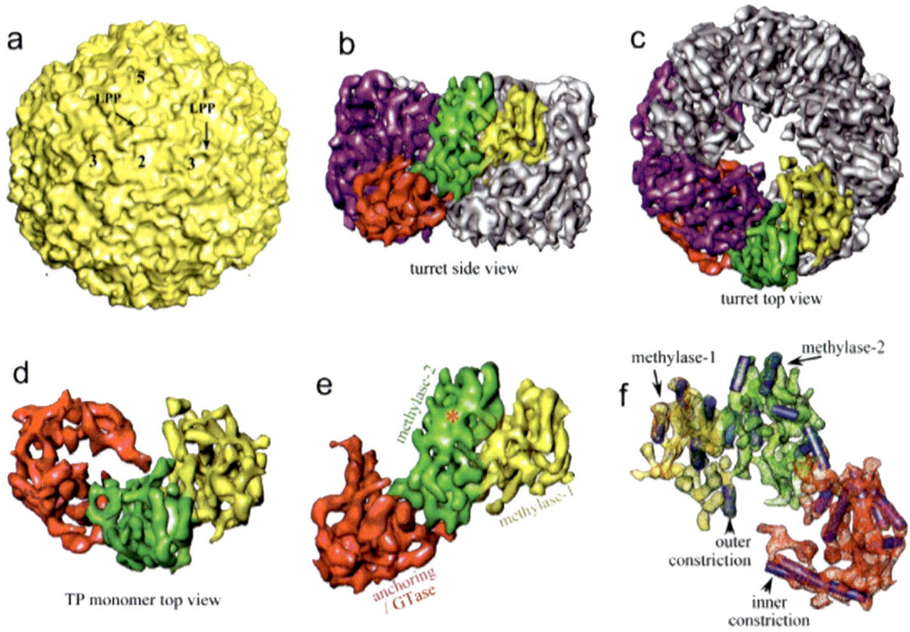

Plate 2.5 Structure of the mRNA capping complex, or turret. (a) Shaded surface representation of CPV with its turrets removed by chemical treatment (Zhang *et al.*, 2002). (b,c) Side (perpendicular to fivefold axis) (b) and top (parallel to fivefold axis) (c) views of a computationally isolated turret. Two adjacent subunits are highlighted in colour. (d,e) Top (d) and side (e) views of an extracted TP monomer. The asterisk indicates the large, twisted β-sheet in the characteristic methylase-like, helix-pair/β-sheet/helix-pair fold of the methylase-2 domain. (f) Wire-frame representation of TP as viewed from the inside the turret channel looking outward. The superimposed 5-Å blue cylinders represent the α-helices identified in each TP. The helices contributing to the formation of the inner and outer channel constrictions are indicated. Adapted from (Zhang *et al.*, 2002) (a) and (Zhou *et al.*, 2003) (b-f) with permission from the publishers. (See also Figure 2.5.)

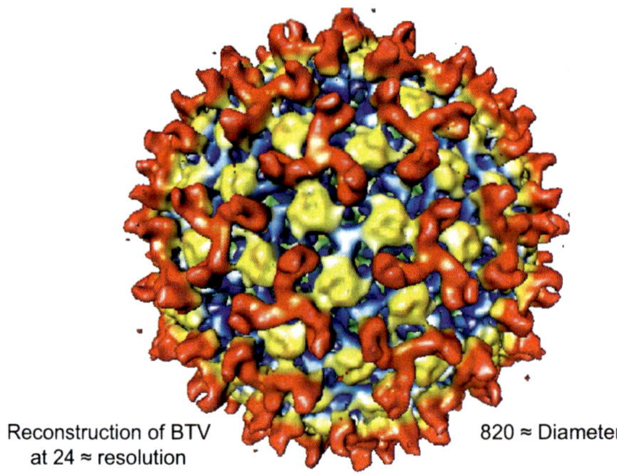

Plate 4.2 Surface representations of the 3D cryo-EM structures (22 Å resolution) of BTV. (A) Whole particle showing sail-shaped triskelion propellers (VP2 trimers) in red and globular domains (VP5 trimers) in yellow (Nason *et al*., 2004). (See also Figure 4.2.)

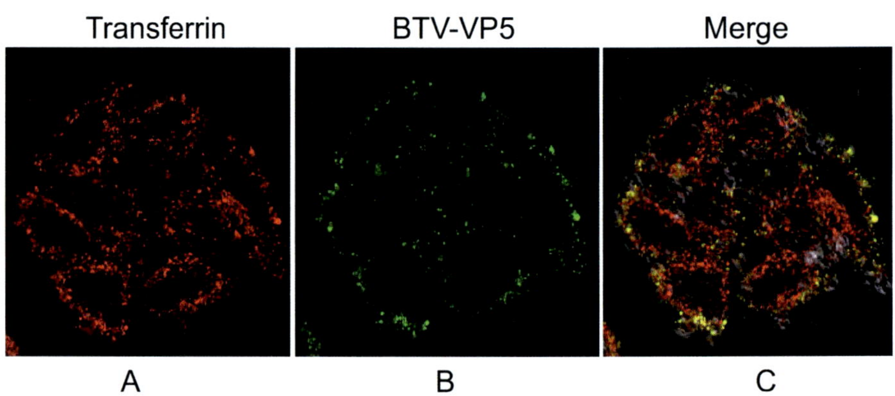

Plate 4.3 BTV entry into the cells via endosome. HeLa cells were infected with BTV and cellular uptake of BTV was monitored by confocal microscopy. BTV particles were stained with a polyclonal anti-VP5 primary antibody and localized by a FITC-conjugated secondary antibody (panel B). Transferrin-TRITC conjugate (Tf-TRITC) was monitored as endosomal marker (panel A). Note that VP5 co-localises with transferrin in the endosomes of infected HeLa cells (Panel C) (Forzan *et al*., 2007). (See also Figure 4.3.)

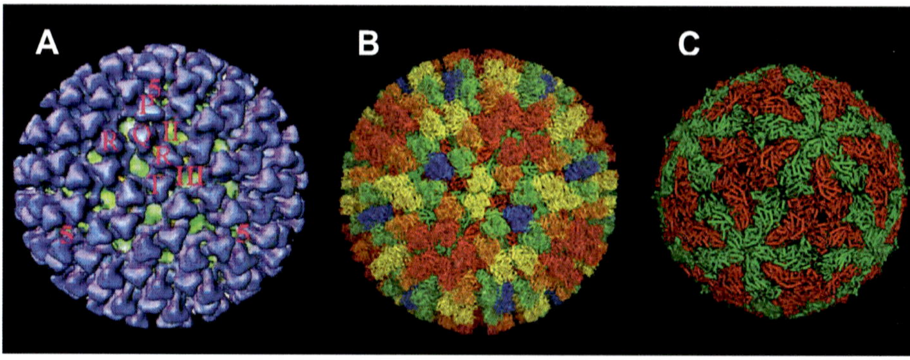

Plate 4.4 3D structures of BTV core. (A). Surface representations of the 3D cryo-EM structures of BTV-10 core (700 Å in diameter) viewed along the icosahedral threefold axis showing the trimers of VP7 (in blue). The five quasi-equivalent trimers (P, Q, R, S, and T) and the locations of channels II and III are marked. (Grimes *et al.*, 1997). (B). The X-ray structure of the BTV-1 core surface showing the detail arrangement of 260 trimers. The structurally equivalent trimers have been given the same colour. (C). The structure of the VP3 layer showing the arrangement of 120 copies of two conformationally distinct types of VP3 molecules, 'A' shown in green and 'B' shown red. 60 copies of each of the two molecules together form 12 decamers that are linked together forming a thin protein shell. The icosahedral fivefold, threefold and symmetry axes are indicated (Grimes *et al.*, 1998). (See also Figure 4.4.)

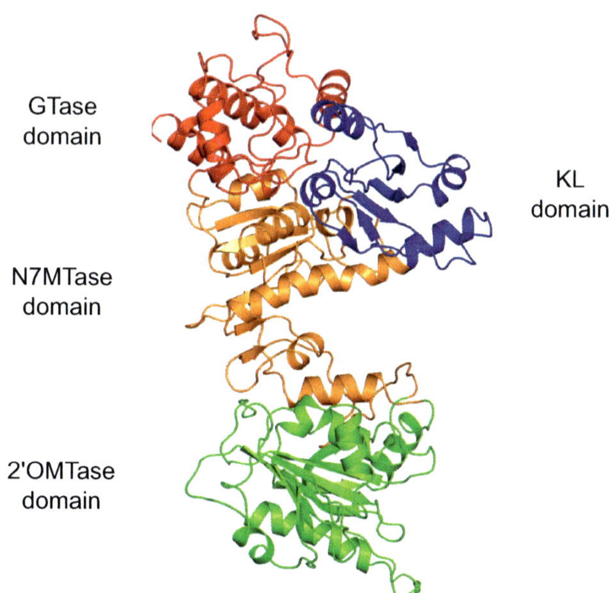

Plate 4.6 Cartoon representation of the structure of the VP4 molecule. Ribbon diagram of the VP4 structure with enzymatic domains coloured: guanylyl transferase (GTase, red) domain, kinase-like (kl, blue) domain, guanine-N7-methyltransferase (N7MTase, orange) domain and nucleoside-2′-O-methyltransferase (2′OMT, green) domain (Sutton *et al.*, 2007). (See also Figure 4.6.)

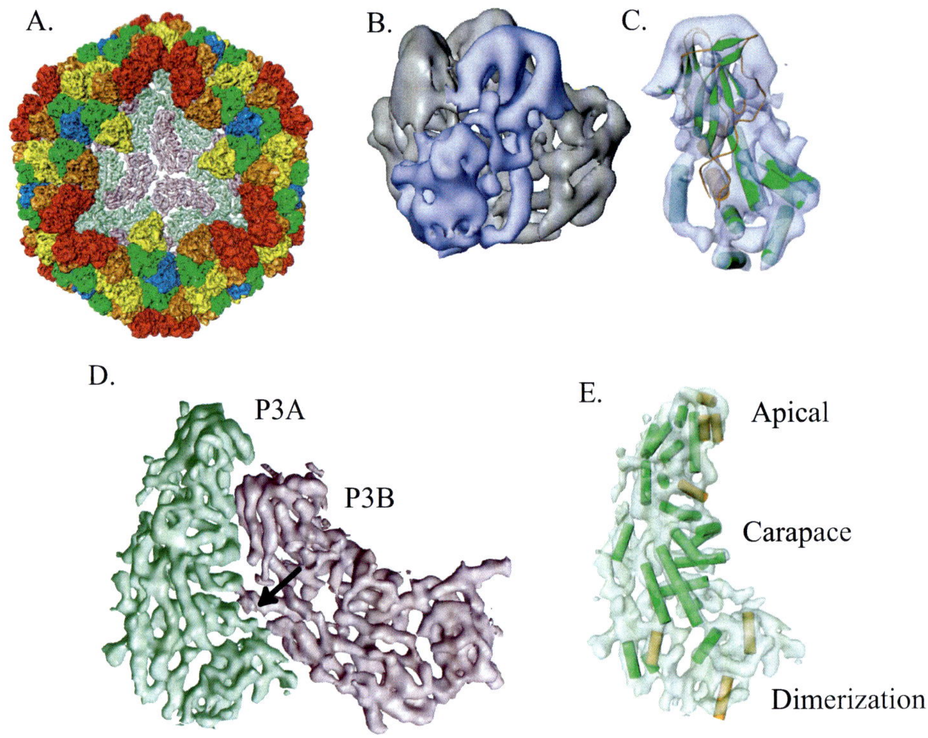

Plate 5.2 CryoEM structure of Rice Dwarf Virus. (A) Structure of the double layered Rice Dwarf Virus at 6.8 Å resolution as determined by cryoEM. An asymmetric unit from the outer capsid shell has been computationally extracted. Individual subunits within the capsid layers are coloured (P3A light green, P3B light red, P trimer red, Q trimer orange, S trimer yellow, R timer green, T trimer blue) (B) Computationally averaged P8 trimer of RDV. A single P8 monomer is highlighted. (C) Pseudo-atomic model of P8 derived from structural analysis. (D) P3 dimer (P3A and P3B). Helix 1 of P3B (light red) is marked with an arrow. (E) Locations of alpha helices of P3A derived from structural analysis. (See also Figure 5.2.)

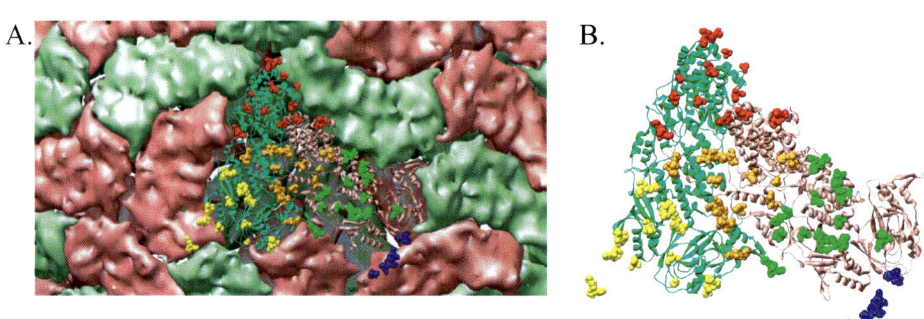

Plate 5.3 RDV protein shell Interactions. Interactions, determined from the X-ray structure, of the outer capsid layer protein, P8, with the inner shell protein, P3, are shown in top views. The atoms shown represent the interacting residues, coloured keyed (same colour scheme as in Figure 5.2) to the corresponding P8 protein trimers. (See also Figure 5.3.)

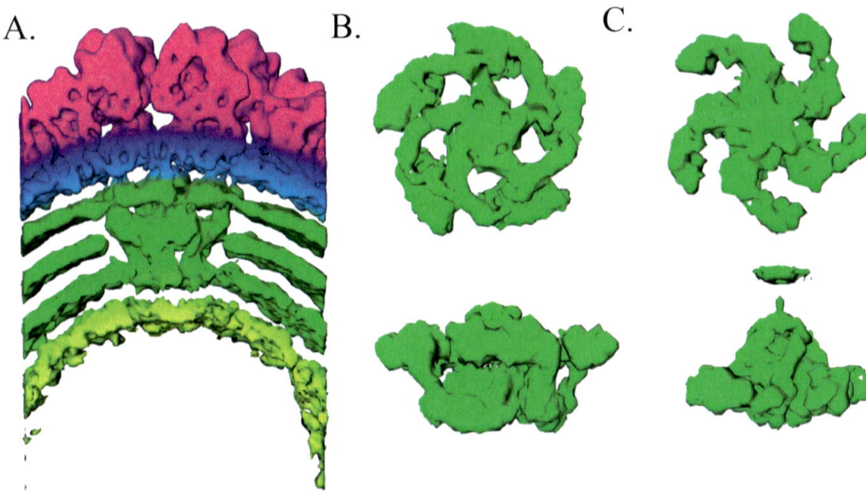

Plate 5.4 Minor proteins of RDV based on cryoEM structures. (A) Radially coloured density at a fivefold vertex. The outer capsid layer is coloured magenta, while the inner layer is coloured blue. The concentric rings of RNA forming the coiled coil structure and the polymerase complex are shown in green. RNA dependant RNA polymerase density at a fivefold vertex after bilateral filtering in the full (B) and empty (C) RDV cryoEM reconstructions. (See also Figure 5.4.)

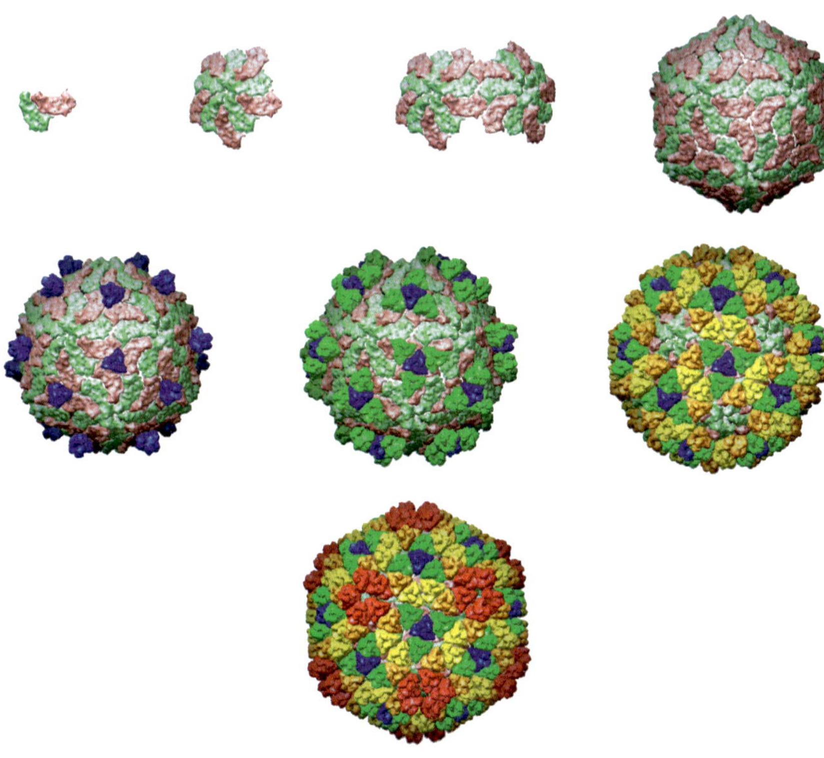

Plate 5.5 Assembly pathway. Assembly of RDV based on a model proposed by Nakagawa and co-workers (Nakagawa *et al.*, 2003). Proteins are coloured as stated in Figure 5.2. (See also Figure 5.5.)

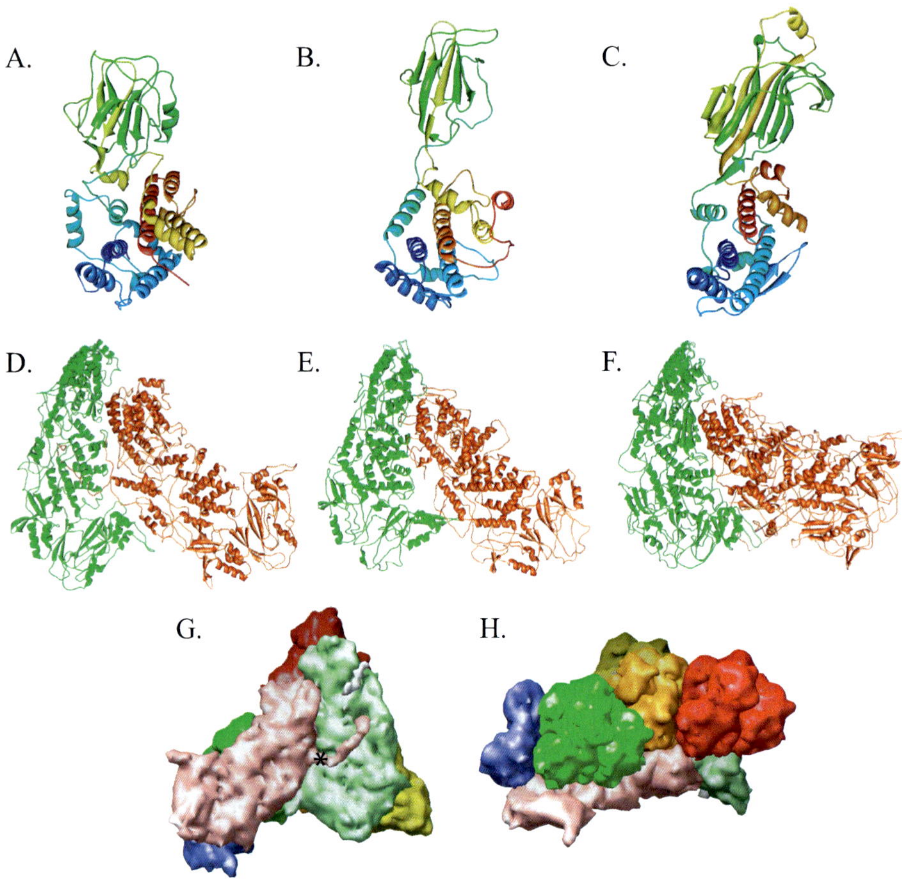

Plate 5.6 Structural comparisons of *Reoviridae* major capsid proteins. (A-C) Comparison of the outer capsid proteins: RDV P8 (A), BTV VP7 (B) and Rotavirus VP6 (C). The ribbon representations of the structures are coloured from the N-terminus (blue) to the C-terminus (red). (D-F) Structural comparison of the dimer of the inner capsid proteins: RDV P3 (D), BTV VP3 (E) and reovirus (F). The subunit directly adjacent to the fivefold axis is coloured green, while the other subunit is coloured red. Asymmetric unit of RDV in a bottom view (G) and side view (H). Helix 1 of P3B is marked with an asterisk in (G). (See also Figure 5.6.)

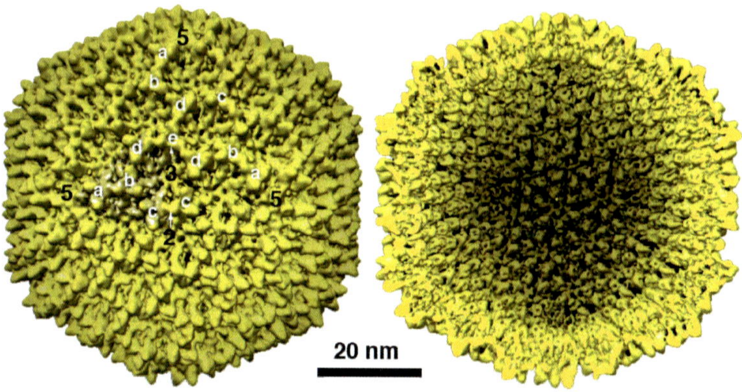

Plate 8.2 Three-dimensional structure of IBDV. The surface-shaded representation, viewed along a threefold axis of icosahedral symmetry, shows a T = 13*l* structure built by five classes of VP2 trimers on the outer surface, labelled a to e (left). A model with the front half of the protein shell removed is shown (right). (See also Figure 8.2.)

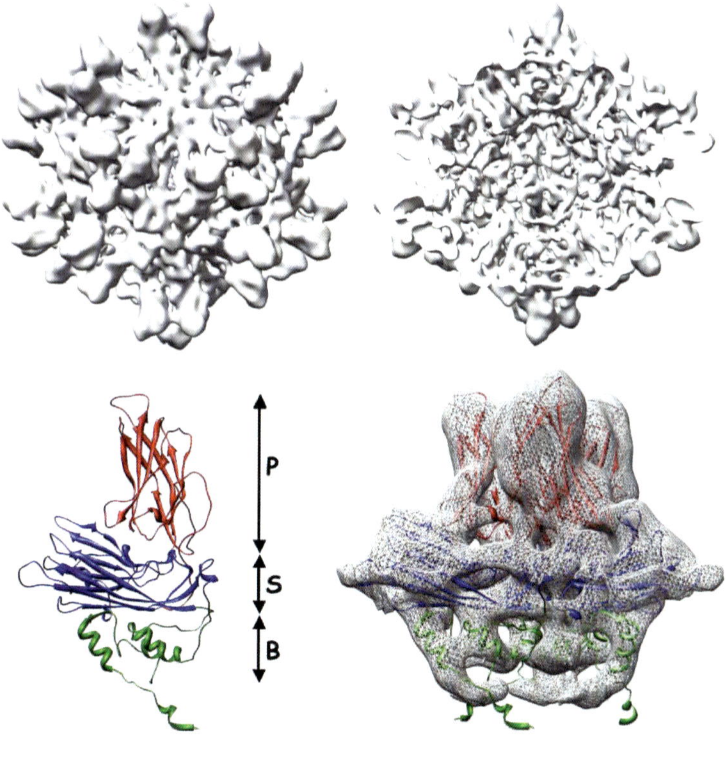

Plate 8.5 Three-dimensional structure of the VP2 subviral particle. Shown are surface-shaded representations of the of the outer (top, left) and inner (top, right) surfaces of VP2 capsids viewed along a twofold axis of icosahedral symmetry. The VP2 atomic model is shown (bottom, left), and the secondary structure elements are colour-coded by domain (red, P; blue, S; green, B). A VP2 trimer atomic model fitted into the corresponding density in the cryo-EM map is shown on the bottom right. (See also Figure 8.5.)

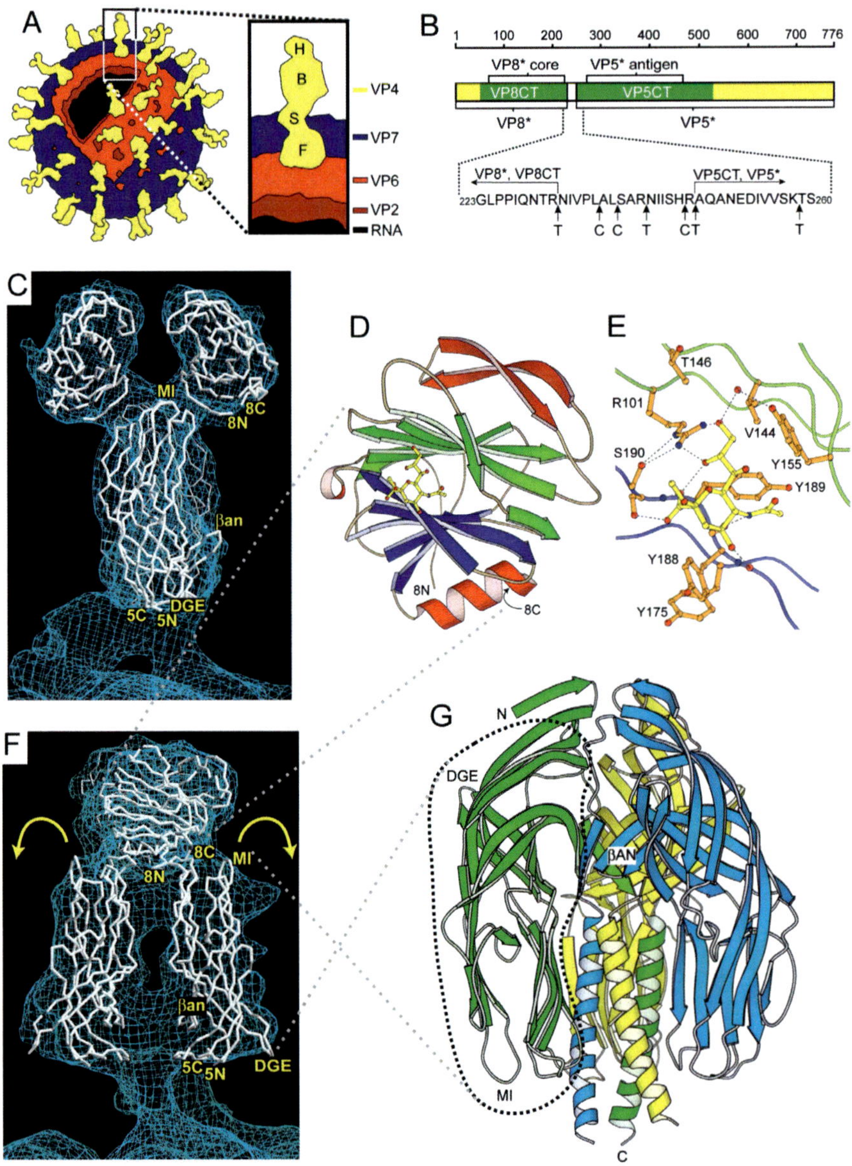

Plate 11.1 Rotavirus VP4. A. Schematic diagram of the rotavirus virion and trypsin-primed VP4 (Dormitzer *et al.*, 2004). The diagram is based on an electron cryomicroscopy-based image reconstruction (Yeager *et al.*, 1990). The proteins that make up each layer of the triple-layered icosahedral particle are indicated next to the expanded view of the spike. The subviral DP contains the RNA, VP2 and VP6 layers. The molecular envelope of the primed VP4 spike has head (H), body (B), stalk (S), and foot (F) domains. B. Linear diagram of VP4 and its fragments. The cleavage products of trypsin digestion of VP4 on the virion, VP8* (residues 1–231) and VP5* (residues 248–776), are given below the bar; the products of sequential chymotrypsin and trypsin digestion of VP4 in solution, VP8CT (residues 46–231) and VP5CT (residues 248-approximately 528), are labelled in the green bars; and the compact globular domains identified by high resolution structural studies, the VP8* core (residues 65–224) and the VP5* antigen domain (residues 267–470), are given above the bar. On the expanded view of the trypsin

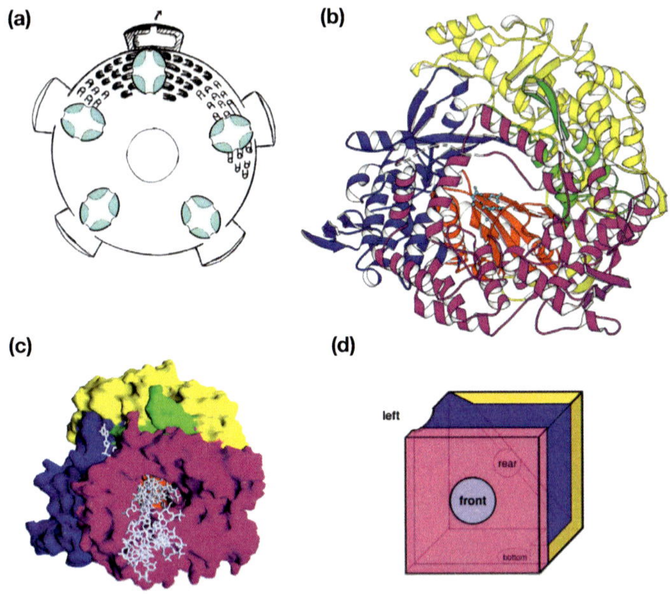

Plate 13.1 Reovirus λ3 (from Tao *et al*, 2002). (a) One λ3 molecule is positioned interiorly near each of the fivefold vertices of the core of reovirus particles. (b) A ribbon diagram of the λ3 crystal structure. The central polymerase domain is shown with its fingers in blue, palm in red and thumb in green. The N- and C-terminal domains are coloured in yellow and magenta, respectively. (c) λ3 represented by its molecular surface, coloured as in (b). A template:product duplex from HIV-RT (Huang *et al*., 1998), shown in grey, is modelled on to λ3 to illustrate the different functions of the four channels. (d) λ3 represented as a cage structure with its N-terminal domain in yellow, the central polymerase domain in blue and the C-terminal bracelet domain in magenta. NTPs and single-stranded templates move to the catalytic core through the rear and left domains, respectively. In its replicase mode, the dsRNA product leaves the polymerase through the front channel. In its transcriptase mode, the mRNA product is believed to leave the polymerase through the bottom channel, with the minus-strand template RNA leaving through the front channel. (See also Figure 13.1.)

---

priming region, the three sites at which trypsin cleaves VP4 on the virion (after arginines 231, 241, and 247) and the additional local site at which trypsin cleaves VP4 in solution (after lysine 258) are indicated by 'T.' The local sites at which chymotrypsin cleaves VP4 in solution (after leucines 236 and 238 and histidine 246) are indicated by 'C.' C. Cα traces of the VP8* core (PDB accession code 1KQR) and a globular domain of VP5CT (PDB accession code 1SLQ) fit to the molecular envelope of the spike from a 12 Å resolution electron cryomicroscopy-based image reconstruction of primed virions (Dormitzer *et al.*, 2004). The N- and C-termini of the VP8* core (8N and 8C), the N- and C-termini of the globular domain of VP5CT (5N and 5C), an exposed hydrophobic potential membrane interaction loop (MI), and the potential integrin binding loop (DGE). D. Ribbon diagram of the VP8* core (Dormitzer *et al*., 2002b). The colours of the ribbons demarcate structural elements. The bound sialoside is depicted with balls and sticks, and the N- and C-termini are indicated. E. Molecular details of sialoside binding (Dormitzer *et al.*, 2002b). Hydrogen bonds are indicated by dotted lines. F. The image of panel C rotated 90° about a vertical axis. The rotation of the body domains during the twofold to threefold reorganization of the protruding regions is indicated by arrows. Dotted lines link the fit Cα traces to the corresponding ribbon diagrams. The lines linking panel F and G are crossed to indicate the domain inversion associated with the transition to the folded-back state. G. Ribbon diagram of VP5CT (Dormitzer *et al.*, 2004). Each subunit is a different colour. The globular domain that is fit to the molecular envelope of the spike is circled with a dotted line. 'βAN' marks the β-annulus. Other labels match those in panel F. (See also Figure 11.1.)

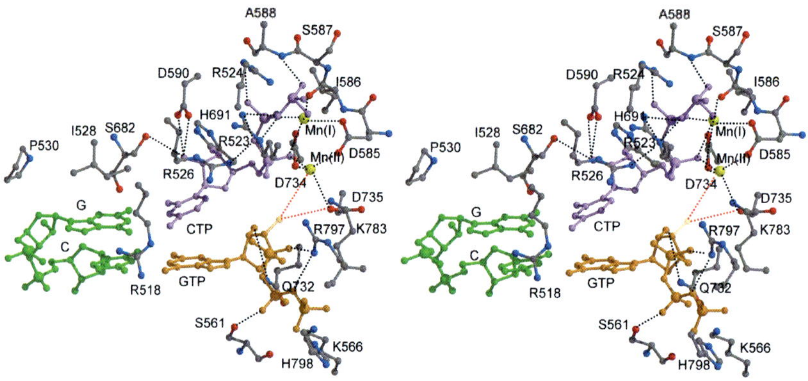

Plate 13.2 Stereo diagram of the active site in the λ3 initiation complex (from Tao *et al*, 2002). Shown are critical amino acids (stick models: blue, nitrogen; red, oxygen), 3′-end of template strand, green; 3′-deoxy rCTP at incoming nucleotide position, purple; 3′-deoxy rGTP at priming position, orange; Mn^{2+} ions, gold spheres; assigned hydrogen bonds and metal ligand interactions, dashed lines. The missing 3′-OH is shown in a lighter shade, and the related hydrogen bonds, in red. (See also Figure 13.2.)

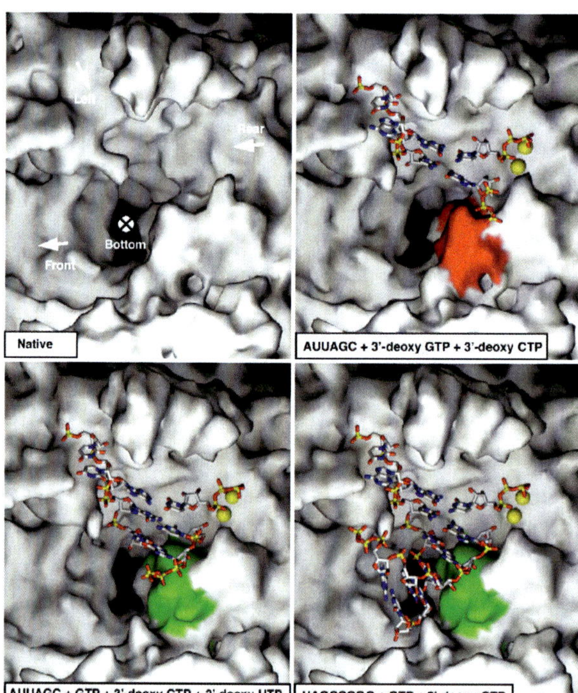

Plate 13.3 RNA synthesis by λ3 (from Tao *et al*, 2002). The active site of λ3 is shown at four progressive stages – as the native apo-enzyme (upper left), an initiation complex (upper right), after synthesis of a 2-nucleotide product (lower left) and after synthesis of a 5-nucleotide product (lower right). λ3 is rendered as a molecular surface; RNA and nucleotides, as atomic models; and metal ions, as yellow spheres. The four channels are indicated on the image of native λ3. Upon formation of the first phosphodiester bond, the loop (residues 658–665) that supports the priming nucleotide changes its conformation (red → green), thus avoiding collision with the dsRNA product. (See also Figure 13.3.)

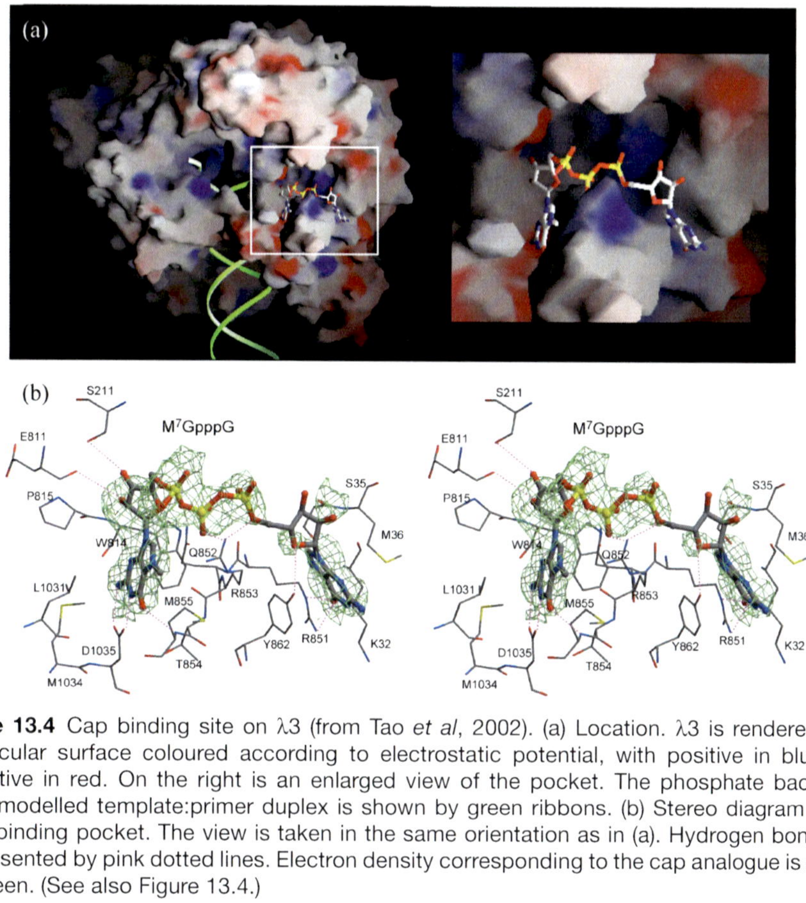

Plate 13.4 Cap binding site on λ3 (from Tao *et al*, 2002). (a) Location. λ3 is rendered as a molecular surface coloured according to electrostatic potential, with positive in blue and negative in red. On the right is an enlarged view of the pocket. The phosphate backbone of a modelled template:primer duplex is shown by green ribbons. (b) Stereo diagram of the cap-binding pocket. The view is taken in the same orientation as in (a). Hydrogen bonds are represented by pink dotted lines. Electron density corresponding to the cap analogue is shown in green. (See also Figure 13.4.)

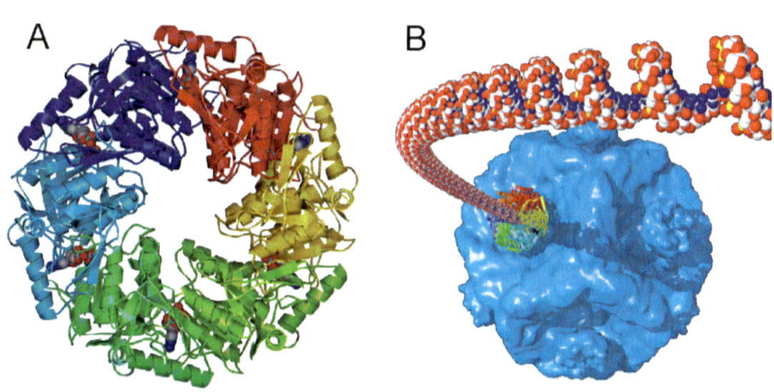

Plate 15.1 (A) Ribbon structure of P4 hexamer. In this representation different colours are used to designate distinct subunits. ADP, binding at the interface between monomers, is shown in a ball-and-stick representation. (B) The P4 hexameric motor is shown in the act of packaging ssRNA into the empty capsid. A 20 Å icosahedral reconstruction of the ϕ6 procapsid (proteins P1, P2, P4 and P7) (de Haas *et al.*, 1999) is shown in a cyan isosurface representation. Density for one of the P4 turrets has been removed and substituted with a ribbon representation of the crystal structure of the hexameric P4. A computer modelled A-form ssRNA is shown in a cpk model (atom colours are: carbon, white; oxygen, red; nitrogen, blue; phosphorus, yellow). (See also Figure 15.1.)

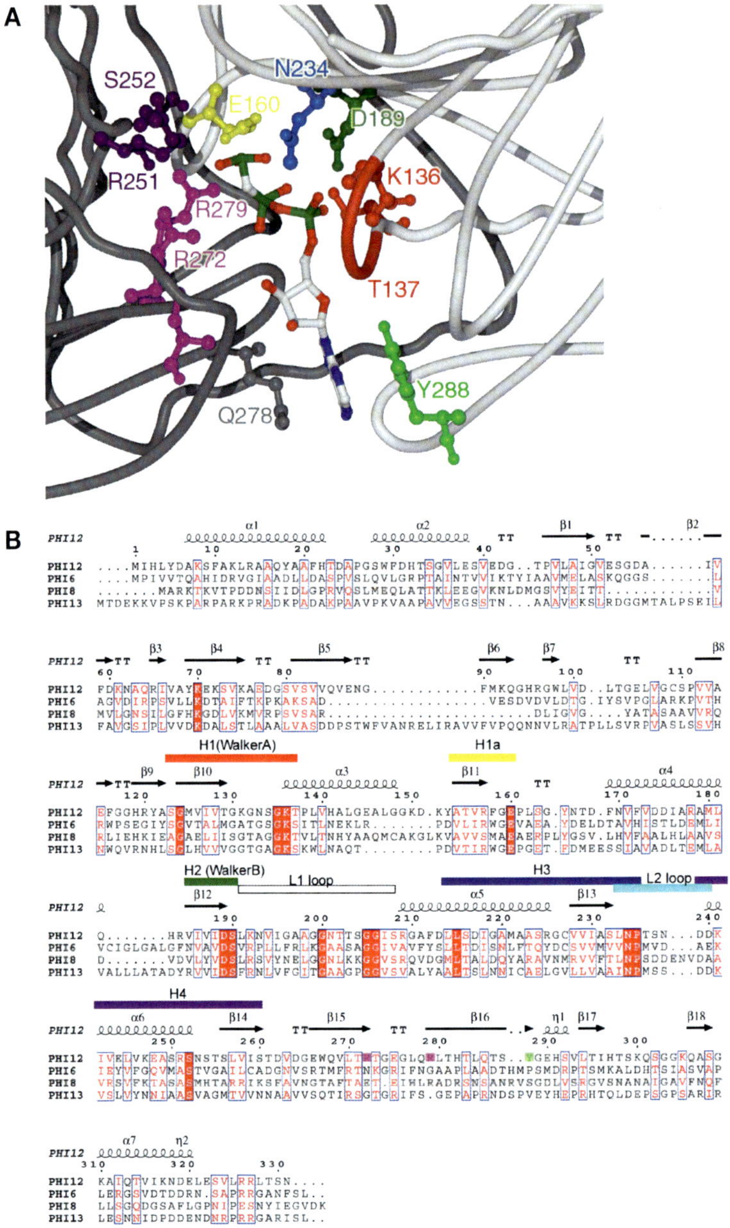

Plate 15.2 Structure and sequence of the conserved helicase motifs in P4 proteins. (A) Detail of the nucleotide binding site using the same colour coding for the conserved motifs as in the bottom panel. AMPcPP is shown in a ball-and-stick representation binding at the interface between two subunits (grey and white). (B) Sequence alignment and assignment of the conserved motifs for *Cystoviridae* P4 proteins (ϕ6, ϕ8, ϕ12 and ϕ13). The conserved motifs are called H1 (red), H1a (yellow), H2 (green), H3 (blue) and H4 (purple) in analogy to the conserved motifs in hexameric helicases. The two loops protruding in the centre of the P4 channel are the disordered loops L1 (transparent) and L2 (cyan). The putative arginine fingers Arg272 and Arg279 are shown in magenta and the adenine binding Tyr288 is shown in light green. (See also Figure 15.2.)

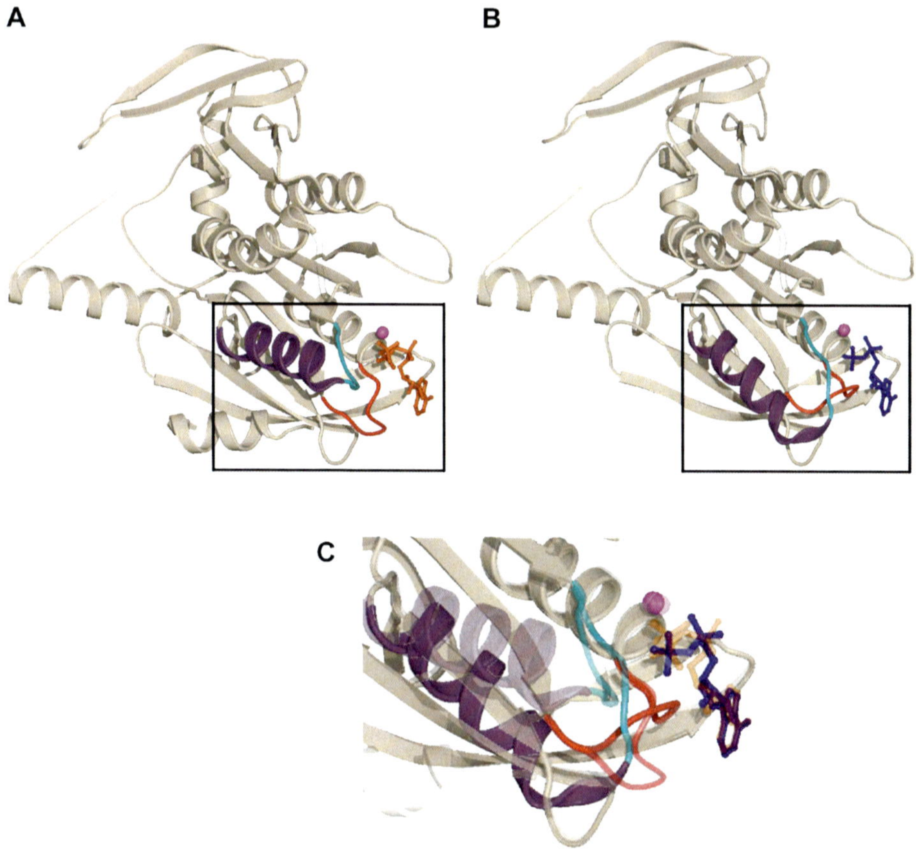

Plate 15.3 Conformational changes induced by ATP-hydrolysis. (A) Conformation of the P-loop (red), L2 loop (cyan) and helix α6 (magenta) in the AMPcPP+Mg^{2+} structure. (B) Conformation of the P-loop, L2 loop and helix α6 in the ADP+Mg^{2+} structure. (C) Superimposition of the two conformations. (See also Figure 15.3.)

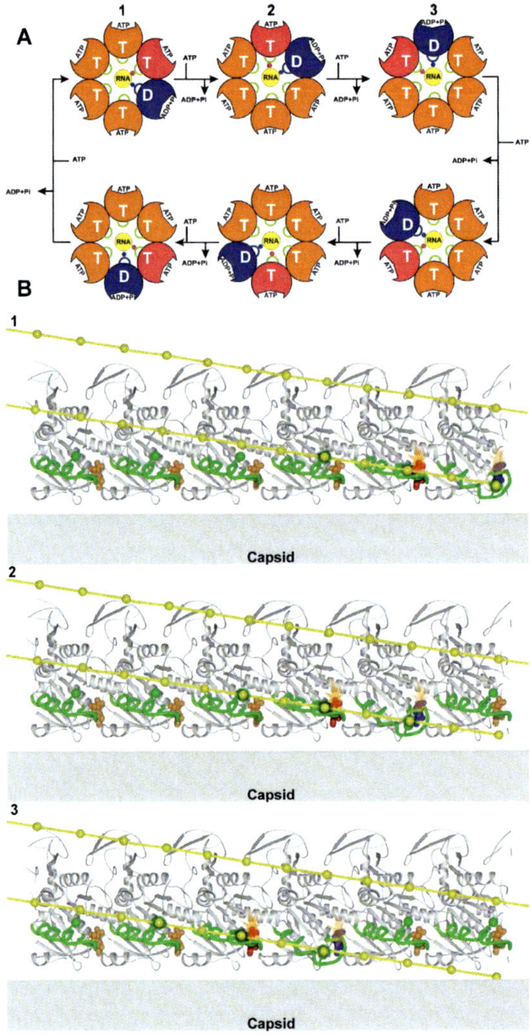

Plate 15.4 Mechanism of ssRNA translocation. (A) Six cycles of ATP hydrolysis are shown, in a schematic illustrating the nucleotide binding state of each subunit during successive cycles. Orange subunits contain bound ATP (T), blue have ADP+Pi (D), and red are hydrolysis competent (or activated) ATP containing sites. RNA binding loops are schematically shown in green (up position) and blue (down position). This is a sequential binding and hydrolysis model where all six bound ATP sites are utilized at different points during sequential cycles. ATP hydrolysis triggers movement of the L2 loop (blue) and the bound RNA backbone (blue dot) down the channel while another backbone phosphate is interacting with the L2 loop of a neighbour (red asterisk) and makes this subunit competent for hydrolysis. These conformational changes propagate as a ripple going counter-clockwise around the ring and do not require a rotation of the ring itself or of the RNA. The notation of the subunits is maintained in panel B. (B) Representation of the hexameric ring structure such that the subunits are laid out flat in two dimensions. The RNA binding loops are indicated in green, the RNA is coloured in yellow with the phosphates bases are represented by small balls. For simplicity, only the first three cycles of ATP hydrolysis are shown. As the hydrolysis wave proceeds from right to left, RNA is drawn inside the viral capsid (in grey) by about 2 subunits per hydrolyzed ATP. (See also Figure 15.4.)

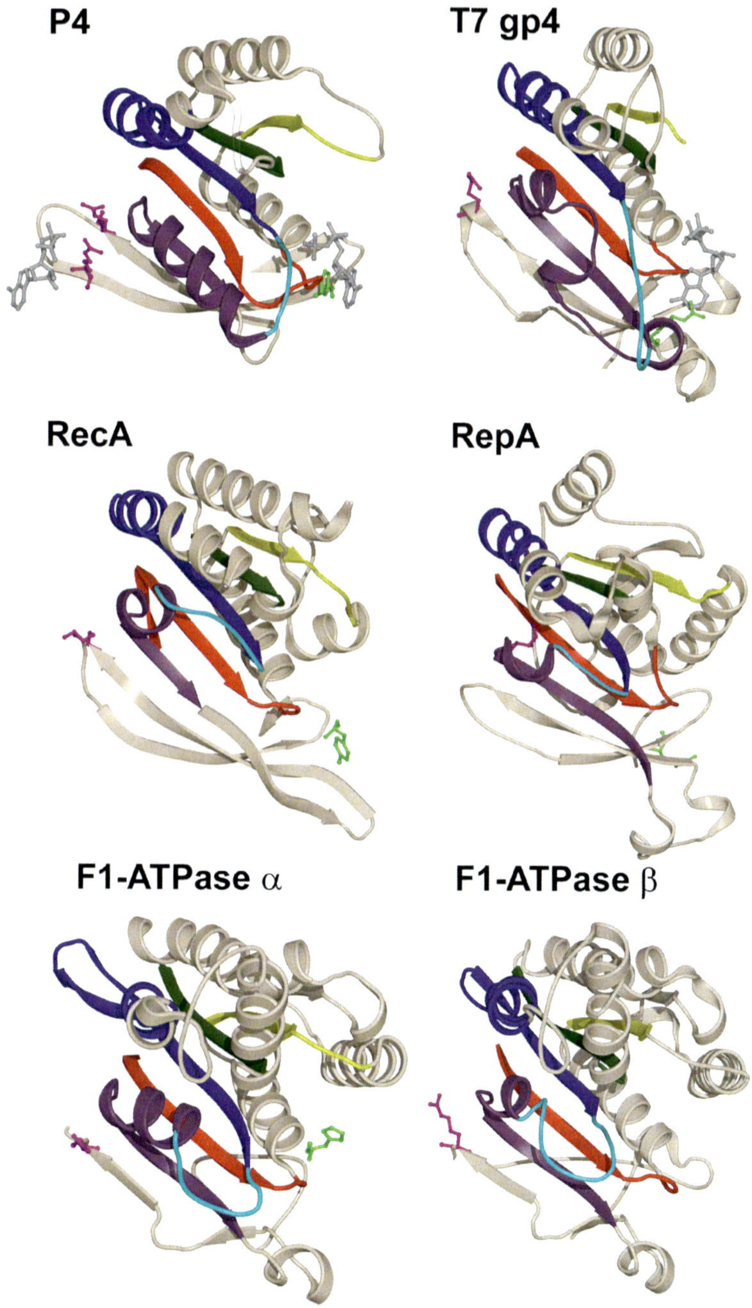

Plate 15.5 Structural superimposition of the nucleotide binding domains of P4 and structurally related proteins: T7 gp4 helicase (T7 gp4), *E. coli* RecA (RecA), regulatory protein RepA (RepA), Thermophilic Bacillus Ps3 F1-ATPase subunit α (F1-ATPase α) and β (F1-ATPase β). The structural superimposition was obtained with SHP (Stuart *et al.*, 1979). The colour scheme for conserved motifs (H1-H4) is the same used in the sequence alignment in Figure 2A. Alignment statistics are reported in Table 15.1 PDB deposition numbers are listed in the legend to Table 15.1. (See also Figure 15.5.)